MW01047796

J. Bien, lith. Photo. by Childs.

J. C. Branner

GEOLOGICAL SURVEY OF MICHIGAN.

UPPER PENINSULA

1869-1873

ACCOMPANIED BY AN

ATLAS OF MAPS.

VOL. I.

PART I. Iron-Bearing Rocks (Economic). *T. B. Brooks.*
PART II. Copper-Bearing Rocks. *Raphael Pumpelly.*
PART III. Palæozoic Rocks. *Dr. C. Rominger.*

PUBLISHED BY AUTHORITY OF THE LEGISLATURE
OF MICHIGAN.

UNDER THE DIRECTION OF THE

BOARD OF GEOLOGICAL SURVEY.

NEW YORK
JULIUS BIEN
1873.

208739

PART I.

IRON-BEARING ROCKS
(ECONOMIC).

BY

T. B. BROOKS.

MEMORANDUM.—It has been deemed advisable that the Appendices, referred to in this Part, should be issued separately as Vol. II.

ATLAS PLATES

REFERRED TO IN PART I.

---

* This map refers also to Parts II. and III.

LIST OF ILLUSTRATIONS.

VIEWS.

PLATES.

FIGURES.

FIGURES—Continued.

TABLE OF CONTENTS.

APPENDICES. (VOL. II.)

Governor J. J. BAGLEY, Hon. W. J. BAXTER, Hon. DANL. B. BRIGGS, } *Board of Geological Survey of the State of Michigan.*

GENTLEMEN,

Herewith I transmit a Report, with maps and illustrations, containing in part the results of my economic survey of the Iron Regions of the Upper Peninsula, made in accordance with a plan approved by your predecessors, under the chairmanship of ex-Governor H. P. Baldwin. I have labored diligently to produce "as complete a manual as possible of information relating to the finding, extracting, transporting, and smelting of the iron ores of the Lake Superior Region;" a book that should possess interest and value to the practical man and capitalist interested in our mines, which have for several years produced nearly one-fourth the ore raised in the United States.

Absence in Europe, on account of ill-health, prevents my giving the book that supervision, in passing through the press, which is essential to accuracy and finish in a work of this kind, especially when, as in this case, the author is not accustomed to bookmaking. Your publisher, Mr. Julius Bien, has promised to perform this duty, which is a guarantee that it will be well done.

I hope that the iron trade of the West will find in this a useful, although incomplete, manual, and that the people of Michigan will approve of the manner in which I have expended the money entrusted to me.

Very respectfully and obediently yours,

T. B. BROOKS.

London, May 1st, 1873.

INTRODUCTION.

IT is customary to preface Geological Reports with a history of the surveys on which they are based; in this case, however, it will be impossible to give more than a brief sketch, without omitting some part of the report itself, the limits of the book, for the publication of which funds were provided, having already been considerably exceeded.

The first survey of the State by Dr. Houghton, which was discontinued on account of his death by drowning in Lake Superior in 1845, is noticed in the first chapter in connection with the discovery of iron ore. The present survey was inaugurated by act of the Legislature in 1869, which appropriated $8,000 per year for the work, one-half of which went to the Upper Peninsula. This amount was again divided equally between the Iron and Copper Regions, which gave $2,000 per year for each to cover all expenses, including salaries, supplies, instruments, travelling, etc. To the $8,000 aggregate for four years from this source, the Geological Board added $1,000 for chemical work, making $9,000 in all received by me from the State for the survey of the Iron Region. In addition to this sum I have expended about $2,000 of my own means, and have not received any compensation for my services.

This small sum would have been inadequate to have accomplished anything worthy of the importance of the work undertaken, had not several corporations and individuals generously come to my relief: indeed on this source of help I counted largely in undertaking the work, and made it an express condition in the arrangement that I should be permitted to avail myself of all the assistance of this kind I could obtain, and also that during the progress of the work I should be free to continue the practice of a profession from which I was sure to obtain further facts bearing on the objects of the survey.

The companies which have contributed valuable data in their

possession, or have instituted special surveys at my suggestions, with the view of furthering the object of the survey, are :—The Marquette, Houghton & Ontonagon Railway, The Portage Lake & Lake Superior Ship Canal Co., The Republic, Washington, Lake Superior, Champion, New York, Spurr Mountain, Iron Cliff, Cannon and Magnetic Iron Companies. E. Breitung bore a part of the expense of making Map No. V., and John Fritz, A. Pardee, and Daniel J. Morrell, of Pennsylvania, S. P. Ely, of Marquette, and A. B. Meeker, of Chicago, contributed generously to the chemical fund, the results of the analyses being given in Chapter X.

The law of 1869 established a Board of Survey, consisting of H. P. Baldwin, Governor; W. J. Baxter, President of the Board of Education, O. Hosford, Superintendent of Public Instruction, with power to select the Geologists, disburse the money appropriated, and perform other necessary duties. Prof. A. Winchell was made Director, who approved the plan for the survey of the Iron Region which I submitted to him, and which is contained in the following letter:

LETTER OF INSTRUCTIONS,

Referred to in Agreement with T. B. BROOKS, *dated Negaunce, Mich., June 5th,* 1869.

" *To Major* T. B. BROOKS, *Assistant of the Geological Survey of Michigan.*

" SIR :—You are hereby authorized and requested to make a Survey of the Marquette Iron District, and to draw up a report on the same, substantially in accordance with the following suggestions:

" 1. By the Marquette Iron District is meant the region embracing all the deposits of iron ore extending from the shore of Lake Superior on the east, through Townships 46, 47, and 48 north, as far as Range 31 west, inclusive, being the region which for the present finds its outlet by railroads through Marquette and Escanaba.

" Your report on this district would appropriately furnish—

" 2. A historical sketch of discovery in the Iron Region of Lake Superior.

" 3. A physiographical sketch of the Marquette Iron District; general topography, hydrography, timber, soil, climate, etc.

" 4. The general geological structure of the district (not entering into details, nor theoretical discussions); identification of iron range stratification; outline description of the rocks; general description of the ores of iron occurring in the district.

" 5. The mines in general; their distribution and grouping.

" 6. Special notices of the mines and mining locations of the District; local structural geology, topography, mineralogical specialties of the ores.

"7. Discovery of ores; geological principles applicable; the use of instruments.

"8. The working of Iron Mines; methods in use here and elsewhere in analogous regions; advantages of each; machinery.

"9. The manufacture of iron and steel; special adaptations of the different varieties of ore in the District; the use of charcoal and mineral coal; resources of charcoal in Michigan; manufacture of charcoal; fluxes; location of furnaces; construction and operation of furnaces.

"10. Transportation of iron ores, and of iron; market; prices.

"11. Commercial statistics of iron ores, and of iron.

"In the discussion of the above topics, it is intended that you make such reference to other iron regions as may be necessary to thorough treatment and illustration of the general subject.

"It is not intended to lay down any stringent rules for your procedure, but only to furnish a general conception of the ground to be worked over. It is desired to produce as complete a manual as possible of information relating to the finding, extracting, transporting, and smelting of the iron ores of the Lake Superior Region, and it is believed that your own experience and the suggestions which may occur to you in the progress of the work will render it proper to deviate from the letter of the foregoing programme, according to the dictates of your own judgment. Specimens are to be collected according to the requirements and provisions of the law of 1869.

"In the prosecution of your field work, it is obvious that you cannot with the money at your disposal enter into detailed and complete examinations of individual properties, but it will promote the interests of the general work, if proprietors can be induced to defray the expenses of such detailed surveys beyond the limits to which you may be able officially to prosecute them; and it is evident that the interests of proprietors, no less than those of the State, will be promoted by committing such detailed surveys to your direction.

"The report, with the requisite maps, plans, and other illustrations, is to be ready for publication by the 31st day of December, 1870.

"(Signed) A. WINCHELL,
"Director Geological Survey,
"Ann Arbor, Mich."

On the completion of this survey of the Marquette Region, the Board decided to extend the work over the Menominee Region as well as further West before publishing, thus embracing all the known iron-fields of the Upper Peninsula. Professor Winchell having resigned in 1871, this part of the work was done under the direction of the Board.

Prof. R. Pumpelly has been engaged, with interruptions, in the Copper Region during the same period I have been at work in the Iron (see his Report, Part II.), and in the spring of 1871 Dr. C. Rominger commenced work on the Palæozoic rocks; his Report on the Silurian rocks of the Upper Peninsula is contained in Part III. of this volume.

The sum appropriated ($20,000) for publishing 2,000 copies of the three reports, with Atlas of maps, enabled the Board to contract for no more than a 500 octavo-page volume, which at the time was deemed sufficient space. I have been generously allowed more than one-half this space, but find that it was not sufficient to contain the material which I had accumulated, and which it seemed to me could be advantageously embodied in the proposed report. It was for some time a question with me, whether I should attempt to consider all the points named in the above scheme (giving each its relative space), which plan would have excluded a large amount of valuable material, or whether I should only attempt to treat each subject in order, as fully as my material would admit and its importance seemed to demand, without attempting at this time to cover the whole ground.

I choose the latter plan, and have in consequence been obliged to entirely omit all consideration of the important subjects of the location, construction and operation of furnaces; of fuels, fluxes, and ore mixtures; of the resources and manufacture of charcoal in Michigan,* as well as the consideration of the question of steel manufacture. The question of the transportation of ore and iron, of markets and prices, was also forced out for want of space. A proper treatment of these subjects would fill a volume.

I trust those gentlemen, who have favored me with lengthy and carefully prepared replies to my numerous inquiries on these excluded subjects, will feel that no injustice has been done them in withholding their papers, until they can be properly presented.

---

* The subject of the resources of Michigan in Charcoal and the location of charcoal furnaces both on the Upper and Lower Peninsulas has been carefully worked up and illustrated by Timber Maps, but there is unfortunately no means provided for their publication.

The following named gentlemen are well acquainted with their respective localities on the Lower Peninsula, and are prepared to give information regarding the timber, etc., which is in many instances unsurpassed:

Joseph Dame, North Port.	O. W. Hart, Torch lake.
E. E. Benedict, Manistee.	A. G. Butler, Frankfort.
E. B. Mills, Mayville.	James Lee, Bingham.
Geo. N. Smith, Bear river.	W. H. Hurlburt, South Haven.
W. H. C. Mitchell, East Traverse bay.	Dennis T. Downing, Little Traverse.
Leroy Warren, Pentwater.	Delos L. Filer, Ludington.
J. S. Dixon, Charlevoix.	William H. Frey, West Olive.

It may be questioned, whether with the purely practical object I have had in view in preparing this report, and the limited space, that so large a place should be given to the subject of Lithology, so ably treated by Mr. Julien, in App. A, Vol. II. The reasons which led to this were my own inability to properly treat this subject, its great relative importance in the study of rocks devoid of fossils, but above all I had collected and catalogued during seven years a more complete suite of specimens from the Azoic of the Upper Peninsula, than had before been got together, which collection I believed worthy the study and paper referred to, and which I saw no better way of utilizing to the public, than as has been done It is open to question whether Mr. Julien's paper should not have been published through some scientific channel, rather than in an industrial report, where it will stand nearly alone as a contribution to science.

Grouping Iron Deposits.—It has been found convenient in this report to disregard such political divisions as counties and towns in designating localities, and to employ instead, either the precise and simple method of U. S. linear surveyors, which can be readily understood by an inspection of Maps II., III. and IV. of Atlas; or, by the use of what may be termed the mineral or industrial geography of the Upper Peninsula, by which it is conveniently divided into regions, districts, groups, etc., which, although not sharply defined, may be considered at present to have the following boundaries: The *Marquette Iron Region* (see Map III., Table XIII., and Chap. IV.) embraces all the developed iron mines of the Upper Peninsula, the ores of which now find their outlets via Marquette, L'Anse and Escanaba by the Marquette, Houghton and Ontonagon and Chicago and Northwestern railroads. This again is subdivided into the (1) Negaunee, (2) Michigamme, (3) Escanaba, and (4) L'Anse districts. These divisions may be conveniently carried still further by a subdivision of the Negaunee District into the Cascade Range, Negaunee Hematite Mines, Ishpeming Group, New England and Saginaw Range; and of the Michigamme District into the Washington, Champion, Spurr and Magnetic Ranges, and Republic Mountain Basin. The S. C. Smith is the only worked mine in the Escanaba District, and no ore has yet been shipped from the L'Anse District or Range. The *Menominee Iron Region* (see Map IV. and Chap. IV.), which as yet has sent no ore to market, is divided into (1) The North Belt in south part of T. 42, (2) The South Belt in Ts. 39 and 40, and (3) The Paint River District. The *Lake Gogebic and Montreal River Region or Range* (Chap. VI.) is so little known that it may be questionable whether it should have a place in this economic grouping; it embraces the country between Lake Gogebic and the west boundary of Michigan, and is 100 miles west of the Marquette Region.

It but remains for me to express my obligations and gratitude to the many gentlemen who have contributed in various ways to the objects of this survey, to officially acknowledge their services and to thank them cordially for myself and on behalf of the Board for what they have done.

To S. P. Ely, of Marquette, the survey is more deeply indebted than to any other person; indeed, I would not have undertaken the work except from assurance of his support, which has been constant and generous from the beginning. To Messrs. H. B. and F. L. Tuttle, of Cleveland, Ohio, I am indebted for a considerable amount of the material embodied on Statistical Tables XII. and XIII. of Atlas, much of which I believe it would have been impossible for me to have procured, except through them; App. J, Vol. II., contains a letter from H. B. Tuttle, who has always, with great promptness and care, answered my various inquiries. To Major Fayette Brown, Cleveland, the survey is indebted for a most valuable paper on the amount of air required by charcoal furnaces and the mode of applying it, based on his experience with the Jackson Co.'s furnaces at Fayette, the almost unparalleled success of which gives his statements great value. S. L. Smith, on the part of the Marquette, Houghton & Ontonagon railroad, placed all the results of that company's explorations, made under my direction, at the disposal of the survey. J. J. Hagerman, Milwaukee, furnished a statement regarding the working of Lake Superior and Iron Ridge, Wisconsin, ores with anthracite and coke, and the successful use of the metal in making rails. John L. Agnew has furnished drawings of the new charcoal furnace, superintended by him at Escanaba, 50 feet high and 12 feet bosh, the largest, so far as I know, in the world. M. R. Hunt, Depere, Wis., has given full details of a remarkable long and successful blast of the First National Iron Co.'s furnaces.

The Historical chapter has been made far more complete and reliable than would otherwise have been possible through the contribution of facts and documents by Messrs. William and John Burt, Messrs. Everett, White, Harlow, Hewitt, and Ely, of Marquette; also by Messrs. Jacob Houghton and Charles T. Harvey. This chapter was rewritten by Charles D. Lawton.

I am indebted to so many persons for the facts embodied in the chapter on Mining, that I can only mention W. E. Dickinson, J. C. Morse, William Sedgwick, A. Kidder, Peter Pascoe, George and Eugene St. Clair, and D. H. Merritt, of Marquette county, and Prof. R. Akerman, of Stockholm, Sweden.

C. H. V. Cavis, S. H. Selden, and George P. Cummings, civil engineers, have greatly aided in the work by their personal efforts in procuring information which is embodied in the maps. The valuable explorations of C. E. and Frank Brotherton, and of A. M. Brotherton, deceased, made for the C. & N. W. and M. H. & O. roads, has been to a large extent placed at my disposal by the officers of these companies.

The nature of the valuable scientific aid given to this work by Alexis A. Julien, Prof. R. Pumpelly, Dr. T. S. Hunt, Prof. George J. Brush, Dr. H. Credner, and Charles E. Wright, are explained in the text of chapters III., V., VI., and in Appendices A, B, and C, Vol. II.

Edwin Harrison, of St. Louis, has given me full and detailed statements regarding the working of his Irondale furnace, which has one of the best records ever made by a charcoal furnace. Robert Wood has prepared most of the manuscript for the press, and, with Mr. Bien, will take care of the publication and indexing.

The survey is indebted to the University of Michigan, Ann Arbor, for the use of rooms without charge, and for the same courtesy (most cordially extended) to the School of Mines, Columbia College, New York, on which institution the survey had no claim.

The Marquette, Houghton & Ontonagon, Chicago & Northwestern, Michigan Central, the Great Western, and Grand Trunk railways have in every instance, when requested, granted passes to persons connected with the survey.

To the gentlemen and companies above named, as well as to Messrs. J. N. Armstrong, American Iron & Steel Association, S. C. Baldwin, William H. Barnum, J. B. Britton, C. M. Boss, J. R. Case, Mr. Childs, Girard Iron Company, C. H. Hall, A. Heberlein, Alexander L. Holley, E. C. Hungerford, Prof. Hayden, Gilbert D.

Johnson, F. B. Jenny, Prof. J. P. Leslie, J. S. Lane, A. W. Maitland, David Morgan, Capt. H. Merry, F. W. Noble, Charles H. Pease, New York Mine, J. R. Orthey, Freiburg Royal School of Mines, James M. Safford, Samuel Thomas, J. M. Wilkinson, H. N. Walker, Walter Williams, Capt. R. D. Weston (deceased), Washington Mine, Dr. White, and Charles R. Westbrook, who have in various ways promoted my work, I am under great obligations. Without their aid this report could not have been prepared. I have forwarded to the Board of Survey a full list of their names and addresses, with the request to furnish each with a copy of this Report and accompanying Atlas.

CHAPTER I.

HISTORICAL SKETCH OF DISCOVERY AND DEVELOPMENT.*

NOTE.—Statistical Tables XII. and XIII. of Atlas contain many facts relevant to this subject, which could not well be incorporated in the text.

MINERAL explorations along the south shore of Lake Superior began at a very early period, and the existence of copper was made known to the world as long ago as 1636, by La Garde, in a book published in Paris. During the subsequent portion of the 17th century frequent mention is made in the "Relations" of the Jesuit Fathers of the finding of this metal.

These Relations† extend from 1632 to 1672, and are made up of the reports or simple narratives of these humble but zealous missionaries, scattered as they were all over the region of the great Lakes, then controlled by the French Government, and are necessarily of inestimable value to the historian and archæologist; and also contain much that is highly interesting to the geologist, as indicating the early discoveries of minerals and the knowledge of their localities and uses, possessed by the natives. In illustration of the allusions to copper found in these reports, we quote simply from one, Claude Allouez, who seems to have been a man of intelligence, as well as one of the most persevering and deserving of these early missionaries. He first visited Lake Superior in 1666, and makes mention of a large mass of copper to be seen near the shore of the lake, with its top rising above the water, giving an opportunity for those who passed that way to cut pieces from it. The writer says, this "rock" has disappeared, having become buried, as he opines, beneath the sands, through the action of the waves. He also states that pieces of copper weighing from 10 to 20 lbs. are frequently found among the savages, who esteem them

---

* C. D. Lawton, Esq., rendered much assistance in the preparation of this chapter.

† These valuable documents have been republished by the Canadian Government.

2

as domestic gods, and hold them in superstitious awe, preserving them, in some instances, time out of mind, among their most precious articles.

In 1672, a map was published in Paris of this region, which was made by these early Jesuits, and on which is represented 1,600 miles of coast and many islands, with what may be considered remarkable accuracy.*

In 1689, Baron La Houtan, in a book relating to travels in Canada, mentions that "upon Lake Superior we find copper mines, the metal of which is fine and plentiful; there being not a seventh part base from the ore."

In 1721, P. De Charlevoix described the native copper deposits, and the superstitions which the Indians had in regard to them, in considerable detail. The occurrence of native copper being so frequent, the wonder of the early voyageurs was naturally excited, being increased also by vague rumors (gathered from the savages) of the existence of gold, silver, and diamonds.

In 1765, Captain Jonathan Carver visited Lake Superior, and in his account dwelt so largely on the abundance of native copper, that a copper company was formed in England in 1771, which actually began mining operations on the Ontonagon river, under the direction of Mr. Alexander Henry, who seems to have been a better historian than miner; for he gives a detailed account of the winding-up of his operations in 1772 and concludes, as the result of his unsuccessful experiment in mining, that the country must be cultivated and peopled before the copper can be profitably mined.

In 1819, Mr. H. R. Schoolcraft accompanied as mineralogist and geologist a government exploring expedition along the south shore of Lake Superior, having for its object the investigation of the copper mines.

In 1823 another government expedition, under charge of Major Long, passed along the north shore of the Lake, having come from the northwest; and mention is made of their having observed copper boulders in the region of the headwaters of the Mississippi.

Steps had been taken with a view to an exploration of this region

---

* A fac-simile of this map, and much other interesting matter relating to the early history of the copper region, may be found in Foster and Whitney's Report, Exec. Doc., 1850, Part I.

during the Presidency of John Adams, but nothing was ever effected. The work of systematic, scientific exploration of the Upper Peninsula of Michigan was first undertaken by Dr. Douglas Houghton, the earliest State Geologist. Dr. Houghton had commenced his examination of this region in 1831, and in his first annual report to the Legislature in 1841 presented the results of his labors up to that period in so able a manner, that the attention of the world became directed to the Northern Peninsula with greatly increased interest. In 1840, Dr. Houghton wrote to the Hon. A. S. Porter, under date December 26th, regarding the mineral wealth of the south shore of Lake Superior: "Ores of zinc, iron and manganese occur in the vicinity of the shore, but I doubt whether either of these, unless it be zinc and iron, is in sufficient abundance to prove of much importance. Ores of copper are much more abundant than either of those before mentioned, and a sufficient examination of them has been made to satisfy me that they may be made to yield an abundant supply of the metal."

In his Geological Report of 1841 Dr. Houghton says: "Although hematite ore is abundantly disseminated through all the rocks of the metamorphic group, it does not appear in sufficient quantity at any one point that has been examined to be of practical importance." At this date Dr. Houghton had traversed the south shore of Lake Superior five times, in a small-boat or canoe, on geological investigations. It is therefore probable that up to 1841 no Indian traditions worthy of credence, in regard to large deposits of iron ore, had come to his knowledge. As there are, so far as known, no considerable outcrops of iron ore, which come nearer than seven miles to the shore of the Lake, it is plain that investigations, based on observations taken along the shore only, could have determined no more than its probable existence, which is plainly indicated in the extracts given. Dr. Houghton was not aware of the existence of iron ore in quantity, until the return of Mr. Burt's party of surveyors to Detroit in the fall of 1844, his examinations in the interior of the country having been confined to the Copper Region. Attention at that early period was entirely directed to searching for ores of more value than iron, and it is worthy of remark, that the Jackson and Cleveland Iron Companies, which were the first two organized, were formed to mine copper, silver, and gold.

The remarkably rapid development of the mineral resources of the Upper Peninsula is largely due, among other causes, to the fact that the United States Linear Surveyors were required to combine geological and topographical observations with their surveys. The use of Burt's solar compass, which permits of rapid and precise observations of local variations (so important in the economic survey of a primitive iron region), served greatly to enhance the value of the results, by making known the position of rocks containing magnetic ore.

The honesty, skill and enthusiasm with which the field-work was executed resulted in the collection of a large amount of geological data, which at the completion of the survey would have left little to be done save the final report, in which the master-mind should classify, group, and harmonize the facts, and thereby develop nature's law from the mass of material collected. Dr. Houghton's untimely death by drowning in Lake Superior, while in the midst of his labors, prevented him from performing the crowning work. Any one familiar with the geology of the Upper Peninsula, who will peruse the manuscript notes * left by Dr. Houghton, will be convinced that his views regarding the geology of the older rocks were far in advance of his time, and such only as geologists years afterward arrived at, and those which are but now, thirty years after he recorded them, universally accepted (see Appendix E, Vol. II.). A brief statement of the origin of a work from which such important results have accrued will be given. In 1843 the financial troubles of the State of Michigan arising out of the "Five Million Loan," as it was called, were of such a character as to cause the Legislature to withhold the annual appropriation for the Geological Survey, which then had been for several years in successful operation under the direction of Dr. Houghton. Thoroughly interested in his scientific work, and believing that the best interests of the State and the cause of science demanded the continuance of the survey, Dr. Houghton asked from the General Government the aid which his own State felt unable to grant, and succeeded in obtaining, in the appropriation for the Public Surveys of

* These manuscript notes are now in the University Library at Ann Arbor, having been presented to that Institution by Dr. Houghton's widow. Dr. Houghton, it will be remembered, was at the time of his death a Professor in the University of Michigan as well as State Geologist.

IVES' MAP, SHOWING IRON HILLS, 1844.

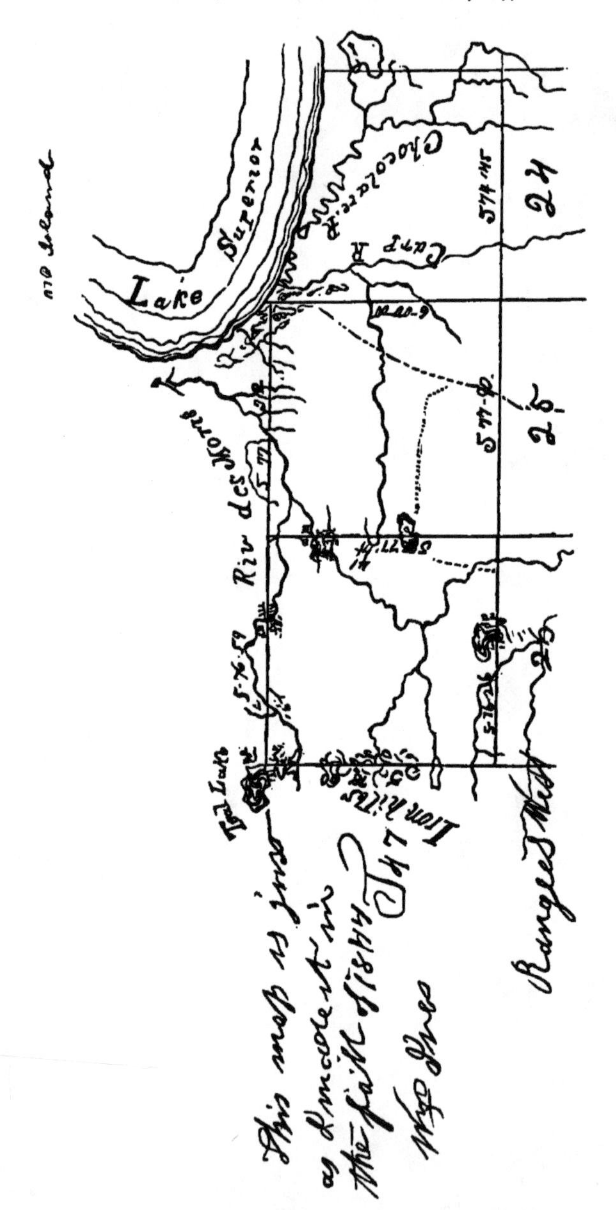

the Upper Peninsula of Michigan, an additional allowance per mile to cover the cost of the geological work. In order to expedite the work and insure the best scientific results from the adoption of his plan, Dr. Houghton himself took the contract from the Government for completing the surveys on the Upper Peninsula, which had been previously begun in 1840 under the direction of the Hon. William A. Burt, United States Deputy Surveyor. In the spring of 1844 Dr. Houghton commenced operations under his contract, the field-work being in charge of Mr. Burt, who received in compensation therefor the allowance granted by the Government. It is proper to add that Mr. Burt entered with deep interest into Dr. Houghton's plans and had, during his survey in the Lower Peninsula, collected for him many specimens and important geological information not required by his instructions.

In 1844 Mr. Burt, with a party consisting of William Ives, compassman, Jacob Houghton, barometer man, H. Mellen, R. S. Mellen, James King, and two Indians named John Taylor and Bonney * was engaged in establishing Township lines and making geological observations, as previously described.

On the 19th September, while running the east line of Town. 47 north, Range 27 west (the great iron Township), they observed by means of the solar compass remarkable variations in the direction of the needle, amounting to 87° from the normal. (See Appendix D, Vol. II.) Ascribing this phenomenon to iron ore, they sought for and found it in the ledges or outcrops at several points. Speci mens † were collected and named by Mr. Burt and Dr. Houghton (See Appendix D, Vol. II.), and were described by them in their official returns; the fact of the great variation and large amount of ore being also especially commented upon. (See Appendix and official notes in Land Office, Washington, Lansing and Marquette.) A map made by Mr. Ives at the time, a fac-simile of which is given in Pl. I., has written along this line the words "Iron Hills." As the Jackson range is not magnetic at this locality, and does not outcrop on the line in question, it is not probable it was seen, but instead one or more of the ranges of flag or soft hematite ore further south.

---

* Bonney's real name was Michael Doner; himself and Taylor are now dead.

† Mr. R. S. Mellen has still in his possession a piece of the ore found that day, which he brought away with him.

In the month of June following, Dr. Houghton and Mr. Burt, with their party, were engaged in subdividing the Township above mentioned (Town. 47 north, Range 26 west), when the former made a personal examination in reference to iron ore, especially at the corners of Sections 29, 30, 31, 32 (see Appendix D, Vol. II.), now known as the Cascade mines, and remarked to Jacob Houghton and others, who were members of the party, that it would some day be very valuable and the basis of an active industry.

It thus appears that the U. S. surveyors, in the fall of 1844, officially established the fact, that iron ore in considerable quantities existed in the Upper Peninsula of Michigan. It is also undoubtedly true, that Indians had previously observed the ore and were acquainted with locations of it, without, however, being able to identify it.

The Jackson Co.—The manner in which this, the earliest developed, and one of the most important of the iron properties on Lake Superior, was discovered (although the enterprise was not mainly undertaken with a view of finding iron), is reliably set forth in the following letter, written by P. M. Everett, now of Marquette, to Captain G. D. Johnson, now of the Lake Superior mine. The letter is dated at Jackson, Mich., Nov. 10th, 1845, and is as follows:—

"I left here on the 23d of July last and was gone until the 24th of October. . . . . I had considerable difficulty in getting any one to join me in the enterprise, but I at last succeeded in forming a company of thirteen. I was appointed treasurer and agent to explore and make locations, for which last purpose we had secured seven permits from the Secretary of War. I took four men with me from Jackson and hired a guide at the Sault, where I bought a boat and coasted up the lake to Copper Harbor, which is over 300 miles from Sault Ste. Marie. . . . . We made several locations, one of which we called iron at the time. It is a mountain of solid iron ore, 150 feet high. The ore looks as bright as a bar of iron just broken. Since coming home we have had some of it smelted, and find that it produces iron and something resembling gold—some say it is gold and copper. Our location is one mile square, and we shall send a company of men up in the Spring to begin operations; our company is called the 'Jackson Mining Co.'"

The actual discovery of the Jackson location was made by S. T.

Carr and E. S. Rockwell, members of Everett's party, who were guided to the locality by an Indian chief, named Manjekijik.*

The superstition of the savage not allowing him to approach the spot, Mr. Carr continued the search alone, resulting in the discovery of the outcrop, which he describes as indicated in Mr. Everett's letter. Previous to the discovery he was led to suppose from the Indians' description, that he would find silver, lead, copper or some other metal more precious than iron, as it was represented and found to be "bright and shiny."

July 23d, 1845, articles of association of the Jackson Iron Company were executed at Jackson, Mich., and by these articles Abram V. Berry was appointed the first *President*, Frederick W. Kirtland, *Secretary*, Philo M. Everett, *Treasurer*, and George W. Carr and William A. Ernst, *Trustees.*

Mr. Berry gives the following account of the early history of his company, in a letter dated at Jackson, Mich., Oct. 21st, 1870:—

"In the summer of 1845, an association was formed in this city, then a village, for the purpose of exploring the mineral region on the south shore of Lake Superior. The company consisted of P. M. Everett, James Ganson, S. T. Carr, G. W. Carr, F. W. Carr, E. W. Rockwell, F. W. Kirtland, W. H. Munroe, A. W. Ernst, F. Farrand, of Jackson, and S. A. Hastings, of Detroit (John Watkins, of Detroit, was interested with Hastings). Eleven individuals of the association procured permits from the War Department to locate one square mile each of mineral land on the south shore of Lake Superior. John Western, of Jackson, was then added to the

---

* In reward for the service of the Indian on this occasion, the officers of the Jackson Company subsequently gave him a written stipulation, of which the following is a copy:—

"River Du Mort, Lake Superior,
May 30, 1846.

This may certify that, in consideration of the services rendered by Manjekijik, a Chippeway Indian, in hunting ores of Location No. 593 of the Jackson Mining Company, that he is entitled to twelve undivided twenty one-hundredths part of the interest of said mining company in said location No. 593.

A. V. Berry, *Superintendent.*
F. W. Kirtland, *Secratary.*"

This agreement on the part of the company was never fulfilled, and Manjekijik finally died in poverty; his relatives, now living in Marquette, are in the same miserable condition, without ever having received, as is averred by those who are cognizant of the facts, any compensation for the services mentioned.

company, making thirteen in all. In the fall of 1845 a company of explorers, consisting of S. T. Carr, P. M. Everett, W. H. Munroe, and E. S. Rockwell, visited Lake Superior, when what is now known as the Jackson location was secured by the permit granted to James Ganson, in the unsurveyed district, the section lines not having been run. The location was described by metes and bounds, commencing at a certain large pine-tree, the position of which was fixed by its course and distance from the corner of Teal lake. When the land was surveyed it was bought at $2.50 per acre. * * *

"In the spring of 1846, another expedition was fitted out, consisting of F. W. Kirtland, E. S. Rockwell, W. H. Munroe and myself, members of the company and several other adventurers; the object being to make further examinations of the iron and to use the remaining permits, by entering other mineral land. * * * * * I found our location much beyond what I had anticipated. After spending twelve days in the woods, exploring the surrounding country, including what was afterwards known as the Cleveland location and building what we called a house, we returned to the mouth of the Carp with 300 pounds of ore on our backs. We then divided; one party was left to keep possession of the location, another went farther up the Lake to use the remaining permits, while I returned to the Sault with the ore. It was my intention at this time to use another permit on the Cleveland location, but on arriving at the Sault I met Dr. Cassels, of Cleveland, agent of a Cleveland company, and having arranged with him that his company should pay a portion of the expense of keeping possession, making roads, etc., I discovered to him the whereabouts of the Cleveland location. He took my canoe, visited the location, and secured it by a permit. On arriving at Jackson we endeavored on two occasions to smelt the ore which I had brought down, in our common cupola furnaces, but failed entirely. In August of the same year, Mr. Olds, of Cucush Prairie, who owned a forge (in which he was making iron from bog ore), then undergoing repairs, succeeded in making a fine bar of iron from our ore in a blacksmith's fire, the first iron ever made from Lake Superior ore. In the winter of 1846-47 we began to get up at Jackson a bellows and other machinery for constructing a forge on the "Carp;" and in the summer of 1847 a company of men commenced building the same, and continued until March, 1848, when a freshet carried away the dam. * *

"—The association was then (1848) merged into an incorporated company, and by some means the pioneers in the enterprise are now all out."

In a book * on the mineral region of Lake Superior, with map by Jacob Houghton, Jr. and T. W. Bristol, published in 1846, only one iron company is mentioned—The Jackson. The description of the company's property is as follows :

Permit No. 593—somewhere in T. 46, N.-R. 27 or 28 W., while on Section 1 of T. 47, R. 27, Permit No. 158 is marked, which was granted to D. Hamilton, of Watervliet, New York. Section 3, same Township, embracing the New York mine, is covered by Permit No. 160, granted to T. Williams, of Newburg, N. Y. Section 10, same Township, embracing parts of the Cleveland and Lake Superior mines, was covered by Permit No. 177, granted to T. Ricket, of Copper Harbor.

In 1846 Fairchild Farrand explored the Jackson location and mined some ore. The company, under the superintendency of Wm. McNair, began, in 1847, the construction of a forge on the Carp river, three miles east of the mine, the first iron being made Feb. 10th, 1848, by forgeman, now Judge, A. N. Barney. Work was stopped in a few days by a freshet which carried away the dam. Mr. Everett came up in the summer of 1848, had the dam repaired and resumed the manufacture of blooms. The first iron made was sold to E. B. Ward, who employed it in the construction of the steamboat "Ocean." This forge was afterwards carried on under leases by B. F. Eaton, and later by the Clinton Iron Co., subsequently by Peter White and lastly by J. P. Pendill ; it made but little iron and no money. The quality varied from the highest (as shown by the experiments of Major Wade, of the U. S. army) to indifferent, the trouble being a lack of uniformity in the blooms. The power was supplied by the Carp river, a dam 18 feet high having been constructed across the stream for this purpose. There were upon either side of the stone arch, and arranged opposite each other, four fires, from each of which a lump was taken every six hours, which was placed under the hammer and forged into blooms

* This little volume, (afterwards revised by Mr. Houghton,) thus early issued, contains much interesting and valuable matter relating to the early discoveries and mining operations of Lake Superior, especially regarding the copper region.

four inches square and two feet in length; the daily product being about three tons, requiring two teams of six horses each to convey them over the intervening ten miles of horrible road to Marquette. These teams, when so fortunate as not to break down, on returning brought back supplies for the men and animals. The same difficulties attended the procuring of the supply of ore and charcoal. The power was also found to be insufficient, owing to a scant supply of water occurring at certain seasons of the year. These difficulties were too numerous and serious for the maintenance of the existence of the concern, and resulted in its abandonment in 1856.

On the 6th of June, 1848, a meeting was called to act on the question of the acceptance of an act of incorporation passed at the preceding session of the Legislature, and it was decided to incorporate the company under the act referred to. The organization was completed under the title of the Jackson Mining Co., of Jackson, Michigan—Fairchild Farrand, *President*, W. A. Ernst, *Secretary*, George Foot, *Treasurer*, F. W. Carr, F. W. Kirtland, Lewis Bascom, and John Western, *Directors*. The capital stock of the company, as also that of the New England Mining Co., organized at this time, was fixed at $300,000, in shares of $100 each; the purpose of each being the mining of copper as well as iron. April 2d, 1849, an amendment to the charter of the Jackson Mining Co., of Jackson, was obtained, when the title was changed to its present form—**Jackson Iron Co.** The first officers under this organization were Ezra Jones, *President*, Wm. A. Ernst, *Secretary*, John Watson, *Treasurer*, S. H. Kimball, James A. Dyer, and James Day, *Directors*.

In 1850, Mr. A. L. Crawford, proprietor of the iron works at Newcastle, Pa., took with him from Lake Superior about five tons of the Jackson ore, and there worked it up. Part of the ore having been made into blooms and rolled into bar-iron, was used for special purposes, and part used for lining in the puddling furnaces. The iron was found to be excellent. About the same time, General Curtis, of Sharon, Pa., proprietor of extensive iron-works at that place, came to Lake Superior to inspect the Jackson and Cleveland locations; his object being to secure an interest, with a view to a future supply of ore for his works, of a better quality than he then possessed. Failing to make an arrangement for the Cleveland, he bought up sufficient stock in the Jackson Co. to give him a controlling interest in the management of its affairs; so that for some years the location was known as "Sharon."

It is proper to remark that General Curtis believed, as did also John Western before him, that, as soon as practicable, the best policy for Lake Superior iron mines to follow would be to sell their ore to the furnaces of Ohio, Pennsylvania, and elsewhere; and in 1852 about 70 tons of the company's ore were taken to Sharon, Pa., and there made into pig-iron in the "old Clay Furnace." There were frequent changes of officers and directors in the Jackson Co. up to 1860, and the history of the company was one of disappointment and financial embarrassment. Between 1860 and 1862 the gentlemen who now compose the Board of Directors came into office, and in 1862 the first dividend was made. The great demand for iron occasioned by the war caused the iron interests of Lake Superior, for the first time, to assume a very successful aspect. The first regular shipments of ore from the Jackson mine were made in 1856, which amounted to about 5,000 tons. Up to this time the different forges in the district had consumed about 25,000 tons of ore. (See Table, Pl. XII. of Atlas.) The Jackson mine, earliest discovered, and first opened and tested, became widely known from the outset, and has ever continued to remain the leading mine in the district. The important village of Negaunee, within whose corporate limits the Jackson mine is situated, dates its origin with the commencement of the company's o erations. As the Chicago and Northwestern and the Marquette, Houghton and Ontonagon railroads form a junction in Negaunee, facilities are thus afforded for shipments over either road—that is, by the way of Escanaba or Marquette. The "openings," or pits, are irregular and numerous, and extend from the west edge of the village of Negaunee west for three-quarters of a mile. The greater portion of the product finds its outlet through a tunnel, which enters the mines from the north side of the hill and is of sufficient size to admit railroad cars and small locomotive engines. From the main tunnel radiate several branches, which extend to, or are being extended to, the different stopes and shafts. The main shafts are supplied with ample steam-power for pumping and hoisting purposes. For details of workings, geological structure, etc., see accompanying maps, tables, and text.

The New England Mining Co. was, like the Jackson, *incorporated* by a special act of the Michigan Legislature passed in 1848. The purpose for which the organization was effected is stated as being the mining and smelting and manufacturing of ores and minerals in the State of Michigan, the language stating the company's objects being identical with that of the Jackson Company; the capital stock was placed at 300,000. It does not appear that anything noticeable was accomplished by this company, thus early organized. The charter came in 1855 into the possession of Capt. E. B. Ward, by whom it is now held.

The Marquette Iron Co.—In the summer of 1848, Mr. Edward Clark, of Worcester, Mass., was sent to Lake Superior by Boston parties, to look for copper, but at the Sault he fell in with Robert J. Graveraet, who induced him to stop at the Carp river and see the iron mines. The Jackson Company's forge was at work and had made a little iron. Clark, on his return to Worcester, carried with him a bloom and some ore from the Jackson Iron mountain, which, on being drawn into wire at a factory, proved excellent. Clark at once proceeded to form an association for the purpose of building a forge on the far-off shore of Lake Superior, assisted by Graveraet, who also appeared in Worcester at this time (having travelled from Marquette to Saginaw on snow-shoes); he succeeded in organizing a company, March 4th, 1849, consisting of E. B. Clark, W. A. Fisher, A. R. Harlow, of Worcester, Mass., and R. J. Graveraet, of Mackinaw; Clark and Graveraet putting in against the capital of the others leases of iron lands of which they claimed to have possession. These iron lands constitute what subsequently became known as the Lake Superior and Cleveland mines, and over which a long controversy arose as to which party should possess the land, and which was finally decided by the Interior Department at Washington in favor of what was known as the Cleveland Company. Mr. Harlow constructed and purchased the necessary machinery to the value of $8,000, and in the spring of 1849 shipped it to Marquette, starting himself with his family on the 11th of June, and arriving in Marquette on the 6th of July thereafter. Graveraet had reached there on the 17th of May previous, taking with him a small party of men, among whom was Peter White, then a lad, but subsequently largely identified with numerous interests in the Iron

Region, and now President of the First National Bank of Marquette. The forge was completed, making the first bloom in just one year from the date of Mr. Harlow's arrival.

The Marquette Iron Co.'s works started with 10 fires, and used Cleveland and Lake Superior ores, mostly the former, making blooms exclusively, which were sold in Pittsburg at prices ranging from $35 to $50. The works were in operation somewhat irregularly until 1853, when the Marquette Company was merged into the Cleveland Company, under the auspices of which the forge continued in operation for a few months longer, and was finally destroyed by fire in 1854. Like all bloomeries started in Marquette County, it was from the first, financially, a failure. The cost of the plant was great, transportation difficult and expensive, and the price of iron during the entire period disproportionately low. There was no dock at Marquette, no canal at the Sault, scarcely a road in the country, no shop for repairs, no skilled labor but what was, together with all supplies, imported "from below," and no regular communication. During the summer of 1849 only three sailing vessels and five propellers arrived at Marquette. The stock of the Marquette Company was bought up by the Cleveland Company, and its property passed to the ownership of the latter.

In 1852 John Downey, Samuel Barney and others began the construction of a forge on the "Little Carp," but after having built some houses, constructed a wheel, etc., permanently abandoned the enterprise.

In 1849 and 1850 a *whetstone quarry* was opened in a bed of novaculite, near the outlet of Teal lake, and Messrs. Smith and Pratt established a factory, for the purpose of sawing these blocks, at the mouth of a small stream near the Marquette landing, and carried on a "thrifty business."

The Iron Mountain Railroad.—The question of transporting the rich ores of Marquette county to the coal of Ohio and Pennsylvania, being one that came to be seriously considered, it naturally suggested the necessity of *a railroad* from the mines (those near the present villages of Negaunee and Ishpeming) to Marquette bay. In 1851 Messrs. Heman B. Ely and John Burt strongly advocated the enterprise, and in the following year Mr. Ely caused a survey to be made; at that period the entire population of Marquette county was less than 150 persons. There being no general railroad law in the State at that time, the construction of the railroad was undertaken by Mr. Ely, assisted by his brothers George H. and Samuel P. Ely, of Rochester, New York, as an individual enterprise, he having previously made a contract with the Jackson and Cleveland Iron Mining Companies and Mr. John Burt, as the representative of other companies, for the transportation of their ores. This contract the two first-named iron companies subsequently attempted to break, and sought to defeat the railroad by constructing a plank-road in opposition to it, thus instituting a serious and embarrassing controversy, which continued until 1855, when all matter of dispute then pending between the Railroad Company, under charge of Mr. Ely, and the Plank-road Company, under charge of Mr. S. H. Kimball, were submitted to arbitration and settled to the satisfaction of both parties—Messrs. C. T. Harvey and Austin Burt being arbitrators. Immediately after the passage of the General Railroad Law of this State in 1855, the Messrs. Ely incorporated the railroad under the title of the Iron Mountain Railroad, and John Burt was first President. A year later the company was strengthened by the addition of Jos. S. Fay, Edwin Parsons, Lewis H. Morgan, and other capitalists; and in 1857 the road was completed and put in operation. Mr. H. B. Ely, to whose foresight and energy the origin and success of the enterprise was largely due, and to whom the interests of Lake Superior became otherwise greatly indebted, died in Marquette, in 1856, before the work upon which he had labored so intently was completed.

The death of his brother, and his own connection with the road, was the occasion of bringing to Marquette Mr. S. P. Ely, who is now more largely identified with the business management of many of the leading enterprises in the Iron Region than any person resident on "Lake Superior." The Iron Mountain Railroad became subsequently a part of the Bay de Noquette and Marquette Railroad, this becoming afterwards, by consolidation, the Marquette and Ontonagon Road, and still later, by further consolidation, a part of the through line of the Marquette, Houghton, and Ontonagon Railroad. The plank-road to which reference is here made was built by the Jackson and Cleveland Companies jointly, but was never used as a plank-road; longitudinal sleepers were laid down and covered with strap-rail, on which horse cars were run. The road was used for two seasons, and cost $120,000, which amount was practically sunk. The cost of transportation was nominally one dollar per ton; each team would make the round trip in a day, bringing four tons of ore. It is proper to add that the rates of transportation fixed by these H. B. Ely contracts, although afterward deemed by the iron companies much too liberal, were lower than any at which ore has ever been carried over the road; the present rates being more than double those agreed upon with Mr. Ely.

Among the most important enterprises early connected with the development of the Lake Superior iron interests was the construction of the **Sault Ste. Marie Ship Canal.** In the St. Mary's river or strait, connecting the waters of Lakes Superior and Huron, occurs, nearly opposite the village of Sault Ste. Marie, a rapid of about one mile in length, and about seventeen feet fall, forming a complete barrier to the communication between the lakes. Some years previous to the construction of the canal this barrier had been overcome partially, by the construction and use of a portage flat-bar railroad, over which all articles of commerce between the lower lakes and Lake Superior were transported and reshipped in both directions. The important and growing interests of Lake Superior demanded more easy and effective means of commercial communication with the lower lakes. The matter being brought before the National Legislature, Congress granted to the State of Michigan, by Act approved Aug. 26th, 1852, 750,000 acres of land for the purpose of aiding in the construction and completion of a ship canal around the falls of Ste. Marie. On the 5th of February following, the State of Michigan, by an Act of its Legislature, accepted the grant of land above mentioned; and to further the objects thereof, authorized the Governor of the State to appoint Commissioners to let the contract for the construction of the canal, and to enter the lands authorized under the grant.

The Commissioners appointed under this legislative act entered into contract with Joseph P. Fairbanks, Erastus Corning and others for building the canal within two years from date thereof; the consideration being the U. S. Government grant of lands. This contract was soon after duly assigned to the Ste. Marie's Falls Ship Canal Co., which company had been organized in the city of New York on the 14th of May, 1853, under an Act of the Legislature of the State of New York, passed April 12th, immediately preceding. At the organization of the company, the following persons were chosen officers and directors of the company: Erastus Corning, *President*, J. W. Brooks, *Vice-President*, J. V. L. Pruyn, *Treasurer* and *Secretary*. *Directors:* Erastus Corning, J. W. Brooks, J. V. L. Pruyn, Jos. P. Fairbanks, John M. Forbes, John F. Seymour, and James F. Joy.

Subsequent to the passage of the grant by Congress, but previous to the acceptance thereof by the State of Michigan, Mr. Charles T. Harvey was authorized by Messrs. Fairbanks and Corning to cause a survey to be made, which he proceeded to do during the month of November, 1852, having secured the services of an experienced engineer from the Erie Canal, Mr. L. L. N. Davis. After the organization of the company, Mr. Harvey was appointed its general agent, and the supervision of the construction placed under his control.

Early in the season of 1853 Mr. Harvey, with 400 men, proceeded to the Sault, and on the 4th of June broke ground for the canal. The remoteness of the locality, and many other unfavorable circumstances, rendered the construction of a work of such magnitude exceedingly difficult, and necessitated at every step of the operations unusual care and energy in the management as well as heavy pecuniary expenditures. Mr. Harvey remained in control of the construction for one year, when he was relieved and placed in charge of the finance, and also appointed agent for the State to select lands under the grant in the Upper Peninsula. Mr. Harvey selected about 200,000 acres of land, 39,000 of which were taken in Marquette county, and were subsequently sold for $500,000 cash, to the Iron Cliff Co. Among the copper land selected was the quarter section on which the Calumet and Hecla Company's mine is situated, and which was sold by the canal company for $60,000, now worth, on the basis of late sales of stock, $13,000,000. The 750,000 acres granted by the General Government were entered by the company as follows: on the Upper Peninsula, 262,283 acres of iron, copper, and timber land, and 487,717 acres of pine land in the Lower Peninsula. A land agency was established at Detroit for the purpose of locating the lands obtained through the grant.

During the summer of 1854 the difficulties necessarily attendant upon building the canal were very much enhanced by disease among the workmen; some 200 of whom died of the cholera, and among them was Mr. Ward, who had charge of the construction. Mr. Harvey was again placed in charge of the work, which, owing to the panic among the workmen, had become nearly suspended; but by the exercise of much skill and energy he succeeded in reorganizing the force, and pushing the work vigorously forward to final completion. On the 19th of April, 1855, the water was let into the canal, and in the following June the work was opened for public use, under the superintendency of Mr. John Burt.

The total cost of the construction of the canal, which includes also the expense attendant upon the selection of lands, as contained in the report of the company under date of January 1st, 1858, was $999,802.46.

The State of New York, by act passed April 15th, 1858, granted a charter incorporating the "**St. Mary's Canal Mineral Land Co.**" Under this act of incorporation, a company was duly organized, and to it was transferred the canal company's lands of the Upper Peninsula. It was soon found that the canal failed to meet the growing wants of the commerce of Lake Superior, owing to the variation in the general level of the Lake Superior becoming somewhat lower than when the canal was completed, thus making a variable difference in the depth of the canal of from one to one and one-half feet; and also that the General Government, by successive appropriations, has caused the channels through Lake George and the St. Clair Flats to be so widened and deepened, that vessels of far heavier tonnage than was originally anticipated could be employed. The Michigan State Legislature adopted a resolution in the session of 1869, offering to cede the canal to the U. S. Government; although Congress has not as yet formally accepted the offer made by the State, nevertheless, under its system of internal improvement, the General Government is now engaged in the enlargement of the canal. The width of the canal is to be increased to 300 feet, and its depth to 16 feet, the locks are to be double, 80 feet in width and 450 feet long. The amount of the government appropriations under which this improvement is being effected is in the aggregate $800,000; and the work, when completed, will be fully adequate to the wants of commerce.

The report of superintendent Guy H. Carleton shows the following to be some of the principal exports and imports through the canal during 1871 and 1872:

	1871.	1872.
Flour, bbls	25,146	42,141
Pork, bbls	8,887	10,306
Beef, bbls	3,054	4,161
Bacon, lbs	163,763	242,475
Lards, lbs	283,141	213,394
Butter, lbs	519,545	559,137
Cheese, lbs	187,340	200,994
Tallow, lbs	104,354	106,170
Soap, boxes	21,799	18,205
Apples, bbls	18,359	20,025
Sugar, lbs	4,062,087	5,454,559
Tea, chests	3,864	7,980
Coffee, bags	5,228	7,815
Salt, bbls	36,199	42,690
Tobacco, lbs	258,179	321,836
Nails, kegs	29,843	34,984
Dried Fruit, lbs	115,366	73,230
Vegetables, bush	27,619	35,263
Lime, bbls	2,338	6,067
Window Glass, boxes	25,226	7,492
Cattle, head	2,639	3,608
Horses and Mules	435	528
Hogs, head	1,625	1,567
Brick, M	1,225	9,067
Furniture, pieces	13,616	44,768
Machinery, tons	1,595	10,593
Engines	18	28
Boilers	17	34
Liquor, bbls	4,366	7,082
Malt, lbs	653,140	1,545,875
Coarse Grain, bush	283,503	444,875
Mdse., tons	23,245	38,215

The following are some of the principal exports from Lake Superior for 1871–72:—

	1871.	1872.
Mass Copper, tons	1,091	1,709
Ingot Copper, tons	7,666	8,547
Stamped Work Copper, tons	5,705	4,365
Iron Ore, tons	327,461	383,105
Pig Iron, tons	23,304	29,341
Fish, half bbls	26,041	14,529
Wheat, bush	1,376,705	567,134
Tallow, lbs	59,225	64,567
Flour, bbls	179,093	94,270
Barley, bush	25,320	898

	1871.	1872.
Silver Ore	464	306
Stone, building, tons	5,528	5,213
Potatoes, bush	.....	636
Copper, manufactured, tons	.....	395
Quartz, tons	.....	591
Wool, tons	.....	30

In 1853 the **Lake Superior Iron Company,** one of the three oldest companies in the district, was formed; articles of association were filed March 13th, capital stock $300,000, in 12,000 shares of $25 each. The capital stock was subsequently increased to $500,000, which has all been returned to the stockholders in dividends. The incorporators were Heman B. Ely and Anson Gorton, of Marquette, Mich.; Samuel P. Ely, George H. Ely, and Alvah Strong, of Rochester, New York. The company commenced operations in 1857 on 120 acres of land in Sections 9 and 10, T. 47, R. 27, which was purchased of John Burt, being a part of the Briggs and Graveraet claim spoken of above under the Cleveland Company. Subsequent purchases enlarged the company's estate to 2,000 acres, its present dimensions. The company's principal openings are upon the land originally purchased. The first shipment of ore (4,658 tons) was made in 1858; since which the increase has been so great that its shipments now exceed those of any mine in the district, as will be seen by reference to the tables. This company have recently constructed, in Marquette, the Grace Furnace, which went into blast in December, 1872, using anthracite coal in the manufacture of pig-iron. The furnace is located on the shore of the bay, within the limits of the city, and is the first anthracite furnace built on Lake Superior. A map of the Lake Superior and Barnum mines accompanies this report.

The Eureka Iron Company was organized October 29th, 1853, with a capital stock of $500,000 in 20,000 shares. The corporators were Eber B. Ward, Harmon De Graffe, Silas M. Kendrick, M. Tracy Howe, P. Thurber, Elijah Wilson, Thomas W. Lockwood, and Francis Choate, with office in Detroit. The organization was effected with a view of mining ore and of manufacturing charcoal pig-iron from Lake Superior ores; preparations were made to build a furnace in Marquette county, but the location was finally changed and the furnace erected where now stands the flourishing city of Wyandotte, becoming the nucleus of the extensive iron works which have since grown up in that locality. The Eureka Company was also the first iron enterprise in which Captain E. B. Ward, subsequently so widely known as a successful iron master, became engaged. The company was formed by Philip Thurber, and a quarter section of land purchased near Marquette of Mr. A. R. Harlow, on which a few hundred tons of ore were mined; but it becoming evident that the ore did not exist in quantity, the work was abandoned. This land was subsequently sold back to Mr. Harlow for his shares of the company's stock, and is now known as Harlow's Mill.

The Cleveland Iron Mining Company filed articles of association March 29th, 1853; capital stock, $500,000, in 20,000 shares. The corporators were John Outhwaite, Morgan L. Hewitt, S. Chamberlain, Samuel L. Mather, Isaac L. Hewitt, Henry F. Brayton, and E. M. Clark, with office in Cleveland, Ohio. The early history of this celebrated mine, one of the oldest and most important in the district, is referred to in connection with that of the Jackson Co.

Dr. J. Lang Cassels, of Cleveland, to whom reference is made in Mr. Berry's letter, visited Lake Superior in 1846, and took, as he expresses it, "squatter's possession" of a square mile for the Dead River Silver and Copper Mining Co. of Cleveland; the property here spoken of includes the mines of the present Cleveland Co. The Jackson Co. had previously taken possession of their lands, and Dr. Cassels obtained guidance thereto from an Indian, there being no white men in the region; the doctor went up from and returned to the Sault in a bark canoe. During the succeeding year, Cassels having left the country, the location was taken possession of by Messrs. Samuel Moody, John Mann and Dr. Edward Rogers. The two former claiming what became the Cleveland mine, and the latter what is known as the Lake Superior. When the Marquette Forge Co. was organized in Worcester, as previously described, Clark had authority from Mann and Moody to lease their location, and Graveraet had similar power from Rogers.

In this manner leases of these lands were put into the organization against $20,000 cash capital, to be paid by Messrs. Harlow and

Fisher. Both the Cleveland Co. and Graveraet, representing Messrs. Moody and Mann, claimed priority of right to the land under a "pre-emptor's mining act." These conflicting claims went before the Department at Washington, where a decision was rendered, which gave the right of purchase to the Cleveland Co. The entries which the Cleveland Co. made did not cover the Lake Superior location, Graveraet still claiming it, in behalf of the Marquette Co., on the ground of the Rogers pre-emption. Previously Isaiah Briggs had been on the land, but, leaving it, Rogers had taken possession. Rogers lost his interest, however, by not being present at the Government sale of lands in November, 1850, and establishing his claim, having been detained by a storm on the lake while endeavoring to proceed to the Sault (where the land office was located) for that purpose. The location was purchased by John Burt, on the basis of the Briggs claim, he having agreed to lease an undivided one-half interest to Graveraet, who was also present in behalf of the Rogers claim. This lease to Graveraet was assigned by him to the Marquette Co., passed with the company's other assets into the possession of the Cleveland Co., and was finally sold for $30,000 to the Lake Superior Iron Co., that company having previously purchased the Briggs title.

The Cleveland association, although formed in 1849, did not do any business in Lake Superior until 1853; at that date the Cleveland and Marquette companies became finally merged by the former company purchasing (including 64 acres of land on which the forge was located) the assets of the latter, and the present Cleveland Iron Co. was formed. The Cleveland Co. continued to run the forge for about two years, until it was burned down. The company mined in 1854, 4,000 tons of ore, which was made into blooms at the different forges in the vicinity. In 1855 they shipped 1,449 tons of ore to the furnaces "below," thus preceding the Jackson Co. one year, and becoming the first to send out of the region any considerable amount of ore. The Jackson Co. had sent a few tons to the World's Fair in New York in 1853, and in 1852 some had been sent to Sharon, as before mentioned. The Cleveland Co. has also an ore dock at Marquette, entirely similar to the docks of the M. H. & O. R. R. Co., of which full descriptions and illustration are given.

On Nov. 8th, 1853, the **Collins Iron Co.** filed articles of association, with a capital stock of $500,000 in 20,000 shares. The corporators were Edward K. Collins, of New York, Solon Farnsworth, Edwin H. Thomson, Robert J. Graveraet, and Charles A. Trowbridge, with office in Detroit.

The company built a forge in 1854, and began to make blooms late in the fall of 1855; Robert J. Graveraet, Supt., and C. A. Trowbridge, Managing Director. E. K. Collins largely interested himself with a view of obtaining a superior quality of iron for the shafts of his ocean steamers. In 1858, about the time the Pioneer Furnace was completed, Mr. S. R. Gay, who had been engaged on that work, leased the Collins Forge and put up a cupola there in which he made some pig-iron. The company immediately thereafter constructed a blast-furnace under the direction of Mr. Gay. This furnace was completed and put in operation December 13th, 1858, with a single stack; all the necessary power being afforded by the Dead river, upon which the furnace is located.

On August 28th, 1854, **the Peninsula Iron Co.** filed articles of association, with a capital stock of $500,000 in 20,000 shares. The corporators were Wm. A. Burt, Austin Burt, Wells Burt, John Burt, Heman B. Ely, Samuel P. Ely, and Geo. H. Ely; the two latter of Rochester, N. Y., the others of Michigan. Office of the company, Marquette, Mich. The company originally owned 800 acres of iron lands, which it sold in 1862 to the Lake Superior Iron Co., and determined on building a blast-furnace at Hamtramck, Detroit, Mich., which furnace was completed in February, 1863, and is still in successful operation. The company also operated a saw-mill for a few years, which they built on the Carp river, a short distance from Marquette.

Oct. 11th, 1854, the articles of association of the **Chicago and Lake Superior Iron Mining and Manufacturing Co.** were filed. Capital stock, $500,000, in 20,000 shares. The corporators were B. S. Morris, Isaac Shelby, Jr., Geo. Staley, Henry Frink, and Samuel S. Baker, all of Chicago, Ill.; and Solomon T. Carr and Fairchild Farrand, of Jackson, Mich. No permanent mining work was ever done by this company.

The Clinton Iron Co. Organized by forgemen from Clinton Co.,

New York, Jan. 20th, 1855. Capital stock, $25,000. *Corporators*, Azel Lathrop, Jr., H. Butler, Chas. Parish, and Daniel Brittol.

The object for which the organization was effected was to lease and operate the Jackson Forge. The company being composed of workmen, who at the time were employed in that concern and were locally styled the "Mudchunk." The market price of blooms being much below the cost of their manufacture, they were enabled to operate the forge but a brief period, and having become hopelessly involved in indebtedness, the company permanently suspended.

The Forest Iron Co. filed articles of association, September 22d, 1855, with a capital stock of $25,000 in 1,000 shares. The corporators were Matthew McConnell, Wm. G. Butler, Wm. G. McComber, M. L. Hewitt, and J. G. Butler. This company was organized for the purpose of putting up a bloom forge on Dead river, and the location became known as Forestville. McConnell, Butler and McComber commenced operations at this point as early as 1852 on their own private account, but becoming financially embarrassed, they sought relief by organizing a company as above indicated, who continued the manufacture of what was called half blooms, the production of which cost them from $180 to $200 a ton. These selling in Pittsburg for $35 to $40, on six months' time, it naturally resulted in the ruin of the company.

To the original projectors of the **Pioneer Iron Co.** belongs the credit of having established the first blast-furnace on Lake Superior; previous to that all the iron manufactured had been made in bloomeries. Mr. C. T. Harvey was the mover of the scheme, and the originator and manager of the company. He induced capitalists (chiefly in New York) to embark in the enterprise, Mr. E. C. Hungerford of Chester, Conn., being chosen Secretary and Resident Treasurer. Although the business was unknown to a single man on Lake Superior, the most sanguine views prevailed from the outset, and a two-stack furnace was constructed near the Jackson mine.

The late S. R. Gay and L. D. Harvey, now Superintendent of the Northern Furnace, were the builders; the work being commenced June, 1857, and completed so as to make the first iron in February of the next year.

Much of the material, including two millions of brick, was brought from Detroit and had to be hauled 13 miles from Marquette by teams; the engines were made at the West Point Foundry. The original stock was $125,000, in 5,000 shares; the articles were filed July 20th, 1857, the corporators being Moses A. Hoppock, Wm. Pearsall and Chas. T. Harvey. Most of the parties interested in the concern were totally ignorant of iron-making and as an instance illustrating the fact, it is related that one of the directors, during the period of construction, inquired when the furnace would be completed so that it might be sent up to Lake Superior; he supposing it was being made in Detroit. These unfavorable circumstances, combined with the financial depression of 1857, at which time the company were obliged to sell their iron for $22, while the cost of its production was $24 per ton, gave no return save anxiety and disappointment.

In the spring of 1860, the furnace was leased for four years to Mr. I. B. B. Case, he agreeing to deliver the pig-iron on board the vessels at Marquette for $17.50 per ton, and paying all the expenses of its manufacture; the company furnishing the timber, standing, for the charcoal, and giving him the advantage of a contract with the Jackson Company for the ore, the royalty for which ($1.00 per ton of iron) he paid. This price proved to be less than the iron could be made for. The furnace was burnt down August 9th, 1864; number two stack was at once rebuilt and put in operation in January following, by Mr. Case.

In 1865, Dr. J. C. McKenzie, then President of the Pioneer Iron Company, entered into negotiations with the Iron Cliff Company, which subsequently resulted, largely through the instrumentality of Major T. B. Brooks, Vice-President of the latter company, in an arrangement (ratified by the stockholders of both companies, March 10th, 1866) by which the Iron Cliff Company came into possession of the furnace, on consideration that it pay to its former proprietors one-third of the profits of the business. Soon after the two companies became practically one, through the purchase of the stock of the Pioneer by the Iron Cliff Company.

The Detroit Iron Mining Company filed articles 15th August, 1857. Capital, $500,000, in 20,000 shares at $25 each, with office in Detroit. Corporators were Patrick Tregent, Guy Foot, Joseph

P. Whittemore, John H. Harmon, John W. Strong, Oville B. Dibble, Nelson P. Stewart, Andrew T. McReynolds, Thornton T. Brodhead, Henry T. Stringham, Henry J. Buckley, Joseph L. Langley, of Detroit, and Edwin H. Thomson, of Flint. The company having ascertained, as they believed, that their lands did not contain sufficient ore for mining purposes, sold them to Mr. J. P. Pendill, and upon them is now built a portion of the village of Negaunee. The McComber mine, which lies at a short distance south of that village, is on this land.

The Excelsior Iron Company filed articles October 6th, 1857. Capital stock, $100,000; 4,000 shares, at $25 each. Corporators were: C. T. Harvey, Sarah V. E. Harvey, E. C. Hungerford, George P. Cummings, and Joseph Harvey, all of Marquette. This company did little but organize. It originated with Mr. C. T. Harvey, and some of the land which it owned has since proved to be valuable mining property, as it embraces the Barnum mine, now owned by the Iron Cliff Company; upon it is also situated a portion of the village of Ishpeming.

The Lake Superior Foundry Company filed articles of association July 14th, 1858. Capital stock (paid in), $10,000; 400 shares, at $25 each. *Corporators:* John Thorn, Isaac Maynard, Thomas Maynard, Nathan E. Platt, of Utica, N. Y., and Charles T. Harvey, of Marquette, Mich. This establishment, which was started in 1858, is now running on a much enlarged scale, under the name of the **Iron Bay Foundry,** D. H. Merritt, proprietor. The location is near the bay, within the city of Marquette.

The Grand Island Iron Company filed articles May 3d, 1859. Capital, $400,000; 16,000 shares, at $25 each; paid in, $110,000. *Corporators:* Thomas Sparks, Henry W. Andrews, William Lippincott, John L. Newbold, John D. Taylor, John R. Wilmer, Samuel Pleasants, William M. Baird, Samuel J. Christian, L. de la Cuesta, William A. Rhodes, Charles Lennig, James C. Fisher, Samuel T. Fisher, Lewis Seal, Coleman Fisher, Henry Maule, William Gaul, J. T. Linnard, Howard Spencer, Caleb Jones, Charles W. Carrigan, of Philadelphia, and Devere Burr, of Washington, D. C., with office in Philadelphia. The property belonging to this company, consisting of 3,000 acres of land, situated on Grand Island harbor, in Munising Township, was sold in 1867 to the Schoolcraft Iron Company, and their operations were confined to some minor improvements in the way of wharves, etc.

The Northern Iron Company filed articles May 16th, 1859. Capital stock, $125,000, in 5,000 shares of $25 each. *Corporators:* John C. Tucker, Moses A. Hoppock, of N. Y., and Charles T. Harvey, of Marquette, with office in Marquette. This company was formed through the efforts of C. T. Harvey, and constructed a blast-furnace at the mouth of the Chocolate river, 5 miles south of Marquette, with a view of making pig-iron with bituminous coal, being the first enterprise of this kind inaugurated in this region. After making about 1,000 tons of iron, the furnace was changed into and run as a charcoal furnace up to June, 1867; since which time it has not been working, and it is now being changed back into a bituminous coal furnace. This is the first charcoal furnace on the Upper Peninsula that has been permanently blown out.

1863.—The great financial prostration of 1857, combined with numerous causes which readily suggest themselves, naturally embarrassed and, in instances, extinguished the new and struggling enterprises of Lake Superior to the extent, that comparatively little was done in the manufacture of iron or the mining of ore up to the opening of 1863. During this interval of time no companies of importance filed articles of association in this region. Very early in the war, however, the greatly increased demand for iron which it occasioned, began to be felt over the country and finally extended its influence to Lake Superior, causing the revival of the languishing enterprises already started and the organization of many new ones. The abundance of ore, together with its surpassing richness in iron and freedom from deleterious substances, the facility with which it could be mined and the greatly improved means of transportation, were becoming generally known, and the strength and exceeding tenacity of the iron manufactured therefrom universally acknowledged. Thus altogether there was opened to the Marquette region an outlook of prosperity, which it had not heretofore experienced, enabling its mining and iron manufacturing companies to assume a basis of more successful operation, and confidently to push forward their improvements.

The articles of association of the **Teal Lake Co.** were filed on the 7th of June, 1863, with a capital stock of $500,000, in 20,000 shares, and an amount paid in of $100,000. The corporators were George A. Fellows, John W. Wheelwright and Charles L. Wright, of New York, with office in New York. Beyond some explorations this company never did any work on Lake Superior, confining its operations chiefly to stock speculations, it being the only iron mining company organized in this region, whose stock was sold at the Brokers' Board in New York.

The articles of association of the **Morgan Iron Co.** were filed on the 1st of July, 1863, with a capital stock of $50,000, in 2,000 shares, and $26,000 paid in. Corporators were Joseph S. Fay, of Boston, Lewis H. Morgan, of New York, Harriet H. Ely, Samuel P. Ely, Ellen S. White and Cornelius Donkersley, of Marquette, with office in Marquette. The capital stock was subsequently increased to $250,000, in 10,000 shares fully paid. The company own 20,000 acres of timber land. In 1863 they constructed the Morgan Furnace, eight miles west of Marquette on the M. H. and O. R. R., and the location has since become known as "Morgan." The furnace was put up under the supervision of Mr. C. Donkersley and has been successful. It went into blast Nov. 27th, 1863, making that year 337 tons of iron, and was the first furnace company in the region to pay a dividend to its stockholders. The extreme high price of iron, created by the war, enabled the company to realize, during the first ten months of the operation of the furnace, a dividend of 100 per cent. over and above the total outlay in its construction. Having exhausted the fuel in the vicinity, the company constructed charcoal kilns upon their lands at a distance of nine miles north from the furnace, and provided for the transportation of the coal by building a wooden railway thereto. The kilns and railway were made in 1869, and most of the coal now used is prepared at these kilns.

In 1867 the Morgan Company built the **Champion Furnace**, which went into blast Dec. 4th of that year. This furnace is located at what is now Champion village, on the line of the M. H. and O. R. R., 31 miles west from Marquette. The ore used is mainly magnetic from the Champion mine, and the record of the furnace is one of gratifying success.

The articles of association of the **Marquette Iron Co.** were filed April 9th, 1864, with a capital of $500,000, in 20,000 shares of $25 each. *Corporators:* George Worthington, Truman P. Handy, Samuel L. Mather, N. B. Hurlbut, Richard C. Parsons, G. D. McMillen, John Outhwaite, of Cleveland, Ohio, and Charles I. Walker, of Detroit, Mich. This company was organized for the purpose of mining iron ore and owns 400 acres of land, lying contiguous to, and south of, the Cleveland mines, 240 acres of which was originally held by the latter company. Its stock is held by stockholders of the Cleveland Company. The year of its organization it shipped 3,922 tons of ore, and has been somewhat regularly in operation since that period.

The Magnetic Iron Co. was organized in 1864; the articles of the company were filed May the 6th of that year, with a capital stock of $500,000 in 20,000 shares. *Corporators:* John C. McKenzie, Alex. Campbell, of Marquette, and Edwin Parsons, of New York. Office in Marquette, but now in Philadelphia, Pa. The property owned by this company consists of 520 acres of land on Section 20, T. 47, R. 30. A shaft 60 feet in depth has been sunk, and other explorations made to test the ore-deposit and the company expect to take out ore, as soon as a branch road is built to the mine.

The Chippewa mining property comprises Section 22, T. 47, R. 30, W., owned by J. S. Waterman, of Philadelphia, and S. S. Burt, of Marquette; considerable exploring has been done on the property and some fair ore found, but no mining done. This property lies on the east side of Michigamme river and opposite the Magnetic and Çannon properties.

The Phœnix Iron Co. filed its articles of association June 7th, 1864. Capital, $500,000, in 20,000 shares, of which $20,000 was paid in. The Corporators were Wm. C. Duncan, Henry J. Buckley and Simon Mandlebaum, of Detroit, with office in Detroit. No mining or manufacturing was ever done in the Marquette Region by this company.

Washington Iron Company filed its articles of association July 30th, 1864. Capital stock, $500,000, in 20,000 shares, at $25 per share; amount paid in, $100,000. The Corporators were Edward Breitung, I. B. B. Case and Samuel P. Ely, of Marquette, Joseph S. Fay, of Boston, and Edwin Parsons, of New York.

This company made its first shipments of ore (4,782 tons) in 1865, and has since been in active operation. The land owned by the company comprises 1,000 acres in the northeast part of T. 47, R. 29, which was purchased of Silas C. Smith, J. J. St. Clair, J. C. McKenzie, and Alexander Campbell, who derived their title from the United States Government. The mine is on the M. H. and O. railroad, at a distance by rail from Marquette of 27 miles. All the company's surplus earnings have been expended in making extensive improvements, of which an adit or tunnel, now over 1,100 feet long, constitutes the chief. Their plans and expenditures have been on an extensive scale, and contemplate operations for a long period to come. The details of the mine, shafts, adit and underground workings, together with the geological structure, are fully shown by the map of the Washington mine, accompanying this report.

The Bancroft Iron Co. filed its articles of association September 12th, 1864; capital stock being $250,000 in 10,000 shares, of which $100,000 was paid in. The Corporators were Wm. E. Dodge, of New York, Samuel L. Mather, John Outhwaite and Wm. L. Cutter, of Cleveland, Peter White and Samuel P. Ely, of Marquette, and Henry L. Fisher and L. S. McKnight, of Detroit, with office in Marquette.

The location of this company is the same as that of the Forest Iron Co., heretofore described; the property of the latter having been purchased by Mr. S. R. Gay, in 1860, he erected on the water-power employed by the old forge a blast-furnace, this being the second furnace he had built on Dead river, the one at Collinsville having been constructed by him the winter before.

Mr. Gay* having died in 1863, his furnace at Forestville passed to the ownership of the Bancroft Iron Co., who have since continued

* It is a fact worthy of note, in connection with the services rendered by Mr. Gay, that he was the first among the iron men who visited Lake Superior to recognize the value of the hematite ores; while engaged in the construction of the Pioneer Furnace, he observed that the Jackson Co. were wasting their soft hematite in large quantities, they supposing it to be worthless. He at once called their attention to its value.

to operate it. The furnace is worked by Mr. L. Huillier on contract, the company paying him a certain price per ton for the iron delivered on the dock in Marquette.

The articles of **The Iron Cliff Co.** were filed September 15th, 1864, with a capital stock of $1,000,000, in 40,000 shares at $25 each. *Corporators:* William B. Ogden and John W. Foster, of Chicago, and Samuel J. Tilden, of New York. Office at Negaunee, Mich. This company in 1864 purchased of the St. Mary's Ship Canal and Mineral Land Co. the 38,000 acres of land which that company owned in Marquette county. Subsequently, as heretofore mentioned, the Iron Cliff Co. came into possession of the Pioneer Co.'s property, thus increasing its estate to over 40,000 acres. The company soon began the construction of a furnace near the Foster mine, which has never been completed. They own and are working the *Barnum* and the *Foster mines*, the latter of which was opened in the spring of 1865. The product is a soft hematite, which forms a good mixture with hard ores. This mine is situated on Secs. 22 and 23, T. 47, R. 27. The first shipment of ore therefrom was made in 1866, and the mine has since been continually worked.

The **Barnum mine** is situated on Sec. 9, T. 47, R. 27, connecting with the Lake Superior Co.'s principal opening. The first shipments of ore were made during 1868, the ore being specular and of excellent quality. The C. and N. W. R. R. has a branch running into the mine, over which shipments are made. The mine is supplied with pumping and hoisting machinery. The map of the Lake Superior mine, which will be found in the accompanying Atlas, embraces the Barnum mine.

On that portion of the estate purchased of the Excelsior Company, in addition to the Barnum, a deposit of specular ore has been found near the corner of Secs. 5, 6, 7, and 8, T. 47, R. 27, which promises well; a branch railroad has been surveyed to it. Besides those already mentioned the company have several other openings. One on Sec. 15, adjoining the Pittsburgh and Lake Angeline Co., opened during the past season, which gives a fine showing of hematite ore. The Cliff-Parsons, also opened during the past season, adjoins the *Old Parsons*, on Sec. 21, T. 47, R. 27.

Another opening is near the quarter-post between Secs. 17 and 18, T. 47, R. 26, from which ore was shipped during the season. A second opening is being made on this same line, at a point farther north, near the section corner. These openings belong to the Negaunee Hematite Group. In addition to their own mines the company are working the Pioneer opening of the Jackson mine on a lease. Near the Foster mine the company have in operation a saw-mill, to which is attached shingle and lath mills.

In 1864 the *Ogden* and *Tilden* mines, situated on Secs. 13, 23, and 24, T. 47, R. 27, were extensively opened, and the branch road, which also extends to the Foster, built to them. The ores, however, proved of too low a percentage to sell in the then existing market, and the work was abandoned. The purchasers of the Iron Cliff estate also controlled the Chicago and Northwestern Railroad, and a short time previous to the purchase effected a consolidation with the Peninsula Road of Michigan, with a view to the future development of iron deposits on this extensive property, and the control of the railroad facilities for transporting the product of these and other mines to Lake Michigan.

The Iron Mountain Mining Co. filed its articles of association Nov. 1, 1864, paid in $100,000. *Corporators:* Geo. E. Hall, of Cleveland, O., Richard Hays, Henry A. Laughlin, and Irwin B. Laughlin, of Pittsburgh, and Gilbert D. Johnson, of Ishpeming. The company own 320 acres of land, being the S. ½ of Sec. 14, T. 47, R. 27. The first shipments of ore were made in 1865, a branch of the C. and N. W. R. R. extending into the mine. All work at this mine has been discontinued, owing to the leanness and refractory nature of the ore, its yield being less than 50 per cent. of iron in the furnace. This mine has been recently leased to Messrs. Clark and Colwell, under whose auspices work will be resumed in the spring of 1873, with the view of finding hematite.

The Michigan Iron Co. filed its articles of association Dec. 30th, 1864. Capital stock, $500,000, in 20,000 shares of $25 each. *Corporators:* Henry J. Colwell, Andrew G. Clark and Samuel P. Ely, of Marquette, with office there.

This company own a large amount of woodlands, two furnaces and considerable other manufacturing property. The Michigan furnace was built by them in 1866, went into blast June, 1867, and has since been in constant operation; it is on the M. H. and O. R. R., 23 miles west of Marquette, and is surrounded by the village of Clarksburgh.

The remaining furnace owned by this company, known as the Greenwood, went into blast in June, 1865, and was purchased by the Michigan Co., together with about 8,000 acres of land of the M. and O. Rd., in 1868. Greenwood is 27 miles from Marquette, on the line of the M. H. and O. R. R., and the furnace has continued in blast since the time of its purchase by the present owners.

In 1864 **The Peninsula Railroad**, from its junction with the Marquette and Ontonagon Railroad at Negaunee to Escanaba (a distance of 62 miles), was completed and put in operation. The project which has resulted in opening this important outlet to the great iron mines was first definitely broached in 1855. In that year meetings were held at Ontonagon, Marquette, and all important points to Milwaukee, with a view to the united action of the people along the route, in the endeavor to obtain governmental aid in the construction of the railroad. These meetings were chiefly initiated by Mr. C. T. Harvey and H. B. Ely. Mr. Harvey, John Burt and others, immediately proceeded to Washington and were instrumental in obtaining from Congress the passage of an act, June, 1856, which donated a large amount of land in aid of railroad enterprises.

Among the projects for which provision was thus made in this grant were the building of a railroad from Marquette to Little Bay de Noquette, and also from thence to Menominee, as well as for the extension of a road from Fond du Lac to this latter point. In 1859, the Chicago, St. Paul and Fond du Lac Railroad Co. (which company had received from Wisconsin the congressional grant), through the agents of its bond-holders, organized under the name of the Chicago and Northwestern Railway Company, and in 1861, under a law of the State of Wisconsin, proceeded to locate a line by the way of Fort Howard to the Menominee river. In 1862 the State of Wisconsin conferred upon the C. and N. W. R. R. Co. all the franchises and rights heretofore granted to the several companies of which it had become the successor; and in the same year the road was extended to Green bay, a distance of 242 miles from Chicago.

The Iron Mountain road was completed and became consolidated with the Bay de Noquette railroad in 1858. The location of the Marquette and State line grant was changed by act of Congress in 1860, so as to extend from Menominee northward along the shore of Green Bay, and thence to Negaunee; and in 1863 the Marquette and State line grant, with the remainder of the Bay de Noquette grant (being coincident with it from Negaunee to Escanaba) having been suffered to lapse, were, by agreement between the grantees, conferred by the State upon the Peninsula Railroad Co., of Michigan. Surveys were made in 1862 (the enterprise being set on foot by C. T. Harvey, who subsequently transferred it to S. J. Tilden, of New York), and work began in the summer of 1863, and in December of the following year the road was opened to the public. During the preceding October, however, the Peninsula road had consolidated with the Chicago and Northwestern, and the line from Marquette to Menominee became known as the Peninsula division of the C. and N. W. R. R. The lands owned by the Peninsula division embrace in the aggregate 1,200,000 acres.

An extensive ore dock was constructed at Escanaba, upwards of 1,300 feet in length, 32 feet in height, and 37 feet in width, capable of receiving in the pockets 20,000 tons of ore at a time, and of shuting it thence into the holds of vessels. This dock was built at an expense of about $200,000. Communication to this excellent and accessible harbor being thus opened, and such ample facilities afforded for the transmission and shipment, large and increasing amounts of ore have since been carried yearly over this route.

Corning Iron Co. filed articles of association March 23d, 1865. Capital stock, $200,000—8,000 shares of $25. *Corporators:* G. C. Davidson, S. Churchill and Chas. T. Harvey, with office in Marquette. This company did nothing worthy of note.

The New York Iron Mining Co. Incorporated April 8th, 1865. Capital stock, $250,000, in 10,000 shares of $25 each. *Corporators:* Samuel J. Tilden, J. P. Sinnett and J. Rankin, of New York.

The mining operations of this company are conducted in the southeast ¼ of southeast ¼, Sect. 3, T. 47, R. 27, being 16 miles west from Marquette and adjoining the Cleveland. The mine is worked under a lease from Mr. A. R. Harlow and the stock is all held by Mr. S. J. Tilden and Messrs. W. L. and F. W. Wetmore. Operations were commenced in the mine in 1864, during which year 8,000 tons of ore were shipped. The statement of its yearly product and other details will be found by reference to the tables in this work; the workings and geological structure are shown by a map. This company is identical with the New York and Boston Iron Mining Co., and also with the New York iron mine, incorporated March 31st, 1865; it soon after changed to the New York Iron Mining Co., as above described.

The Pittsburgh and Lake Angeline Iron Co. was incorporated Nov. 11th, 1865. Capital stock, $500,000, in 20,000 shares of $25 each. James Laughlin, *President*, T. Dwight Eels, *Secretary* and *Treasurer*. The company own 1,376 acres of land, situated in T. 47 and 48, R. 27 and 28, of the former Town., and R. 31 of the latter. They also hold a lease of about 300 acres, on which is located the Edwards mine. The company's mines consist of the Lake Angeline and Edwards; the *Lake Angeline* mine is situated on the south shore of Lake Angeline and on the line of the M. H. and O. and C. and N. W. R. Rs., 17 miles from Marquette and 66 miles from Escanaba, and produces both specular and hematite ore, the latter of first quality.

The **Edwards mine** lying contiguous to the Washington, is also on the line of the M. H. and O. R. R., distant from Marquette 28 miles, and produces only magnetic ore. Work was commenced in 1865, the first shipments being made in the following year. The mining is all conducted underground, the ore being raised to the surface through shafts and is the only mine in the Iron Region which has been exclusively worked in this way. The results of this company's operations are shown in the accompanying tables and the mine workings by maps and illustrations.

The Schoolcraft Iron Co. filed articles of association April 8th, 1866. Capital stock, $500,000, in 20,000 shares of $25 each. Paid in, $250,000; the remaining 10,000 shares being held by the com-

J Bien, lith. Photo. by Child.

JACKSON IRON CO'S. FURNACES AND COAL KILNS

pany. *Corporators:* Hiram A. Burt, Peter White and H. R. Mather, of Marquette; office at Marquette, Michigan.

A furnace was constructed by this company at Munising, Schoolcraft county, on Grand Island bay, which went into blast in June, 1868, and was blown out in about six months thereafter. The furnace continued "in and out" of blast somewhat irregularly, until the company went into bankruptcy. In 1871 the furnace and other property, including 40,000 acres of hard wood land, which had belonged to them, passed into the hands of Peter White, Esq., by whom it was transferred to the Munising Iron Co., an organization effected for the purpose of owning and operating this estate, which is now being successfully done. Mr. Peter White, of Marquette, is managing director.

The Marquette and Pacific Rolling-Mill Co. filed its articles of association Oct. 1st, 1866. Capital stock, $500,000, in 20,000 shares of $25 each. The corporators were John Burt, Samuel P. Ely, Wm. Burt, Edward Breitung, Timothy T. Hurley, Cornelius Donkersley, W. L. Wetmore, Peter White and Alvin C. Burt, of Marquette. Office in Marquette, Mich.

The company has constructed at Marquette a bituminous blast-furnace, with rolling-mill connected therewith. The works are located near the lake shore, at a short distance south from the city, went into operation in the summer of 1871, and are connected with the M. H. and O. R. R. by a branch track. Upon their land at Negaunee, the company have opened a mine of manganiferous hematite ore, to which a side track has been extended, connecting it with both railroads; from this mine the company's furnace at Marquette is in part supplied. This rolling mill is the first erected on Lake Superior, and the furnace the first which has continually used bituminous coal. H. A. Burt is superintendent.

The *Fayette Furnace* was constructed and put in operation in December, 1867, the enterprise originating with Major Fayette Brown, general agent of **The Jackson Iron Co.** It is located at "Snail Shell Harbor," in Big Bay de Noquette, 20 miles east of Escanaba, and about it has grown up the beautiful village of Fayette. It is owned by the Jackson Iron Co., with general office in Cleveland, Ohio. The company own 16,000 acres of land, excellently well timbered with hard wood, and generally adapted to agricultural purposes, the soil being of limestone formation. From the ledges of limestone, which exist in the immediate neighborhood, material for the necessary flux is obtained, as well as for the manufacture of all the lime used by the company. They possess a full complement of charcoal kilns, and a large portion of the necessary wood is purchased, the company preferring to save their own timber as long as possible. This wood is delivered by the parties of whom it is bought at the furnace, or along the line of the company's railroad, of which they have constructed for this purpose six miles, laid with T-rail, and operated with two small locomotive engines, it being the only furnace on the Upper Peninsula that operates a locomotive railway for the exclusive purpose of transporting fuel. The company have also a saw-mill, machine-shop, etc. The furnace, as originally started, consisted of a single stack, which is shown in the accompanying illustration. A second one was subsequently erected, and both stacks have since been in operation with results more favorable, than any other charcoal furnaces using Lake Superior ore. The extraordinary favorable working of these furnaces will be fully realized from the following statements, furnished from the company's reports: During the 73 days immediately preceding April 13th, 1872, there were made in the No. 1 stack an average of $27\frac{6}{10}$ tons per day, using 94 bushels of charcoal and 125 lbs. of limestone per ton, the ore being from the Jackson mine and yielding from 62½ to 64½ %. On August 4th following, the same stack again went into blast, making, during the first quarter, a period of 91 days, 2,258 tons of iron, an average of 27^{8} tons per day, using by measure 92 bushels of charcoal per ton. No. 2 was also in blast during a portion of the same period with corresponding results. On December 14th No. 2 stack had produced, during the previous four weeks, an average of $26\frac{39}{100}$ tons per day, and on January 18th, 1873, had produced, during the previous five weeks, an average of $29\frac{34}{100}$ tons per day; the charge used during this time was 26½ (called 30) bushels of charcoal, 1,000 lbs. of ore (⅓ soft and ⅔ hard specular Jackson), 35 lbs. of limestone and 10 lbs. of clay.

These results require no comment relative to the efficiency of the management. The coal is of the best quality, kept dry under shelter, as is also the ore, which is crushed finer than is customary. The stacks are each 42 feet high inside and 9 feet 6 inches bosh;

4 feet 8 inches, and 5 feet 8 inches diameter, 3 feet below the top, and 4 feet and 5 feet at the top respectively. The hearths are 4 feet diameter battering from the bottom; the tuyeres, three in number, with 3½ inch nozzle, are placed 40 inches above the bottom of the hearth. Two blowing engines are used, the cylinders respectively 36 and 48 inches in length, with diameter of 50 and 44 inches. The engines make from 24 to 28 revolutions per minute, and both of them are only run when the two stacks are in operation. The temperature of the hot blast averages in one about 600° and in the other 750°. Originally No. 2 stack had a five-foot cone, but did not make as much iron, nor as cheaply, as the other, until the cone was reduced in height to 4 feet 4 inches, since which time it has worked equally well with the other. The total product of these furnaces during 1871 and '72 was 19,117 tons, which were used as follows:

For Bessemer Steel..............	17,465	tons.
" Malleable Iron..............	88	"
" Wheels.....................	787	"
" Foundry, etc................	400	"
" Forge purposes..............	377	"

Genl. Agt., Major Fayette Brown, Cleveland, Ohio. Local Agt., C. L. Rhodes, Fayette, Mich. Founder, Jos. Harris, Fayette, Mich.

The Deer Lake Iron Company.—Articles of association were filed July 9th, 1868. Capital, $75,000—3,000 shares at $25 each. *Corporators:* George P. Cummings, of Marquette, Edward C. Hungerford, of Chester, Conn., Gardner Green, Caleb B. Rogers, Moses Pierce, Samuel B. Case, Theodore T. McCurdy, John E. Ward, James Lloyd Greene, James C. Colby (Ex'r), Daniel T. Gulliver, William R. Potter and Enoch F. Chapman, of Norwich, Conn.; Giles Blague, Jr., New York, Geo. Smith, New York, G. F. Ward, E. R. Ward, Old Saybrook, Conn., and James H. Mainwaring, of Montville, Conn., with office at Marquette, Mich.

This company organized for the purpose of smelting iron ore, and immediately constructed a furnace, which went into operation in Sept., 1868. This furnace, the smallest in the district, is located at Deer lake on the Carp river, two miles north from the village of Ishpeming on the M. H. and O. R. R., with which place it is connected by a tram railway. The stack is 33 feet high and 7 feet 8 inches bosh, thus making it perhaps the smallest furnace which has been built in the United States during the past 7 years. Another peculiarity of this furnace is the comparatively enormous size of its hot-blast oven, to which is doubtless due in part the favorable results, which, considering its small size and peculiar management, the furnace has accomplished. The oven, on the Pleyer plan, contains 45 tons of metal, which is 50 per cent. more than that contained in the ovens of our largest charcoal furnaces; having twice the capacity of the Deer lake stack. The furnace is driven by water, employing an 18-inch turbine wheel under 35 feet head, thus leaving all the gas available for heating the blast, which is brought to an extremely high temperature. It runs but six days in the week, "banking up" Saturday night and starting again on Sunday night. Notwithstanding an arrangement necessarily disadvantageous to the greatest production, the furnace has averaged during several consecutive weeks 11 tons of pig-iron per day, using 110 bushels of charcoal to the ton, one-half of which is made from pine slabs,—the ore used being hard ore from the New York mine, averaging 66 per cent. The origin of this enterprise is due to Mr. E. C. Hungerford, who also determined its unusual size and the peculiar policy under which the furnace has been managed. Near the present one the company are now building a new iron shell furnace, 9 feet bosh.

The Cannon Iron Company.—Articles filed July, 1869. Capital, $500,000; 20,000 shares, $25 each. *Corporators:* Bernard A. Hoppes and Wm. H. Berry, of Philadelphia, and Samuel S. Burt, of Marquette, with office in Philadelphia. This company organized for the purpose of mining iron ore, but beyond making explorations on their lands with this view, nothing has as yet been done.

Bay Furnace Company.—Articles filed July 19th, 1869. Capital stock, $150,000; 6,000 shares at $25 each. *Corporators:* William Shea, of Munising, Mich., George Wagner, Jay C. Morse, Frank B. Spear and James Pickands, of Marquette, John Outhwaite, of Cleveland, and John P. Outhwaite, of Ishpeming, Mich., with office in Marquette.

This concern organized for the purpose of smelting iron ore, and immediately proceeded to the construction of a blast-furnace for that purpose. This furnace was completed and went into operation on the 6th of March, 1870. It is located at Onota, in Schoolcraft county, on Grand Island bay, 40 miles from Marquette. But one stack was originally constructed; a second one, however, has since been erected and put in readiness for the blast. The ore used is from the Cleveland and McComber mines, received by the way of Marquette. This company own about 20,000 acres of land, mostly hard wood timber, from which the fuel for the furnace is obtained.

The Whetstone Iron Company.—Organized Aug. 20th, 1869. Capital stock, $150,000, in 6,000 shares of $25 each. Office at Marquette. This company have not commenced operations. Corporators were William Burts, Samuel Peck, A. A. Cole, Thomas O. Hampton, Clark Stratton, A. S. Harvey and A. G. Benedict.

Champion Iron Company.—Organized August 23d, 1869, with a capital stock of $500,000, in 20,000 shares of $25 each. *Corporators:* Joseph S. Fay, of Boston, Edwin Parsons, of New York, Thomas C. Foster, of Cambridge, Mass., and Samuel P. Ely and Peter White, of Marquette. The company own about 1,600 acres of land, but their mining operations are conducted on that portion of their land comprising the south half of Sec. 31, T. 48, R. 29, being 32 miles by railroad from Marquette. The ore is principally magnetic, though a large amount of slate ore is obtained. The Champion mine is upon the south outcrop of the magnetic ore basin, which underlies Lake Michigamme, and near the village of Champion, about half a mile distant from the furnace of that name. The company are now working chiefly underground, as is fully shown in Map VII. of Atlas, where the geological structure and all other important details will also be found.

The Lake Superior Foundry Company filed their articles of association Sept. 2d, 1869, with a capital stock of $50,000—2,000 shares at $25 each. *Corporators:* Daniel H. Merritt, Lotan E. Osborn, Henry J. Colwell, William L. Wetmore, Jay C. Morse, Alfred Kidder, James Pickands and Thomas Fitzgerald, of Marquette, Mich.; Gilbert D. Johnson, Seymour Johnson, Harvey Diamond and Robert Nelson, of Ishpeming. The works (located at

Ishpeming) are quite extensive and adapted to general and particular foundry and machine work. (See Iron Bay Foundry, p. 33.)

Silas C. Smith Iron Company.—Articles of association filed Jan., 1870. Capital, $500,000, in 20,000 shares at $25 each. *Corporators:* Silas C. Smith, of Ashtabula, O., Oliver F. Forsyth and Wm. H. Lyons, of Flint, Mich., with office at Ashtabula, O.

The property of this company consists of 703 acres of land in Sections 18, 20, and 28, T. 45, R. 25, upon which have been made numerous openings, showing soft hematite ore in quantity, the main one being near the E. ¼ post of Sect. 18. A tunnel is being driven into the deposit, of sufficient size for the admission of railway cars from a branch road five miles in length, which connects with the Chicago and Northwestern railroad. The ore at present is loaded into the cars from temporary docks, provided with pockets for that purpose. The principal stockholders are Silas C. Smith, the discoverer, General James Pierce, of Sharpsville, Pa., and Henry Fassett, of Ashtabula, O. The shipments of ore and other details will be seen by reference to the mining tables.

The Pittsburgh and Lake Superior Iron Co. filed articles of association June 28th, 1870. Capital stock, $500,000, in 20,000 shares of $25 each. *Corporators:* James McAuley, C. T. Spang, C. G. Hussy, Thos. M. Howe and James M. Cooper, of Pittsburgh; Sherman J. Bacon, of New York, Joseph G. Hussy, of Cleveland and W. M. Sinclair, of Philadelphia; with office at Pittsburgh, Pa. The company own 2,691 acres of land in Towns. 47 and 48, Ranges 25 and 26, their title to which was derived direct from the United States Government. Work was commenced on their property near the Cascade mines in Sept., 1872, houses, etc., were erected, a railroad side track built and a pit opened on Sec. 32, which is called the Hussy mine, and from which about 2,000 tons have been shipped.

The Republic Iron Co. was organized Oct. 20th, 1870. Capital stock, $500,000, in 20,000 shares. Office in Marquette. *Corporators:* E. Breitung, S. P. Ely and Ed. Parsons. This company own 1,328 acres of land, being in part in Sections 6, 7, and 18, T. 46, R. 29, comprising what was formerly known as Smith mountain, which is unquestionably one of the largest deposits of pure specular and magnetic ore on the Upper Peninsula, if not in the United States. The great extent and value of this deposit was observed and commented on by the early United States surveyors, when engaged in running the township lines in that locality in 1846. The property was explored and selected by Silas C. Smith, of Marquette, and entered in the name of Dr. James St. Clair, in 1854 and 1855. A branch from the M. H. and O. R. R. has been constructed to the mine, over which the shipments of ore are now being made. See Tables, Plts. XII. and XIII. of Atlas. A complete map of this property, based upon careful surveys, exhibiting the topography, geological structure, magnetism and other important details, will be found in the Atlas accompanying this work, together with full descriptions.

The Cascade Iron Co. is an association of Pittsburgh men, owning 3,120 acres of land in Sections 19, 20, 29, 30, 31, and 25, T. 47, Ranges 26 and 27. These lands were entered by Waterman Palmer and purchased by the present company in 1869. An examination of the iron deposits in this locality was made by Dr. Douglas Houghton, in 1845, while engaged in running the interior section lines. (See Appendix D., Vol. II.)

The company's mines are provided with side tracks, connecting with a branch road of six miles in length to the C. and N. W. R. R. Mining operations commenced in 1871, and the openings (including the leased mines) are seven in number. There are other improvements, such as a saw-mill run by water, a store, sufficient number of dwellings, barns, repair-shop, etc. The expenditure which these improvements (including the branch railroad and side tracks) have necessitated has been very large, and future operations are contemplated upon a scale of considerable magnitude. (See Statistical Tables.)

The Cascade Company, under another organization, to wit, **The Escanaba Iron Co.**, are constructing a blast-furnace at Escanaba, to consist of two stacks, one of which will go into operation in January, 1873; the height of stack, 56 feet; diameter of bosh, 12 feet. The entire structure is built in the most complete and substantial manner, and when finished, will probably not be surpassed, if equalled, in capacity, durability, or beauty, by any similar furnace in the United States. The principal owners are Joseph Kirkpatrick, William Bagaley, James Lyon, William Smith, Samuel Riddle and Samuel Hartman; Joseph Kirkpatrick, *President*, James Lyon, *Treasurer*, and John L. Agnew, *General Superintendent*.

The Emma Mine, one of the Cascade openings, is on the E. ½ of E. ½ of N. E. ¼, Sec. 31, and is being worked under a lease from the Cascade Company by an association of Pittsburgh gentlemen, who are represented at the mine by Mr. James E. Clark. They commenced shipping ore in 1872.

The Bagaley Mine, likewise one of the Cascade openings, is also worked under a lease from the Cascade Company, by Messrs. Wilcox & Bagaley, and its total product is about 6,000 tons.

The Gribben Iron Co., having a capital stock of $500,000, in 20,000 shares of $25 each, was organized 1872. The mining property comprises a lease on the S. E. ¼, Sec. 28, T. 47, R. 26, being on the Cascade range. Mining and exploring operations during the season have resulted in taking out considerable ore, some of which has been shipped for testing. The company have built a side track, which connects with the Cascade branch of the C. and N. W. R. R. Officers of the company are: W. C. McComber, *President*, C. H. Hopkins, *Secretary*, and James Mathews, *Treasurer*; all of Negaunee, Mich.

The Carr Iron Co. was also organized in the summer of 1872, with a capital stock of $250,000. Its real estate comprises forty acres of land, situated on Sec. 33, T. 47, R. 26, being also in the Cascade range. The officers are Amos Root, *President*, Jackson, Mich.; E.W. Barber, *Secretary*, Jackson, Mich.; and W. H. Maynard, *Managing Director*, Marquette.

Negaunee Hematite Mines. A large number of new companies have recently been organized for the purpose of mining hematite ore in the vicinity of Negaunee. These new locations, which have been and are in process of being developed, are situated in Sections 6, 7, 8, and 18, T. 47, R. 26, and comprise what are known as the McComber, Grand Central, Rolling Mill, Himrod, Ada, Negaunee, Calhoun and Spurr, Green Bay, Allen, the Iron Cliff "Sec. 18," and other mines. The McComber mine, opened by William C. Mc-

Comber in 1870, is worked on a lease from J. P. Pendill, of Negaunee, at a royalty of fifty cents per ton for ore. The mine has been worked for the past three seasons, and in the spring of 1872 the lease was sold to parties interested in the Cleveland mine, who in July organized a company. The Rolling Mill mine, heretofore spoken of, is worked in part under a lease from A. L. Crawford. The company, however, own the greater portion of the land.

All these workings, except Sec. 18 and the McComber, are worked on leases from Edward Breitung, at 75 cents per ton royalty, he having leased from the owners, Messrs. Harvey and Reynolds, at 50 cents per ton royalty. Some of these pits have been worked during the past season, and nearly all of them are prepared for active operations during the coming year. Railroad side tracks are either completed, or in process of construction, to the several mines; dwellings and other improvements have been made, or are contemplated at each, and several of the locations bid fair to be the scene of active mining operations. The product is for the most part a soft hematite, containing usually from one to five per cent. of manganese, which renders the ore more easily worked in the furnace and is probably beneficial to the iron. The yield of metallic iron of the best of these ores is 50 per cent. and upwards, the average, however, is below that. See Map No. V. and Table Pl. XII. of Atlas.

Among the promising iron properties upon which work has been commenced during the present season, and from which large shipments may be reasonably anticipated, are the Michigamme and Spurr Mountain mines, at both of which work has actively commenced; side tracks are being constructed at both places, connecting with the M. H. and O. R. R. The mines are situated upon the same magnetic range and are about two miles apart.

The property of the **Spurr Mountain Co.** (which company was organized in September last) comprises 160 acres of land, and the point at which mining operations have been commenced is at what is known as Spurr mountain. The preliminary work has uncovered the south side of a very large mass of magnetic ore of a great degree of purity; rising at the highest point to a height of 60 feet above the surface of the ground at the base of the hill. This remarkable outcrop of ore is situated (as will be seen by reference to the accompanying map) 900 feet east and 700 feet north from the west and south boundaries respectively of the company's property. It was first discovered to the public in 1868. The examinations which have been made, established beyond any reasonable doubt the presence of the ore in a very large quantity and of a uniform purity and quality. The natural facilities afforded at Spurr mountain for commencing mining operations are excellent, and with the exception of Republic mountain there is, so far as known, no other locality in Marquette county where occurs so large an exposure of pure ore, rising at so great an elevation above the general level and at which there is apparently so little preliminary work necessary.

This range has been explored to a considerable extent in either direction; westerly, across the east half of Sec. 23, owned by the M. H. and O. R. R. Co., the examinations show the presence of the ore, but to how great an extent the deposit exists future workings alone can determine; easterly, as is elsewhere more fully related, the range has been traced along the north side of Lake Michigamme for several miles. The officers of the Spurr Mountain Co. are: H. N. Walker, Esq., of Detroit, *Prest.;* Col. Freeman Norvell, *Supt.* and *Sec.* The distances from the mine to the ports of L'Anse and Marquette are respectively, by rail, about 24 and 39 miles.

The Michigamme Co. was organized in the winter of 1870–71, the organization being effected mainly by persons already largely identified with Lake Superior iron interests. The land owned by the company comprises 1,400 acres, situated on the north side of Lake Michigamme. Preliminary work was begun in the spring of 1872, and prosecuted during the summer. The point selected for the commencement of mining operations is near the shore of the lake, and upon each side of the line between Sections 19 and 20, the developments resulting from this work thus far being of the most promising character. Improvements, not previously indicated, consist of a large, substantial steam saw-mill, with other machinery attached thereto, an office, dwellings, etc. At a short distance south and west from this location the company have laid out a village plat, to be called "Michigamme," and which promises to be built up with considerable rapidity. The distance to L'Anse is about 26 miles, and to Marquette 37, by rail. The officers of the company are: William H. Barnum, of Lime Rock, Conn., *Prest.;* James Rood, of Chicago, *Sec.* and *Treas.;* and Jacob Houghton, *Supt.*

The Keystone Iron Co. also organized in the fall of 1872, with capital stock of $500,000, in 20,000 shares of $25 each. The property comprises the southeast ¼ of southwest ¼, Sec. 32, T. 48, R. 29, distant from Marquette, by rail, 29 miles, from Escanaba 77, and from L'Anse 35. The company are at work preparing for mining the ensuing season. A. P. Swineford, Marquette, *General Agent.*

A number of mining enterprises, comprising **The Albion, Saginaw, Lake Superior Company's new openings, The New England, Winthrop, Shenango, and Parsons,** in Secs. 19, 20, 21, 16, T. 47, R. 27, are situated east and west, parallel and contiguous ranges of specular and hematite ore, are all connected by branches with the M. H. and O. R. R., and soon to be with the C. and N. W. Road.

The Albion mine, opened in 1871 by the brothers St. Clair, who hold the property comprising the northeast ¼ of the northwest ¼, Sec. 19, on a lease from Messrs. E. Breitung and S. L. Smith, at a royalty of 75c. per ton; up to the present time but a small amount of ore has been mined. The opening is immediately west of the Saginaw mine and on the same ore belt.

The Saginaw Mine, situated on the northwest ¼ of the northeast ¼ of Sec. 19, T. 47, R. 27, was opened in 1872, and during the same season shipped (via M. H. and O. R. R.) 19,000 tons of specular ore. The mine was worked on a lease by Messrs. Maas, Lonstorf and Mitchell, of Negaunee, on a royalty of 50c. per ton for the ore. During the fall of 1872 the lessees sold out to parties representing the Cleveland Rolling Mill Co. for $300,000, and immediately thereafter the Saginaw Mining Co. was organized with a capital stock of $500,000 in 20,000 shares. A. B. Stone, of Cleveland, *Prest.*, and A. G. Stone, of Cleveland, *Sec.* and *Treas.* A side track has been surveyed, to connect with the Chicago and N. W. Railroad, and the grading finished to the Winthrop mine. The land on which the Saginaw mine is located was purchased of the State of Michigan, with four other contiguous "40's" situated about the centre of same section, seven years ago, by Messrs. Heater, Elison and Conrad; the latter having made the selections.

Between the Saginaw and New England mines, on Sec. 20, the Lake Superior Iron Co. have a very promising opening, from which a considerable shipment of specular slate ore was made in 1872.

The New England Mine, on same range, is situated on the east ½, northeast ¼, Sec. 20, T. 47, Range 27. The shipments from this mine commenced in 1866, and up to the present time about 60,000 tons of ore have been mined and shipped via Marquette. The property is mainly owned by Captain E. B. Ward, of Detroit, and the mining operations are conducted by H. G. Williams under a contract. The principal part of the product is a hematite ore. A very narrow bed of excellent specular slate ore was worked several years, but not proving sufficiently profitable, work was discontinued. The ore is chiefly consumed at the extensive works controlled by Capt. Ward at Chicago, Milwaukee, and Wyandotte.

Adjoining the New England is the **Winthrop Mine**, situated in the southwest ¼, Sec. 21, T. 47, R. 47, owned by A. B. Meeker and A. G. Clark, of Chicago, and H. J. Colwell, of Marquette, and opened in 1870 by Messrs. Richardson and Wood, who work the mine on contract. Up to the close of 1872 about 25,000 tons of ore have been shipped, and the indications are favorable for increased shipments during the coming year. The product is a hematite ore, one of the richest of the class in the district. A. B. Meeker, of Chicago, is *Prest.*, A. G. Clark, *Sec.* and *Treas.*, and H. G. Colwell, Clarksburgh, *Gen'l Agt.*

The Shenango Iron Co. was organized in September, 1872, with a capital stock of $500,000, in 20,000 shares of $25 each. The land worked by the company comprises the north-west ¼ of south-east ¼ of Sec. 21, T. 47, R. 27, and adjoins the Winthrop, the deposit being a continuation of that mine.

The officers are C. Donkersley, of Appleton, Wis., *Prest.*, and H. D. Smith, *Sec.* and *Treas.*; in addition to these, E. Decker, Charles Reis and George L. Hutchinson, constitute the Board of Directors. A small amount of ore was shipped during the fall of 1872, and the company are erecting machinery, including the sinking of a shaft 60 feet in depth, with the view of doing considerable mining the coming season. The land is leased of the Williams Iron Co., who in turn lease of the Pittsburgh and Lake Angeline Co., who are the owners of land. The ore is mined by Messrs. Hurd and Orthey, part owners, on contract.

The Boston Mine, situated on the southwest ¼ of the northeast ¼ of Sec. 28, was organized in 1872, and a lease of the property above described secured by Messrs. Day, Anderson and others, with a view of mining operations. The lease of these parties is the same as that of the Shenango.

The Parsons, or "**Old Parsons,**" mine is located between the New England and the Lake Superior Companies' opening on Section 16, northeast of the Winthrop. Several thousand tons of specular slate ore were shipped from each of these mines, but work has been discontinued.

The Kloman Iron Co. was organized in December, 1872, with a capital stock of $500,000, in 20,000 shares. The corporators were Andrew Kloman, William Coleman, Thomas M. Carnegie, Jacob Houghton and T. B. Brooks. The company own 437 acres of land adjoining and northwest of the Republican mountain, being in part in Sec. 6, T. 46, R. 29, on the west side of the Michigamme river. The company have commenced mining on the continuation of the Republic mountain deposit and are building a short railroad to connect the mine with the Republic branch.

The Howell Hoppock Iron Mining Co. filed articles of association January 13th, 1873. *Corporators:* Lewis J. Day, Wm. R. Bourne, Wm. Rice, James S. Ward and Frank Austin. Office in Ishpeming, Mich. Organized to mine on the northwest ¼ of northeast ¼ of Sec. 28, T. 47, R. 27. Capital stock, $500,000, in 20,000 shares.

The Watson Iron Co. filed articles of association January 16th, 1873, with capital stock fixed at $500,000, in 20,000 shares of $25 each. *Corporators:* C. J. Hussey, E. T. Daro, Thomas M. Howe, M. K. Moorhead, George F. McLeane, W. J. Moorhead, Charles F. Spang, John W. Chalfant, Campbell B. Herron and James W.

Brown, all of Pittsburgh, Pa., and James W. Watson, of Marquette county, Mich. The property of this company comprises the northwest ¼ of Sec. 32, T. 47, R. 26 and which constitutes $325,000 of the capital stock. This ¼ section is a part of the estate of the Pittsburgh and Lake Superior Iron Co. and is on the Cascade range. Operations were commenced in September last by this latter company, of which mention has already been made under the Hussey mine.

In the **Menominee Iron Region** two companies, called respectively the Breen and Ingalls Iron Mining Companies, have been organized and are engaged in explorations, and in addition to the operations inaugurated by these companies, explorations are being made by private parties. The completion of the Peninsula railroad from Escanaba to Menominee, affording better promises for transportation, will stimulate operations of this character, which have heretofore been deferred from want of railroad communications.

The **Breen Mining Co.** owns 120 acres of land in Sec. 22, T. 39, R. 28, distant from Escanaba by proposed road 35 miles, from Menominee 55 miles and from Deer river 28 miles. The ore is chiefly flag, with some hematite. The property is being explored by Capt. E. B. Ward, J. J. Hagerman and J. W. Vandyke, who have an option of leasing or purchasing the mine. The officers are E. S. Ingalls, *Pres.*, T. B. Breen, *Sec.*, S. P. Saxton, *Treas.*, Thomas Breen, Bently Breen, and S. P. Saxton, *Directors*—all of Menominee, Mich.

The **Ingalls Mining Co.'s** property constituted 240 acres of land situated in Sections 8 and 9, T. 39, R. 29. The distance from Escanaba by proposed road is 44 miles and from Menominee 64 miles. The officers are E. S. Ingalls, *Pres.*, C. L. Ingalls, *Sec.*, and F. S. Mullburg, *Treas.*

An effort has been made to manufacture pig-iron by using *peat as a fuel*, but has not as yet proved in the requisite degree successful. A **peat furnace** was constructed at Ishpeming and went into operation early in the year 1872, but very soon went out of blast; subsequently it started again and made about 200 tons of iron and again stopped, it being the intention to alter and enlarge the stack, the better, it is thought, to adapt it to the peculiarities of the fuel. The peat is prepared from a bed of the material which exists in proximity to the furnace.

The Ericson Manufacturing Co. was organized in April, 1872, to conduct general manufacturing operations, with a nominal capital of $150,000. *Corporators:* Peter E. Ericson, John Carlson, A. J. Burt and Wm. Burt.

The company are operating a foundry and machine-shop, which they have built on Whetstone brook, within the city of Marquette. The machinery is driven by water-power.

Mr. Jno. Burt commenced, in September, 1872, the construction of a charcoal furnace, on the lake shore, at the mouth of the Carp river, south of Marquette. The stack is being built of stone, with a nine-foot bosh, and the whole is to be completed and put in operation in the spring of 1873. It is intended to supply the fuel from points along the lake shore, transporting it to the furnace in boats in the same manner that the wood for the Burt furnaces in Detroit is obtained, of which latter furnaces the one being built at the Carp will be a duplicate, and will be the first built on the Upper Peninsula based on this plan of obtaining fuel.

Very recently **The Carp River Iron Co.** has been organized, and own the furnace and about 500 acres of land at that point, including the water-power on the Carp, etc. The business office will be in Marquette.

SANDSTONES.

The Lake Superior sandstones are very carefully described by Dr. Rominger in his accompanying report, commencing with page 80, and the results of his observations, as therein described, are of great practical and scientific interest. There are two organized companies now engaged in quarrying and marketing sandstone within the limits of the city of Marquette, the locations being contiguous.

The Marquette Brown Stone Co. was organized in August, 1872, with a capital stock of $500,000, in 20,000 shares. The corporators were Peter White, Wm. Burt, F. P. Wetmore, S. P. Ely,

Sidney Adams, J. H. Jacobs, H. R. Mather and Alfred Green. In addition to quarrying stone, the company's franchises include the mining and smelting of ore, etc. Office in Marquette, Mich.

This company's property was previously known as the Wolf Quarry, located on the farm formerly owned by J. P. Pendill, and has been worked for some time past, the stone being principally used in Chicago. It is of a uniform dark-brown color, free from pebbles and clay holes. It apparently exists in great quantity, and is readily quarried and transferred to vessels. Mr. Peter White is constructing in Marquette a fine business block with a variety of stone from this quarry, which is variegated and striped with different colors, giving to the building a unique and pleasing appearance.

The articles of association of **The Burt Free Stone Co.** were filed Oct. 3d, 1872. Capital stock $500,000, in 20,000 shares of $25 each. The corporators were John Burt, William Burt, Hiram A. Burt, A. Judson Burt and Wm. A. Burt. Office in Marquette.

This company have opened a quarry of sandstone adjoining the one described above and the deposit is similar, the stone being lighter colored.

Both companies are prepared to furnish stone in large quantities. For full description of the sandstone found in these quarries, see Dr. Rominger's report, pages 90 and 91.

In addition to the above, **The Lake Superior Stone Co.** has been more recently formed with the amount of capital stock and number of shares as the preceding. The company own and hold in lease about 296 acres of land, situated on the west side of Keweenaw bay and on the north side of Portage Entry. The stone outcrops horizontally in a bluff, which rises from the water of the bay and is thus readily accessible for removal from the bed to vessels.

It is intended to begin operations in the spring. The corporators are H. H. Stafford, V. B. Cochran, W. S. Dalliba, E. J. Mapes and A. Kidder. Office, Marquette, Mich. See Dr. Rominger's report, page 95.

The fine new Court-House at Milwaukee is built with sandstone obtained from Bass island, near Bayfield, on Lake Superior, at which point stones have been quarried for several years.

The quarry described by Dr. Rominger, page 89 of his report, is now owned by Messrs. Winty and Mossinger, of Chicago, and Thomas Craig, of Marquette.

ROOFING SLATE.

There are three companies which were organized for the purpose of quarrying and selling roofing slate; but one of them, however, has actually commenced operations and is now at work on explorations.

The Huron Bay Iron and Slate Co. filed articles of association January 19th, 1872. Capital stock, $500,000, in 20,000 shares. The corporators were Peter White, W. L. Wetmore, F. P. Wetmore, J. C. Morse, James Pickands, A. R. Harlow, M. H. Maynard, D. H. Ball, Wm. Burt, D. H. Merritt, Sidney Adams and H. R. Mather. Office, Marquette, Michigan. The company own 2,000 acres of land in T. 51, R. 31.

The Huron Bay Slate and Iron Co. was organized subsequently, with same capital stock and number of shares. The corporators are W. L. Wetmore, Peter White, M. H. Maynard, Wm. Burt, Thomas Brown, J. J. Williams, S. L. Smith, Alex. McDonald, John H. Knight, W. C. Wheeler, H. R. Mather, Jas. D. Reid, F. P. Wetmore and R. C. Wetmore. Office in Marquette. The company own 1,100 acres of land in T. 51, R. 31, and have commenced work near Slate river, about four miles south of Huron bay, on the northeast quarter of section 33 in the above town. The slate apparently exists in very large quantities.

The Stafford Slate Co., an association comprising H. H. Stafford, V. B. Cochran, E. J. Mapes, A. Kidder, J. M. Wilkinson, Wm. Burt A. J. Burt and W. S. Dalliba, own 1,900 acres in T. 51, R. 31. The operations of this company thus far consist in having cut out a road from L'Anse to their property on Section 27, in the above town, a distance of 15 miles.

The color of the slate found in T. 51, R. 31, is somewhat varied, the green, purple and gray are found on Sections 14, 15, and 16. South of this are found large deposits of black slate, extending several miles east and west, with an apparent thickness of several hundred feet, the cleavage planes dipping to the south.

SAW-MILLS.

The following saw-mills are now in operation, all of which, with the exception of the ones at Whitefish Point, at Onota and Fayette (the two former of which are in Schoolcraft county and the latter in Delta), are in Marquette county:

Name of Firm.	Location.
Decker and Steele	Eagle Mills.
Edward Fraser	Cherry Creek.
George Wagner	Laughing Whitefish Pt.
A. R. Harlow	Little Presque Isle.
H. A. Stone	Bancroft.
Jackson Iron Co	Negaunee.
Iron Cliffs Co	"
Mr. Jackson	Palmer Falls (Cascade).
Hartman and Connelly	Little Lake.
Cleveland Iron Co	Ishpeming.
Lake Superior Iron Co	"
Deer Lake Iron Co	Deer Lake.
Michigan Iron Co	Clarksburg.
Michigamme Iron Co	Michigamme.
Edward Breitung	Republic Mt.
C. T. Harvey	Chocolate.
Bay Iron Co	Onota.

These mills produced in the aggregate, during the year 1872 (besides shingles, laths and a small amount of hard wood), thirteen and a half million feet of pine lumber, all of which, excepting the product of the three mills above designated, was, or will be, consumed in Marquette county. The total product during the coming year, if the winter is favorable, will be much greater, as most of these companies are preparing to get in a larger amount of logs. The Michigamme mill, which has a nominal capacity of 4,000,000 feet, has but recently started, and thus did not contribute to the total product of 1872.

COMPLETION OF THE RAILWAY SYSTEM.

Marquette, Houghton and Ontonagon R. R.

Among the most important events affecting the interests of this portion of our State, which transpired during the year 1872, was the extension of the C. and N. W. R. R. from Menominee to Escanaba, the consolidation of Marquette and Ontonagon Railroad with the Houghton and Ontonagon, and the completion of the line to L'Anse, thus making complete railroad communication from the head of Keweenaw bay to Chicago, a distance of 462 miles.

The development of the mineral resources of a country are so intimately blended with the improvement of its facilities for transportation, as to render it essential in considering the progress of the former, to give due credit to the latter. Iron ores having a low value per ton must be reached by rail or water before their value can be realized; differing in this particular from the ores of the precious metals, which will bear wagon or even pack-mule transportation. Especially is this true with reference to an isolated region like the Upper Peninsula, which is as yet a comparative wilderness, possessing but a small population, a rigorous climate, few thoroughfares and with a surface so rough and rocky in portions of its territory, as to render their construction a matter of much difficulty. It naturally follows, that the addition of two so important avenues of communication to the railroad facilities of the Peninsula becomes in a pre-eminent degree a matter of congratulation and importance. The history of the enterprise, which has thus resulted in the connection of the bays of Marquette and Keweenaw, is in brief as follows:

As has been previously related in speaking of the Peninsula road, the United States granted to the State of Michigan, by an act passed on the 3d of June, 1856, every alternate section of land for six sections in width, designated in odd numbers, to aid in constructing a railroad from Little Bay de Noquette to Marquette and thence to Ontonagon, and from the two last places to the Wisconsin State line. The State, by an act passed Feb. 14th, 1857, conferred this grant upon the Little Bay de Noquette and Ontonagon Railway

Co., and two other railroad corporations, all of which lines were required to be completed within ten years, a condition with which neither of the companies complied.

In 1863 the State conferred the forfeited franchises and grant previously given to the Marquette and Ontonagon *Railway* Co., upon the Marquette and Ontonagon *Railroad* Co., under certain conditions. Congress in 1864 extended the grant five years, in the subsequent year added four sections per mile thereto, and in 1868 fixed the time for a full compliance with the conditions of the grant until Dec. 31st, 1872. During the period of its existence, the company built twenty miles of main line of railroad, commencing near the Lake Superior mine at the terminus of what was formerly the Bay de Noquette road, and extending to a point on the south side of Lake Michigamme.

In 1870 the State decided that the company, by reason of its failure to complete any extension of their lines, had forfeited the greater portion of the grant. On the 24th of Jan., 1871, the Legislature confirmed the action taken by the State Board of Control during the month of April previous, which conferred the forfeited or unearned lands upon the Houghton and Ontonagon Railroad Company, a new organization, incorporated Jan. 15th, 1870, and of which the following Michigan men were among the principal stockholders: H. N. Walker, *President*, S. L. Smith, Chas. H. Palmer, Geo. Jerome and S. F. Seager. The conditions of the act of Congress required the completion of thirty miles of road before the close of the year 1872, which fortunately this company have succeeded in accomplishing. Jacob Houghton was chosen Chief Engineer; and having located the line from Champion to L'Anse during the winter, the construction was begun in the spring of 1871 at the L'Anse terminus, and on the 16th of Dec., 1872, the first train passed over the entire line to Marquette, sixty-four miles; the whole having been placed under one management by the consolidation of the two companies effected during the previous summer. The completion of the road to L'Anse, exclusive of innumerable other advantages, opens to market the products of several iron mines, among the most promising of the region.

In anticipation of future shipments of ore from L'Anse, the company have constructed at this terminus of the road an extensive dock, a full representation of which from careful drawings

J. Bien, lith. Photo. by Childs.

ORE DOCK, MARQUETTE

M.H & O.Rd. (Looking East) Vessels loading- Break water.

is herewith presented.* They have also built, at this point, in a very substantial manner, a round-house, turn-table, machine-shop, etc.

The charter of the company and the grant of lands provide for the extension of the road to Ontonagon, and it is but reasonable to assume that the energy, which has characterized the prosecution of the enterprise thus far under its present efficient management, will result in the accomplishment of the work before the expiration of the time fixed by law. The length of the main line is 62 miles, of branches 20 miles and of sidings 18 miles, making 100 miles of road now constructed and in operation.

The dimensions and capacity of the company's railroad dock at Marquette, a representation of which is given in the accompanying view, are as follows :—Total length, 1,222½ feet; working length, 720 feet; height above water, 38 feet, and width of top, 53 feet, on which are four tracks for cars. Whole number of pockets, situated on both sides, 136, of which 120 have a capacity of 55 tons each, and 16 (steamboat-pockets) of 100 tons each. From both sides 8 vessels can be loading at the same time, and 6,000 tons have been loaded in a single day. Three vessels arrived on Saturday, after 8 o'clock in the evening, and were loaded and gone early Sunday morning. Vessels with a capacity of 476 tons may be loaded in one hour and fifteen minutes; vessels of 683 tons, in one hour and thirty-five minutes; the average time is three hours. The average capacity of vessels is about 650 tons, ranging from 400 for the smallest to 1,100 for the largest. Total amount of ore shipped over the dock from May 12th, 1872, to the following Nov. 25th, 301,210 tons, of which 75,000 tons were taken by steam, and 225,-000 by sail-vessels; the estimated capacity of the dock, with a sufficient number of vessels to receive the ore, is 500,000 tons.

The working capacity is indicated by the amount of rolling stock, which at the opening of navigation, 1873, will consist of 1,600 ore-cars, 50 box and platform-cars, 7 passenger and baggage-cars and 28 locomotives. The present officers are: H. N. Walker, of Detroit, *President*, S. P. Ely, Marquette, *Vice-President*, Moses Taylor, New York, *Treasurer*, Freeman Norvell, Detroit, *Secretary*, Jacob Houghton, Michigamme, *Chief Engineer*.

Directors: H. N. Walker, Detroit, C. H. Palmer, Pontiac, S.

---

* Appendix F., Vol. II.

P. Ely, Marquette, John Steward, New York, Alexander Agassiz, Boston, S. L. Smith, Lansing, George Jerome, Detroit, Moses Taylor, New York, C. Francis Adams, Jr., Boston.

By the Peninsula division of the **Chicago and Northwestern Railway** the distance from Escanaba to Lake Angeline is $67\frac{50}{100}$ miles, and the branches completed and in course of construction, $37\frac{80}{100}$ miles; sidings, $15\frac{80}{100}$ miles; making a total length of track between these points of $121\frac{10}{100}$ miles.

The total amount of track between Escanaba and Menominee is $65\frac{70}{100}$ miles, of which $2\frac{30}{100}$ are side-track, making a total amount of track between Menominee and Lake Angeline, inclusive of sidings and lurches, $186\frac{80}{100}$ miles.

Estimated amount of rolling stock, which will be necessary and available for the business of 1873, between Escanaba and Negaunee:

Number of locomotives	33
" ore-cars (750 of them 6-wheeled)	3,000
" other cars	100

For the estimated business between Escanaba and Menominee:

Number of locomotives	6
" cars (exclusive of ore-cars)	100

S. C. Baldwin, *Div. Supt.*
Marvin Hughitt, *Gen. Supt.* } C. & N. W. R. R.

Statistics showing past production, with present condition and capacity of the mines and furnaces of the Upper Peninsula, might properly follow this historical sketch, thus bringing it to date and supplying facts, which could not well have been incorporated into the text. It was thought better, however, to arrange such information in tabular form, which has been done on Plates XII. and XIII. of Atlas, to which attention is here again called.

The Marquette Mining Journal, of Marquette, Mich., publishes an interesting yearly exhibit of the product and condition of the mines and furnaces.

In Appendix G, Vol. II., will be found statistics of population for the whole Upper Peninsula, from the United States Census for 1870.

CHAPTER II.

GEOLOGICAL SKETCH OF THE UPPER PENINSULA.

(*Where to Explore.*)

1. GEOGRAPHICAL DISTRIBUTION OF THE ROCK SYSTEMS.

IN prospecting for valuable minerals the intelligent explorer should constantly observe several kinds of phenomena. If his search degenerates into a simple blind hunt for ore, he would deserve the success of a hunter who went into a gameless region, or who hunted for game whose habits he did not understand. The following general geological facts and laws will possess value to the explorer in enabling him to wisely select his field of labor and in prosecuting his work.

As all the *sandstone* suitable for building, which has yet been found in the Lake Superior region, belongs to a system of rocks named by geologists Lower Silurian, and all the workable deposits of *iron ore* have been found in another system called the Huronian, while all the *copper* and workable *silver*, in a third system appears known as the Copper-Bearing Rocks; and as no workable deposits of useful minerals have yet been found in the fourth and oldest system, the Laurentian or granitic rocks, it follows, that it is of the utmost importance to the explorer that he be acquainted with the boundaries of these several fields and not waste his energies on unproductive ground. I do not mean to assert that iron ore will not be found in the Silurian sandstones, for in St. Lawrence County, N. Y., and in the Maramec district, Missouri, valuable deposits of ore exist in rocks of this age. Large deposits of iron ore also occur in the Laurentian (granite) rocks of Canada and Northern New York, and again, the iron ores of Thunder bay are contained in rocks which the Canadian geologists declare to be the equivalents of our Copper series; but at this date it is a fact, that no workable deposits of iron ore have been found in the Upper Peninsula in rocks of these systems, and an explorer or miner would not be considered wise, who should search for iron outside the Huronian limits. It is not only important that he be acquainted with the boundaries of the four great rock systems, but also with their leading characteristics. We will therefore first sketch in some detail the geographical distribution of these systems, as developed on the south shore of Lake Superior, beginning with the youngest and uppermost. The reader should have before him the map of the Upper Peninsula Pl. I. of the Atlas. The boundaries marked are not always exact, but embody the best information available and are not far wrong.

I. *Lower Silurian.*—The Lower Silurian system, the youngest or lowest division of the Palæozoic rocks represented on the Upper Peninsula, is made up of various sandstones and limestones which are fully described in Dr. Rominger's Report, Part III. The entire Peninsula, east of the meridian of Marquette, is underlaid by Silurian rocks and the " Copper range " is flanked by a Silurian flat on the south side, which separates it from the iron series, until the two, together with the South copper range, come together west of Lake Gogebic.

About two-thirds of the whole area of the Upper Peninsula, or 9,982 square miles, is underlaid by this system.

II. *The Copper-bearing Rocks*, corresponding with the upper copper-bearing rocks of the Canadian geologists, occupy a narrow belt on the northwestern edge of the Upper Peninsula. These rocks have less superficial extent than either of the other formations, underlying only about 1,186 square miles, or, say 7 per cent. of the whole surface. For descriptions of them see Prof. Pumpelly's Report, Part II.

III. *The Iron-bearing Rocks*, corresponding, it is assumed, with the Huronian system of Canada, consist of a series of extensively folded beds of diórite, quartzite, chloritic schists, clay and mica slates, and graphitic shales, among which are intercalated extensive beds of several varieties of iron ore. The same rocks occur on the east and north shores of Lake Superior, where they also contain iron. The Huronian area represented on the map equals about 1,992 square miles, or nearly one-eighth of the whole area of the Upper Peninsula.

IV. *The Granitic Rocks*, which so far have produced no useful minerals, and which are believed to be the equivalents of the Laurentian of Canada, are represented as underlaying about 1,839 square miles, equal to 12 per cent. of the total area.

As our examinations in the southwestern part adjoining the Wisconsin line have not been thorough, there is considerable uncertainty regarding some of the lines dividing the Huronian and Laurentian rocks, and a portion of this region, equal to about 668 square miles, or 4 per cent. of the whole area, is left blank on the map.

While, as has been stated, it is not proven that iron ore may not exist in the other great systems in workable quantities, there is every reason to believe, that by far the greater part, if not all the workable deposits, are contained in the Huronian area above described. It must not, however, by any means be understood, that all of this area is iron-bearing. The several iron districts, which have been more or less explored, will be described in another place; they will be found to cover not more than about one-fifth part of the Huronian area, or, say one-fortieth of the whole area of the Upper Peninsula, and on less than one-half of this area have the ores been proven to have commercial value.

Recapitulation.

I.	Lower Silurian area, about	9,982	square miles.
II.	Copper-bearing area, about	1,186	"
III.	Huronian or Iron-bearing area, about	1,992	"
IV.	Laurentian area, about	1,839	"
	Unknown area, about	668	"
	Total area of Upper Peninsula, exclusive of islands, about	15,667	"

In a complete and systematically arranged geological sketch the lithology of the four systems would properly belong here, but what is written on this subject necessarily pertains almost entirely to the Huronian, the whole matter will therefore be considered in Chapter III., following, and in Appendices A, B and C, Vol. II.

II.—TOPOGRAPHY.

It is of importance to the prospecter to carefully observe the topography or form of the surface, for it is well known that useful minerals generally occur in corresponding topographical positions over considerable areas; again, the topography is the very best key to the nature of the underlying rocks, if these be concealed by earth, as is often the case. As the human physiognomy indicates the fundamental characteristics of the man, so the earth's physiognomy suggests the forces and materials lying beneath. It is safe to assert that within certain limits an experienced topographical geologist can, from a correct topographical map, judge of the nature of the rock underlying the surface represented; and conversely, from a geological map, he can predict the general form of the surface. In the same way, an experienced explorer does not hesitate to express an opinion as to whether he is on the "mineral range," from the form of the ground. We will now sketch in some detail the characteristic topography of the four great systems.

I. *Silurian.*—The prevailing surface characteristic of the Silurian region is a nearly level plain, underlaid by horizontal sandstones and limestones, often swampy and sometimes, where fire has destroyed the timber, a desert. The tame, flat, sandy and swampy country along the line of the Chicago and Northwestern Railroad, between Escanaba and Negaunee, is underlaid by Silurian rocks, but is far below the average in the value of its timber. Where rivers or water-courses have cut into these rocks, or waves wasted them, perpendicular bluffs are presented, which afford an excellent opportunity to explore and study the formation. The famous "Pictured Rocks" are bluffs of this character, from 50 to 200 feet high. From the top of these bluffs the country is flat, proving that they are the results of the action of water cutting its way into a horizontal plane, and are not, so to speak, built up and completed hills like those of the older rocks.

There is one apparent exception to this general flatness of the Silurian topography. Many of the highest hills and mountains in the Menominee iron region are capped with horizontal sandstone and limestone, which is never found in the valleys; the base, however, embracing the great mass of these elevations is always an old rock, and in the iron fields always Huronian. There is no doubt but that the sandstone once filled the valleys, extending in an unbroken bed of irregular thickness across the whole of the Menominee region, covering the older rocks, just as it now covers them further east. Since its formation it has here been mostly eroded, but still caps the elevations as described. If it were all gone, the hills, made as they are, largely of highly inclined beds of quartzite, marble and ferruginous rocks, would remain, but with somewhat diminished heights.

Should the eastern part of the Upper Peninsula be elevated at any future time, so as to bring the underlying azoic rocks above the lake level, the Silurian rocks may there also become so eroded as to only cap the Huronian hills, as they now do in the region described. That the older rocks extend eastward under the Silurian, is, I suppose, a geological necessity, and is, I think, directly proven by the existence of local magnetic attractions in this Silurian area, which are undoubtedly due to the existence of beds of iron ore in the underlying Huronian. The explorer in the Menominee region finds these beds of sandstone much in his way, covering, as they do, in some instances, the ores.

Small lakes of clear water, with sandy bottoms but no outlets, are a characteristic feature of the Silurian area. The U. S. Survey maps represent about one-half of the whole surface of these rocks, which underlie the central and eastern portion of the Upper Peninsula, as swamp; the solid rock has often been found within a few feet of the surface in the swamp region. The western Silurian area being the prolongation of the Keweenaw Bay valley west, and embracing in part the Sturgeon, Ontonagon, Presquisle and Black rivers, has fewer lakes, much less swamp, and is more broken, than the eastern part already described.

Soft woods, including pine, are more prevalent on the Silurian rocks than on the older series; but on the other hand, some of the finest bodies of sugar-maple and beech found on the Upper Peninsula, are on these rocks. Beech has not, so far as I know, been found growing on the older rocks; whether this be due to climatic or soil influence has not been determined.*

The water divide, or height of land, of the central and east part of the Peninsula, is much nearer Lake Superior than Lake Michigan. It is an irregular line, approximately parallel with the shore of the lake, having an elevation where it crosses the Peninsula railroad of about 650 feet. See Map, Pl. I.

II. *Copper-bearing Rocks.*—There is probably no more striking

---

* A timber map has been prepared, but could not be published for want of means.

topographical feature in Michigan, than the "Mineral" or Copper range, including Keweenaw Peninsula, of which it is the backbone. Ranges would better express the fact, for west of the Ontonagon river there are three; the Main or central Range which extends from Keweenaw Point far into Wisconsin, being flanked on the north by the Porcupine mountain range and on the south by the South copper range, each separated from the other by broad Silurian flats. The general trend of the three ranges is north, 60° east, and south 60° west, but they are not quite straight, as may be seen on the map. The ridge is broad, generally more than three miles, and the crest quite even, but is cut down to lake level at Portage lake, and further west is deeply eroded by the Fire steel, Flint steel, Ontonagon and other rivers. The surface of the ridge or plateau is from 500 to 600 feet high in the vicinity of Portage lake, and rises to a height of 884 feet at Mount Houghton, near Keweenaw Point. Between the Ontonagon river and Lake Gogebic the Central range attains, in isolated peaks, an elevation of 1,100 feet, and the Porcupine mountain range is over 900 feet high; the range is more broken towards the west, and in the vicinity of Rockland presents a series of oval mammillary hills with steep escarpments on the south side. This is also the character of the South copper range, between Lake Gogebic and Montreal river.

The iron range immediately south of the South copper range, and west of Gogebic, is lower, the hills having more gentle slopes; the range being in places obscured by low ground. As this is the only part of the Upper Peninsula, so far as I know, where the iron explorer may come in contact with copper rocks, it is important to observe the topographical differences above noted, especially as the copper traps in some places resemble the diorites or greenstones of the iron region. Lakes and swamps, so numerous in the iron and granite regions, are infrequent on the copper belt, as must follow from the form of the surface. The reason for the striking regularity in the leading topographical features of the copper range is to be found in the great uniformity in the strike and dip of the rocks, as is explained under Stratigraphy. The timber of the copper range is generally sugar-maple, is abundant and of excellent quality; very little pine or other soft wood occurs here.

III. *Iron-bearing Rocks.*—The topography of the Huronian rocks differs essentially from that of either the Silurian, or the copper series. It is almost everywhere hilly and often mountainous, forming peaks higher than any in the copper range; but instead of a continuous range, or series of parallel ranges, it is rather a broad belt or irregular area of mountains, hills, swamps and lakes. It may be said, that the ruling topographical features, especially the mountains, have a general east and west trend, but there are numerous exceptions to this law; for example, the Michigamme river, from the lake to Republic mountain, runs northwest to southeast; and Michigamme lake itself has a north-south arm, nearly as long as the main lake, which runs east-west. The ridges west of Paint river, in T. 42, R. 33, run north-south, conforming with the bedding of the rocks.

Probably one of the most persistent ridges in the Marquette region is formed by the "lower quartzite," which outcrops on the shore of Lake Superior just south of Marquette, and rising rapidly from the lake it forms Mt. Mesnard on Sec. 34, T. 48, R. 25; from this peak it extends westerly, crossing the railroad at the Morgan furnace, then by way of the old Jackson Forge and along north side of Teal lake to south side of Deer lake, it holds its westerly course for a total aggregate distance of over 15 miles. The Chocolate and Morgan flux quarries and the Teal lake whetstone quarry are in this range. More persistent and conspicuous, and nearly as long, is the Greenstone ridge, which skirts the north side of the Michigamme and the Three lakes extending from the Bijiki river to the west end of the First lake, a distance of eleven miles:—points on this range are three hundred feet above Michigamme lake, which is 950 feet above Lake Superior. Summit mountain, one mile easterly from the Foster Mine, is one of the prominent landmarks of the region, looking as it does from an elevation of about 1,300 feet over the flat granite and Silurian region to the south. It forms one of a chain of hills which extend from the south end of Lake Fairbanks westerly for about 10 miles, but which form in no sense a ridge.

The mountains, or hill ranges, above described are exceptional in their regularity and continuity. Broken chains of irregular hills and short ridges of various sizes, separated by lakes and swamps, is the prevailing character; the highest hills are seldom over 300 feet above the low grounds at their base and about 1,300 feet above Lake Superior. Outcrops of rock, forming often perpendicular ledges of moderate height, are more numerous in the iron-bearing rocks, than in either of the systems described, except in the westerly part of the copper range. Although the relief of the surface is considerably modified by drift, it is generally plain that the strike, dip, and texture of the underlying rock has determined the general outline or contour; we should therefore expect that the great variation in these rocks, hereafter to be described, would produce this varied topography.

The topography of the Marquette region is very like the iron region of southern New York and northern New Jersey, except in its smaller elevations; a profile running north and south through the Jackson Mine, Marquette, would closely resemble a profile running northwest and southeast through the Sterling Mine, New York, platted say to half the scale.

Passing to the Menominee iron region, we find greater simplicity in the geological structure and a correspondingly less varied surface.

Obeying the influences of the great rock beds beneath, the elevations there have a tolerably uniform east-west trend and consequent parallelism. The south iron range, of which the Breen Mine is the east end so far as known, can be traced through a greater part of its course by a ridge, often bold, which crosses Town. 39, R. 29, and T. 40, R. 30, for a distance of over 15 miles, the bearing being west-northwest. The north iron range, about 12 miles from the other in the south part of Town. 42, Ranges 28, 29 and 30, is in places a prominent topographical feature. The capping of horizontal sandstones, which has already been mentioned as characterizing the Menominee hills, gives a somewhat more even character to the crest lines, and in places produces a strikingly different profile.

The Gogebic and Montreal river range, above referred to, is better marked by its running parallel with and lying south of the South copper range, than by any essential character of its own.

IV. *Laurentian.*—The surface of the granite country south of the Marquette region, at the same time the most extensive and best known, is not unlike that of the iron-bearing rocks on a much smaller scale. There are no mountains, the hills are lower, being usually mere knobs, seldom exceeding 50 feet in height; the ridges shorter and swamps more numerous. A coarse pitting of the surface, or promiscuous sprinkling of little hills, and low, short ridges may convey the idea. Sometimes the knobs range themselves in lines constituting low ridges, with jagged crest line; these ridges, when near the Huronian rocks, are usually parallel with them; if they have any prevailing direction, it is east and west.

Perpendicular walls of granitic gneiss 15 to 40 feet in height sometimes face the ridges for several hundred feet in length, constituting the most regular topographical feature within the Laurentian area.

Small beaver meadows are common here as in the other rocks, and sometimes a succession of dams, one above the other, forms a long narrow meadow, which produces considerable quantities of wild hay.

This region was once heavily timbered, largely with pine, which has been prostrated by a hurricane, and since burned over several times. The soil, naturally light, has burned up and so washed away, as to expose the white-gray, pink and dark-green rocks in every direction, affording an unsurpassed opportunity to study this series; the boulders are very numerous and often of great size. The light colors of the rock, scarcity of vegetation and an abundance o. standing trunks of dead trees give the landscape a peculiar aspect; but a second growth of poplar and wild cherry is rapidly changing this dismal character.

The fallen timber, swamps, steep bluffs and ledges, and numerous boulders, make travelling through the Laurentian area difficult and laborious in the highest degree. Florida swamps have denser vegetation and are much larger; sea-coast marshes often have more mud; the highlands of the Hudson present more formidable elevations, but, all in all, the writer believes it requires more physical exertion to travel 5 miles per day (all a man can accomplish with a pack) through Lake Superior granite windfall, than in any other region east of the Mississippi. The trees were prostrated by northwesterly winds, judging by the direction in which they lie; persons have travelled in a southeasterly direction on the trunks of fallen trees (mostly pine) for over a mile without once touching the ground.

III.—STRATIGRAPHY.

Scarcely second to the two classes of phenomena already mentioned is the observance of the rock masses, or strata, as to their direction or strike, and inclination or dip; the order of their superposition and thickness; but more important than either is to ascertain between what rocks the mineral sought for occurs. Useful minerals which occur in beds, like the iron ores of Lake Superior, will usually be overlaid and underlayed by rocks, having different characters and which maintain those characters for considerable distances. Next to finding the ore itself, it is desirable to find the hanging or footwall rock. Whoever identifies the upper quartzite in the Marquette region, or the upper marble in the Menominee region, has a sure key to the discovery of any ore that may exist in the vicinity.

With few exceptions, all the rocks in the region we are describing are stratified—that is, arranged in more or less regular beds or layers, which are sometimes horizontal, but usually highly inclined. This stratification or *bedding* is generally indicated by a difference in color of the several layers, oftentimes by a difference in the material itself, but occasionally the only difference is in the texture or size and arrangement of the minerals, making up the rock. Thus, rocks made of quartz, sand and pebbles, may vary from a fine sandstone to a coarse conglomerate. In general, a *striped rock*, whether the stripes be broad or narrow, plain or obscure, on fresh fracture or weathered surface, is a stratified rock. Usually rocks split easier on the bedding planes, than in any other direction; but the converse is true in the case of most clay slates and in some other rocks, which split more easily on their *joints* and *cleavage* planes, the direction of which seldom coincides with the bedding and is often at right angles with it. If a rock splits most easily along its striping, it is always safe to assume, the true bedding planes have been found. Such planes are supposed to have had their origin in the original deposition of the mud and sand, of which most rocks are made. Similar marks can be seen in excavations in sand and clay, which may be regarded as unconsolidated rocks. The cleavage and joint planes above indicated, which are always more regular in strike and dip, than the others, are supposed to have originated from pressure, subsequent to the formation of the rock.

The term plane, as used in describing bedding, must not be understood to signify a straight-line surface; on the contrary, they are usually curved planes, sometimes folding and doubling on each other, so as to produce a very intricate structure. Not only do these plicatures take place on the small scale, as shown in hand specimens, but precisely similar folds exist in masses of rock, which may be hundreds of feet thick. The resulting curved strata take the name of troughs or basins, if the convexity is downward, the general term *synclinal* structure being applied to this form. Connecting the synclinal troughs and basins are *anticlinal* domes and saddles. The whole may be described as rolling or wave-like forms. Sometimes the power which produced the *folds* seemed greater than the rocks could bear, and cracks or breaks, and *faults* or throws, are the result, though these are not numerous in the Lake Superior region. Cracks so produced and filled with material, other than that constituting the adjacent rocks, are called *dykes*; or if the material be crystalline and metalliferous, *veins*. As iron ore in workable quantities does not occur in this form in this region, vein phenomena will not be considered here.

An examination of the four great rock systems will illustrate and prove the above remarks on stratification.

I. Beginning, as before, with the uppermost or youngest, which is at the same time the softest and lightest rock, the *Silurian* brown and gray sandstones and limestones, so well exposed on the south shore of Lake Superior, we have a perfect illustration of the regular and horizontal bedding, without folds, faults, or dykes. An inspection of the Marquette quarry, or any of the numerous natural exposures, will convince any one that these rocks are but consolidated sandbanks.

II. *The Copper-bearing Rocks.*—Some beds of this series are sandstones nearly or quite identical with the Silurian in appearance, but the great mass is made up of different varieties of copper trap, which are often amygdaloidal; interstratified are beds of a peculiar conglomerate. The stratification of these rocks, considered in large masses, is nearly as regular as the sandstones, and differs only in the fact that the layers are inclined, dipping northwest and north toward Lake Superior at a varying angle, which seems to be greatest on the south side of the range, and is there often vertical. It is least at Keweenaw Point, where it is as low as 23°.

III. *The Iron-bearing* or *Huronian Rocks* are immediately beneath, and are exposed to the south of the copper rocks. This series are, on the average, heavier and harder, than either of the others and folded to a far greater degree. The prevailing rock is a greenstone or diorite, in which, like the copper traps, the bedding is usually obscure; but the intercalated schists and slates which usually bear strong marks of stratification, make it usually not difficult to determine the dip of the beds at any point. This dip varies both in amount and direction, but is generally at a high angle, and is more apt to be to the north or south than in any other direction.

IV. Descending to the oldest or bottom rocks of the Lake Superior country, the granites and associated beds (*Laurentian*), we find the bedding indications still more obscure and often entirely wanting. Here there is, if possible, more irregularity in strike and dip, than in the Huronian.

IV.—BOULDERS (FLOAT ORE).

Fragments of iron ore which have been detached from the parent ledge and are found loose on the surface, or in the drift beneath, possess great interest to the explorer, and are among his most important helps and guides. The same remarks are applicable, but to a less extent, to boulders of other rocks. As a rule, in the iron region of Lake Superior, it is safe to assume, that when boulders of a particular variety of rock are abundant on the surface, a ledge of the same will be found in place very near—if not immediately under the boulders, then up hill from them, or perhaps a little to the north or east; the more angular or sharp-cornered the boulders, the nearer we would expect to find the ledge.

In the Menominee region it may almost be said, that this rule is invariable, as there seems to have been less movement of the drift material here than farther to the north.

In the Michigamme district a large amount of float ore is found some distance south of the iron range, part of the fragments being very large and containing at least 100 tons of ore. Sections 19, 29, and 30 of T. 48, R. 30, and Sections 25, 36, and 35 of T. 48, R. 31, contain many such boulders, which were probably derived from the Michigamme range. Considerable digging has been done at several of the larger boulders, which has failed to find the ore in place, and the magnetic attractions are of a character which indicate detached boulders and not a continuous ledge. For mode of distinguishing boulders of magnetic ore, see chapter on use of the magnetic needle.

These Michigamme ore boulders are all found south of the iron range which produced them, and but few at a greater distance than two and one-half miles, most of them being much nearer. This southerly and westerly direction of the drift is, so far as I know, universal in the iron region of the Upper Peninsula, and it is fully confirmed by the direction of the drift scratches in the solid rock, which vary from north to east, averaging about northeast and southwest.

Therefore, if iron boulders be found in considerable abundance, the explorer may assume, especially if they are angular, that he has iron underneath the surface; if rounded or abraded, the ledge may be to the north or east. If the boulders be magnetic, the place of the ledge should be found, with comparative ease, by means of the needle; but if specular, it may be an expensive and difficult work. Soft hematite, from its nature, can never occur in the form of boulders, as it would weather into a reddish soil. Iron boulders are often met with in digging test-pits and shafts; in such instances, if near the ledge, I have generally found the ore in place very near; if considerably above it in the drift, the same rules would apply as to surface boulders.

Attention should be given to the character of boulders other than iron, which may be associated with it, or found where there is no iron. Occasional granite boulders occur everywhere in the Lake Superior iron region and have no economic significance. I have never seen an abundance of granite boulders, however, except over granitic rocks, and so far, these rocks have not produced workable deposits of iron.

Boulders of quartzite, diorite and slate usually accompany those of iron in the Marquette region, and marble boulders, as well as quartzite, are most significant in the Menominee region.

The above laws, regarding the occurrence of *iron boulders*, give the facts regarding their geographical distribution great importance in iron explorations. If, where there are iron boulders, we may confidently look for iron, then conversely, where there are none, we should not expect to find iron. I do not assert that every deposit of hard ore is marked by float or boulders, but, so far as the facts have come to my knowledge, this is the case in the region under consideration.

Except in one or two instances, which have not been verified, I have heard of no iron boulders in the so-called silver-lead region, which extends north from the Marquette iron region to Lake Superior, which would lead one to believe, that merchantable hard ores will be found there. And except the L'Anse range in north part of T. 49, R. 33, this is true of the belt of country, west from the so-called silver-lead region. The region, without iron boulders, may be briefly described by saying, that it is bounded west and south by the line of the Peninsula division of the Chicago and Northwestern, and by the Marquette, Houghton, and Ontonagon railways. In other words, a person travelling by rail from Escanaba through Negaunee to L'Anse would have the region of iron boulders on the left, and the boulderless region on the right hand, or towards the lake.

Limiting their distribution still further, we may say, that iron boulders have only been found in quantity and quality, which would point toward economic importance in (1.) T. 45, R. 25, in the vicinity of the S. C. Smith mine, which is the most easterly locality in which they have been observed on the Upper Peninsula of Michigan; (2.) the Negaunee and Michigamme iron districts, extending in belts of irregular width from Negaunee west to the First lake in S. 17, T. 48, R. 31; (3.) the L'Anse iron range, in north part of T. 49, R. 33; (4.) south and southwest from Michigamme lake, embracing wholly or in part Towns. 44 to 47 north, and Ranges 39 to 32 west; (5.) the Menominee iron region, embracing wholly or in part Towns. 39 to 42 north, and from Range 28 west to the Menominee and Brulè rivers, but not west of Range 33; (6.) the Lake Gogebic and Montreal river iron belt, south of the South copper range.

Hunting for boulders is something like hunting game; when on the ground the best woodsman, the most active and observant will be the most successful, assuming, of course, that he knows at sight what he is looking for. (See chapter on Explorations.) I have found Indians good help in this kind of work, and believe that the incentive of a bonus in money for boulders or outcrops is often good policy. The best places in which to observe boulder phenomena is in the beds of rapid streams and under the roots of trees, the latter, probably, having been the most fruitful field. A windfall is as good as five thousand dollars' worth of test-pits to the section.

With boulder phenomena may be classed the reddish or ***brownish earth***, which comes from the disintegration of iron ore rocks of a hematitic character, and ***magnetic sand***, which is very generally distributed, and which comes from the disintegration of magnetic ore. Such material may, for our purposes, be regarded as made up of minute boulders and the same remarks will apply, except that I should not expect to find red earth far removed from the ferruginous rock which produced it. Minute quantities of magnetic sand can be found almost everywhere in this region.

CHAPTER III.

LITHOLOGY.* (*Mineral Composition and Classification of Rocks.*)

In the preceding sketch the terms sandstone, limestone, conglomerate, trap, diorite, granite, etc., occur. It is evident that no satisfactory and useful progress can be made in geological field-work, which includes prospecting, until one has learned to recognize and name the more common varieties of rock. For this purpose we have to give attention to their mineral composition, that is, we must ascertain of what simple mineral or minerals the rock in question is chiefly made up and to observe, whether such minerals are angular, presenting bright facets (crystalline), or whether they are rounded like sand and gravel (fragmental). Not only must the prospector be able to recognize at sight the mineral he is seeking, but in case it is not exposed, which often happens, then those rocks, which are known to indicate its presence or absence. Experienced prospectors will not spend much time in looking for iron among granite rocks, nor in the copper traps, nor yet in the region of horizontal sandstones and limestones.

The mineral composition of rocks, by which they are identified, described and named, constitutes the science of Lithology, one of the most abstruse departments of Geology. A high authority on this subject has remarked :—" In all attempts to define and classify rocks, it should be borne in mind that they are not definite lithological species, but admixtures of two or more mineralogical species, and can only be arbitrarily defined and limited." When rocks present recognizable crystalline minerals, the task of describing and naming is comparatively easy ; but when the constituent minerals are obscure, as is often the case in the rocks we are considering, the attempt to employ specific names, which shall define such vaguely compounded aggregates, will be exceedingly difficult.

* The stratigraphical order of the rocks here considered will be found in the succeeding chapter.

The difficulty may be illustrated by supposing, were an attempt made, to give such name to a common brick, as will designate its composition and structure. Bricks are made in general of sand and clay, but several varieties of sand, and as many of clay, are employed in different localities, which, being mixed in various proportions and differently burned, give rise to a wide variation in composition and appearance and could not be expressed by a single word or term. In the case of rocks we have, of course, no previous knowledge of the numerous ingredients employed in their composition, by which the difficulty is greatly increased. It may seem at first sight, as if chemical analysis should form a reliable basis for rock nomenclature, but this is not the case. Van Cotta asserts, that a rock containing 72 silica, 11 alumina, 2.8 oxide of iron, 1 lime, 1.2 magnesia, 1.2 potash, 2 soda and 0.4 water, may be either a granite or a gneiss, protogine, granulite, quartz-porphyry, felsite, petrosilex, pitch-stone, trachyte-porphyry, obsidian, or pearlstone; and by giving a little range in the percentages of some of the constituents, half a dozen other rock names could be added. Here we have eleven different rocks, having precisely the same chemical composition, but widely different in physical character.

It must be borne in mind, in studying this subject, that the solid crust of the globe is almost entirely made up of ten or eleven simple *chemical elements*, which variously combined, according to the laws of chemistry, produce the few *minerals* which in turn, mechanically mixed, constitute ordinary *rocks;* hence we should expect, that the average chemical composition of a series of rocks, wherever found and of whatever character, would nearly agree.

The materials of the first formed rocks, whatever their origin, have been worked over and over by rains and waves and chemical forces, distributed over sea-bottoms, consolidated and elevated, to pass again through the same process by just such means, as are now at work in producing similar results.

The reader who may not be familiar with the physical characters and composition of the minerals—quartz, feldspar, hornblende, chlorite, talc, argillite, mica and the oxides of iron and manganese, which make up the great bulk of the rocks herein described, is advised to refer to some elementary work on geology or mineralogy.

Extensive rock formations are now generally named after the locality, where they were first thoroughly studied, or are best exposed, and their minor beds and layers are often named according to their peculiar mineral composition, or with reference to their relative age, that is, order of superposition. The names Laurentian, Huronian and Silurian are geographical names of the first class. No attempt will here be made to describe the lithological character of either the Copper bearing traps, conglomerates and sandstones, nor the Silurian sandstones and limestones ; these will be fully treated by Prof. Pumpelly and Dr. Rominger, respectively. What has been and will hereafter be said of the geographical distribution and topographical and stratigraphical character of these rocks was considered necessary, to acquaint the prospector and explorer with those general principles of geology, which lie at the foundation of intelligent and successful work. Whoever would become thoroughly acquainted with these systems is referred to Parts II. and III. of this volume. A number of specimens from the Laurentian are described in Appendix A, Vol. II. (see descriptions 252 to 299) ; but they do not cover all the lithological families represented in that system.

In subdividing the Huronian or iron-bearing series, which we have particularly to study, the rocks have been grouped (1) *lithologically*, *i.e.*, according to their mineral composition, and (2) *stratigraphically*, *i.e.*, according to relative age. As this system was first described and named by the Canadian geologists, their names have been employed as far as possible in the body of this report ; the identity in composition of many of our rocks with theirs, having been established by an examination of a large number of Marquette specimens by Dr. T. Sterry Hunt.

Alexis A. Julien, A.M., of the School of Mines, New York, has made careful studies, both in the field and laboratory, of a large number of specimens from the Lake Superior region, his results being in part given in Appendix A, Vol. II. As his paper was not obtained in time to modify this chapter and the geological descriptions which follow, in accordance with Mr. Julien's nomenclature and orthography, what follows may be regarded as an independent and popular presentation of this subject, which is scientifically and more fully treated in the Appendix, the practical needs of the explorer and miner being here chiefly considered.

The specimens examined by Mr. Julien are in part from the Marquette region ; the L'Anse, Menominee, and Gogebic districts are also well represented, thus embracing an area over 125 miles long and having an extreme width of 60 miles. The specimens described belong to a catalogued collection, numbering over 2,500 specimens, being probably the most complete suite of rocks from the Azoic of the Upper Peninsula yet collected. Those from the Montreal river and Gogebic district were collected by Prof. R. Pumpelly and myself, and are believed to be the first described from that region. Prof. Pumpelly took very full lithological notes in the field, but has not yet, so far as I know, made them public. Dr. H. Credner's publications are very full on the lithology of the Menominee region, he having spent two seasons in that field.

Appendix B, Vol. II., contains a list (named by Mr. Julien) of the specimens constituting the State collection, over thirty duplicate suites of which were collected and have been distributed among the incorporated colleges of Michigan and other leading institutions and cabinets, of this country and Europe.

Appendix C, Vol. II., contains a list of 76 specimens, number 1,001 to 1,076, determined by the microscope by Chas. E. Wright, under the direction of the Faculty of the School of Mines, Freiberg, Saxony. A suite of these rocks is at Freiberg and others in Michigan.

The several beds or layers of the Huronian system, as developed in the Marquette region, are numbered upwards from I. to XIX., always written in Roman numerals. These strata being particularly described as to thickness, geographical extent, etc., in following chapters, it need here only be said in general that I., II., III., IV. are composed of beds of silicious ferruginous schist, alternating with chloritic schists and diorites, the relations of which have not been fully made out; V. is a quartzite, sometimes containing marble and beds of argillite and novaculite; VI., VIII. and X. are silicious ferruginous schists; VII., IX. and XI. are dioritic rocks, varying much in character; XIII. is the bed which contains all the rich specular and magnetic ore, associated with mixed ore and magnesian schist; XIV. is a quartzite, often conglomeritic; XV. is argillite or clay slate; XVI. is uncertain, it contains some soft hematite; XVII. is anthophyllitic schist, containing iron and manganese; XVIII. is doubtful; XIX. is mica schist, containing staurolite, andalusite and garnets. This classification, it will be borne in mind, applies only to the Marquette region, the equivalency of the rocks of the Menominee and other regions not having been fully made out.

These beds appear to be metamorphosed sedimentary strata, having many folds or corrugations, thereby forming in the Marquette region an irregular trough or basin, which, commencing on the shore of Lake Superior, extends west more than forty miles. The upturned edges of these rocks are quite irregular in their trend and present numerous outcrops. While some of the beds present lithological characters so constant, that they can be identified wherever seen, others undergo great changes. Marble passes into quartzite, which in turn graduates into novaculite; diorites, almost porphyritic, are the equivalents of soft magnesian schists. In this fact is found the objection to designating beds by their lithological character, while to numbers or geographical names no such objection exists. The total thickness of the whole series in the Marquette region is least at Lake Superior, where only the lower beds exist, and greatest at Lake Michigamme, where the whole nineteen are apparently present, and may have an aggregate thickness of 5,000 feet.

Near the junction of the Huronian and Laurentian systems, in the Marquette region, are several varieties of gneissic rocks, composed in the main of crystalline feldspar, with glassy quartz and much chlorite. Intersecting these are beds of hornblendic schist, argillite and sometimes chloritic schist. These rocks are entirely beneath all of the iron beds, seem to contain no useful minerals or ores and are of uncertain age. No attempt is here made to describe or classify them.

The following description and classification has resulted from an examination of a large number of specimens of "ore and rock," collected with the view of embracing all varieties found in the iron-bearing series of the Marquette region, together with a study of the parent masses in the field, which latter is of great importance on account of the variations in composition of the same bed, to which attention has been directed.

The *specific gravity* of over five hundred specimens, weighing from 3,000 to 10,000 grains, was determined by a balance, which turned when loaded, by the addition of two grains. The magnetic properties were carefully examined and are given in part in the chapter on the magnetism of rocks. Most of the specimens examined were arranged into ten *lithological groups* (having no reference to age), which are designated in what follows by the first ten letters of the alphabet. When a specimen represented a very small and unimportant layer, it was thrown out as exceptional and not important to the object of this report.

It must be constantly borne in mind, that the divisions between these ten lithological groups or families are not sharply marked; one passes into the other by insensible gradations, thus producing many intermediate varieties, which it was difficult, if not impossible, to classify or describe. The first family, A, will include all valuable iron ores, the remaining nine (B to J) will include "rocks." But as iron ore, in large masses, has all the geological characters of the associated rocks, the popular general classification of minerals into "ores" and "rocks" will be disregarded except as above mentioned. Except in a few instances, where Mr. Julien's collection was incomplete, all minute lithological descriptions have been omitted, for such, frequent reference will be made to his paper; and for the reason that he had not access to maps and sections, which gave the stratigraphical distribution of the various rocks, this part has been made quite full in that respect.

In a few instances reference is made to the full suite of Marquette rocks, numbered 6,000 to 6,222, deposited by me in the cabinet of the University of Michigan, at Ann Arbor.

A. Iron Ores.

(Occurring in formations X., XII., XIII. and below V.)

Only such ores as are now employed in the manufacture of iron will be described under this head. They are in order of present supply, the (a) specular hematite or ***red specular ore***, as this class is designated in the iron trade; (b) ***the magnetic***; (c) the "mixed" or ***second-class ore***, which may be either specular or magnetic; (d) the ***soft hematite***, and (e) ***the flag ores***. Another variety, the magnetic specular, might be added, which, as the name implies, is a mixture of the black and red oxides, which gives a purple streak. The local terms "hard," embracing both the magnetic and specular ores, and "soft," for the soft hematites, are convenient.

The commercial statistics, modes of mining, and composition will be considered under their proper heads,* attention being directed here chiefly to the mineralogical and physical character of each

* See Chapters IX. and X., Plate XIII. of Atlas, and Appendix J, Vol. II.

ore. Under Woodcraft and Surface Explorations, Chapter VII., are given some brief practical rules for distinguishing iron ores, for the benefit of those, who know little or nothing of rocks.

All the specular, magnetic, and mixed ores, and a part of the soft hematites, are found in one formation; bed XIII. of my arrangement, which has its most easterly exposure near the Jackson mine and extends irregularly and indefinitely westward, embracing all the mines now producing rich hard ore.

It may be said of these ores in general, that they are essentially oxides of iron, with a few per cent. of silica added, and generally contain minute quantities of sulphur and phosphorus, but no titanium. Alumina in quantity not exceeding two and one-half per cent., with one-fourth as much manganese, is sometimes found, together with alkalies, which seldom aggregate over one and one-half per cent. The soft hematites are in part hydrated sesquioxides, hence contain water and usually more silica, than the hard ores; traces of organic matter are sometimes found, and manganese is almost exclusively confined, to the soft ores. Many specimens of specular and magnetic ore have been analyzed, which gave ninety-eight per cent. of oxide of iron, the balance being nearly pure silica. For numerous analyses of all the ores, see Chapter X., Appendix J, Vol. II., and Plate XIII. of Atlas. Weathering has no appreciable effect on the hard ores, except to crumble and cover with soil the more granular varieties. The exposed surfaces of the compact ores (by far the most prevalent variety) are of almost as high lustre as fresh fractures, and are often highly polished, showing no weathered coating like almost all other rocks. In the "mixed ores" the jasper bands are sometimes slightly elevated on the weathered surface, due to their greater hardness.

a. Red Specular Ores.—Miners divide these into *slate* and *granular*. The former resembles closely in its structure the soft greenish chloritic schists, commonly associated with it. The slabs, into which the slate ore easily splits, are not uniform in thickness like roofing-slate, but taper always in one and often in three ways, producing elongated pieces often resembling in form a short, stout, two-edged sword-blade, with surfaces as bright as polished steel, but striated and uneven. See Specimens 46, 47, 48, State Collection, Appendix B, Vol. II., and 1,050 Appendix C, Vol. II. Thin edges of such slates can be pulverized into a bright scaly powder by the finger-nail, and occasionally the whole mass is too friable for economic handling. The magnet will generally lift one or two per cent. of the powdered ore, and occasionally one-fourth of the whole, in which case the streak is purple. These last, constituting magnetic slates, are more friable than the pure red specular slates, due in some way to the larger admixture of magnetite. See Specimen 49, State Collection, Appendix B, Vol. II.

The *granular* or massive specular ore shows no tendency to split in slabs, and is made up usually of minute crystalline grains, which are sometimes, however, so large that their octahedral form can be easily recognized without the aid of a lens; fine specimens of this variety occur at the Cleveland and New York Mines. Mineralogists apply the name *martite* to the red oxide of iron, when it has the crystalline form of the octahedron, which belongs to magnetic ore. See Specimens 2, 43, 44 and 45, State Collection, Appendix B, Vol. II. It is not improbable, that all of the granular specular ores under consideration may have once been magnetic and in some way have gained the two per cent. of oxygen necessary to change them from black to red oxides. See Dana's System of Mineralogy, 5th ed., p. 142.

The granular ore is generally firm in texture and never friable, like the granular magnetic. Some highly compacted varieties, which contain a little silica, are very hard, constituting the hardest rock to drill which the miner encounters. This variety is called the "fine-grained steely ore;" some specimens of it possess almost the highest specific gravity observed, 5.23, while the rich softer ores of the same class averaged about 4.85. See Spec. 45, State Collection, Appendix B, Vol. II.

From the examination of a considerable number of specimens of red ore, it was found that the magnet would usually lift an appreciable portion of the powder. In the case of one coarse-grained specimen of pure ore from the New York mine, one-third of the pulverized ore was removed by the magnet. Spec. 1060, App. C, Vol. II. The percentage of powder lifted by a magnet in twenty-one specimens, together with color of powder, is given in Table, App. H, Vol. II. Numerous specific-gravity determinations of this variety of ore will be found in App. B, Vol. II.

b. Magnetic Ore.—The description given above of the granular specular ore applies with equal force to this class, except that the latter is more of granular and often friable, has the magnetic property and gives a black or purple powder instead of red. Sometimes the rich magnetites crumble easily into grains, like some Lake Champlain ores, to which the term "shot ore" is applied; again, it is very hard, as in Pit No. 8 of the Washington mine. See Specs. 39, 40, 41 and 42, State Coll., App. B, Vol. II. The compact tabular form so frequent in the magnetic ores of New Jersey and Southern New York is not common in the best ores of the Marquette region, nor are the latter ores as highly magnetic as the former, or at least good loadstones are not so common; the ore from the Magnetic mine (see Spec. 17, State Coll.) has most of this tabular character.

Typical *slate* ores occur with the magnetites, but they are of the character already described, that is, mixtures of the two oxides, the magnet not removing over one-fourth of the powder, while it takes all in the case of the granular variety. The specific gravity of the granular magnetic ores, as will be seen in Appendix B, Vol. II., varied from 4.59 to 5.01, the average of many specimens being 4.81. Specs. 1,054 and 1,059 of Appendix C, Vol. II., are also varieties of this ore.

The following minerals and rocks are most commonly associated with hard ores: a soft grayish-green *chloritic schist*, which sometimes, owing to bad sorting, goes to market in sufficient quantity to perceptibly reduce the furnace yield. The magnesia it contains might tend to stiffen the slag, otherwise it can have no effect in the furnace further, than what is mentioned above. This rock is described under Group D. See Specs. 53, 54, and 55, State Coll., App. B, Vol. II.

Micaceous red oxide of iron often occurs in scales and bunches, particularly in proximity to jasper. It has been improperly called plumbago, but is in reality in no way related to it, being chemically pure oxide of iron, having the crystalline structure of mica. A soft whitish mineral, often called *magnesia*, and appearing not unlike flour, occurs occasionally in specular ore and frequently in "soft hematite." This substance is usually most abundant in the more jaspery varieties of specular ore; an examination by Prof. Brush determined it to be *kaolinite*, a hydrated silicate of alumina (clay) in minute crystalline scales. The presence of this clay in small quantity could not but help the working of the furnace, by forming a more fusible slag, but it would of course diminish the yield of iron, if in quantity.

The needle and velvety forms of the mineral *Göethite* (a hydrated oxide of iron) are not uncommon at the Jackson mine, and "*Grape ore*" (botryoidal limonite), sometimes finely colored with yellow ochre, is found at several of the mines, but always in soft hematite. Fine specimens of crystallized quartz are rare, and no form of lime has been observed, although analyses show minute quantities. Bunches of *iron pyrites* are occasionally found, especially in the magnetic mines. At the Champion mine a thin layer containing this mineral occurs next the hanging wall, but it is easily separated from the ore, and is not sent to market. Hornblende, so generally present in the magnetic mines of New York, New Jersey and Sweden, is rare in the Marquette mines, of XII. and XIII.

c. Second-class Ore.—By far the most abundant, and commercially objectionable ingredient in the Marquette ores of all kinds, is the so-called jasper, a reddish ferruginous quartz, which is invariably found associated with the best ores, usually in thin seams or lamina conforming to the bedding, but sometimes in a form approaching a breccia. In the hard ores this impurity can usually be readily distinguished, but in the soft hematites it is often only found by analyses. As this rock possesses considerable scientific as well as commercial interest (the better varieties constituting the second-class ores), I will attempt to describe and illustrate it somewhat minutely. It consists of jasper, varying from bright red to dull reddish-brown, with occasional seams of white quartz, and usually pure specular or magnetic ore of high lustre. These materials are arranged in alternating lamina, varying in thickness up to one inch. These lamina are often highly contorted, zigzagging, and turning sometimes in opposite directions within a few inches. The jasper bands are in places broken up into little rectangular fragments, which are slightly thrown out of place, as it were, by tiny faults; the ore fills the break, so that the whole mass has the appearance of a breccia. There can be little doubt, but that the true breccia at the east end of the Jackson mine has this origin, and it would be interesting to consider whether this idea might not be extended to other conglomerates in the Huronian series. The contorted laminated structure, with the striking contrast of colors, is beautiful, and affords fine miniature examples of the anticlinal and synclinal folding and faulting of large rock masses. Sometimes the lamina are very irregular and indistinct, and one or the other of the minerals greatly preponderates. When the jasper layers all thin out (as they usually do somewhere), the ore becomes first class. Some phases of this interesting rock, with descriptions, are given in Appendix K, Vol. II., Figures 19 to 29. See Specs. 36 and 37, State Coll., Appendix B, Vol. II.

The miners call this material "mixed ore;" and those varieties in which the jasper does not constitute over 20 per cent. of the whole, are sold as second-class ore, yielding about fifty per cent. in the furnace; for rail-heads and some other uses requiring a hard iron, the presence of silica in the ore is not objectionable. The quantity of "mixed ore" is greatly in excess of the pure ore, and it will some time undoubtedly have considerable commercial value. Its nature is such, as to admit of the ready mechanical separation of the pulverized ore from the jasper by jigging, a process now employed in separating ores in the Lake Champlain region. For fixing puddling furnaces, or for any branch of iron industry which may demand pulverized ore (as the Elerhausen process promised to), it is very probable that this method may advantageously be employed, and a cheap ore produced.

"Mixed ore" is seen in outcrops far oftener than the purer ores, the softer character of which has caused their erosion, whereby they had become covered with soil; but as the mixed ores are usually associated with the pure varieties, their outcrops possess great significance in prospecting. It is important in this connection not to confound the "flag ores," (e) to be described, which they sometimes closely resemble, with this variety. The quartz of the magnetic mixed ore is usually white, or lighter colored than the red mixed ore.

d. The *soft hematites* of the Marquette region differ entirely from the ores above described, and are closely related to the brown hematites of Eastern Pennsylvania and Connecticut. In color they are various shades of brown, red and yellow, earthy in form, and generally so slightly compacted, as to be easily mined with pick and shovel. They are invariably associated with, or rather occur in, a limonitic silicious schist, from which they seem to have been derived by decomposition and disintegration. These ores occur in two distinct formations, X. and XII., and probably in others, in irregular bunches or pockets, surrounded by the schist and passing by gradations, often abrupt, into it. Scattered through the ore, and conforming in their positions with the original bedding of the rock, are fragments of the schist. When the ore shows stratification, which it often does not, it also conforms with the bedding of the schist. The specific gravity of the soft hematite ore varied from 3.50 to 3.81, the average of five specimens being 3.59, and specimens of the schist varied from 2.80 to 3.38. Strictly this schist should be described under the next group of rocks, B, to which it belongs, but its assumed parentage of the hematite ore, here considered, has led to the digression. See Specs. of soft hematite 1,067, 1,077, 1,079, and of schist 1,040, 1,065, and 1,069, Appendix C, Vol. II.; also, Specs. 25 and 26, State Coll., App. B., Vol. II.

The following analyses of the schist and ore, from the Foster mine, by Dr. C. F. Chandler, will help to make their relations better understood:—

		Schist.	Ore.
Sesquioxide of iron		44.33	79.49
Alumina		2.14	1.19
Oxide of Manganese		.16	.25
Lime		.36	.27
Magnesia		.13	.33
Silica		47.10	9.28
Phosphoric Acid		0.13	0.19
Sulphuric Acid		0.17	0.17
Water		5.19	8.74
		99.71	99.91
Equivalent to	Iron	31.03	55.64
	Sulphur	.068	.068
	Phosphorus.	.057	.083

It will be observed that the essential difference is in the amount of silica, of which the schist has over 47 per cent., while the ore has less than 10 per cent., and again the ore has 25 per cent. more metallic iron than the rock. The one would evidently be converted into the other, both as to its chemical and physical characters, by the abstraction of the greater part of its silica. It is not at all improbable, that this change may have been brought about by the alkaline waters of former thermal springs, such as are now producing similar results in other parts of the world. There seems to be very little sand or clay in this ore, and washing has not appeared to improve its quality, as is the case with the eastern ores which it resembles. If the fragments of silicious rock, which are scattered through it, are carefully picked out by the miner, an ore uniform in character is obtained. Except the ever-present silica, there are only two minerals, which it is necessary to mention as being generally associated with this variety of ore. 1st. The *white clay* (kaolinite), above described, which is far more abundant in this ore than the hard ores; bunches as large as a hen's egg being sometimes seen. There can be no doubt but that the kindly working of the furnace usually obtained by using the best quality of this ore, is due in part to this clay as well as to the porous character of the ore. (Calcining the ore would expel the water, of which it contains from 2 to 9 per cent., and should also cause it to reduce more easily in the furnace.) The second and most important mineral to be mentioned is the *oxide of manganese*, usually if not always in the form of Pyrolusite; minute quantities of this metal, always less than one per cent., are sometimes found in the hard ores, but from 1 to 4 per cent. is constantly present in several of the hematite deposits, which is so important an element in their value, as to almost warrant the subdivision of the soft hematites into two classes, the *manganiferous* and *non-manganiferous*.

The recently developed hematite mines near Negaunee, belonging to formation X., contain most manganese; others contain little or none. Scarcely enough of the ore has been worked to determine its place in the market; but there can be no doubt, that when equally rich in metallic iron, the manganese would give this ore the advantage, as a mixture for the furnace, over the non-manganiferous varieties. See Spec. 25, State Coll., App. B, Vol. II.

The hematite ores now in the market, as a class, vary greatly in richness, from an average of not exceeding 40 per cent. of metallic iron for some deposits, to at least 55 per cent. in the case of others. This difference is in part brought out in Chapter X.

Passing from the Marquette region to the undeveloped districts, we find on the L'Anse range, at the Taylor. mine, a large deposit of hematite of excellent quality. At the Breen mine, on the south belt of the Menominee region, is also a good "show" of hematite. Promising indications of this ore were also found between Lake Gogebic and Montreal river; all of these localities and their ores will be described hereafter.

e. The last variety of merchantable ore, to be described in this report and designated *Flag*, has been in use so short a time, that but little can be said of its metallurgical character. It corresponds more nearly with the second-class ores (*c*), than with either variety described, differing from it more in structure than in composition. The ores embraced under this head are abundant and have received various local names, which will be found significant and convenient, as lean ores, iron slates, magnetic slates and silicious ores. They have also been called "lower ores," in reference to their subordinate geological position, being older than the rich ores of formation XIII., already described. Flag ores are in reality only varieties of the ferruginous schists, constituting Group B, next to be described, which are sufficiently rich in iron; to possess market value. The percentage of metallic iron in these ores and the associated schists varies from say 5 to nearly 60, those above 50 now constituting a merchantable ore. The remaining material is generally silica, always silicious, but sometimes contains more or less chlorite, manganese, argillite, mica, garnet, or hornblende added. This ore is always flaggy in structure, the layers being occasionally thin enough, to warrant the application of the term slate. All forms of the oxide of iron can be observed, a mixture of the black and red prevailing. The hydrated oxide, producing limonitic silicious schist, has been described above, as the rock from which the soft hematite ore seems to have been derived, and an analysis is there given, to which nothing need be added here.

Stratigraphically these rocks are older than the ores described under *a* and *b*, and constitute at least four beds, X., VIII., VI., and below V., separated by diorites, chloritic schists, quartzites and argillites. Like the mixed ores (*c*) they are banded, but the marking is seldom bright and often obscure, produced by the interlamination of a dull reddish or whitish quartz, with dull *silicious* instead of *pure* ore. There are exceptions to this rule, but they are not numerous in this region. As this is a point of much importance to iron prospectors, it may be asserted, that when white or red quartz (jasper) is found banded with an ore which can be scratched with the knife, it is in all probability the "mixed ore," which accompanies the pure ores of bed XIII.; but if the quartz be dull and not sharply defined in its layers, and particularly if the knife marks the ore layers like a pencil, instead of cutting them, then we probably have one of the flag-ore formations. It is difficult to say, whether the red or black oxides prevail in many flag ores; hence whether particular varieties should be described as hematitic or magnetic.

All ores and ferruginous rocks become more magnetic as they are followed west in the Marquette region, the maximum amount of magnetite occurring in the Michigamme district. The ferruginous schists of the Republic Mountain series are among the most highly magnetic rocks in the whole region. At the Ogden mine, Section 13, T. 47, R. 27, the abrupt transition of the hematitic into the magnetic variety can be plainly observed, by following the *strike* of the beds less than 200 feet. This transition probably often occurs in the same bed, and, of course, might occur still oftener in crossing the formations, that is, in passing from one bed to another.

Several varieties of *flag ore* will now be described, showing a wide range in lithological character, which we should not be warranted in grouping together in a strictly scientific classification; but our arrangement of rocks, as has been stated, is rather economic and for the use of practical men.

(1) A showy, granular, chloritic, specular ore was found in a small pocket-like mass at the north ¼ post of Sec. 26, T. 47, R. 26, at locality known as the Gillmore mine. A specimen having a specific gravity of 4.28 gave Dr. C. F. Chandler metallic iron 60.46, alumina 3.49, lime 0.60, magnesia 1.33, silica 7.05, sulphur 0.30, phosphoric acid 0.08, water and alkalies not determined 0.77.

A similar ore, but containing some magnetite and peculiar white glistening spots, which appear to be mica scales, is found at the Chippewa location, Sec. 22, T. 47, R. 30. A specimen of this gave Prof. A. B. Prescott metallic iron 53.17, and insoluble silicious matter 20.20. Neither of these varieties are flaggy. See Specs. 6,156 and 6,206, University of Michigan cabinet.

(2) A specular slate ore, holding reddish specks on freshly fractured surfaces, is found at the Cascade location, bedded with layers of jasper, having the local significant name of "Bird's-eye Slate." A specimen of this gave J. B. Britton metallic iron 59.65, insoluble silicious matter 12.24, alumina 0.88, lime 0.14, magnesia 0.08, oxide of manganese 0.02, water 1.08, with traces of sulphur and phosphorus. See Spec. 6,190, University of Michigan cabinet, and Spec. 6, State Coll., App. B, Vol. II.

(3) South of the Cascade range is a flag ore, beautifully banded with

red jasper and silicious iron ore, closely resembling some of the mixed ores of Bed XIII. above described, and interesting on this account.

(4) Northeast of the Cascade location, and near the centre of Sec. 29, T. 47, R. 26, is a granular slate ore showing on fresh fracture a peculiar fine reticulated appearance and indistinct octahedral forms. A specimen of this gave Mr. Britton 59.42 per cent. of metallic iron. See Spec. 6,191, University of Mich. cabinet. Since the foregoing was written, shipments of flag ore have been made from the Cascade mines (see Plate XII. of Atlas), and with it a considerable amount of a good quality of specular ore.

(5) At the Tilden mine, while the prevailing ore is a 40 per cent. ordinary red flag ore, there are seams or layers of bright steely ore, very hard and heavy, which yield, according to analyses made by Dr. Draper, 62 per cent. metallic iron. This ore possesses particular interest from its close resemblance to the Pilot Knob ore, Mo.

(6) While the most abundant ore at the Iron Mountain mine, Sec. 14, T. 47, R. 27, is much like the Tilden and Ogden ores already mentioned, there is a peculiar variety, containing manganese, which is also found on the hills south of Negaunee and on the lands of the Deer Lake Company, north of the New York mine. This ore is a very dark-colored silicious hematitic schist, containing on the average several per cent. of manganese, single specimens of which have proved to be nearly pure oxide of manganese. Some of this ore from Iron Mountain was tested in the furnace as a mixture, but was found to be silicious. The need of ferro-manganese in steel-making would make ores of this character a legitimate object of exploration. An experienced iron-master recently expressed the opinion that a 30 per cent. iron ore, with 12 to 20 per cent. of manganese, would soon have commercial value. It is possible that such a variety may exist in some of the beds under consideration. The soft or hematitic variety of this ore has already been mentioned.

(7) Passing from the Negaunee to the Michigamme district, we find two flag ores worth noticing. On the Magnetic Company's property, Sec. 20, T. 47, R. 30, is a large amount of a very compact, hard, heavy, highly magnetic ore, laminated with a greenish horn-blendic mineral, producing an unusual banded structure. A piece of one of the layers of ore gave Mr. Britton 56.78 metallic

iron, 19.44 insoluble silicious matter, less than one per cent. of alumina, lime and magnesia, and a trace of phosphorus. See Spec. 18, State Coll., App. B, Vol. II.; also Chapter X. Recent explorations have developed a workable deposit of this ore.

(8) Adjoining this property, to the southeast is Sec. 28, owned by the Cannon Iron Co., on the north side of which is a thin layer of micaceous specular ore, closely resembling that described above under A, but containing more silica. A specimen of this afforded Professor Prescott 55.12 metallic iron, 19.80 insoluble silicious matter, with traces of sulphur and phosphorus. This and the banded ore associated with it, has a closer resemblance to the slate and "mixed ore" of some of the old mines, than any place I have seen in the flag-ore series, to which it seems to me geologically to belong; its relation to the associated mica schist is interesting. See Group H below. The Chippewa ore, near the Cannon, has already been mentioned above in connection with the Gillmore.

The foregoing brief descriptions of several varieties of flag ore embrace all those, which have come under my notice in the Marquette region and give promise of having early commercial value.

As will be elsewhere (Chapter V.) more fully described, the hard ores found in the Menominee region up to October, 1872, are more nearly allied to flag ores than to either of the first-class ores of the Marquette region. Flag ores of a low grade have also been found in the L'Anse and Gogebic districts, as will be mentioned hereafter.

A very limited experience in working these ores, together with the little I have been able to learn from others, leads me to believe, that they require more limestone and coal and produce a harder metal, having comparatively little strength, but which is probably well adapted to making rail-heads. I think a large mixture of manganiferous hematite might help the working of a furnace consuming flag ore. Precisely the same remarks may be made of the second-class ores (*c*); indeed, these two classes are to all intents and purposes identical in their metallurgical character, and are only separated here because of their different geological occurrence. The second-class ores are, it will be remembered, simply inferior grades of the rich hard ores of XIII.

The flag ores have here received relatively far more attention, than their present commercial importance warrants, for the following reasons:—1st, Their quantity, so far as can now be judged, is greater by tenfold than the first-class hard ores, and for this reason they must, at some future time, constitute a large part of the total production of the region. 2d. Very serious disappointments and losses have occurred in the past, and are likely to be repeated in the future, from mistaking flag ore for first-class ore. This arises from the fact, that the better varieties of flag ore closely resemble the poorer varieties of the rich ore. So close is this resemblance, that the best judges of ore in the Marquette region have erred. It is doubtful, if the matter can be settled definitely, except by thorough explorations, aided by the well-known laws of the geological occurrence of the two ores, which will be more fully brought out in succeeding chapters.

It is not asserted that first-class hard ores may not be found associated with the flag ores, hence below and older than formation XIII.; but it is a fact, that over one million dollars have been sunk in such search, and excepting the West End mine of the Cascade range (if that is an exception), no workable deposit of strictly high grade hard ore has been found in the flag-ore series.

B. Ferruginous, Silicious, and Jaspery Schists.

(Occurring in formations XII., X., VIII., VI., and below V.)

The best general idea of the character of the rocks embraced here can be conveyed by saying, that they are identical with the flag ores last described, except in containing less iron and usually more silicious matter. On geological grounds, as has been remarked, the flag ores should be embraced under this head and described as a subclass, rich in iron. It remains therefore for me to mention briefly, a few of the remaining varieties of this series, which are so poor in iron as to render it highly improbable that they will ever possess value as ores: I design to embrace in this group Mr. Julien's quartz schist, silicious schist, and jasper schist, Appendix A, Vol. II. For minute lithological descriptions of numerous varieties see Specs. 154 to 173, App. A, Vol. II.

At Republic Mountain are three highly magnetic beds of silicious, chloritic and hornblendic schists, numbers VI., VIII., and X. See Map No. VI. of Atlas. The peculiar striping—whitish, greenish, brownish, and yellowish—exhibited in the large outcrops suggested the name "rag-carpet schist." A specimen made up of numerous chippings of this rock gave 31 per cent. of metallic iron; this is believed to be above the average. Both the red and black oxides are present, and some of the layers hold an ore, which, if it could be separated, might yield 50 per cent.

South of the Washington mine these rocks contain the minimum amount of iron, a specimen of which gave Charles E. Wright less than 5 per cent. Garnets and anthophyllite, or mica, seem to replace the iron, producing a grayish and brownish schist, the mineralogical character of which is obscure. See Group I. The old Michigan mine ore, Section 18, T. 47, R. 28, seems to be a variety of this peculiar schist, but much more highly charged with metal, specimens of which, I should judge, would afford 30 to 40 per cent. of metallic iron.

Passing to the Negaunee district we find in the railroad cut at the northwest end of Lake Fairbanks a chloritic, magnetic, silicious schist of a brownish gray color, faintly banded and very hard; it is aphanitic in character, and shows no disposition to split on the planes of bedding. In the railroad cut near the centre of Section 8, one mile and a half southeast of Negaunee, is a soft variety of ferruginous rock, affording some good red chalk. The rock seems to be chloritic, layers of which are impregnated with red oxide of iron. A similar material was found in numerous test pits in the east part of Section 18, T. 47, R. 26. Recent explorations in this vicinity prove this rock to be associated with the Negaunee hematites, which are fully described in Chapter IV.*

One of the best characterized and abundant varieties of this group is the banded ferruginous jaspery schist, which constitutes in the Michigamme district the whole of formation XII., and is also abundant in parts of ore formation XIII. Such varieties of "mixed ore," as contain too little iron to give them commercial value (unfortunately the greater part), would be classed here. The full descriptions and illustrations already given of "mixed ore" under A, will make any further description unnecessary, for this is a similar rock with little or no iron. See Spec. 32, State Coll., App. B, Vol. II., and for several other varieties of this group see Specs. 1,026, 1,034, 1,061, and 1,064, Appendix C, Vol. II. The Felch mountain series contain a large amount of a similar rock.

---

* It is questionable whether this rock should be classed under D or G.

C. Diorites, Dioritic Schists and Related Rocks (*Greenstones*,)*

(constituting formations XI., IX., VII., and one or more beds below them.)

These obscurely bedded rocks, locally designated greenstones and sometimes traps, are co-extensive with the ferruginous rocks A and B, very abundant, outcropping throughout the Huronian region, and present much variety in appearance. They range in structure from very fine-grained or compact (almost aphanite) to coarsely granular and crystalline, being sometimes porphyritic in character. The color of the fresh fracture is from dull-light to dark or blackish green, the weathered surface being usually lighter and of a grayish green or brownish color, not unfrequently spotted or mottled, showing a dark-green, or black, lamellar mineral (hornblende), set in a whitish, and sometimes reddish, softer mineral (feldspar). The rock is exceedingly tough, powdering under blows of the hammer rather than break. It can be scratched by the knife, giving a light grayish-green powder, and is fused without difficulty before the blow-pipe. On the one hand, it graduates into a heavier, tougher, blacker variety, which is unquestionably hornblende rock, with some feldspar, well shown at the Greenwood Furnace quarry, on Sec. 15, T. 47, R. 28. See Specs. 1,018 and 1,020, App. C, Vol. II. On the other hand, it passes into a softer, lighter colored rock of lower specific gravity, which, while it has the same streak, weathers similar to the true diorite, is eminently schistose in character, splitting easily, and appearing more like chloritic schist than any other rock. The Pioneer Furnace quarry at Negaunee contains this schist and several transition varieties, some of which approach the granular massive rock. See Specs. 1,001, 1,005, 1,006, and 1,015, App. C, Vol. II. On the north side of Lake Michigamme, and west, varieties occur having a true slaty structure in appearance, although not splitting easily. See Spec. 1,028, App. C, Vol. II.

At several points dioritic schists, semi-amygdaloidal in character, were observed, and in one instance the rock had a strong resemblance to a conglomerate. See Spec. 1,024, App. C, Vol. II.; and

---

* See Dr. Houghton's Notes on Diorites, Appendix E, Vol. II.

Spec. 71, State Coll., App. B, Vol. II. It is of much practical importance to distinguish between the schist of this group and the true chloritic schist to be described under the next head, D, which is usually found associated with the pure ores of Bed XIII.*

At Republic mountain a dioritic schist graduates into black mica schist, and large garnets are there found in typical diorite. Iron pyrites are usually seen sprinkled through the rock, and epidote is sometimes observed. Dr. Hunt found chromium in two specimens. South of the Old Washington mine, in Bed XI., occurs a variety, which in places may almost be described as hornblendic schist; that in other parts of the same bed, near at hand, graduates into the above-described dioritic schist.

In the railroad cut at the foot of Moss Mt., west of Negaunee, is an exposure of soft dioritic schist, in which are imbedded rounded lumps of diorite, which, when broken, show a crystalline reddish feldspar. See Specs. 1,001 and 1,002, App. C, Vol. II. Spec. 77, App. B, Vol. II., is another beautiful and rare variety, in which the feldspar is red. On the south side of Sec. 9, T. 49, R. 33, is a heavy bed of coarse-grained friable diorite, which has in places disintegrated into sand. Mr. Julien regards this and the associated dioritic rocks of the L'Anse range as possessing such distinctive characteristic as to warrant him in describing them as a distinct variety. See Specs. 342 to 353, App. A, Vol. II. He also classes the well-known peculiar serpentine rock of Presque isle with the diorites. See Spec. 321, App. A, Vol. II, also App. E.

The magnet usually lifts less than one per cent. of a powdered diorite, but in one case it took nearly all, and the specimen attracted the needle. This piece was from the ridge south of the New England mine; it had the essential character of a compact, perhaps hornblendic diorite, but its magnetic property and very high specific gravity, 3.29, prove that it is exceptionally rich in iron. It will be shown below, that in addition to the magnetite, seventeen per cent. of metallic iron exists in some diorites in the form of combined protoxide, which does not attract the needle. The specific gravity of the typical rock varied from 2.84 to 2.96, the average of six specimens being 2.91. The hornblendic varieties ranged as high as 3.01, while the schistose variety fell as low as 2.70,

---

* See Julien's remarks under Chloritic schist, App. A, Vol. II.

averaging 2.82. A garnetiferous specimen, from Smith Mountain, gave 3.02, while a peculiar variety from north of Greenwood Furnace, which appeared to be feldspathic in character, gave but 2.71. Numerous additional specific gravity determinations are given in App. B, Vol. II. The precise character of the constituent minerals of this rock is obscure. Mr. Julien has minutely described numerous varieties in App. A, Vol. II., Specs. 302 to 353.

The following analysis of a specimen from bed XI. is from Foster & Whitney's Report, Part 2d, p. 92. The specimen was from Sect. 10, T. 47, R. 27, on south side of the Cleveland and Lake Superior ore deposits:—

		OXYGEN.
Silica	46.31	24.06
Alumina	11.14	5.21
Protoxide of iron	21.69	4.82
Lime	9.68	2.76
Soda	6.91	1.78
Water	4.44	
Magnesia	trace.	
	100.17	

From this it is deduced that the rock is a mixture of labradorite feldspar with hornblende or pyroxene. Regarding the presence of water, numerous analysis of similar rocks in Canada show the same result. See Geology of Canada, pages 469, 604, 605, and 612. Dr. Hunt expresses the opinion, that in the case of the Marquette diorites, the hornblendic mineral often becomes softened and hydrated, passing into a degenerate form more nearly allied to chlorite or delessite (in which water is an essential constituent), than to a true hornblende. This chloritic mineral is sometimes seen scattered through the body of the rock, and very often near the weathered surface.

The absence of *magnesia*, which is regarded as an essential ingredient of chlorite and delessite, and as very rarely absent from hornblende, as shown by the above analysis, deserves notice. Dr. Hunt remarks that the hornblendic element may very likely be the iron hornblende described by Dana, System of Mineralogy, 5th ed. p. 234, under the name grünerite. The unusually large amount of iron shown by Whitney's analysis and the high specific gravity observed would favor this view. The conversion of this non-magnesian diorite into a magnesian schist (chloritic or delessitic) would require the introduction of the magnesian element under some law of pseudomorphism, the possibility of which is proven by chemical geology.

Magnesia is not, however, absent from all varieties of the diorite. A chromiferous specimen from near the centre of Sec. 36, T. 48, R. 28, was found by Dr. Hunt to be rich in magnesia, containing more of this element than of lime; the specimen was not a typical one, but showed a tendency to pass into a steatitic rock, which might be expected to contain magnesia. Until, however, the presence of magnesia in the schists and its absence from the diorites is proven by more analyses, it is not worth while to conjecture in the matter, and I here digress only to record a few facts, bearing on an interesting and unsettled question in chemical geology. In the absence of any additional light, we adopt the hypothesis that the Marquette "greenstones" are diorites, composed essentially of a non-magnesian iron hornblende and some feldspar other than orthoclase.

It is of great importance that the prospector should have a good practical acquaintance with this rock, for it is everywhere associated with iron ores in the Upper Peninsula. He should be able to recognize it at sight, to distinguish its varieties, and especially he must not confound the Huronian diorite with a similar rock, found in the Laurentian, nor with Copper trap. More than one piece of land has been bought for iron on the Laurentian area, because "greenstone" was found on it.

The bedding of these rocks is generally obscure, and in the granular varieties entirely wanting. It is usually only after a full study of the rock in mass, and after its relations with the under and overlaying beds are fully made out, that one becomes convinced, whatever its origin, it presents in mass precisely the same phenomenon as regards stratification, as do the accompanying schists and quartzites.

I have nowhere seen the granular diorites show more unmistakable evidence of bedding than on the small knob southwest of Bear Lake, Republic Mountain, shown in Fig. 1, scale $\frac{1}{16}$th. The cross shading represents massive diorite, and the parallel shading a slaty silicious iron ore.

No reference is hëre made to the false stratification or joints, which are numerous and interesting, but which, unfortunately, for want of space, can receive no other attention here, than to warn the observer against mistaking *joint* planes for *bedding* planes, which is sometimes done, even by experienced observers.

This description, as has been stated, is intended to apply to the diorites of the iron-bearing or Huronian series, and more especially

Fig. 1.

Stratification of Diorite.

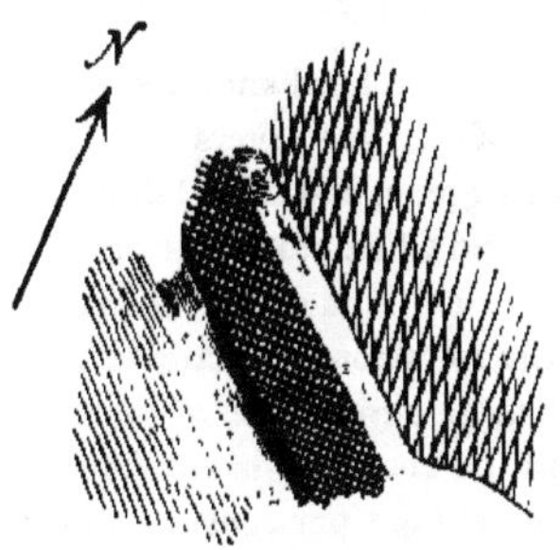

to the Marquette region; but a similar rock, as has been observed, occurs abundantly in dykes or veins, and probably in beds in the Laurentian rocks. A fine example of such a dyke can be seen penetrating a granitic gneiss, near the northeast corner of Sec. 7, T. 46, R. 29. At other points in the Laurentian area immense masses of a dioritic rock were observed, the stratigraphical relations of which to the gneiss and granites was not made out. The average specific gravity of the dyke diorite was 3.03. Mr. Julien describes some specimens of diorite from the Laurentian in App. A, Vol. II.

The following designated specimens, in addition to those already referred to, constitute a tolerably full collection of the more important varieties:—Granular diorites, 1,007, 1,008, 1,009, 1,010, 1,011, 1,012, 1,014, and 1,016; Dioritic schists, 1,001, 1,019, and 1,023 of App. C, Vol. II. The State Collection, App. B, Vol. II., also contains a large number of specimens of diorite of several varieties.

The distribution of this rock in the Huronian of the Upper Peninsula is interesting. It is far more abundant in the Marquette region and contiguous to the ore deposits, than elsewhere. The related rocks in the L'Anse region are abundant; but in the West iron district, and on its prolongation into Wisconsin, where it forms the Penokie range, diorites are rare. In the Menominee region they seem to be replaced to a great extent by chloritic schists and hornblendic schists, as described in Chapter V. Whether future explorations will prove that the best ores are always associated with the typical diorite, remains to be seen.

D. Magnesian Schists (*mostly chloritic*).

(See Mr. Julien's description, Specs. 179 to 188, App. A, Vol. II.)

Intercalated with the pure hard and mixed ores, at all the mines worked in formation XIII., are layers of a soft schistose rock, of some shade of grayish green, and often talcy in feeling. The Cleveland, Lake Superior and Champion mines are good localities for an examination of this rock. It is unquestionably a magnesian schist, varying from chloritic to talcose in character, and sometimes apparently containing a large percentage of argillite. In places, as at the Old Washington, its character is unmistakably talcose. Specimens obtained there held 4.2 per cent. of water, and had a specific gravity of 2.81, with light grayish-green color, and other characteristics of talcose schist. See Specs. 1,046, App. C, Vol. II. The corresponding schist at the Champion mine is also decidedly talcy. On the same magnetic range, but further west, at the Spurr Mountain, the equivalent schist is unmistakably chloritic. See Specs. 179 to 181, App. A, Vol. II. A rare variety of talc schist is represented by Spec. 74, App. B, Vol. II., obtained at the Grace furnace, Marquette.

In the Lake Superior and Barnum mines this rock is, in places, of a light green color, less soapy in feel, has a higher specific gravity and is of uncertain composition. See Spec. 55, State Coll., App. B, Vol. II. At this locality it has a marked cleavage structure, the planes of which trend east and west, and are nearly vertical, being distinct from its bedding, which latter is very obscure. Its structure bears a striking resemblance to that of the specular slate ores, noticed under A, even to the presence in both of minute octahedral crystals. Prof. Pumpelly has suggested, that one may be a pseudomorph after the other. In this connection it may be remarked, that no gradual transition of one into the other was observed, the division planes being in each instance sharply defined.

Specimen No, 1,043, App. C, Vol. II., from the Washington mine, is grayish, less schistose in structure than the last described variety, and gave up, when pulverized, one-third its bulk to the magnet. A similar massive variety from the same mine, which contained three per cent. of water, held black hard scales, which Prof. Brush decided had the character of ottrelite.

A reddish gray variety of this rock (see Spec. 6,164, University of Mich. Cabinet), holding grains of vitreous quartz, is from a heavy bed on the northeast side of the S. C. Smith soft hematite ore deposit, on Sections 17, 18, and 20, T. 45, R. 25.

South of the Edwards mine, at the Republic Mountain, and at other places in the ferruginous schists, occur bunches and thin irregular beds of a pure chlorite, often micaceous, which always contain garnets. See Spec. 6,097, University of Mich. Cabinet. This specimen shows, under the lens, minute elongated crystalline faces, closely resembling those seen in the diorite. Spec. 184, App. A, Vol. II., is garnetiferous. The "keal" or red chalk, found at several mines, is a variety of this schist impregnated with oxide of iron. See Spec. 6,183, University of Mich. Cabinet.

A very peculiar occurrence of this rock are the so-called "slate-dykes," which can be seen at the New England, Lake Superior and Jackson mines, but still better in the quartzite ridge, just north of the outlet of Teal lake. These dykes are often several feet in width, cut across the stratification, and are filled with a magnesian schist. If space permits, this subject will be more fully considered elsewhere. See Specs. 1,053, 1,068, App. C, Vol. II.

The Lower Quartzite bed V. often contains talc in bunches, small beds and disseminated, producing in places a talcy rock. The *novaculite* of that formation is due to the presence of talc and argillite. These rocks will, on account of their association, be more fully described in the Quartzite group.

It would be difficult for a skilled lithologist, and impossible for me, to draw the line between the chloritic schists here considered and the dioritic schists mentioned under Group C. So far I have chiefly noted occurrences of the magnesian schists, in formations XIII. and V., where they are not associated with true diorites. But at the Marquette quarries we find what may be called typical chloritic schists, bedded with granular diorites. See Specs. 182 and 183, App. A, Vol. II. At this locality the planes separating the two kinds of rock are well defined; at others, which have been designated, the transition is gradual.

Along the north border of the Laurentian area, which lies south of Lake Gogebic (see Map I.), are numerous exposures of a chloritic schist (see Specs. 187 and 188, App. A, Vol. II.), which in places becomes massive and granular, a form designated "greenstone" by the United States Linear Surveyors, and so marked on their maps. See Specs. of Diorite, 309 and 212, App. A, Vol. II.

The specimens of Laurentian Gneiss, 275 and 299, App. A, Vol. II., contain chlorite as an essential ingredient, proving this mineral to be as widely disseminated in the Laurentian as Huronian. An examination of Prof. Pumpelly's very exhaustive chapters on the lithology of the copper-bearing rocks, will show chlorite to be of frequent occurrence in that system; demonstrating it to be next to feldspar and quartz, one of the most universally diffused minerals in the Azoic of the Upper Peninsula.

E. QUARTZITE—*Conglomerates, Breccias, and Sandstones.*

(Principal development in Formations V. and XIV. See Mr. Julien's descriptions, 126 to 140, and also 358 and 359, App. A, Vol. II.)

After diorite and the ferruginous schists, no rock is more abundant in the Marquette region, and none more frequently found in outcrops, than the different varieties of this group. Two extensive beds exist—XIV. lies immediately over the ore formation, and V. near the base of the series. The last appears to be the most persistent and wide-spread member of the Huronian system. It can be traced from the shore of Lake Superior, near Chocolate river, westward for 40 miles, and possesses unusually economic interest from its affording the marble, used to a limited extent as furnace flux, and the whetstone rock (novaculite), which was at one time quarried for market. This quartzite has also recently been successfully employed as lining for Bessemer converters.

The Upper Quartzite (XIV.) is co-extensive with the ore formation XIII.; it is seen as the hanging wall of the most easterly point, at which rich hard ore is mined, and overlays the most westerly deposit yet explored. Between these is a third bed, seen in the railroad cut near the west end of Lake Fairbanks, the extent of which has not been made out. See Spec. 21, App. B, Vol. II.

At the west end of Lake Michigamme, near the centre of Sec. 25, T. 48, R. 31, is a large mass of quartzite, which appears to be a ledge, but if so, the bed is concealed to a greater extent than usual, for it has not been observed elsewhere. No. XVIII. is assigned for this quartzite, or for whatever rock may be found in the gap between Beds XVII. and XIX. The Cascade iron range is divided by a thin bed of quartzose rock, which varies from a quartzite to the coarsest conglomerate I have observed in the region, but which, like the two last-mentioned beds, seems to be local. At the Greenwood furnace is a heavy and persistent bed of quartzite, in which are intercalated layers of clay slate; its age has not been determined; it resembles the lower quartzite.

The extreme hardness of quartzite (the knife makes no impression on it, and it will readily scratch glass), and its general dissimilarity to the other members of the series, renders its recognition easy and much description unnecessary.

Vein quartz, occurring in bunches, seams and veins, in nearly all rocks, is not embraced in this description; nor are those slightly ferruginous quartz schists, already described in Group B, which a strictly scientific classification would place under this head. Quartzite is seldom white, often light-gray, or dark-gray and sometimes reddish or greenish. The effect of weathering does not penetrate the rock beyond a mere film, dulling the lustre and color of a fresh fracture rather, than changing it; but the latter effect is sometimes produced in the impure varieties. Broken pieces often show grains of glassy quartz; and the arenaceous character is sometimes so plain, as to leave no doubt in the mind, that the rock is a metamorphosed sandstone or conglomerate (see Fig. 2). Again, the whole mass is compact, having much the appearance of vein-quartz. In structure it is usually massive, and the bedding obscure; but in places, as at the northeast corner of Teal lake, it is banded, presenting a flaggy structure, like the ferruginous schists. The mean specific gravity of a large number of specimens was 2.69. See App. B, Vol. II.

The foregoing description applies in general to all the beds; but as it is often of importance to the explorer to distinguish the Upper bed on account of its relation to the ore formation, a few points of difference will be noted. As has been remarked, the Lower bed is often calcareous, turning in places into a true marble, as at the Morgan Furnace; and the same formation is often talcy in character, containing in certain localities bunches and beds of a talcy material and in other places beds of argillite. An intimate mixture of these minerals with the quartzose material produces novaculite, which was formerly quarried just east of Teal Lake outlet. See Spec. 13, State Coll., App. B., Vol. II. Red oxide of iron in grains and small bunches, is not infrequent in the Lower bed, as can be seen in northeast quarter of Sec. 22, T. 47, R. 26.

So far I have seen neither marble, talc, nor novaculite in the Upper Quartzite, and only once, at the Lake Superior Mine, have I seen argillite associated with it. As this exception has much interest, it will be fully considered in another place. The Lower Quartzite is seldom conglomeritic, the upper one often so, and in places on the Spurr Mountain range it is a true conglomerate, containing pebbles of white and glassy quartz and jasper. See Specs. 115 to 118, App. A, Vol. II. At Republic Mountain large fragments of ferruginous schist are seen in the base of the Upper bed. Southwest of the Old Washington mine it is a coarse conglomeritic rock, which is in places schistose or slaty. See Spec. 122, App. A, Vol. II.

The matrix of this variety (See also Spec. 6,085, University of Mich. Cabinet) is a soft, micaceous, slaty material, containing fine grains of specular ore and holding pebbles of white quartz. The Upper bed overlying the east end of the Jackson, and that over the New York mine, also hold pebbles. Mica scales and epidote were found in the same bed at the Republic Mountain, and in places it had almost the appearance of fine-grained granite.

As if to leave in our minds no shadow of doubt, as to the sedimentary origin of this rock, nature has, in addition to the conglomerate on the Spurr Mountain range, given us a variety of the Upper Quartzite, which can only be described as a fine-grained, friable, banded *sandstone*. See Specs. 358 and 359, App. A, Vol. II. The alternations of magnetic sand with quartz sand, producing the stripes, is very interesting in connection with the origin of these ores. It is doubtful if any true breccias (conglomerates with angular pebbles) occur associated with the rocks here described, if at all in the region. The brecciated rocks, a variety of "mixed ore" found in formation XIII., is believed to have had the origin ascribed under Group A.

Specimens of University of Mich. Cabinet, Nos. 6,193, 6,084, 6,180, 6,211, 6,219, and 6,122 are from these quartzite beds. Specs. 8 to 14, State Coll., App. B, Vol. II., are from the Lower bed, and Specs. 50, 51, and 52, same Coll., are from the Upper. The extensive beds of quartzite, which occur in the Menominee region, will be fully considered in Chapter V. This rock is also of frequent occurrence in the L'Anse range and toward the Montreal river, as will appear in following Chapters. A beautiful example of false stratification, or discordant parallelism, was observed in this last-named region, as is shown by Fig. 2, sketched near the south quarter post of Sec. 10, T. 47, R. 45. It was a true granular quartzite, but showed deposition marks almost as plainly as a fresh-cut sandbank.

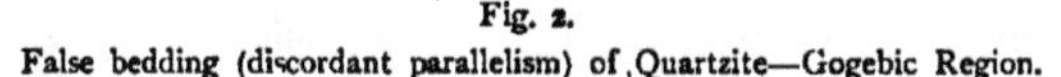

Fig. 2.
False bedding (discordant parallelism) of Quartzite—Gogebic Region.

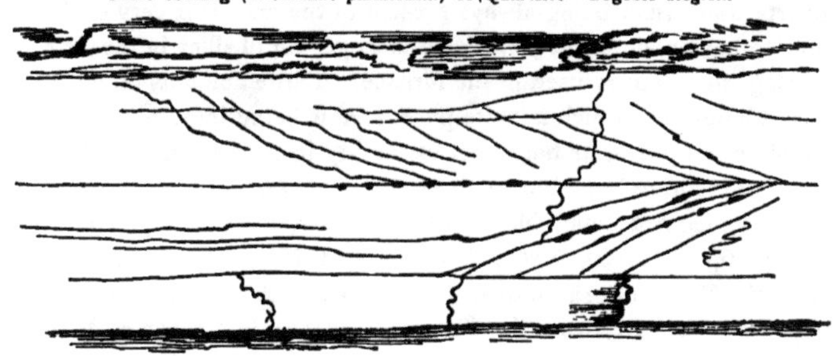

F. Marble (*Limestone and Dolomite*).

(See Mr. Julien's descriptions, 101 to 113, App. A, Vol. II.)

The association of this rock with the Lower Quartzite, or rather the transition of the latter into marble, has been mentioned. This transition is seldom complete, the marble being always more or less silicious. As is usual in such cases, the change is gradual, producing all varieties, from calcareous quartzite to silicious marble. The prevailing colors are light gray, salmon and reddish. The purest varieties often present a sparry structure, with large lamellar facets like orthoclase feldspar, with which it is often confounded, but from which it can readily be distinguished by its softness. Beds of argillite are invariably associated with the marble. See Fig. 19, App. E, Vol. II. Outcrops often present minute ribs or ridges of the more silicious layers, left by the weathering away of the purer marble.

The mean specific gravity of a large number of specimens averaged 2.82. See App. B, Vol. II. Pure marble has the same composition as pure limestone, of which it is simply a crystalline or highly altered form, that is, it is a carbonate of lime;—if carbonate of magnesia is present in considerable quantity, as is often the case on the Upper Peninsula, the rock becomes a *dolomite*. Marble is readily distinguished from its effervescing with acids, when pulverized.

Marquette marble has been considerably used as a blast furnace flux, for which purpose it only answers passably well, on account of the silica so generally present; silica, in the form of quartz, and jasper being always present in the *ores*, it is very desirable to have none in the *flux*, for it is to get rid of silica in the form of slag, that lime is used in the furnace. Large amounts of Kelly island limestone, which is quite pure, is now being imported. For building purposes, its hardness, variability in texture and the difficulty of securing large blocks, have so far prevented its use; beautifully variegated small blocks can, however, be easily procured. Specs. 6,198, 6,199, 6,200, University of Michigan Cabinet, are from the Morgan Furnace quarry, and Specs. 106 to 113, State Col., App. B, Vol. II., from the Chocolate quarry, just south of Marquette, all belonging to formation V., represent the chief varieties of this rock.

No marble has been observed in the L'Anse district, nor between Lake Gogebic and Montreal river, but it is one of the most abundant rocks in the Menominee region, where it occurs in a much purer form than in Marquette, usually more dolomitic. See Chapter V, and Specs. 102 and 103, App. A, Vol. II. Marble of similar quality is also abundant in the vicinity of Fence and Michigamme rivers, in Towns 44 and 45, R. 31. See Spec. 105, App. A, Vol. II.

G. Argillite or Clay Slates and Related Rocks.*

(Constitutes bed XV., and occurs in bed V. and elsewhere.)

It was previously mentioned under Groups E and F, that beds of clay-slate were sometimes interstratified with layers of quartzite and marble. Fine examples of this, in the case of both rocks, can be seen respectively at the Greenwood and Morgan furnaces. In addition to these, at least two distinct beds of argillite have been made out; one immediately beneath the ferruginous schist of formation X., to be seen in outcrop on the south shore of Teal lake, near west end, and in the railroad cut about one mile east of Negaunee. See Spec. 20, App. B, Vol. II. Another and far more extensive bed is XV., which forms the stratum next above the Upper Quartzite; boulders of this bed, which had the appearance of being near the parent ledge, were found in the railroad cutting, near the pockets at the Washington mine. At the Champion this formation is exposed in the branch railroad, and it is found at numerous points on the north shore of Lake Michigamme.

The prevailing color of this rock is usually dark brown or blackish, but where associated with the marble it is sometimes reddish. It has a true slaty cleavage, distinct from the bedding, but seldom splits in sufficiently large or regular slates to warrant us in supposing it may in places produce roofing slates, although experienced persons express the belief, that good slates will yet be found in the Marquette region. Black carbonaceous matter is often present in this slate, a preponderance of which produces the rock which will be described hereafter under J. A variety at the Greenwood furnace contains a large amount of iron-pyrites; and the first stack built of it had to be taken down, from the decomposition of this mineral. The slate in the branch railroad cut, at Champion, shows a slight tendency to be micaceous and holds garnets. See Spec. 56, App. B, Vol. II. Silicious bands often exist

* Mr. Julien has in App. A, Vol. II., given the results of much study of these rocks, and has divided them into the true argillites and several other varieties possessing a different composition. See descriptions 189 to 225. As this difference cannot readily be made out by the unscientific, and as it is not important to the practical man, it will not here be attempted to separate these varieties.

in this rock, faintly marking its bedding at an angle with the cleavage, as can be seen in Spec. 20, App. B, Vol. II.

Overlaying the Lake Superior and Barnum ore deposits, hence occupying the place of the Upper Quartzite, is a greenish-gray schist, obscure in its composition, and somewhat like the magnesian schists D, but apparently of the same general character as this group. See Spec. 55, App. B, Vol. II. This rock may very properly be regarded as the connecting link between Groups D and G, which evidently graduate into each other, as did C and D. It is frequently stained reddish-brown along the seams and cracks, proving the presence of protoxide of iron, and shows in places beautiful dendritic delineations of manganese. This formation does not show the cleavage structure, so conspicuous in the schists of Group D, which are bedded with the pure ore at these mines. At the most westerly opening of the Lake Superior, thin beds of quartzite appear, indicating that the presence of argillite in this bed is probably only local. See Map No. IX.

An example of a magnesian schist (D) graduating into an argillaceous variety can be seen in the slate which overlies the specular ore of No. 1 pit, New England mine, which, by its high specific gravity (3.03), evidently contains considerable iron. Another ferruginous and probably chloritic variety occurs on N. W. ¼ Sec. 31, T. 47, R. 25, where explorations for iron have been made by the Morgan Iron Co.

The average specific gravity of a number of typical specimens of argillite was 2.75. See App. B, Vol. II. The rocks above described are illustrated by Specimens 1,039, 1,072, and 1,036, App. C, Vol. II.

Beyond the limits of the Marquette region, we find in the recently explored Huron Bay district, particularly in the south part of T. 51, R. 31, the finest clay slates so far discovered in Michigan. Several competent experts have examined this district, and pronounced the slates of the best quality for roofing and other purposes, and in immense quantity. See Spec. 81, App. B, Vol. II. Companies are now at work in this district, the organization of which is given at the end of Chap. I. For an account of the clay-slates in the Menominee region, see Dr. H. Credner's papers (Leipsic).

This rock also occurs west of Lake Gogebic, as will be mentioned hereafter.

H. Mica-schist.

(Formation XIX. contains the principal development of this rock. See Mr. Julien's description, No. 301, App. A, Vol. II.)

There appears to be but one extensive stratum of this rock, the character of which is unmistakable, which is at the same time the youngest and one of the thickest beds of the whole Huronian series. This formation, which I have numbered XIX., forms the surface rock along the south shore of Michigamme lake, among its islands, along the outlet for several miles, and westward from the lake through the southern parts of T. 48, Ranges 31 and 32, as shown on Map III. The rock is sometimes so silicious as to be rather a micaceous quartzite, but usually its true character is very plain. It frequently contains seams and bunches of white quartz, occasionally seams of black hornblende, and often holds numerous imperfect crystals of a delicately pink-colored, coarsely fibrous mineral, which Prof. Brush decided was andalusite, and brownish, smaller, and more perfect crystals of staurolite.

Andalusite and staurolite have not been observed elsewhere in the Marquette region in rocks of any age. Imperfect small reddish garnets are sometimes abundant, but they were not observed at the same places as the first-named minerals, and seemed to be nearer the base of the formation. The mica, which usually holds but little quartz, is of a brownish color on fresh fracture, weathering more grayish; its scales show a constant tendency to bend themselves around the imbedded crystals, like the fibres of wood around a knot. The projecting rounded crystals give the weathered rock a warty look, having somewhat the appearance of a conglomerate, as can be seen on the most southerly islands in Lake Michigamme. The specific gravity of this porphyritic mica-schist varied from 2.81 to 2.89, the mean being 2.84. See Specs. 1,031, App. C, and 61, App. B, Vol. II.

Descending in the series, the next mica-schist to be noticed is entirely different from the above, in being black, and decidedly dioritic in its affinities. It occurs in the upper part of diorite bed XI. at Republic Mountain. The deposit is not extensive, and its relations with the diorite indicate that it is a local variety, apparently graduating into dioritic schist.*

* The local micaceous character of bed XV. has been noticed.

One other mica-schist, that associated with the Cannon ore on Sec. 28, T. 47, R. 30, deserves notice. This rock resembles XIX. only in the brownish color of its mica; it contains no crystals of other minerals, and is always quartzose, sometimes to the point of becoming a micaceous quartz-schist. The age of this rock has not been satisfactorily determined, but it is near the base of the series. The striking peculiarity of this variety is the fact, that in places the mica is replaced by micaceous specular iron ore, thereby becoming a specular schist, a rock very nearly related to the itaberite of some writers. The Cannon Iron Company's explorations, in which a fair specular slate ore has been found, are located in a highly ferruginous part of this bed. See Spec. 16, App. B, Vol. II. The relations of this rock with the lower quartzite of the North belt, Menominee Iron region, is fully discussed in another place.

I. ANTHOPHYLLITIC SCHIST.—(in bed XVII. and others.)

(See Mr. Julien's descriptions 174 to 178, App. A, Vol. II.)

Immediately below the great mica-schist bed, XIX., and probably separated from it by a stratum of quartzite, XVIII., is a well-defined stratum of a slightly magnetic rock, varying in color from brownish-black to dull slate on fresh fracture, and grayish to blackish in outcrop. It often shows manganese,* and always a fibrous, light-brown mineral, which Prof. Brush, from the examination of some imperfect specimens, decided to be anthophyllite,† a variety of hornblende, and suggested the name here employed for this group.

Numerous outcrops of the rock occur along the north shore of Michigamme lake, and a fine development at the mouth of the Bi-ji-ki river, as well as at the Champion furnace, where layers rich in manganese occur. A specimen afforded Dr. C. F. Chandler 25.2 per cent. of metallic iron, and 4.37 per cent. of metallic manganese. See Specs. 58 and 59, App. B, Vol. II., and 178 App. A, Vol. II.

Below the ore formation XIII., at the Spurr Mountain, are layers of schist of a similar character, a specimen of which afforded Mr. Britton 45.21 metallic iron, 1.78 metallic manganese, 26.36 silica.

---

* This variety resembles plumbago, and may contain carbon.

† Prof. Dana now regards anthophollite as a distinct mineral.

A moderate increase in the percentage of iron and manganese therein found (which may very likely take place in some part of the bed) might render this rock a workable ore, particularly as the associated mineral is an easily fusible hornblende instead of the silica so common in the other ores. Ores containing 12 to 20 per cent. of manganese need not be rich in iron, to give them merchantable value.

Underlying this formation (XVII.), or perhaps forming its base, is a rock, numbered XVI., which at Champion and on Sec. 26, T. 48, R. 31, shows a tendency to pass into a *limonitic schist*, and may very likely afford workable soft hematite ore in some part of its course. The propriety of giving this rock, about which so little is known, a distinct stratigraphical designation, may be questioned; but its ferruginous character, pointing toward the possibility of commercial value, led to this course.

South of the Washington mine, and therefore stratigraphically below the ore formation,—for the whole dips north,—there is an obscure schistose rock of a gray color, weathering brown, and containing very little iron, often garnets, but made up chiefly of a light brownish fibrous mineral, which is probably anthophyllite, but which in places resembles mica. These rocks are extensive, stretching from the Champion mine eastward to the old Michigan mine. They are generally slightly magnetic, and unquestionably occupy the place of the silicious ferruginous schists of Group B. The diorites associated with them are also peculiar, the two sometimes resembling each other. This obscure series is well illustrated by Specimens 6,086 to 6,099, University of Mich. Collection. See also Specs. 174 and 175, App. A, Vol. II., and 27, App. B, Vol. II. Their affinities are apparently with this group.

J. Carbonaceous Shale.

(See Mr. Julien's descriptions, 246 to 251, App. A, Vol. II.)

The presence of plumbago or graphite (a form of carbon) was noticed in the anthophyllitic schists, last described. Carbonaceous matter has also been observed in various clay-slates, as was noticed in describing the Argillite Group, and we could have placed this rock there as a variety of clay-slate, very rich in carbonaceous matter. It is of a bluish-black color, but burns white before the blow-pipe, marks paper like a piece of charcoal, is soft and brittle, slaty in structure, and is the lightest rock yet found, having a specific gravity of but 2.06.

This rock has been found in the Marquette region only at two localities: 1. The S. C. Smith mine, T. 45, R. 25, where it seems to bound the iron-ore formation on the northeast. See Spec. 6,163, University of Mich. Collection.

(2.) On the south side of Sec. 9, T. 49, R. 33, along Plumbago brook, as will be fully described in the account of the L'Anse Iron range, is a large deposit of carbonaceous shale, a specimen of which gave Prof. Brush—carbon, 20.86; earthy matter, 77.78; moisture, 1.37. Another sample from same locality gave Mr. Britton—moisture and carbonaceous matter, 22.51; oxide of iron, 4.37; earthy matter, 73.12. See Spec. 64, App. B. Vol. II. These analyses prove the material to have no commercial value, but possess scientific interest as proving the existence of a large amount of carbon in the Huronian rocks. The equivalency of these shales with the members of the Marquette series has not been established; they are undoubtedly Huronian, and are, I suppose, younger than the ore formation XIII.

CHAPTER IV.

GEOLOGY OF THE MARQUETTE IRON REGION.

1. MICHIGAMME DISTRICT.

IN describing the geological structure of the Marquette Iron series, I shall begin with the Michigamme district, because its structure is simplest, the iron ranges easily followed on account of their magnetism, and because my explorations and surveys have there been more thorough than in either of the other districts.

The **Champion mine**, 33 miles west of Marquette, is at one of the most extensive, regular and typical deposits of ore in the whole region (see Map No. VII.). The strike is a few degrees south of west, and dip north at an angle of 68°. The extent and nature of the workings at the date of the survey may be seen by reference to the map. Up to this time the mine has produced an aggregate of 225,000 tons of magnetic and slate ore of first quality. The general form of the ore mass is that of a huge irregular lens, or flattened cylinder-shaped mass, which thins out to the east and west to so narrow a width, as not to be workable. The easterly portion of the deposit is black, fine and coarse-grained magnetic ore; the westerly portion is specular slate ore, with a small admixture of magnetite. The local magnetic attractions are very strong and are fully considered in Chapter VIII. The position of the plane dividing the two varieties is approximately shown in the sketch of workings on Map No. VII. The whole mass here described is not, however, pure ore, as may be seen by inspecting plans of the first and second levels on the map. Minor irregular lens and pod-shaped masses of pure ore, "mixed ore" (banded ore and quartz), together with whitish and greenish magnesian schists, alternate like the muscles of an animal, forming, as a whole, a comparatively regular deposit. Overlying the ore on the north side is a hanging wall of gray quartzite, the thickness of which is considerable, but could not be accurately determined on account of the drift. Immediately south of the ore, if it may not be regarded as a part of the ore formation, is a banded jaspery or quartzose rock, containing some iron. Next south, and underlying the whole ore formation, as may be seen by an outcrop near the east end of the mine, is a bed of diorite ("greenstone"); this rock in places becomes schistose and chloritic in character. South of the diorite is a silicious schist and then a swamp. The arrangement of these beds may be seen in geological section A—A," on the map, where they are numbered in Roman numerals X. to XIV., the latter designating the quartzite.

Following the Champion range east one mile, we arrive at the **Keystone Company's mine,*** where but little work has been done, and the arrangement of the rocks in consequence not so easily made out. A small bed of magnetic ore was opened at this locality two years ago, and what is said to be a large deposit of specular ore has but just been discovered on the same place. Five hundred feet north are a number of outcrops, indicating the presence of a heavy bed of conglomeritic schist, which holds masses of quartzite, varying in size from pebbles to others two feet by one thick, and even larger. It also contains flattish fragments of various schists and slates. Further north it passes into a brownish schist, containing pebbles of quartzite. This rock is believed to correspond with the overlying quartzite of the Champion, and is marked XIV. on the map and sections. North of this, and exposed in the railroad cut, is a micaceous slate, containing garnets, marked XV., and represented by Specimen 56, State Collection, App. B, Vol. II.

North and west of this locality, about one-fifth of a mile, are a number of test-pits, in many of which is exposed a soft, brownish, ferruginous rock, which affords hand specimens of soft hematite ore. This rock is marked XVI., and is represented in the State Collection by Specimen 57, App. B, Vol. II. Immediately south of the Keystone workings is a specular schist or conglomerate, in which flattened pebbles, or very uneven lamina of quartz, are contained between thin layers of micaceous specular ore. This formation is believed to be the equivalent of XII. of the Champion mine section, and is so numbered on the map.

West and south are numerous extensive outcrops of a brownish banded magnetic schist, marked X. on Section C—C", Map VII.

---

* Late "Parsons Mine."

The arrangement and character of the rocks along the intermediate section, B—B,' will be sufficiently understood from the above descriptions and an inspection of the map. The other formations represented will be considered in another place.

At the **Spurr and Michigamme mines** we find rocks identical in their general character and sequence, although the order is reversed, this series being on the opposite side of the basin from the Champion. Projecting all the facts observed along the north shore of Michigamme Lake on one plane, which we will assume to pass north and south through the Spurr Mountain mine, the following Geological Section is easily made out:

Commencing at the most southerly and uppermost bed (the whole series dips to the south), we have, first, a comparatively soft, grayish and blackish flaggy rock, containing considerable iron, a little manganese and often made up largely of a hornblendic mineral, which occurs in needle-shaped crystals. Professor Brush calls this rock anthophyllitic schist. See Specimens 58 and 59, State Collection, App. B, Vol. II., and Chap. III.

This rock is numbered XVII. on geological section No. 9, map of the Marquette Iron region, which see. It is also well exposed at the mouth of the Bi-ji-ki river, in the railroad cut just east, at the Champion furnace, and at numerous projecting points along the north shore of the lake.

The next rock to the north, in descending order, (numbered XVI. on the map and section,) on account of its tendency to decomposition, has never been seen in outcrop; it is exposed by the explorations for ore, made on the north side of Sec. 26, T. 48, R. 31, and at the Champion; its character was indicated in describing the Champion series, and need not be repeated here. As will be seen, this rock has the same number in each section, and the two exposures are believed to belong to the same bed. It is not improbable that future investigations may prove it to be a variety of the ferruginous anthophyllitic schist XVII., already described, a point which was considered in Chapter III., Group I.

Next below is a dark-colored clay-slate, which also, on account of its softness, is seldom seen in outcrop. It is, however, exposed on the point in northeast part of Section 29, and at other places along the north shore of the lake. On the Spurr mountain, geological section No. 9, this formation is numbered XV., and is

9

believed to underlay the swamp and creek immediately south of the mountain which finds easterly prolongation in Black bay. As will be seen by reference to the Champion sections, this rock is regarded as the equivalent of the micaceous clay-slate XV., there described.

North of this clay-slate, and immediately overlying the ore at both the Spurr and Michigamme mines, is a quartzose rock numbered XIV., which is in places a hard conglomerate, and again, especially when in contact with the ore, a fine whitish sandstone. See Specimen 52, State Collection, App. B, Vol. II., and Julien's descriptions, Specs. 358 and 359, App. A, Vol. II. This rock is unquestionably the equivalent of the upper quartzite XIV. of the Champion section, which, on the whole, it closely resembles in its lithological character. See also Group E, Chapter III.

The prevailing variety of ore of the mines on this range is a fine-grained, somewhat friable, rich, blackish magnetite. See Specimens 40 and 41, State Collection, App. B, Vol. II., and also Iron Ores, Chap. III. There is also at the Michigamme mine a hard, fine-grained, steely magnetic ore, in considerable quantity. Analyses of these ores will be found in Chapter X. The surface indications, magnetic attractions, explorations and mining operations but just commenced, point unmistakably to large deposits of high grade magnetic ore at both localities.

The **Spurr Mountain** is an east and west ridge, the summit of which is 118 feet above Lake Michigamme and 75 feet above the creek, which passes south of it. This ridge terminates abruptly to the west near the centre of the northwest ¼ of the southwest ¼ of Sec. 24, T. 48, R. 31, where there is a natural exposure of merchantable ore 40 feet thick horizontally, being the largest outcrop of pure magnetic ore I ever saw. Mining operations, just begun, have demonstrated the thickness to be still greater, and the deposit to extend at least several hundred feet east and west, with a probability, based on magnetic attractions, of its extending much farther. The bold face, small amount of earth covering, softness of the ore, its apparent freedom from rock, convenience of the railroad and accessibility, present facilities for mining and shipping, which could not well be surpassed. The magnetic observations made at this locality, where the attractions were remarkably strong, are given with illustrative diagrams in the special chapter devoted to that subject. It is easy by means of the dip compass, to follow this iron range two-thirds of the distance along the north side of Michigamme lake, and west-northwest from the Spurr to the First lake, an aggregate distance of over nine miles, as may be seen by the map of the Marquette Iron region, No. III. It must not by any means, however, be supposed, that here is a workable deposit of ore nine miles long; this has not been proven, but on the contrary, it has been proven that for a considerable portion of this distance the ore is not workable, having altogether too large an admixture of rock. Therefore, while it may be confidently asserted, that all of the rich hard ore which will be found in this vicinity, will be in or near the belt of magnetic attraction already described, it may be asserted with equal truth, that at least three-fourths of the whole length of this belt is barren ground, according to the present standard of merchantable ore. The law of the distribution of the rich "chimneys," "shoots," or "courses of ore," as they are designated in different mining regions, along a given iron range, has not been made out. The subject is more fully considered in Chapters VII. and IX.

Besides the deposits already described on this range, one other has to be mentioned, that on the east side of railroad Sec. 23, adjoining the Spurr on the west. The magnetic attractions here are remarkably strong, and explorations have revealed the existence of a small workable deposit of first-class magnetic ore. Whether this deposit connects with the Spurr or not, was not fully determined.

As has been remarked, both the granular and compact varieties of magnetic ore occur at the **Michigamme mine.** The explorations on this location, which were conducted by the writer, developed in a distance of 1,200 feet, east and west, seven places, where pure ore existed of a thickness of from seven to thirty-five feet, rendering it probable, that the ore deposit is continuous and workable for the whole of this distance. Mining operations, which have commenced at this location, confirm these results. Pure ore was found in place at two points on same range west, on Sec. 19 of the Michigamme Company's property, but not enough work was done to prove their extent. Eastward the ore can be traced by the magnetic needle into Michigamme lake, on the south side of Sec. 20.

There can be no doubt these deposits and the Champion belong to the same horizon, being the opposite croppings of the synclinal basin, which passes under Michigamme lake; although the Champion deposit has not been traced westward, nor the Michigamme range eastward, to points where they come directly opposite each other. Whether the specular slate ore found so abundantly at the Champion will be found on the north side of the lake, remains to be seen. I see no reason why it should not; the explorations, so far, have been based entirely on magnetic attractions, and would therefore not be likely to result in finding specular ore.

Underlying the pure ore here, as at the Champion, is a ferruginous quartzose rock, which has an immense development on the Spurr-Michigamme range, where it is a well-characterized reddish quartz schist (jasper), containing thin layers of pure specular ore; these layers being occasionally thick enough to afford hand specimens. See Specimen 33, State Collection, App. B, Vol. II. A similar rock is found, as will be seen hereafter, at the Republic mountain, where it has the same relative position and number, XII.

Underlying this iron series we find, as at the Champion, a diorite (greenstone), but which here has a much greater development, forming a conspicuous ridge which borders the Michigamme and Three Lakes valley on the north, and which has already been described under Topography in Chapter II.

This greenstone ridge is separated from the granite region to the north by a valley about half a mile wide, which is underlaid by various schists and quartzites, about which little is known. Two are marked X. and V. on the Spurr-mountain section No. 9.

The most easterly developed mines in the Michigamme district are the **Washington** and **Edwards**, represented by map No. VIII. The general structure, which we are now considering, can be easiest made out at the Edwards and "old mine," which are adjacent, and about three-fourths of a mile west of the Washington mine proper. The general character and order of the ore and accompanying rocks at this locality is so similar to that of the mines already described, that a careful inspection of the map and accompanying sections leaves but little to be said. The Upper Quartzite XIV. is fully exposed in outcrop, as well as in the railroad cut, just west of the mines, where it is a coarse conglomerate, often schistose, as is shown by Specimen No. 51, State Collection, App. B, Vol. II.

The same formation is a compact gray quartzite at the Edwards mine, and at other points in the vicinity.

The ore formation XIII. affords at this group of mines all the varieties, already designated as being found at the Champion, Spurr, and Michigamme mines. Like the Champion, here are intercalated beds of magnesian schist, the arrangement of which are shown on the sections of workings given on the map already referred to, as well as in the plan of the Edwards mine, by A. Kidder, Plate XIX., Chap. IX., where the subject of detailed structure is more fully considered. One of these schists, of a decided talcy character, is represented by Specimen 54 of State Collection, App. B, Vol. II.

The underlying ferruginous quartzose rock, XII , has a large development south of the Edwards mine, and to it probably belongs the "red ore" of the old Washington. Southwest of the latter mine are large exposures of the peculiar conglomeritic specular schist, mentioned as occurring on the Keystone property, east of the Champion.

The dioritic formation, XI., is represented by a large outcrop of a greenish schistose rock, apparently chloritic, which can be seen immediately south of the old mine. Below this formation are alternating schists and diorites of different varieties, which are sufficiently well shown on the map and sections. One of the most interesting varieties is represented by Specimen No. 27, State Collection, App. B, Vol. II., procured 500 feet south of Pit No. 9, Washington mine.

The Washington mine proper presents some of the most complicated structural problems, to be found in the Marquette region, and I will not here either attempt their solution, or even advance the hypothesis which I have formed. Suffice it to say that, in general, the mine is a monoclinal deposit, dipping away from the St. Clair mountain (which term I apply to the high ground to the south) to the north and under the great swamp. The minor rolls, the peculiar faulting at the East Hill, and the trap dykes, would, if fully considered, occupy a chapter.

I cannot, however, pass to another mine, without noticing the singular manner in which the mass of ore, known as Anderson's cut, or Pit No. 1, is terminated in its downward course, as shown by Figs. 3 and 4. It will have been observed, that the usual form of ore masses is *lenticular*, *i.e.*, they generally terminate by *wedging out* more or less gradually each way. This exceptional mass, as will be seen, is obliquely and abruptly cut off, the bottom rock be-

Fig. 3.—Looking East.

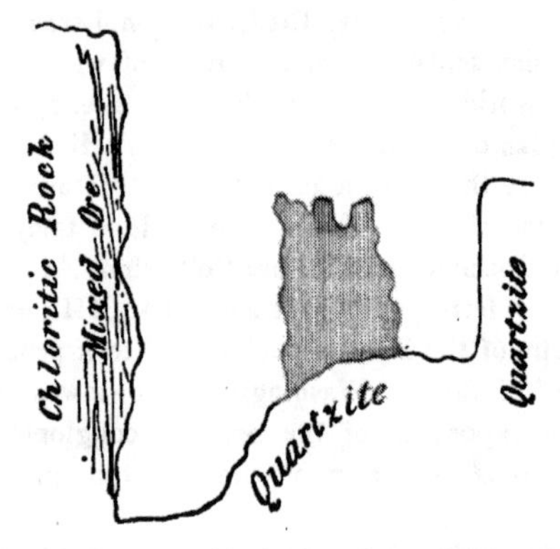

ing a quartzite of the same kind, that bounds the deposit on the north, and there is no evidence of faulting on the plane of this floor,

Fig. 4.—Looking East.

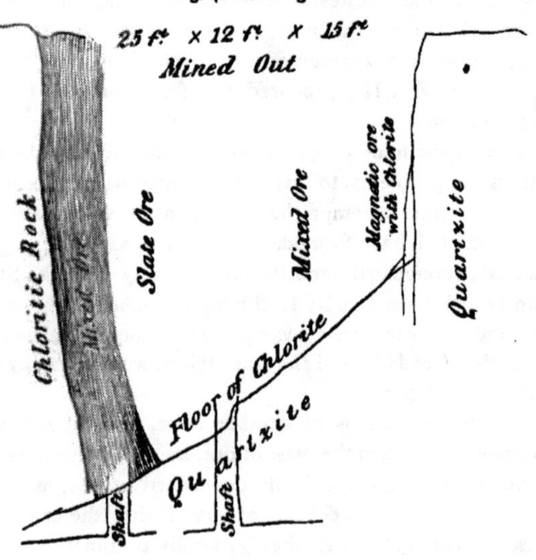

or along the quartzite wall. An hypothesis to account for this phenomena, based on a sedimentary origin for these rocks, will readily suggest itself and need not be stated.

The **Republic mountain** and its prolongation on the Kloman lot, is the only remaining ore deposit of the class under consideration, which remains to be described in the Michigamme district. See map No. VI. The **Magnetic mine** group, embracing the Cannon and Chippewa locations, belong to a different geological horizon, produce different ores, and will be considered hereafter.

The immense mass of pure specular ore, which was naturally exposed near the centre of the north ½ of the southeast ¼ of Sec. 7, T. 46, R. 29, could leave no reasonable doubt in the mind of the experienced observer, that this deposit of ore was one of the largest, if not *the* largest, in the Marquette region. This outcrop, the extent of which is shown on the map of the Republic mountain, being there marked "pure specular ore," is, so far as I know, the largest outcrop of any equally rich ore, ever found in the United States.

The elevation of the ore, 120 to 150 feet above Michigamme river, gives an unsurpassed opportunity for mining operations, which began in the spring of 1872, and confirm, as far as they extend, the "surface show." Several other small outcrops of pure ore occur in the iron belt, one of the largest of which is near the centre of the Kloman mine lot, in southwest fractional ¼ of Sec. 6, same Township.

The numerous outcrops of rock and ore at this mountain, the strong magnetism possessed by three of the beds, the remarkable uniformity in thickness of the several formations, and the bold topographical features presented, all of which were carefully surveyed and are faithfully represented and explained on the accompanying topographical, geological, and magnetic maps and charts (Plates VI. and XII. Atlas), leave but little more to be said in this place, regarding the general structure of the Republic mountain.

The lithological character of the rocks and ores will also be fully understood from the 14 specimens from this locality, which are embraced in the State Collection, App. B, Vol. II. The ten formations represented by colors on the map, as composing the Huronian series, will now be enumerated, commencing with the lowest, which reposes non-conformably on the Laurentian granites and gneisses.*

The lowest bed of the series will be numbered V., for reasons which will hereafter appear.

V. A quartzose rock, which is exposed at but a few points, and is best seen near 4,600 southwest and 6,200 southeast (see rectangular ordinates on map), from which locality Specimen 8, State Collection, App. B, Vol. II., was obtained.

VI. Is a magnetic, bright, banded, silicious and chloritic schist, containing considerable iron. See Specimen 15, State Collection, App. B, Vol. II., from near locality of Specimen 8. Very large exposures of this schist occur on the northeast side of the mountain, and southeast of Bear lake. The regular, various-colored stripes,

---

* This sketch (6,100 southeast and 4,700 southwest, Map VI.) represents outcrops of Huronian quartzites and schists dipping north-northwest, and the Laurentian gneisses, *a a*, dipping northeast, the latter being within 50 feet of the former. The actual contact is not seen, but the stratigraphical relations indicated, in connection with the wide difference in

Fig. 5.

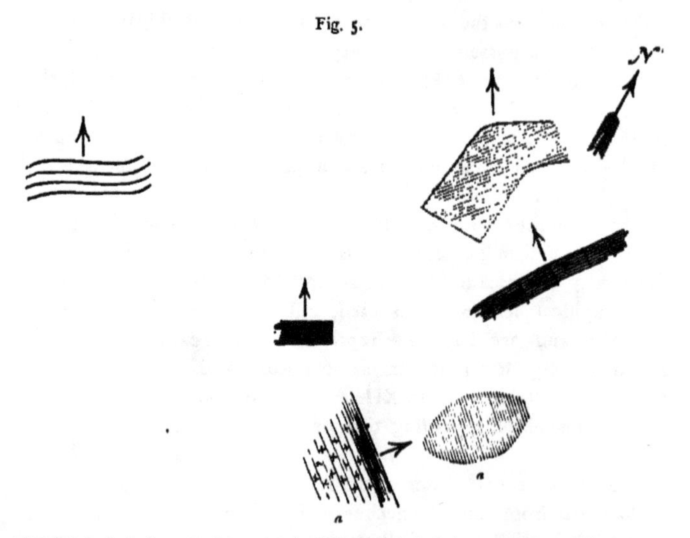

their lithological character, leaves no doubt in my mind of the non-conformability of the two systems, the Huronian being the youngest. This non-conformability can also be observed on the L'Anse Range. See page 156.

which this formation, as well as VIII. and X. displays, strongly suggests a rag carpet. The greenish layers are apparently chloritic, the whitish and grayish are quartz, and the brown and dark gray are silicious layers of the red and black oxides of iron. Some of these lamina are quite pure iron ore, and the whole mass may contain from 15 to 30 per cent. of metallic iron. The magnetic power displayed by these schists is remarkable, as will be seen by inspecting the charts and explanatory text already referred to.

VII. Is a diorite of the general character of those, so fully described by Mr. Julien in App. A, Vol. II., as will be seen by reference to Specimen No. 18, State Collection, App. B, Vol. II.

VIII. This magnetic silicious schist in its lithological character differs in no essential particular from No. VI., already described. See Specimen No. 19, State Collection, App. B, Vol. II. This formation is noticeably thin, not exceeding 40 or 50 feet, the other beds being from three to five times this thickness, as can be seen on the map.

IX. Is a Diorite similar to VII. See Specimen No. 22, State Collection, App. B, Vol. II.

X. A magnetic silicious schist similar to VIII. and VI., but containing in places more iron, as at 5,600 southeast and 2,500 southwest, from which locality Specimen 23, State Collection, was obtained. This, it will be observed, is a fair specimen of magnetic flag ore, containing probably 45 per cent. of metallic iron.

XI. This formation is made up of a coarse-grained diorite, in which a light grayish and reddish feldspar is a conspicuous ingredient, as may be seen on the Kloman lot, as well as at the knob southwest of Bear lake, from which Specimen No. 29, State Collection, App. B, Vol. II., was obtained.

A schistose variety, containing considerable black mica, occurs in the same formation, at 3,400 southwest and 5,300 southeast, where Specimen No. 30, State Collection, was obtained, although it does not truly represent the prevailing variety at this locality.

XII. This is a reddish quartz or jasper schist, containing thin lamina of specular ore, and very similar to the corresponding formation of the Spurr mountain series already described, as will be seen by an examination of Specimen 32, State Collection, App. B, Vol. II.

XIII. We have now reached the iron-ore formation, the principal outcrops in which have been enumerated. Four varieties of material chiefly make up this formation, which in the order of apparent quantity are as follows:

a. A banded rock made up of alternating layers of red quartz or jasper and specular ore, designated by the miners as "*mixed ore*," the richer varieties of which are now shipped as second-class ore. See Specimens 36 and 37, State Collection, App. B, Vol. II. The contorted and plicated lamina of this rock, brought out by the alternating bright red and steely bands, and which could be but poorly illustrated in Figs. 19 to 29, App. K. Vol. II., are very beautiful, being often contorted and plicated in a striking manner. See Iron Ores, Chapter III. It may be remarked in passing, that such contortions in the constituent lamina of rock formations generally indicate the presence of great folds in the whole formation, as is plainly the case at this locality.

On the southwest side of the basin, at points in the ore formation marked "specular conglomerate" on the map, occurs a true schistose conglomerate, in which pebbles, chiefly quartz, are bedded in a matrix of silicious ore. On the supposition that this rock may be a secondary form of the laminated or mixed ore, and from a desire not to multiply subdivisions in this connection, it will at present receive no further consideration.

b. Next to the mixed ore in quantity, so far as can be judged by what can be seen, is the pure *specular* ore. See Specimen 46, State Collection, App. B, Vol. II. The specific gravity of these specimens varied from 5.09 to 5.56, the average of four being 5.24, or greater than that of any other ore in the region, which should indicate a somewhat greater richness in metallic iron; whether furnace work will confirm this, remains to be seen.

c. The next in supposed order of quantity is a rich, black, *magnetic* ore, similar to the Spurr and Champion ores, but much coarser in its grain. See Specimen 39, State Collection, App. B, Vol. II.

d. Dividing the specular ore below, from the magnetic ore above, can be seen, in cut No. 1, Republic mine, a bed several feet in thickness of a *magnesian schist* similar to that previously mentioned, as being found in the Washington and Champion mines. See Specimen 53, State Collection, App. B, Vol. II.

XIV. The Upper Quartzite at Republic mountain is a gray massive rock, sometimes banded, and, near the contact with the iron, sometimes conglomeritic, containing large and small flattened fragments of flaggy ore. The prevailing variety is represented by Specimen No. 50, State Collection, App. B, Vol. II.

XV. Near the south point of Smith Bay is a considerable outcrop of what appears to be a dioritic schist, not unlike Specimen 31, State Coll., containing mica and garnets. It has some resemblance, as will be seen by the description, to the micaceous clay-slate of corresponding number of the Champion section, Specimen 56, App. B, Vol. II.

The horse-shoe form of the surface rocks, as indicated by outcrops, which is so conspicuous a feature on the map, taken in connection with the dip of the strata, as indicated by the arrows and geological section, leave no doubt whatever as to the structure of Republic mountain. It is evidently the south-east end of a synclinal trough with Smith's Bay in the centre, under which, at an unknown depth, all the rocks represented would be found and in the same order. The conjectural division plane, dividing the quartzite and ore (see section), may be regarded as hypothetical, only as to its position, which of course can finally be determined by boring.

It will be observed, that where the northeast side of the horseshoe crosses the river, there is an offset of about 250 feet to the right, and that where the southwest arm of the shoe should cross the river, but very little appearance of Huronian rocks can be discovered on the west side, the Laurentian rocks to a great extent taking their place. These facts can be best explained, by supposing a *fault* to follow the line of this portion of the river, the east being the down side. On this supposition the Huronian rocks on the west side would have been eroded to a much greater extent than on the east, leaving as a consequence the narrow and incompleted series, shown on a section through the Kloman mine.

The proximity of the Champion ore deposit to the Laurentian, it being only about 400 feet distant, while at the Keystone (three-fourths of a mile east) the distance is three or four times as far, leaving room for a greatly increased thickness of vertical brownish banded magnetic schist (see Map VII.), can be best explained, by supposing a *fault*, similar to that just described, but having a direction nearly at right angles ; that is, east by south.

These two instances are the best established cases of faults on a large scale, that have come under my notice, in the whole region.

Calling to mind the series of rocks, which have been described as occurring at the Spurr, Michigamme, Champion, Keystone, Edwards, Washington, and Republic mines, we are irresistibly led to the conclusion, that they are equivalents of each other, belong to the same series, and are of the same age. This hypothesis has already been introduced and carried through the descriptions by the corresponding numbers, which have been attached to equivalent formations in each section; it will no longer be regarded as an hypothesis, but accepted as a demonstrated theory. The Republic mountain section, it will be seen, is most complete for the rocks immediately below the iron, and the Spurr mountain section for those above. The latter embraces one formation of great extent and interest, which was not described, viz.:—XIX., which is made to include the several varieties of mica schists, so extensively developed on the south shore and among the islands of Lake Michigamme. This schist is often very silicious, and, in places, contains numerous crystals of garnets, andalusite and staurolite. See Specimen 61, State Collection, App. B, Vol. II., and Group H, Chapter III.

Near the centre of Sec. 25, at the west end of the lake, is a large mass, probably a ledge, of light-gray quartzite, which may fill in part at least, what appears to be a blank between the anthophyllitic schist XVII. and the mica schist XIX., just described. The number XVIII. is provisionally attached to this quartzite.

We have now described fifteen members of the Huronian series, from V. to XIX., both inclusive. This mica schist is the youngest member of the series, so far as my observations extend, to be found on the Upper Peninsula. It is proper to remark, however, that equivalency, member for member, of the Marquette rocks with the L'Anse, Gogebic and Menominee series, has not been established; they are all Huronian, and it is doubtful if any are younger than XIX.

With regard to the strata below V., there is less certainty as to their order and equivalency. I believe, that the iron ore and associated rocks, to be seen at the **Magnetic, Cannon, and Chippewa** locations, belong here. They are in any event the equivalents of each other, and are very near the base of the Huronian series. See Geological Section, No. 10, map of the Marquette iron region, which extends from the Cannon to the Chippewa. At the latter location is a considerable deposit of ferruginous, silicious schist, or lean flag ore, in which occurs, in what I understand to be an irregular pocket-like mass, a peculiar specular ore of fair percentage, greenish-gray color, and containing numerous bright facets, which resemble scales of mica. This is in comparatively low, wet ground, and the extent of the deposit has not been determined. It resembles the Gilmore ore at north side, Sec. 26, T. 47, R. 26, Cascade range, the two being unlike any other ores in the region.

About 100 tons of 55 per cent. ore was taken from the latter location several years since, but work was not continued. The Gilmore deposit, as well as the Chippewa, is nearly in contact with the Laurentian.

At the **Cannon** location is a banded jaspery rock, holding thin layers of specular ore, which bears a striking resemblance to the rock of formation XII., and even to some varieties of "mixed ore." See Specimen 16, State Collection, App. B, Vol. II. A seam, several inches thick, of pure specular ore, was found here, but did not enlarge on being followed downward. The remarkable characteristic of this schist is the fact, that on following the range northwest and southeast, mica replaces the ore, and we have a micaceous quartz schist, or mica schist depending on the quantity of the latter mineral. These facts, already noticed, possess interest in their bearing on the nature of the Felch mountain ore deposit of the Menominee region, hereafter to be considered.

By far the most promising mine of this group, so far as existing explorations reveal, is the **Magnetic**, in south ½ of northwest ¼ of Sec. 20, T. 47, R. 30. The existence of a workable deposit of magnetic ore of medium richness has been proven. This ore, although highly magnetic, differs entirely in its character from those already described, as will be seen by inspecting Specimen No. 17 of State Collection, App. B, Vol. II. It is very hard, exceedingly fine-grained, and breaks into cubic or tabular pieces. Its structure is more like the flag ores than the first-class magnetites. It should yield about 55 per cent. in the furnace, although none has as yet been worked. The gangue is largely actinolite, instead of the more common quartz, which will help the reduction of the ore.

The relative geological position of this ore is shown in the accompanying north and south section, in connection with Map No. III., already referred to. As to the age of the series represented, I have but little doubt on account of their proximity to the Laurentian, and on lithological grounds, that they are the equivalents of the lowest rocks of the Republic mountain series, and are probably older than the lower quartzite V.

FIG. 6.

Geological Section (looking west).
Magnetic Mine. Sec. 20, T. 47, R. 30.

Level of water in Lake Michigamme 950 feet above L. S.

SCALE OF FEET.

A. Granite. B. Micaceous Quartz Schist. C. Quartzite and Quartz Schist.
D. Banded Magnetic Schist (ore). E. Greenstone or Diorite. F. Dioritic Schist.

B, C, D are undoubtedly the equivalents of the specular and micaceous schists of the Cannon series.

The line of magnetic attraction, running southwest and south, and finally south by east from the Magnetic mine, which has been traced to Sec. 9, T. 45, R. 30, is one of the longest and most persistent belts of attraction in the whole Lake Superior region. The maps of the United States Linear Surveyors mark its position very plainly, as is shown in the chapter on the Magnetism on Rocks, Plate v. Comparatively little exploration has been made on this range; but I see no reason why deposits of the character and equal in value to the magnetic, may not be found along it.

A large amount of very poor ore, and a small amount of very good ore, has been found in south part of Sec. 7 and the north part of 18, T. 47, R. 28; and quite recently a workable deposit of first-class specular ore is reported to have been found there, the locality being known as the **Michigan Mine.** Specimen No. 2, State Collection, App. B, Vol. II., is from this deposit.

Clarksburg, Geological Section No. 6, map of Marquette iron region, records the leading facts to be observed in this vicinity. The Roman numerals marked on the several formations express their *relative* ages correctly; whether they also express the equivalency of these rocks with the Washington and other series previously described, I am not quite certain. Specimen No. 3, of State Collection, from formation marked III., possesses lithological interest, as being a Huronian rock allied to the Laurentian gneisses.

2. NEGAUNEE DISTRICT.

Following the same principle here that guided us in describing the mines of the Michigamme district—that is, beginning with those simplest in geological structure—we find on the **Saginaw and New England** range of mines (being the most westerly of this district), a structure almost identical with that of the Champion and Spurr mines. Referring to Geological Section No. 4, map of Marquette iron region, the rocks in the vicinity of the New England mine are represented as follows:—The ore formation XIII. is made up, as at the Republic mountain, of "mixed ore" (banded ore and jasper), magnesian schist and pure specular slate ore; magnetic ore being absent here, as in all the mines of this district. The quantity of specular slate ore at this mine is, so far as known, small; the small lens-shaped mass, that was formerly worked, having been abandoned.

Overlying the ore formation is the Upper Quartzite, XIV., dipping at a low angle to the north, as may be seen just north of the Parsons mine. This quartzite again comes to the surface about half a mile north, in a flat synclinal, where it again dips north and does not rise until we reach the New Excelsior mine, owned by the Iron Cliff Co., which is shown on the section.*

Returning to the **New England mine**, we find between the ore XII. and the quartzite XIV., a mass of specular conglomerate, somewhat similar to that described as existing at the Republic mountain, where it was regarded as belonging to the ore formation. The fact that it overlays the pure ore at this locality, and has lithological affinities with some of the conglomeritic varieties of the Upper quartzite, leads me to doubt in which formation it should be included. I incline to the view, that it belongs in XIV.

* This general section was constructed more than a year before ore was found at this locality, but it has not been found necessary to make any changes in it.

Formation XII., underlying the ore, is here widely different in its lithological character and economic value from the corresponding formation of the Michigamme district, where, it will be remembered, it was a valueless reddish quartz schist, containing thin lamina of iron. If we suppose tepid, alkaline waters to have permeated this formation, and to have dissolved out the greater portion of the silicious matter, leaving the iron oxide in a hydrated earthy condition, we would have the essential character exhibited by this formation as developed on the New England-Saginaw range, and as will be afterward seen at the Lake Superior mine. This is not offered so much as an hypothesis to account for the difference, as to illustrate the facts observed. The prevailing variety of rock in this formation is a brownish silicious schist, containing a considerable amount of iron (Specimen 26, State Collection, App. B., Vol. II.). Scattered through this formation are here and there large and small pockets of soft earthy hematite ore, having usually the most irregular forms, that can possibly be conceived. This subject was discussed under iron ores, Chapter III. Specimens 34 and 35, State Collection, are ores of this class.

The **Winthrop and Shenango mines** are in this formation, and are producing hematite ores as rich as any now worked in the district, and excepting perhaps the Lake Superior and McComber, richer than any other of this class, as indicated by analyses, Chapter X.

Underlying this hematite formation is a diorite, XI., similar in its general character to the rock, having a corresponding number in the Michigamme district; below this and south, are various ferruginous schists and diorites, corresponding in a general way with the Michigamme series, but which have not been carefully examined in the vicinity of the New England mine. Recent explorations afford opportunities for study, which did not exist when this section was made.

The series at the **Saginaw** and intermediate mines, as well as further west, is so near an exact duplicate of what has been given above, as to require no further mention than to state, that the deposits of specular ore are larger than at the New England, which has been mentioned as being rather small for profitable working. There has been too little work done at these new mines, to determine the extent of the deposits, but I see no reason to suppose that any of those now worked will prove very large. The fact that Sec. 16, the Parsons and New England mines, have produced specular ores and have been abandoned, is significant. No doubt, considerable amounts of first-class ore will be taken out on this range at a profit. The only question is, whether they will continue to produce such ore in quantity for a series of years, at a fair cost for mining.

This range of ore has been traced westerly into the northeast ¼ of Sec. 24, T. 47, R. 28; west of this the drift becomes very deep and the ore range is lost. A shaft 67 feet through the sand in this vicinity found no ledge. Whether there is any stratigraphical connection between this ore formation and the Washington, six miles distant west by north, is not determined. So far as is now known, it is economically a blank in the Marquette iron belt. Work now in progress at the new Michigan mine, already noticed, may throw light on this interesting and important question. It is not at all improbable, that the Negaunee and Michigamme districts may be independent ore basins, in which case the intervening rocks, which are all Huronian, would consist of the lower members of the series, that is below XIII. Even should this be the case, valuable hematite and flag ores may be found in this now barren district.

The new **Excelsior Mine**, previously mentioned and shown on the New England section, is near the southeast corner of Sec. 6, T. 47, R. 27, and is, as will be seen, the opposite cropping of the basin. There is so much drift between these ranges, that not much can be said definitely about the nature of thc intervening rocks; but it seems probable that we have here a great basin, underlaid by ore at an unknown depth, and that the New England and Excelsior deposits are related to each other in the same way, as it was assumed are the Champion and Michigamme deposits. This could be cheaply tested, and possibly an important discovery of ore made, by a drill-hole through the quartzite, near the railroad on the west side of Section 16. All efforts to find an extension of the Excelsior deposit east and west have so far failed.

Returning to the New England range and following it eastward, we find that near the south ¼ post of Section 16, it bends suddenly to the northeast, making its way diagonally across this section to the **Lake Angeline Mine**, which produces specular ore, having such admixture of jasper, as to cause it to rank intermediate in the market between first and second class ores. Whether the deposit worked at this mine belongs to bed XII. or XIII., I have not determined, the ore partaking somewhat the character of each. The overlying rocks on the north are covered by the waters of Lake Angeline.

To the south is a high ridge of diorite, XI., on the south side of which is an extensive deposit of soft hematite, owned and worked in part by the Lake Angeline and Iron Cliff Companies.

I suppose this hematite to belong to formation X., and therefore of the same age as the Negaunee and Foster hematites, which will be fully described below. It will be borne in mind, that the hematite ores on the Saginaw range occur in formation XII.

Without attempting to point out at present the structural relations of the Lake Angeline and Lake Superior ore deposits, we will pass at once to a consideration of the latter mine, one of the most extensive, productive and geologically interesting in the Marquette region.

The accompanying map, No. IX., representing the **Lake Superior** specular and hematite workings, together with the **Barnum mine**, is intended to give the geological facts to be observed in considerable detail, as well as the condition of the workings in 1870. The structure of the east half of this mine is more complicated, than that of any other in the district, and some questions connected with it remain unsolved.

Regarding for the present the west half of the mine only, we find presented on a small scale about the same structural phenomena, which is so prominent a feature in the Republic mountain rocks. The basin, or trough, in this case, however, abruptly narrows up, the sides and bottom being as it were gathered in, as if to be tied, at a point just south of the engine-house; to the west the outcropping edges of the basin diverge rapidly, and its bottom sinks into the earth in the same degree. If we suppose the frustrum of a hollow cone, lying with its axis horizontal and its small end towards the east, to be cut in two by a horizontal plane, representing the surface of the ground, the lower half will represent my conception of the form of the Lake Superior-Barnum ore basin. Conceive now this cone to be made of sheet-lead, and to be considerably bent and dented, and the illustration will be still more applicable.

A study and comparison of sections D—D′, C—C′, B—B′, and A—A′, in connection with the plan of the mine (Map IX.) will, I think, render it plain that this conception of the structure is in accordance with the facts; although the minor folds and faults con

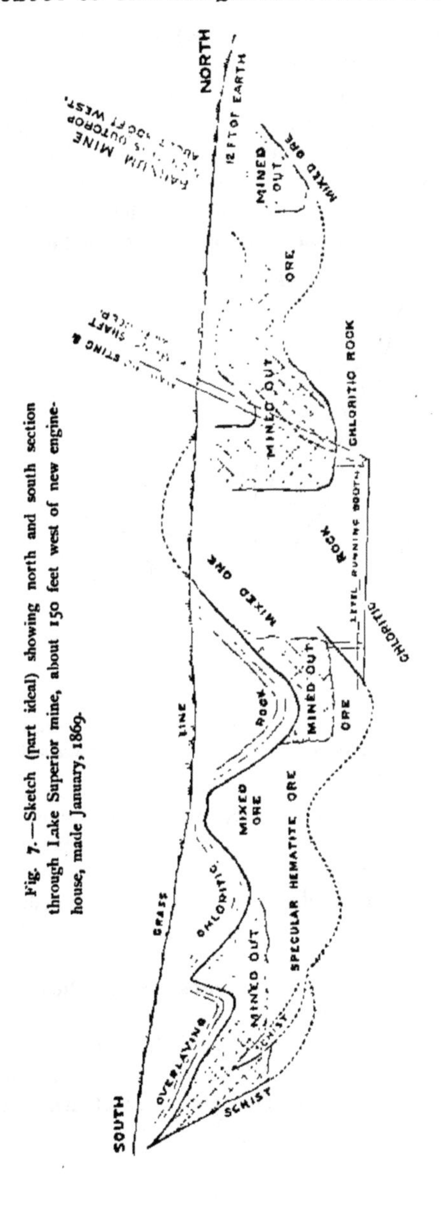

Fig. 7.—Sketch (part ideal) showing north and south section through Lake Superior mine, about 150 feet west of new engine-house, made January, 1869.

siderably obscure and confuse the general structural question. Of course, it is not absolutely proven, that the Barnum deposit dipping south, and the continuation of the main Lake Superior deposit, now worked in Pit No. 25, which dips north, are opposite croppings of the same bed, and that the intervening space is underlaid by the ore formation, and that, therefore, if work continue long enough they will eventually connect under ground; but certainly all the facts point to this conclusion. The importance of this theory in

Fig. 8.

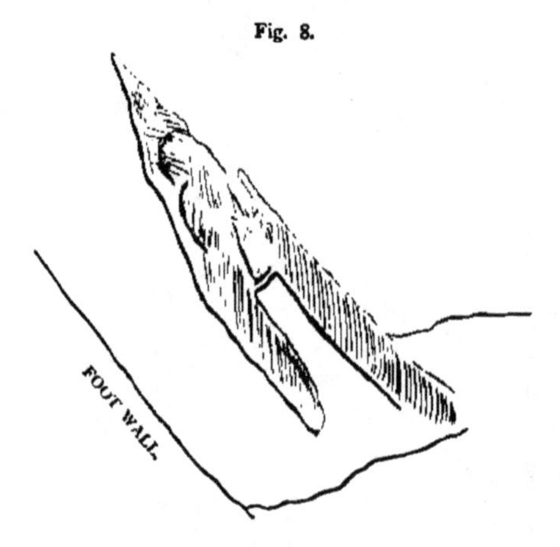

Fig. 8, represents on a large scale the south or left-hand end of the section represented in Fig. 7, and brings out the peculiar form of the "horse" of magnesian schist, which is shaded, the ore being white.

its bearing on explorations for ore, mining and valuing ore deposits, is very apparent. It shows, that such formations are not vein or dyke-line deposits, but true stratified beds, like the rocks by which they are enclosed. Their structure is therefore essentially the same as the coal, limestone, sandstone, and slate-beds, which are regarded as sedimentary deposits from water, subsequently more or less altered by heat pressure, and chemical waters acting during immense periods of time.

The Lake Superior-Barnum deposit evidently has a *bottom*, which will be reached within a period, of which it is worth while for the present generation to take some heed. So of many other deposits in the region.

As we go westerly from these mines the basins become, as we have seen, wider and correspondingly deeper. A depth of 300 feet in the Edwards mine reveals no essential change in the dip of the deposit, as will be seen by reference to the plans of the mine. The same is true of the Champion mine.

The time may come when, having worked out the steep upturned edges of the basins, and the flatter or deeper portions of the deposit are reached, ore properties will be valued somewhat according to the number of acres *underlaid by ore*, as coal now is.

Passing to the east portion of the Lake Superior mine, I confess myself unable to give any intelligent hypothesis of its structure. The facts observed are in part recorded on the Map of the mine on section E—E′, and on the accompanying sketch, in part ideal, which represents on a small scale a section near E—E′. There seems to

Fig. 9.—Sketch showing Geological Section of the Lake Superior mine (looking west), near Sec. E—E′, Map IX.

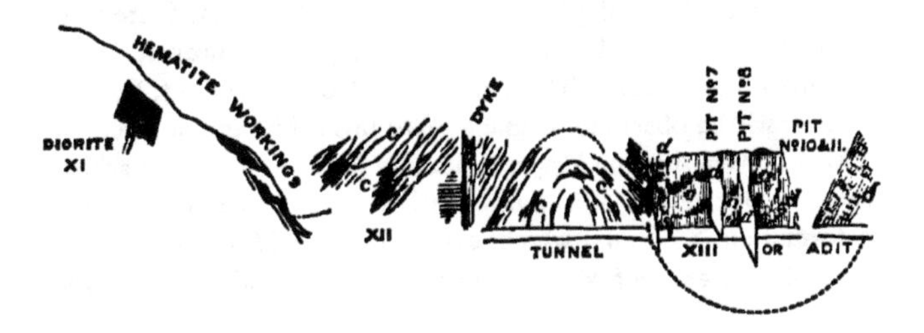

a. Chloritic schist. *b.* "Mixed ore." *c.* Limonitic schist (hematite rock). *d.* Pure Ore.

have been such a gathering together, crumpling, squeezing and breaking of the strata, as to nearly obliterate the stratification. An attempt has been made to represent the present condition of things, so far as revealed, by the workings. The remarkable features are the great masses of light grayish-green chloritic schist, having a vertical east and west cleavage, no discernible bedding planes, and holding small lenticular masses of specular ore, which conform in their strike and dip with this cleavage, and which seem to have no structural connection with the main deposits. They appear like dykes of ore, squeezed out of the parent mass, which we may suppose to have been in a comparatively plastic state, when the folding took place; or they may have been small beds, contained originally in the chloritic schist, and brought to their present form and position by the same causes, which produce the cleavage in the schist. A comparison of these sections, showing effect of the folding on a large scale, with the figures (19 to 29, Vol. II.) representing the contorted lamina of the mixed ore of Republic mountain, will be found instructive. Indeed the same phenomena may be observed abundantly at the Lake Superior mine, and still better at the Cleveland knob.

Lake Superior mine sections E—E', and Fig. 9, may almost be said to represent a huge breccia.

The peculiar nature of the hanging wall of the Lake Superior mine deserves further notice. Instead of the quartzite, which we have hitherto found overlying all the deposits of rich ore, we have here a magnesian schist very similar to, if not identical with, that already mentioned as being associated with the ore, as will be seen by reference to the geological sections, and to Spec. 55, State Collection, App. B, Vol. II. These rocks are given, however, different colors on the maps. The hanging wall of pit No. 25, Section A—A', it will be observed, is made up of this schist and of layers of quartzite. Whether the Upper Quartzite is replaced by this schist, making it belong to XIV., or whether it is a member of the ore formation XIII., in which case XIV. would be wanting at this locality, I am not able to determine, but incline to the first opinion.

The hematite formation XII. is fully developed at this locality, producing an excellent ore which is extensively worked. The relation of this formation to the overlying and underlying rock is obscure, as has already been pointed out. This relation was very plain, it will be remembered, on the Saginaw-New England range.

The structural hypothesis by which I have attempted to connect the Lake Superior deposit with the Lake Angeline on the south, and Marquette, Cleveland and New York mines on the east, need not be further described here, but will be understood I think, by those interested in the question, from an examination of the following figure in connection with the maps.

Fig. 10.—Sketch (part ideal) showing position of ore basins at Ishpeming.

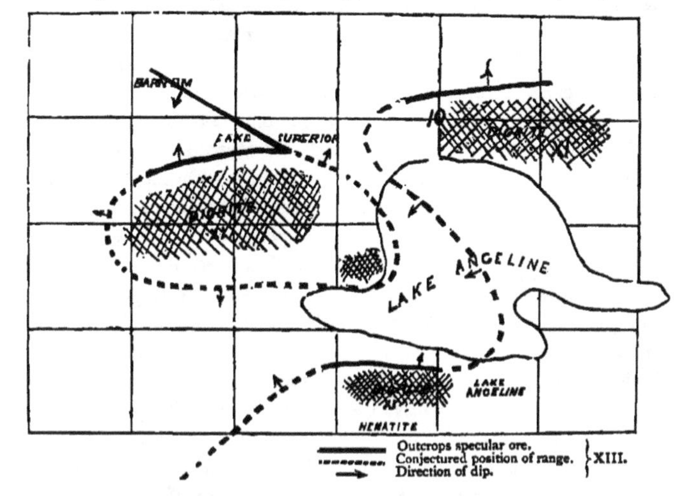

New York, Cleveland and Marquette Mines.

The geological facts to be observed, the general structure, nature and extent of the workings of the **New York mine**, which is one of the most regular deposits in the district, are so plainly set forth on the accompanying Map, No. X., that but few words of description are necessary. It will be seen to be a monoclinal deposit, in every essential particular, like the Barnum, Champion and Spurr. Two interesting facts will be observed: 1st. The absence of formation XII.; the pure ore, with its associated chloritic schists, seems to occupy the whole space between the Upper Quartzite, XIV., and the diorite, XI. It may be here observed that, as a rule, the purest ores are found in the upper part of the ore formation, that is, nearest the Upper Quartzite; the New York mine presents an exception. 2d. The deposits on the north side of the railroad, worked by Pits No. 3 and 4, have a striking resemblance to the small deposits, Pits 16 to 21, of the Lake Superior mine, just described. The facts to be noted at the Collins location, just east, taken in connection with Pits 3 and 4 of the New York, point plainly towards the existence of a small independent trough, north of the Cleveland-New York deposit. Explorations and mining operations so far, do not indicate the presence of a large amount of first-class ore here.

I made no special survey of the **Cleveland mine**, the fund at my disposal not permitting it; the main object of the survey in this direction being, to represent in detail a sufficient number of typical mines, to cover the various structural phenomena to be found in the district. The sketch of the Cleveland and Marquette mines, Plate II., from A. Heberlein's map, in connection with the New York mine (Map No. X.), will give a good general idea of this group. It will be seen, that the most northerly pit (Gents, No. 3) of the Cleveland mine, is a continuation of the New York deposit, having the same strike and dip. Gents pit is in one of the largest deposits of pure specular ore in the whole Lake Superior region. It dips south, forming the northerly edge of a narrow synclinal basin, which immediately comes to the surface again in the Swedes pit, where the ore has a northerly dip. These two pits produced in 1872 over 100,000 tons of ore. The ore basin widens and deepens to the west in a similar manner to the Lake Superior, and undoubtedly underlays the swamp, on which the village of Ishpeming is built. The connection of these deposits with those worked in the more southerly Cleveland and Marquette openings, has not received that attention which would enable me to express an opinion on the subject.

There can be little doubt, but that the Cleveland mine promises as well, if not better, for the future production of first-class specular ore, than any one of the older mines.

Jackson Mine and Negaunee Hematite Deposits.

No special survey was made of the **Jackson mine**; but the accompanying Plate (iii.), from O. Dresler's map and Atlas map of the iron Mines at Negaunee (No. V.) will make known the general structure of the mine, which is essentially similar to that of the Cleveland and Lake Superior. This mine, although it produces first-class specular ore, will be here considered in connection with the hematite deposits, because they are adjacent, and their geological structure can be most conveniently described together. The Jackson mine, so far as is known, is the extreme east end of the

SKETCH OF CLEVELAND AND MARQUETTE MINES.

Marquette Iron Region.

A. HEBERLEIN, M.E.

1869.

1 Swedish Church.	7 Barn.
2 Saw-mill.	8 School-house.
3 Blacksmith's shops.	9 Carpenter's shop.
4 Engine-houses.	10 Captain's house.
5 Shaft-houses.	11 Round-house.
6 Store and office.	12 Ore pockets.

□ Shaft. .... Drifts. —·—·— Boundary line Marquette property.

C & N.W.R.

M.H.& O.R.R.

MARQUETTE MINE

LAKE SUPERIOR COMPANY.

SCALE 480 FEET TO 1 INCH.

PL. III.

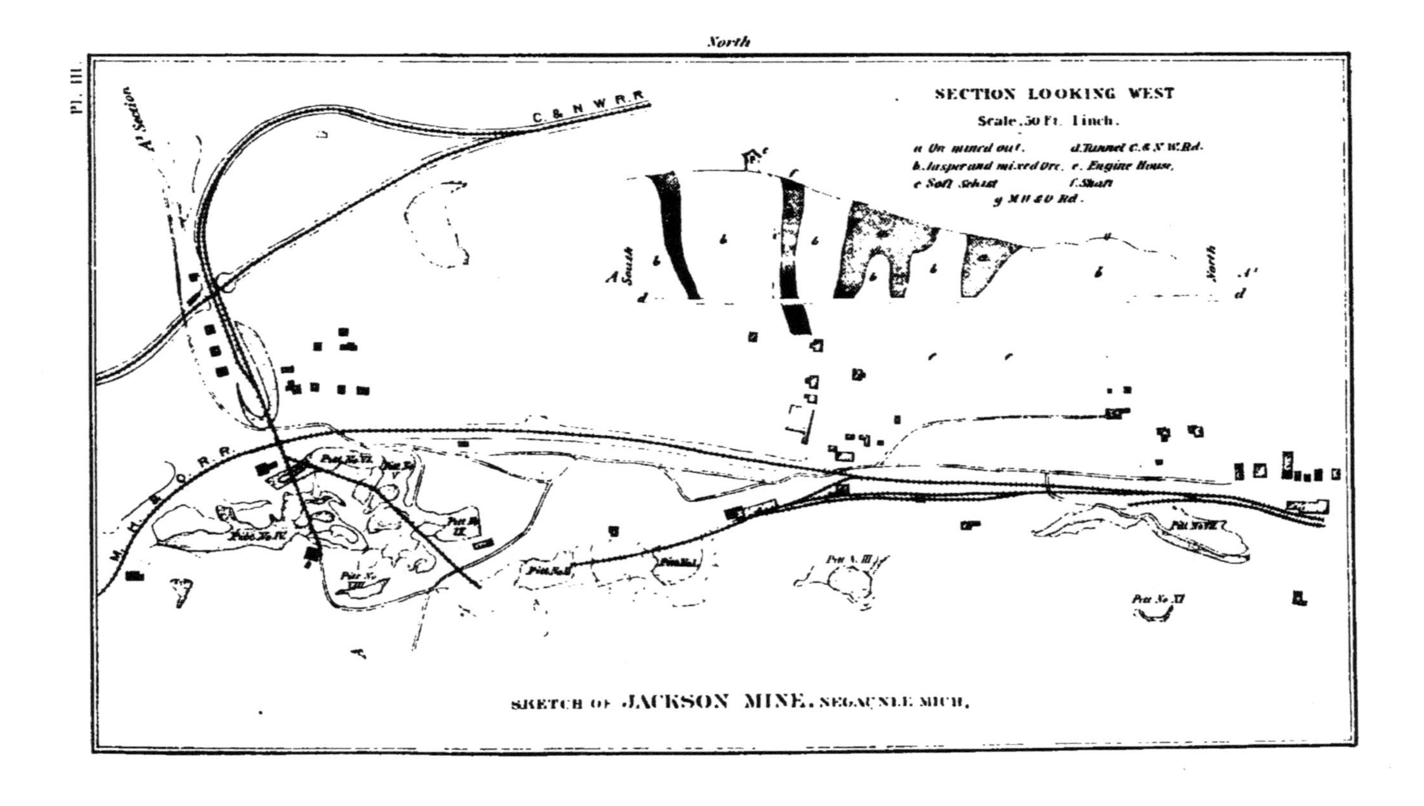
North
SECTION LOOKING WEST
Scale, 50 Ft. 1 inch.
a Ore mined out.
b. Jasper and mixed Ore.
c Soft Schist
d. Tunnel C. & N. W. Rd.
e. Engine House,
f. Shaft
g M H & O Rd.
South
A¹ Section
C. & N. W. R. R.
M. H. & O. R. R.
Pitt No. VIII
SKETCH OF JACKSON MINE, NEGAUNEE MICH.

rich ore basin formed by bed XIII. No workable deposit of ore of any kind has been found north and east from this locality, and the ores to the south are believed to belong to a lower horizon, and to be, on the whole, inferior in quality.

Looking back over the field we have now hastily surveyed, and assisted by the map of the Marquette iron-region, it will be seen that, while there are many minor irregularities, on the whole the ore basin gradually widens towards the west, from a mere point at the Jackson mine to a width of fully five miles at the west end of Michigamme lake, beyond which too little is known, to enable us to accurately define its limits. It follows, therefore, that all the Huronian rocks north, east and south from the Jackson mine, are below, or *older than the ore formation* (XIII.) and all the rocks to the westward and inside of the ore-basin are *younger*, hence above it.

The large amount of exploration work, done in the vicinity of Negaunee, in searching for hematite within the last few years, has aided greatly to develop the geological structure of that locality. But unfortunately, the money I had to expend here was more than exhausted, before this work began, so I have been enabled only in part to avail myself of it.

The facts observed are mostly recorded on the local map, mentioned above, and on the general map of the region. By reference to the former it will be seen, that a belt of country, about one mile wide, extending southeast from the Jackson mine, is dotted over quite irregularly with hematite workings, which are mostly on lands leased from Edward Breitung, as is explained in a note on the map. These mines produce dark-colored earthy hematite, containing metallic manganese, often up to an average of 5 per cent., varying considerably in the amount of metallic iron, but on the whole averaging lower, than the hematite ores heretofore mentioned, as will be seen from the chapter on analyses. I believe these ores all belong to one formation, No. X., in which, up to this time, no merchantable ores, except the Lake Angeline hematite, have been mentioned as occurring; it is at least certain, that they are older than formation XII., which embraces the Lake Superior and Winthrop deposits.

The geological sections A—A′ through the Himrod and Green Bay mines, and B—B′ through the Jackson Company's new hematite and old specular ore workings, fully illustrate the hypothesis of structure adopted. It will be seen, that the ore is contained between two beds of diorite, IX. below and XI. above, and that there is associated with the ore, chloritic schists and various ferruginous schists and flag ore. These last-named rocks, it will be remembered, made up this entire formation in the Michigamme district, where hematites are wanting, as are magnetic ores in the district we are describing. Underlying the lower diorite mentioned, is a clay slate, which is in turn underlaid by a gray quartzite, to be seen outcropping near the centre of the north half of Sec. 8, and represented in Sec. A—A′ under the number VIII. This is undoubtedly the same quartzite to be seen in the railway cut near the northwest end of Goose lake, where it is overlayed by a soft schist. See formations VIII. and XI., Geological Sec. No. 1, Map III. The clay slate on south shore and near west end of Teal lake, and exposed in railroad cut one mile east of Negaunee, is also believed to be of the same age.

The lithological character of the several formations, mentioned above, will be better understood by an examination of the following specimens of the State Collection: No. 21 quartzite from VIII.; No. 20 is a clay slate also from VIII.; No. 31 is from diorite IX.; Nos. 24 and 25 are hematite ores from formation X.; No. 26 is a specimen of ferruginous silicious schist from the Foster mine, which is also regarded as belonging to the same formation (X.); Specimen 28, from the same formation, is a magnetic, chloritic, silicious schist.

Referring again to Map No. V., it will be observed, that the Jackson Company's hematite workings, the McComber, Maas and Lonstorf's most northwesterly opening, the Rolling Mill, Himrod, Spurr and Calhoun, and Iron Cliff Co.'s Sec. 18 mines, are all in a rude curve, skirting the great development of diorite, which seems to limit these deposits on the southwest, and under which they all dip. The remaining openings are mostly contained in a narrow belt, which extends east-southeast from the Grand Central, diverging from the other range, which curves to the south. The diorite ridge which runs through the centre of the latter range is apparently a synclinal ridge underlaid by ore, which should therefore dip towards it from all directions, as is the fact so far as known. Undulations in the bed now unknown, may very likely bring the ore to the surface at several other points.

There can be no doubt of the great extent of this ore; it certainly can be on the average more cheaply mined and shipped than any other ore in the region, except perhaps the hematites of the Taylor and S. C. Smith mines. Location at the junction of two railroads, and contiguity to a prosperous village, are additional advantages, which will go a long way towards offsetting the disadvantages of lower percentage. The presence of several per cent. of manganese in this ore helps its working in the furnace, rendering it a desirable mixture. The McComber mine was first opened, and its ore is well and favorably known to many furnacemen. My analyses indicate, that this is a richer ore than the other mines of this group, but this cannot be established without further developments, as work has but just begun at most of them.

The **Teal Lake** ore deposit belongs to the same formation, as may be seen by an inspection of the map and sections. I have not been able, however, to find any good hematite in the old exploration pits, now nearly filled ; a lean flag ore is very abundant.

The **Foster mine,** near southwest corner of Sec. 23, T. 47, R. 27, is another hematite deposit belonging to formation X. It has produced a considerable amount of hematite ore of medium grade, which contains no manganese ; the deposits, or rather pockets, are pre-eminently irregular in form and uncertain in extent. The geological position of the Foster range is shown on Map No. III. and accompanying sections.

The Cascade Range.—The deposits on this range are the only ones now wrought, which remain to be described in the Marquette region. Like nearly every other described in this report, this ore was known to the United States linear surveyors, and afterwards examined and commented upon in considerable detail by Foster and Whitney. The range extends east and west through the south part of T. 47, R. 26. See Map III. The locality known as the **Gilmore mine,** at ¼ post between sections 23 and 26, is the most easterly point at which ore has been seen in quantity. This, it will be observed, is about three and one-half miles east, and two miles south, of the Negaunee hematite mines. The range has been traced west by south from this place for five miles, or to a point just four and one-half miles south of the Jackson mines. This country has recently been opened up by a branch of the C. and N. W. road, which closely follows the ore range. The principal openings have been made by the **Cascade, Pittsburg and Lake Superior, Carr and Gribben** Iron companies, who shipped an aggregate, in 1872, of over 40,000 tons, nearly all of which was by the first-named company and its lessees. The last two named companies —Carr and Gribben—have done too little work, to enable us to speak with much certainty about their deposits. (See tables, on Sheets XII. and XIII., Atlas.) By reference to the chapter on analyses, which is quite full regarding these ores, it will be seen that they have, on the average, less metallic iron and more silica, than the standard hard ores of the district. The West-End mine, however, worked by the Cascade company, and which produced last year about one-third of their product, appears to be an exception to the above rule, and to rank nearly with the first-class specular ores ; certainly considerable amount of high grade ore was taken from this pit last year, but whether it was kept separate from the leaner varieties in the shipments I do not know. The ore which largely prevails is a silicious or quartzose, or jaspery (practically these words have the same import) red oxide, having a characteristic coarse, slaty, or *flaggy* structure ; hence the name by which they are known throughout this report. They correspond nearly in composition, although not in their appearance and geological position, with the second-class ores of the old mines, as the analyses referred to prove. See Iron Ores, Chap. III. Some varieties closely resemble, if they are not identical with, certain varieties of the high grade ores ; but as a rule they are lighter in weight, duller in color and lustre, are harder under the knife, and pre-eminently flaggy or slaty in structure. I have not been able to obtain a statement of the working of these ores in the furnace. Further information regarding their lithological character may be obtained from descriptions of Specimens 5 and 6 of the State Collection, App. B, Vol. II. ; the latter is the beautiful "Bird's-eye" slate ore from the Bagaley and Wilcox pit. Specimen 7 is from the diorite bed, which overlays the West-end mine, and is interesting from its resemblance to granite in outcrop. *

The structural position which these ores seem to me to occupy is shown on geological section No. 2 of Map No. III. They are near the Laurentian, and the whole series is overlaid by a talcy quartzite, which I believe to be the equivalent of No. V. of the Re-

* Mr. Julien has determined the feldspar in this rare variety to be orthoclase.

public mountain series, and to be the same bed, which outcrops so conspicuously on the north side of Teal lake, and is calcareous at the Morgan furnace and at the Chocolate flux quarry, where it strikes the shore of Lake Superior. This rock varies more widely in its lithological character, than any other in the region, as will be pointed out elsewhere. If this hypothesis is correct, it will follow, that these ores are the equivalents of the Michigan and Magnetic ores of the Michigamme district, and are older than any iron bed made out in the Republic mountain series. The fact, that no iron in quantity has been found north of Teal and Deer lakes under quartzite V., where we should expect to find the opposite cropping of the Cascade series, is to be regarded in considering this question. The shortness of this range, which appears to terminate abruptly to the west, has not been found far east, and has altogether a local and isolated character, is significant. A hasty examination will satisfy any one that the *quantity* of ore in these deposits is very great, and that it is very favorably situated for mining and transporting. The accompanying north and south sec-

Fig. 11.—Geological section across Cascade range, looking west.—Part Ideal.

a. Flag ore or silicious hematite schist, in places quite rich. *b.* Banded jasper and specular ore with flag ore. *c.* Hematite rock or hematitic silicious schist. *d.* Diorite and dioritic schist. *e.* Quartzite. *f.* Conglomeritic and brecciated quartzite.

tion represents the different rocks to be seen outcropping on this range, projected on one plane. No attempt has been made to group them under formations I. to IV., to which they are supposed to belong. The general section No. 2, Map III., which has been mentioned, should be examined in connection with this sketch.

The **Iron Mountain, Ogden and Tilden** mines, not now worked, produced flag ores similar to those of the Cascade range, but not so rich on the average. These deposits belong, as will be seen by Map No. III., to formation X.; the Iron Mountain and Tilden mines being in opposite croppings of the same basin. The **Foster mine**, as has been observed, is also in the same formation, being overlaid and underlaid by flag ores. The Negaunee hematite and Teal lake ores being also in X., make that formation remarkably fruitful in the quantity and variety of ore, which it contains; but it does not, so far as known, hold the high grade specular ores in quantity.

Lower Quartzite, embracing Marble and Novaculite.

A brief consideration of the question of materials for *furnace flux* may come within the limits determined for this report. The subject, so far as the Silurian limestones are concerned, has been fully considered by Dr. Rominger, in Part III., who gives many analyses. The Menominee marbles will be mentioned in Chapter V. on that region. No calcareous, or other rock suitable for flux, has yet been found in the Laurentian system of the Upper Peninsula, although in Canada large beds of marble occur in this oldest series. It remains only for us to consider the silicious variegated marbles, found in the eastern part of the Marquette region, none having been worked west of Goose lake, which happens to mark the most easterly show of iron. The purest stone is found at the Morgan furnace, seven miles west of Marquette, where a heavy east and west bed of silicious marble, with vertical dip, and having associated with it clay slates, is prominently exposed. The prevailing colors are light-gray and pink. Specimens 11 and 12, State Collection, are from this locality; and Specimen 70, from the Gorge, represents the chloritic schist, which underlies the marble on the north.

The Chocolate Flux quarry on the shore of Lake Superior, three miles south of Marquette, is another locality, from which a small amount of furnace flux has been obtained. But the admixture of quartzose matter is here so great, that its use has been abandoned. Specimens 9 and 10, State Collection, represent the so-called "marble" and slate from this locality. It and the associated rocks are fully described in the extract from Dr. Houghton's unpublished notes, given in Appendix E, to which a sketch is appended. Mr. Julien examined a full suite of specimens from this locality, which are described in App. A, Vol. II., Nos. 106 to 113. No other marble locality possesses sufficient interest, to warrant mention, although flux has been quarried at several points near

Goose lake. It has been mentioned that the ***novaculite*** quarry, just east of Teal lake, from which whetstones were taken more than twenty years ago, is in the same formation. These stones are not now worked. See Specimen 13, State Collection, App. B, Vol. II.

During the past season several car-loads of quartzite were quarried in the same vicinity, and used as lining for Bessemer steel converters, at Capt. E. B. Ward's works, for which purpose it answered well.

The various marbles, slates, and quartzose rocks described above, are all believed to belong to one and the same formation, the Lower Quartzite (No. V.), which, it will be remembered, underlies the Republic mountain series, and overlies the Cascade series. This formation is one of the most interesting, geologically, in the Marquette region, and is worthy of a far more careful study than I have been able to give it. Specimens 8 to 13, inclusive, State Collection, App. B, Vol. II., represent several varieties of rock from this formation; as many more varieties could easily be procured, including some very fair specimens of iron ore from south and east of Goose lake.

A brief description, in addition to what has already been given, of the great geological basin formed by this quartzite, which embraces within its folds the great mass of the Huronian rocks, and nineteen-twentieths of all the ore, will possess interest. Like the ore horizon XIII., which we saw came to a point at the Jackson mine, and widened to the west, so the opposite croppings of this quartzite converge to the east and come together at the Chocolate Flux quarry, already described. From this starting-point the ***south rim*** of the basin bears away towards Goose lake, where some minor folds and low dips make it the surface rock for a large area northeast of the lake. From the south end of the Lake west, the formation has a prevailing talcky character, often argillaceous and sometimes conglomeritic; it has a great thickness and strikes west by south. West of the Cascade it seems to assume more the character of a chloritic gneiss and protogine, or at least a well-defined bed of protogine rock occupies the position in which we would expect to find the quartzite. See Map No. III. and sections.

The ***northerly rim***, starting also from the Chocolate quarry, maintains a nearly due west course, crossing the railroad at the

Morgan furnace (where it holds the maximum amount of lime), forms the barrier rock in the Carp at the Old Jackson forge, passes north of Teal lake and south of Deer lake, occasionally at various points further west, and last, so far as I know, north of the Spurr mountain, nearly 40 miles west of Lake Superior.

3. ESCANABA DISTRICT.

The most southeasterly deposit in the Marquette region, and one which is entirely isolated from the localites already described, is the **S. C. Smith Mine**, producing soft hematite ore; it is located on Sects. 17, 18, and 20, T. 45, R. 25, and connected by a branch with the C. and N. W. railroad. It is but 42 miles from Escanaba, giving it a great advantage in distance over any mine, now shipping ore through that port. The geographical position is less remarkable than what might be called its geological isolation, for it appears to be in a small patch of Huronian rocks, in the midst of a great area of barren territory, underlaid by the Laurentian and Silurian systems. See Map III. The discovery of this deposit, a few years since, by Silas C. Smith, Esq., reflects great credit on his knowledge of the nature and distribution of ore deposits, and his perseverance in searching for them. Mr. Smith also first directed attention to the Republic mountain, which was, until within a few years, called by his name; he also made the first explorations in the Menominee region.

The few outcrops about the S. C. Smith mine, and the small amount of work done, when my examinations were made, enable me to say very little about its geological structure. The ore range runs northwest and southeast, approximately parallel with the Escanaba river, and cuts the southwest corner of Sect. 17. Contiguous on the northeast (whether underlying or overlying I am unable to say) is a bed of black clay-slate, in places identical with the so-called "plumbago" of the L'Anse range, which has been heretofore considered. Numerous fragments of a similar slate, probably belonging to the same formation, are found on the east side of Sec. 29. Laurentian granite is seen on both sides of the river, just east of this locality, away from which we have a right to assume the slate dips, rendering it probable, that the whole series dips southwesterly, in which case the slate would form the foot-wall of the ore deposit, as on the L'Anse range. On Section 20, west of the river, a talcky schist, holding grains of quartz, was observed, but its relations with the other rocks were not determined.

Near the west ¼ post of Section 20, and at other points in the vicinity, a flag-ore of good quality has been found; a specimen from one of the test-pits gave Mr. Britton 56 per cent. of metallic iron; whether there is any considerable amount of ore of this degree of richness has not, I think, been determined. Hand specimens of very fair specular ore could be found, but, as a whole, it seemed to me to be much more closely allied to the flag ores. Small boulders of this kind of ore had been found in this vicinity by C. E. Brotherton, some years ago.

Lapping over the upturned edges of the black slate on Sec. 17, and extending towards the east, is a horizontal Silurian limestone, which is, however, cut off by the river, beyond which numerous outcrops of granite and gneiss rear their heads above the flat sand plain. Silurian rocks are also seen on parts of Sec. 19, but west and northwest the country is all Laurentian, so far as I have been able to learn. South and east is a great plain, undoubtedly underlaid by Silurian rocks, but affording no outcrops, except near Little lake, where an isolated hill, apparently Huronian, rises out of the plain; I have not learned that any indications of iron have been found there.

I regret not having had the time and means to make a re-examination of this interesting and important district, after last season's extensive developments, and reluctantly present this imperfect sketch for want of fuller and more complete data.

4. L'Anse District. (See Plate IV.)

The United States surveyors marked "iron ore" in two places on the line between Sects. 4 and 9, T. 49, R. 33. A quartzose or silicious brown and red ore can be seen outcropping, at several points in this vicinity. These facts early drew the attention of explorers to this district, and a considerable amount of land was bought from the government, for iron, as early as 1864. The fine harbor at the head of Keweenaw bay, only seven miles distant, and the abundance of excellent hard wood, tributary to this bay, have long caused it to be regarded as one of the best points in the northwest, at which to make charcoal pig-iron, and establish other manufactories related thereto. The soil along the protected shores of Keweenaw bay is good, which led to the establishment of Indian missions there many years ago. A circle having the village of L'Anse as a centre, and a radius of 35 miles, would embrace the Washington, Edwards, Champion, Republic, Michigamme, Spurr Mountain, Magnetic and Taylor mines, with others less promising, together with all the copper mines in the Portage Lake district, the Hecla-Calumet mine, as also the principal mines in the Ontonagon district. It would also embrace all the roofing slate territory to which attention has already been directed, and an immense sandstone area, about which little is known. The amount of hard wood within the circle would be surpassed by very few equal areas on the Upper Peninsula, and the quantity of pine is large. A railroad running west, tapping the Ontonagon copper region, and continuing through the Gogebic and Montreal river mineral region, so as to connect with the Northern Pacific road, would, with existing roads and the excellent water communication, make the greater part of the area described easily accessible from L'Anse. If the advantages of the geographical position of L'Anse have not been here overstated, it is somewhat remarkable that the locality should have remained so long undeveloped. The want of railroad communication with the outside world was, undoubtedly, the main reason. What effect the very heavy grades, encountered within ten miles of the town, will have on the amount of ore which will be carried there from the Michigamme district, remains to be seen. The ore from the Taylor mine, and others that may be opened on the L'Anse range, can be put on board vessel at L'Anse at less cost for transportation, than any equally good ores with which I am acquainted, on the entire chain of the Great Lakes.

As has been before remarked, the L'Anse iron range, so far as made out, lies in the north part of T. 49, R. 33, the best ore being in Secs. 9, 8, 4, and 5; it has a general easterly and westerly trend, like nearly all of the iron ranges of the Upper Peninsula.

The **Taylor Mine**, the only point where the existence of a workable deposit has been demonstrated by actual exploration, is near the centre of the northeast ¼ of northwest ¼ Sec. 9, T. 49, R. 33.

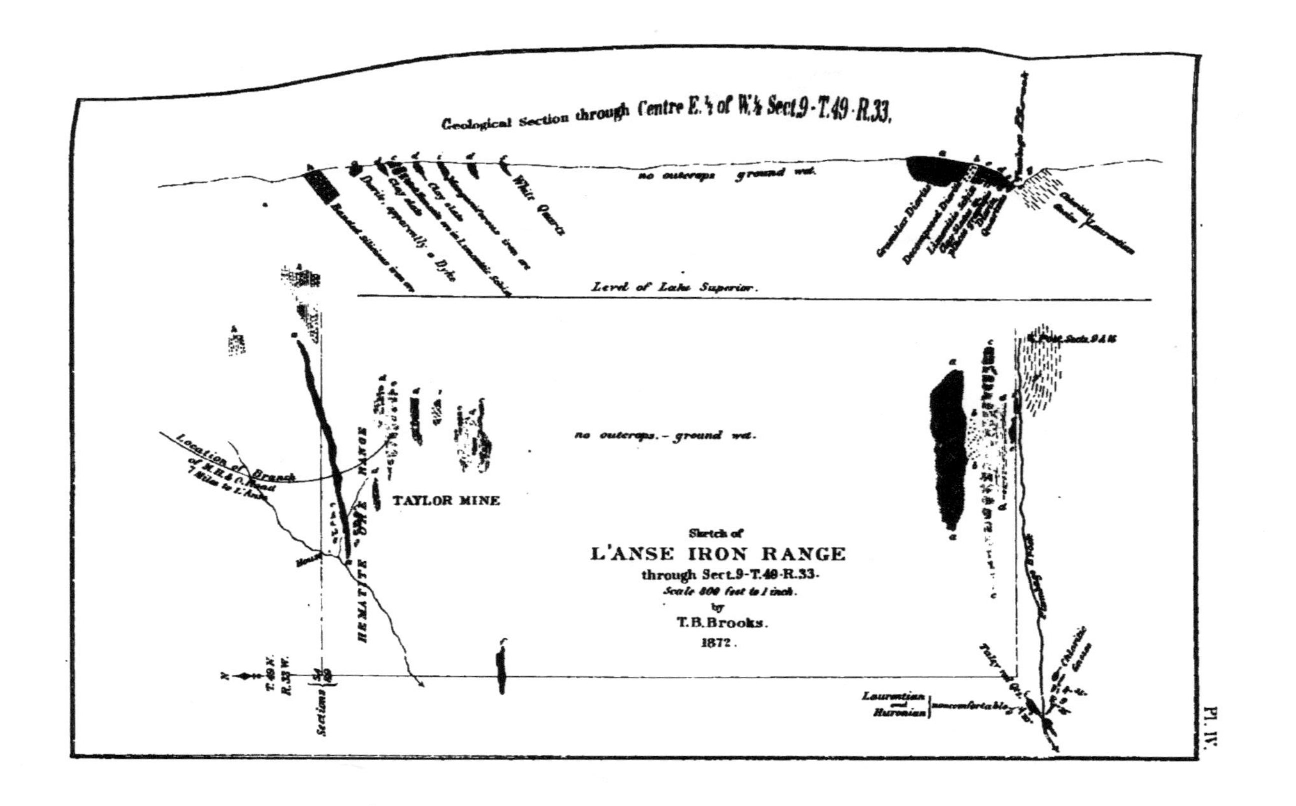
Geological Section through Centre E. ½ of W. ½ Sect. 9 · T. 49 · R. 33.
no outcrops ground wet.
White Quartz
Clay slate
Clay slate
Diorite, apparently a Dyke
Banded Siliceous iron ore
Granular Diorite
Level of Lake Superior.
no outcrops. - ground wet.
TAYLOR MINE
HEMATITE ORE RANGE
Location of Branch of M.H.&O. Road
Sketch of
L'ANSE IRON RANGE
through Sect. 9 · T. 49 · R. 33.
Scale 400 feet to 1 inch.
by
T. B. Brooks.
1872.
N
T. 49 N.
R. 33 W.
Laurentian and Huronian
Chloritic Gneiss

This ore deposit is 950 feet above the surface of Lake Superior, and seven miles from L'Anse by railroad, built or building. The ground slopes gently to the west, affording an excellent opportunity for attacking the ore, which is covered by but a few feet of earth. The timber in the vicinity is first-rate hard wood.

The prevailing variety of ore at the Taylor mine is a soft hematite, similar in character to that of the Lake Superior and Winthrop mines. A number of analyses of average specimens, the results of which are given in full in Chapter X., varied from 44 to 57 per cent. metallic iron, with a remarkably small percentage of silica for an ore of this class. I see no reason to doubt but that a hematite can be mined here, which will yield an average of 55 per cent. of pig-metal in the furnace. Cross trenches and drifts show the deposit to have a maximum thickness 20 to 25 feet free from rock, and three or four times this thickness of such mixtures of ore and rock, as usually occur at hematite mines. The distance between the most easterly and westerly points at which ore has been found, is about 1,000 feet, but up to this time the explorations made have not demonstrated the deposit workable, as to quantity and quality, for more than about one-fourth of this distance. The oft-mentioned irregular pocket-like character of these deposits makes it difficult to predict, with any degree of certainty, regarding them, beyond what can be actually seen. But the heavy bed of hematitic rocks, which show a constant tendency by their decomposition to pass into ore, together with what has been actually developed by the workings, leaves no reasonable doubt but what there is here a large workable deposit of ore.

About 200 feet south of this ore deposit, and overlying it (the whole series dip south), is a bed of highly manganiferous iron ore, average specimens of which have yielded as much as 44 per cent. of the oxide of manganese; such ore must, of course, be comparatively poor in iron; this subject was considered under iron ores in Chapter III. The deposit is of uniform quality for a thickness of ten feet, and was penetrated by a shaft for the same distance. One per cent. of oxide of manganese was reported in some of the analyses of soft hematite mentioned above, showing the general dissemination of this substance, which seems to have its greatest concentration at the point we are describing. Whether this ore would possess value in the manufacture of metallic manganese, I am not able to say, but its presence, undoubtedly, gives additional value to iron ores, in improving the quality of the metal produced, and causing the ore to work more easily in the furnace, besides especially adapting the metal for steel manufacture.

Several other "shows" of iron in this vicinity are worth mentioning. Near the south ¼ post of Sec. 4, being on the north face of a high hill, is an extensive outcrop of several varieties of flag ore, more or less mixed with rock, in the vicinity of which considerable exploration work has been done. Some rich hand specimens of specular ore have been procured at this locality, but the great mass of the material to be seen is made up of layers of silicious ore, banded with quartzose material, the latter greatly predominating. The indications of hematite to be seen here are not promising. I see no reason why a flag ore yielding from 40 to 50 per cent., may not be sought for with reasonable chances of success. A similar ore was found several hundred feet farther north. The quantity of this mixed material existing in the S. ½ of S. ½ Sect. 4 is undoubtedly very great.

In the S. ½ of the N. E. ¼ of Sect. 8 are outcrops of hematitic rocks, which point towards the continuation of the Taylor mine series, making this a promising ground for exploration. Further west and southwest the ground falls off, the drift deepens, and no outcrops of any rock, so far as I know, are to be found, except in the immediate valley of Plumbago brook, where in Sect. 13, Town 49, R. 34, is an outcrop of argellite, which suggests a possibility of there being roofing-slate in the vicinity. Three miles west of the Taylor mine is the east edge of a treeless, sandy plain, which occupies nearly the whole of T. 49, R. 34, and extends into the townships south and west.

A similar desert country is passed through by the Peninsula Railway, commencing 7 miles from Negaunee. This latter, however, is underlaid chiefly by Silurian rocks, while the other is believed to be Huronian.

On the south side of Sect. 9, between Plumbago brook and the diorite ridge, which extends easterly and westerly more than one-half way across T. 49, R. 33, is a range of hematitic rock, similar to that at the Taylor mine, but which is not so promising for ore, so far as explorations have revealed. It has been traced for a distance of more than half a mile, and is the rock which immediately underlies the diorite, being itself in turn underlaid by clay-slate, the whole series dipping to the north, as will be seen on Plate IV.

Before dismissing the economic consideration of this district, it would be proper to notice the so-called "plumbago," found so abundantly in the north bank of Plumbago brook; but as this subject has been fully treated under the head of Carbonaceous Shale, Chap. III., it need not be further referred to here.

The **Huron bay slates** with associated rocks, may be regarded as belonging to the L'Anse series, although more than ten miles away in a northeasterly direction.

This district, which is now being explored for roofing-slate, affords indications of iron at several points, which I have not had such opportunity to examine, as would enable me to make any definite statement about them. So far as I can learn, those best acquainted in the district are not sanguine as to the existence of workable deposits of merchantable ore. At the end of Chap. I. will be found brief statements, regarding the slate companies now at work in this little-known district.

An inspection of Plate IV., in connection with what has been said, makes it necessary to add very little, regarding the structure of this range. The absence of outcrops through the central portion of Sec. 9, leaves the geological section quite incomplete. There can be little doubt, however, but that the quartzites, diorites, clay-slates and hematitic schists, so well exposed on the north side of Plumbago brook, where they dip north, are the equivalents of the Taylor mine series, which dip south, although the sequence is not exactly the same; and the diorite, so conspicuous on the south rim, is not exposed on the north side of the basin, unless the dyke-like mass of greenstone north of the Taylor mine represents it, which I do not think probable. The absence of outcrops also makes it impossible to determine whether there are any minor folds between the two croppings of the basin. If there are no such folds, then there is room for a considerable series of rocks above or younger, than those enumerated; and among them should occur, if it exists here at all, the rich hard ore of the Marquette district. It is assumed in this hypothesis, that the rocks to be seen are the equivalents of formations I. to X. of the Marquette series; this assumption is based chiefly on lithological grounds. Any rich hard ores found must be specular or red oxides, as there is an entire absence of magnetic attraction in the L'Anse district. Magnetic ores have not as yet been found associated with soft hematites, so far as I am aware, in the Upper Peninsula.

The diorite immediately north of the Taylor mine has been mentioned as *dyke-like*. Whether it actually cuts the series of clay and ferruginous slates and schists at an acute angle, was not determined, but in places it certainly has that appearance. If it does so, it is the only case that has come under my observation, in which the Huronian diorites (often termed greenstones and traps) do not conform with the schistose and slaty strata, with which they are associated. This locality, in connection with others which show *unmistakable dykes* of magnesian *schist* cutting various rocks, is worth the study of the geologist, but is comparatively not of much importance to the explorer and miner. Mr. Julien, as will be seen by reference to App. A, Vol. II., Specs. 342 to 353, regards the L'Anse greenstones as a peculiar variety of diorite.

Another point of considerable interest, in connection with the diorites of this locality, is the *dioritic sand*, which forms the base of the great south bed, and separates it from the underlying hematitic schist on the south. This material is an angular, coarse, dark, greenish sand, and has evidently been produced by the disintegration of the rock, which is in places quite friable.

But by far the most interesting geological fact to be observed at this locality, and one, the importance of which can scarcely be overestimated in considering the grand subdivisions of the Azoic rocks, is the *nonconformability* of the Huronian, or iron-bearing series, with the older Laurentian, which can be observed in the gorge formed by Plumbago brook, about 400 feet southwest of the southwest corner of Sec. 9, T. 49, R. 33 (See Plate IV.). Here a talcky, red, quartzose rock, dipping at a low angle northwest, and which is unmistakably Huronian, is seen nearly in contact with a Laurentian chloritic gneiss, which dips at an angle of about 35° south-southwest. The same phenomena can be noted at a point near the Republic mountain (see page 126); and the nonconformability is further proven by the fact that the Laurentian generally abounds in dykes of granite and diorite, which are almost entirely absent from the Huronian.

CHAPTER V.

MENOMINEE IRON REGION.*

THE centre of this region is about 40 miles west by north from Escanaba, 50 miles south-west from Marquette, and 50 miles north from Menominee, as the bird flies. (See Map, No. II.) The area known to bear iron is embraced within a square of 16 miles, being portions of Towns 39, 40, 41 and 42, Ranges 28, 29, 30 and 31. This does not include the iron deposits west of the Paint river, nor the Michigamme mountain, owned by the Republic Iron Co., in Sect. 4, T. 43, R. 31.† The iron ores in the Menominee region occur in two approximately parallel E. and W. *belts*, each probably composed of two distinct *ranges* or horizons of ore; these belts are separated by a broad granite area, in which a little unpromising iron has been found on Sects. 10 and 15, T. 41, R. 29.

This granite area narrows towards the west, caused by the convergence of the iron belts, and has nearly the shape of a flat-iron. The region is drained by the Menominee river, which skirts its W. and S. sides, and by the Sturgeon, a branch of the Menominee, which winds through the eastern part of the iron-fields.

---

* The facts contained in this chapter, as well as on Map No. IV. of Atlas, are largely from the Surveys and Explorations of Prof. R. Pumpelly and his assistant, Dr. H. Credner, made for the Portage Lake and Lake Superior Ship Canal Co. Prof. Pumpelly placed his private notes and sketches at my disposal, and added most valuable explanations. A valuable paper on this region is "The pre-Silurian formation of the Upper Peninsula of Michigan, in North America, by Dr. Herman Credner, Leipsic, illustrated by maps, diagrams and geological sections found in Plates VIII. to XII. (from the Journal of the German Geological Society)." Prof. Pumpelly and Dr. Credner are not in any way responsible for the hypothesis of structure here employed, nor for the views expressed as to the quality of the ores.

† A large amount of silicious iron ore occurs at this locality on the S. W. side of a high hill. Marble is found south and west, but in greatest abundance to the north, between Deer and Fence rivers, and on the upper waters of those streams. This district possesses much geological interest, and quite possibly economic importance, but means were not available for its examination.

1. SOUTH IRON BELT.

The South and, geologically, uppermost iron range of this Belt is probably the most regular and one of the most extensive iron deposits on the Upper Peninsula. The most easterly exposure of ore in this range is at the Breen mine on N. ½ of N. W. ¼ of Sec. 22, T. 39, R. 28. This location is 34 miles from Escanaba, and 45 miles from Menominee, in a bee line. The air-line distance from the elbow of the C. & N. W. R. R., now in operation, is 12½ miles.

Travelling from the Breen mine on a course N. 74° W., which is parallel with the general course of the river, we find on S. ½ of Sects. 11 and 10, N. ½ of Sect. 9, and S. ½ of Sect. 6, T. 39, R. 29, large natural exposures of ore, which have been still farther developed by recent explorations.

In the N. ½ of Sect. 2, T. 39, R. 30, are boulders of iron-ore, and near the S. ¼ post of Sec. 34, T. 40, R. 30, magnetic attractions, which indicate the presence of the iron range. Near the S. ¼ post of Sec. 30, T. 40, R. 30, is a large exposure of ore; thence, following a line of magnetic attraction which leads about W. by N., we find in the centre of the S. E. ¼ of Sec. 25, T. 40, R. 31, another exposure of ore, and a continuation of the local magnetic variations, westerly towards the Menominee river, two miles distant. A range of iron ore, corresponding with this and probably its continuation, has been made out in Wisconsin, between the Brulé and Pine Rivers. Here are no less than nine large exposures of ore, the extreme ones 16 miles apart, which lie in one straight, narrow belt.

Immediately N. of this iron range is a broad belt of impure marble, equally regular, of greater thickness, but which apparently widens towards the W.

North of this, in the vicinity of the Sturgeon River, on Secs. 7 and 8, T. 39, R. 28, and Sec. 12, T. 39, R. 29, are local magnetic attractions and iron boulders, which are believed to mark the position of another geologically lower iron range, although no outcrop has been seen in this vicinity; but near the centre of N. ½ of Sec. 20, T. 40, R. 30, just N. of Lake Antoine, is an outcrop of silicious ore.

Strong magnetic attractions can be observed near the S. W. cor. of Sec. 22, and iron boulders in Sec. 27, and also on north shore N. of Lake Fumée, in T. 40, R. 30.

These indications make certain the presence of a second iron range, although it cannot be demonstrated that these several shows belong to one horizon.

These two ranges, separated by the marble, constitute the South iron belt. North of and underlying both, is an immense bed of quartzite, which is well exposed at the falls of the Sturgeon river, Sec. 8, T. 39, R. 28; also on Sec. 1, T. 39, R. 29, and Sec. 28, T. 40, R. 29, and at the southwest ¼ of Sec. 23, T. 40, R. 30, as will be seen by the map. This quartzite, although believed to be geologically conformable with the ore formations, is not parallel with them, running more northwesterly, and dividing in T. 40, R. 30, into two and perhaps three ranges.

North of this quartzite, and underlying the whole series already described, are the Laurentian, granites, gneisses and schists, which make up the *granite area*, already referred to as probably being barren in workable deposits of ore, and which, therefore, our investigations do not embrace.

South of the south iron range, already described, is a bed of chloritic schist, well exposed on the south shore of Lake Hanbury, Sec. 15, T. 39, R. 29, and on the Sturgeon river in Sec. 13. Immediately south is a second quartzite, which is quite different in its character from the bed already described.

Next south is a broad belt of argillaceous slate, running parallel with the iron range, and exposed at several points in T. 39, Ranges 28 and 29. (See map.) South of this, and embracing portions of the Menominee river, is a broad well-defined belt of chloritic, hornblendic and dioritic rocks, running parallel with the iron range, the harder members of which form the barrier rocks of all the falls in this part of the Menominee, and probably those of Pine river in Wisconsin. This series are perfectly exposed at Sturgeon Falls, Sec. 27, T. 39, R. 29, and at the great and little Bequensenec Falls, and Sand Portage, in T. 39, R. 30.

2. North Iron Belt.

The North iron belt or range has a course nearly due east and west, and is all embraced, so far as known, in the south tier of Secs. of T. 42, Ranges 28, 29, and 30. The most easterly discovered exposure of ore, known as the Felch mountain, is in the N. ½ of Secs. 32 and 33, T. 42, R. 28, and is sixteen miles north and three miles west from the Breen mine, the position of which has been defined. Travelling due west, fragments of iron ore are found in N. E. ¼ of Sec. 31, T. 42, R. 28; after which no absolute proof of the presence of iron is found (although it is probably continuous) until we reach Sec. 31, T. 42, R. 29, where, in the centre of the section, is an immense exposure of iron ore in an E. W. ridge, which can be traced westerly half-way across Sec. 36 of the next Township. The natural exposure of ore on Sec. 31 is larger than at any other point in the Menominee region, and the quality is as good, if not better, so far as can be judged by surface indications. Magnetic attractions and iron boulders, found farther west and southwest on this range, prove its extension in that direction. Whether the westerly course continues, or whether it curves to the southwest, as seems probable from the position of the lower quartzite and local magnetic attractions in the northwest part of T. 41, R. 30, has not been determined. The latter hypothesis is most in accordance with the known facts, although the southeast dip of the quartzite on Secs. 17 and 18, observed by Dr. Credner, is not explained. If this hypothesis is true, the iron range should cross the Menominee somewhere in Secs. 24 or 25, T. 41, R. 31, into Wisconsin. There can be little doubt but that the North and South belts belong to one geological horizon, hence somewhere come together.

The existence of two distinct iron ranges in the North belt, does not admit of so easy proof as in case of the South belt. The facts which point towards this are the following: About one-fourth of a mile north of the iron range, already described as existing on Sec. 36, T. 42, R. 30, is a bed of marble running east and west, parallel with the iron, on both sides of which are slight magnetic attractions. Prof. Pumpelly found, "about 80 paces south of this marble, an outcrop of strata made up of layers of quartz, magnetic iron and chlorite," probably of no economic value.

Again, in the E. ½ of Sec. 35, are two parallel lines of feeble magnetic attractions, several hundred feet apart, and to the north are some large, angular boulders of magnetic ore; similar smaller boulders are found between Secs. 33 and 28, still farther west.

South of the iron deposits on Secs. 31 and 36, is a bed of marble, somewhat similar to the one already described as underlying the south iron range of the South belt, and possibly the equivalent of it, as the two have the same relative geological position. Farther south, immediately adjacent to, and overlying the granitic rocks, is a heavy bed of quartzite, which is undoubtedly the equivalent of the lower quartzite, already described as forming the base of the South belt. This quartzite at the S. ¼ post of Sec. 31, T. 42, R. 29, is characterized by the presence of mica scales in the bedding planes, and might be denominated a micaceous quartz schist. It has considerable resemblance to the rock, associated with the Cannon ore in the Marquette region. This fact possesses considerable geological interest in connection with the relative age of the Felch mountain ore deposit, which, I think, belongs in this lower quartzite. See Chap. III., Group H. mica schists, and below.

The Huronian rocks in the N. ½ of Sec. 31, are covered with horizontal layers of Silurian sandstone, hence cannot be seen. North of the iron on Sec. 36, is the marble already mentioned, which is peculiar in being filled in places with crystals of kyanite, giving the gray weathered surface of the rock a rough jagged character, like a coarse rasp.

Just N. of the N. ¼ post of Sec. 31, T. 42, R. 29, is an east and west range of gneiss rock, and still farther north a heavy bed of hornblendic schist. At numerous points east and west, through the centre of T. 42, Ranges 28, 29, and 30, are outcrops of similar hornblendic rocks, together with beds of mica schist and gneiss, traversed in places by dykes, and perhaps by beds of granite. This broad belt of hornblendic rocks is apparently represented in its westerly extension, where it crosses the Michigamme river, by the mica and chloritic schists and gneisses, so well exposed at the Falls of the Michigamme, Cedar Portage, Long Portage, Norway Portage and intermediate points in Towns. 41 and 42, R. 31. Similar rocks cross the Paint river, a few miles farther west. This series would correspond in their geological position, as they do partially in their lithological and topographical characteristics, to the hornblendic and chloritic series, already described as forming the southernmost formations of the South belt, and which there, as here, produce numerous waterfalls.

Near the centre of this hornblendic belt, in the north part of Secs. 21, 22, 23, and 24, T. 42, R. 29, is a line of comparatively feeble magnetic attractions, which seems to have no equivalent in the

South belt, unless it be in Sec. 28, T. 39, N. R. 18, E. Wisconsin; or in one of the beds of hornblendic rock at Little Bequensenec Falls, to be described hereafter, which contains many specks of sulphuret of iron and of magnetic ore.

This line of attractions, noticed in T. 42, R. 29, may represent the north edge of a basin, of which the North iron belt, already described, is the south edge; but I incline to the hypothesis, that it is an independent ferruginous range. No outcrop or boulder of iron has been seen upon it in Michigan, and it is doubtful if it is of any economic importance, although of much geological interest, as helping to elucidate the structure.

Returning to the most easterly exposure of iron on the North belt, the Felch mountain, we find a different and less complete sequence of rocks. Except some boulders about one mile west, no marble can here be seen. The Felch mountain ore rests immediately upon, and is bounded on the south by hornblendic, micaceous and gneissoid rocks, which are undoubtedly Laurentian, thus shutting out the marble and quartzite, already described as existing under the iron to the west. No indications, which would suggest the presence of a second iron range, can be found here. Within half a mile north the hornblendic schists are to be seen. At the N. ¼ post of Sec. 31, about 1½ miles westerly, is a large exposure of quartzite, running east and west, and apparently dipping to the north, although the bedding is indistinct. This may be the equivalent of the north marble range, Sec. 36, T. 42, R. 30, for quartzites sometimes pass into marbles in the Marquette region.

The Felch mountain ore, so called, is in reality a dull red jasper-like quartzite, containing numerous thin lamina and minute gash veins of very pure specular ore. It has somewhat the appearance of the "mixed" or second class ore of the Marquette region (see Chap. III. A), differing in containing less iron, and in the fact, that the ore lamina have less continuity. Considerable amount of a similar rock can be seen on the Penokie iron range, Wisconsin. I have a two pound specimen of specular ore from the Felch mountain, which is as rich as any I ever saw. The deposit is somewhat magnetic, the east and west belt of magnetic influence having considerable breadth.

It is not at all improbable, that better ores may be found adjoining this on the north, or possibly still further north, in a geological position corresponding with the ore on Sec. 31, T. 42, R. 29.

In the south half of Sec. 36, T. 42, R. 29, about two miles west of the Felch mountain, Prof. Pumpelly and Dr. Credner observed a variety of the lower quartzite, the character of which is important in connection with the age of the Felch mountain deposit. It has been described as containing mica enough on its planes of stratification, to make it semi-schistose, is porous, and contains thin streaks of magnetic iron in crystals, with here and there cubes of iron pyrites.

The above facts lead me to accept the hypothesis already advanced, that the Felch mountain ore deposit is itself in the Lower Quartzite. If we suppose the mica contained in the quartzite exposed at S. ¼ post of Sec. 31, and in the S. part of Sec. 36, to be replaced entirely by specular ore, a Felch mountain ore would be the result. This hypothesis is supported by the fact, that the Cannon ore, Sec. 28, T. 47, R. 30, is a quartz schist, having specular ore in its bedding planes, and which in a short distance changes into mica. (See Chap. III., Mica schist.) It should be noted, however, that while the Cannon ore is micaceous, the Felch mountain is eminently granular. The Cannon, like the Felch deposit, is at the base of the Huronian series, resting immediately on the Laurentian.

It has already been mentioned that *Silurian sandstone* capped the iron bearing rocks on N ½ of Sec. 31, T. 42, R. 29; the same is true in places on Sections 34, 35, and 36, in same Township, as also in Sections 31, 32, and 33, in the Township east. Passing to the South belt, we find the sandstone covering the iron series in Section 25, T. 40, R. 31, in Secs. 30, 29, 23, and 36, T. 40, R. 30; also in Sections 9 and 10, T. 39, R. 29, and in Sec. 15, T. 39, R. 28 immediately north of the Breen mine, as well as at numerous other points, which it is not necessary to mention.*

Explorations eastward on the two iron belts of the Menominee region, reveal the presence of this sandstone and its accompanying overlying limestone (calciferous sand rock), in greater quantity, even to the point of entirely covering up the Huronian and Laurentian rocks, which is done, so far as known, from near the east side of the Menominee iron region, all the way to the Canadian line at the Sault Ste. Marie. Local magnetic attractions, discovered by

---

* These irregular patches of sandstone are not represented on the maps.

United States surveyors at various points in this Silurian area, render it likely that the iron-bearing or Huronian rocks extend far to the eastward, connecting probably with the similar rocks of the north shore of Lake Huron, where they were first studied and named by the Canadian geologists. Pine explorers inform me, that they have observed dark-colored heavy rocks, which were somewhat magnetic, in the eastern portion of the Upper Peninsula. These may have been Huronian islands in the sea, in which the sandstones were laid down. This subject is discussed in Chap. II.

Like their equivalents in the Marquette region, the ore strata and accompanying rocks of the Menominee region usually conform in their strike with the general trend of the belts and ranges, and dip at high angles, thus presenting their upturned edges to the observer, and affording, where exposed, the best possible opportunity to observe the thickness of the beds and their mineral composition. But highly inclined strata, especially if they should be overturned, as is occasionally the case, are not favorable for making out the structure and sequence of the various beds. This question is farther complicated by the difficulty of distinguishing, in the case of the clay and chloritic slates, between the cleavage and bedding planes. The latter are sometimes very obscure, and have been confounded with the other, thus leading to erroneous results.

The geographical distribution of rocks in the Menominee region which has already been given in a general way, in connection with what has been said in Chapter II. concerning the structural relations of the Laurentian, Silurian and Huronian systems, leaves but little more to be said regarding the structure. The Laurentian area is the broad backbone of the great E. and W. anticlinal, on and against the north and south sides of which the iron series repose, dipping away from the axis; that is, the South belt south and the North belt north. This general structure, it will be observed, is similar to that presented by the Michigamme district on the south and the L'Anse-Huron bay districts on the north of the Marquette region, separated as they are by a great Laurentian anticlinal. It is probable that the Laurentian area of the Menominee region may wedge out at a point just west of the Menominee river, in the same way as do the Laurentian rocks of the Marquette region in the west part of T. 49, R. 33. (See Map I.)

In order to bring out the structure more fully for the information of the explorer and miner, three geological sections will be given, two on the South and one on the North belt. Like most geological sections, they are to a certain extent ideal, but are intended to correctly present the facts, together with such inferences as seem to be warranted. I should note that Dr. Credner's corresponding sections differ considerably in the hypothetical parts from mine, as will be seen by reference to his paper already mentioned.

Geological Sections, Menominee Iron Region.

Section A.

Projecting the more important rock exposures of the eastern portion of the South belt on one plane, which may be taken at right angles with the strike of the rocks, that is, N. 16° E, through Sturgeon Falls, Sec. 27, T. 39, R. 29, the following series will be found (See Map No. IV.) :—

At the falls of the Sturgeon, Sections 8 and 9, T. 39, R. 28, is a group of strata, which divide rocks unmistakably Laurentian on the N., from the lower Huronian quartzite on the S., and which Prof. Pumpelly and Dr. Credner regard as of Laurentian age, but which seems to me to admit of some doubt, as they conform with the bedding of both systems (all being conformable) and have lithological affinities with both.

Prof. Pumpelly describes them as follows, beginning with the uppermost strata :—

1. Talcose slates, soft, light-greenish, gray, with distinct ripple-marks.

2. Four beds of conglomerates, consisting of more or less rounded fragments of quartz, granite and gneiss, 15 to 30 feet wide. See Spec. 65, State Coll., App. B, Vol. II. This conglomerate has not been observed elsewhere, although a somewhat similar rock outcrops on Sec. 10, T. 42, R. 28.

3. Underlying the series are two beds of protogine gneiss, of reddish color, separated by a bed of chloritic schist; the upper one of the beds of protogine encloses a segregated vein, two feet wide, of a mixture of magnetic iron and sulphuret of iron, which does not promise to make a workable deposit.

North of this series, at the head of rapids on Sec. 9, T. 39, R. 28, unmistakable Laurentian rocks occur, but which appear to be conformable with the Huronian. The chief varieties found here as well as elsewhere in the Menominee region are,—a granite (in places porphyritic) syenite, mica-gneiss, with some mica-schist, hornblendic-gneiss and schists, chloritic and talcose gneiss, with some chloritic and talcose slates.

I. The lowest, geologically, and most northerly formation which is unmistakably Huronian in the South iron belt, is a *quartzite*, which outcrops conspicuously at the Falls of the Sturgeon river, Sec. 8, T. 39, R. 28 (not Sturgeon Falls), where it is not far from 1,000 feet thick, and rises to an elevation of over 200 feet above the river. It is usually light-gray, massive, compact, and often semi-vitreous, with indistinct bedding; has more the appearance of vein quartz than the Marquette quartzites. In places it shows ripple-marks with great distinctness; the weather has no appreciable effect on it.

This formation outcrops conspicuously, forming high ledges on Sec. 9, T. 39, R. 28, on Sec. 1, T. 39, R. 29 and Sec. 28, T. 40, R. 29. A quartzite, believed to be the equivalent of this, outcrops near the N. W. cor. Sec. 26, T. 40, R. 30. The Felch mountain iron deposit is also supposed to belong to this formation, as has already been explained.

II. A quartzose *sandstone* and conglomerate rock, which has a lithological character more allied to the Silurian than the Huronian, seems to overly this quartzite on the S., outcropping near the S. W. cor. of Sec. 2, T. 39, R. 29, and on the E. bank of Sturgeon river, on Sec. 8, T. 39, R. 28. But little is known about it, and its existence as a member of the iron series is not absolutely proven. From its soft, friable character it would more likely be found under swamps than on elevations.

The marble outcropping in Sections 24 and 25, T. 40, R. 30, would appear to occupy the same horizon. The same marble may exist on this geological section, but it has not been seen; the formation we are describing may be its equivalent.

III. The existence here of a range of slightly *magnetic* ore is indicated by angular boulders of lean ore in the valley of the Pine river, Sec. 12, T. 39, R. 29, and by magnetic attractions, Secs. 7 and 8, T. 39, R. 28. It does not, however, outcrop in this vicinity.

The hypothesis assumed for the structure of the South belt would make this ore the equivalent of the range known to exist north of Lakes Antoine and Fumeê, in T. 40, R. 30. It is possible, as will be seen hereafter, that this conjectured iron range may be the equivalent of the main iron deposit of the North belt.

IV. Crystalline *limestone or marble.*—This formation has an immense development in the South belt, far greater than in the other, its thickness being probably greater than that of the quartzite I. It is generally thinly bedded, and usually of a light-gray color, but is sometimes reddish, yellowish, or bluish.* The upper portion contains thin bands of slate, in which it resembles the marbles of the Marquette region, but differs from them in being freer from silica, less variegated in color, having fewer joints, as well as in being immensely greater in its extent, and more dolomitic. The Marquette marbles are indeed but calcareous beds in the Lower Quartzite (V.) of that series, there being no proper marble formation in the rocks of that region.

A piece of marble from near the Breen mine gave Dr. Rominger carbonate of lime, 61 per cent.; carbonate of magnesia, 34 per cent.; hydrated oxide of iron and manganese, 1 per cent.; and silicious matter, 0.25; which composition would make the rock rather a dolomite than a limestone. Specimen No. 66, State Collection, App. B, Vol. II., came from Sec. 11, T. 39, R. 29. Five specimens from this locality gave an average specific gravity of 2.81, approximately determined. Dr. Rominger gave attention to the value of this rock for building. (See his Report, Part III.) Large outcrops of marble occur on the south side of the Pine river on Secs. 11 and 12, T. 39, R. 29, and on the Sturgeon river, Secs. 17 and 18, T. 39, R. 28.

V. The principal *iron ore formation* of the South belt overlies, on the south side, the formation just described. It is made up chiefly, so far as is now known, of silicious specular slate ores, corresponding nearly with the so-called flag ores of the Marquette region. There is generally such admixture of magnetite as to produce moderate variations in the needle, but no evidence of the existence of a large body of magnetic ore. Specimen 68, State Col-

* The weathered surface is often rough, from minute ridges, caused by the more silicious layers, which best resisted the weathering.

lection, App. B, Vol. II. is from Sec. 11, T. 39, R. 29. At the Breen mine some very good soft hematite occurs in the same formation, which promises to be in workable quantities. See Specimen 67, State Collection, App. B., Vol. II. This ore would probably be found elsewhere if sought for, but it never outcrops. A blackish, porous ore, hematitic in its character, containing 56 per cent. of iron and nearly 1 per cent. of manganese, was found in a pit at the ¼ post between Sections 9 and 10, T. 39, R. 29, but its extent was not determined. Boulders of the same ore were seen in other places on the range.

The best exposures of the hard ores of this formation in the vicinity of the Sturgeon river, besides the Breen mine, are in Secs. 11, 10, 9 and 6, T. 39, R. 29. These ores will be described more fully, and analyses given hereafter.

VI. On the south shore of Lake Hanbury, which lies in Secs. 9, 10, 15 and 16, T. 39, R. 29, is an extensive outcrop of *chloritic schist*, the most easily splitting planes of which strike west by north, and dip south at a high angle. A similar rock, believed to be the same bed, can be seen on the Sturgeon river, near centre of Sec. 13, T. 39, R. 29. South of Lake Hanbury, 200 steps, is a rock partaking of a dioritic character, but which is probably a harder granular form of the same schist. Such rocks often graduate into each other in the Marquette region (Chap. III.). This schist may probably underlie Lake Hanbury and the swamps easterly and westerly from it.* It is represented on the section as following in its foldings formations VII. and VIII., described below. It is at least possible that this formation may be the same as the Menominee river diorites and chloritic schists, IX. and X., there brought to the surface by another series of more southern folds. But this hypothesis is not assumed in this discussion.

VII. *Clay-Slate.*—At 350 steps south of Lake Hanbury, on lines between Secs. 15 and 16, T. 39, R. 29, is a bluish and greenish gray slate, showing indistinct contorted bedding, with prevailing dip to *north;* the cleavage planes of which strike about north 70° west, and dip 80° to south. Veins of white quartz occur in

* Since the above was written Professor Pumpelly has informed me that he observed a large outcrop of marble south of the iron formation III., in T. 40, R. 30, which will be described below under Section B. This marble may fill the apparent blank existing at Lake Hanbury.

these planes. At 550 steps south of the lake, a similar slate is found dipping *north* under the quartzite VIII., next to be described. It is believed that these two outcrops of slate, are the opposite sides of a synclinal trough, which holds the quartzite.

In the N. E. ¼ of Sec. 20, T. 39, R. 29, is an outcrop of talcose clay-slate. In Secs. 29 and 39, T. 39, R. 28, are several outcrops of dark colored, finely cleavable, but indistinctly bedded clay-slates. It is assumed that all these outcrops are parts of bed VII., which is folded into a synclinal and partial eroded anticlinal, as represented on section A of Map IV.

I am not in possession of sufficient facts to demonstrate the precise relations of these beds to each other, but the general fact is established by the northerly dips observed by me on Secs. 14, 15 and 16, that there are at least two folds between the iron range and the Menominee river, which probably reduces the estimated total thickness given in Dr. Credner's paper (18,000 feet), one-third. See page 175.

VIII. Associated with the clay-slates south of Lake Hanbury, is a bluish gray *quartzite*, which weathers into a brown, friable sandstone,* and in places reticulated with fine veins of quartz. At 550 steps south of Lake Hanbury, on line between Secs. 15 and 16, T. 39, R. 29, this quartzite is underlaid, as has been mentioned, by the clay-slate, VII., the division plane dipping plainly to the north at an angle of from 45° to 75°; the same rocks with the same northerly dip were observed farther east, on Secs. 15 and 14. This quartzite may be simply a local bed in the clay-slate formation, hence not entitled to a distinct number. The marked contortions both in the clay slate and quartzite are noticeable, and point unmistakably to the presence of a great fold. The cleavage planes maintain their east-west strike and southerly dip.

IX. This number is intended to include the soft *magnesian schists* (chloritic, talcose, and probably argillaceous) occurring so abundantly along the Menominee river, in the vicinity of the mouth of the Sturgeon, as well as at the several falls above. They will be more particularly described under geological section B.

---

* "Iron slate" is marked on the United States plats at this locality. The brown color of the quartzite has something the appearance of iron rust. The very feeble magnetic attractions existing along this range, indicate the presence of magnetite.

X. This formation is designed to embrace the granular *dioritic* rocks which form the barrier of the Sturgeon and other falls above, for 20 miles. It varies considerably in character, but on the whole bears a strong family resemblance to the granular diorites of the Marquette region. A peculiar gray variety, occurring at Sturgeon Falls, Sec. 27, T. 39, R. 29, is illustrated by Specimen No. 65, State Collection, App. B, Vol. II. This is the formation, it will be remembered, which in its supposed westerly prolongation into Wisconsin, produces the falls in the Pine river, and near them becomes iron-bearing. If the hornblendic schists mentioned as occurring in T. 42, are Huronian, they are probably the equivalents of this formation.

XI.—South of X., on or near the Menominee river, in south part of T. 39, R. 29, are several exposures of what appear to be *magnesian schists* and *protogine*, the structural relations of which to the rocks already described have not been made out. A rock similar to the protogine was observed in Sec. 13, T. 42, R. 30, and would there seem to have about the same relative position to the North belt that this has to the South belt.

Geological Section B runs northeast by north, across T. 40, R. 30, cutting Lake Antoine, and passes near the head of Great Bequensenec Falls. (See Map IV.)

I. Lower *quartzite*.—This formation appears far more conspicuously in this section than in A, owing to the double fold hypothetically introduced to cover the facts observed in the N. ½ of T. 40, R. 30. The large exposure of quartzite lying against the Laurentian, on Secs. 1 and 2, and the numerous angular boulders on Secs. 7 and 8, with the outcrop of quartzite near S. W. cor. of Sec. 23, taken in connection with the granite exposures on Secs. 4 and 9, lead one to the conclusion that one bed of quartzite, forming a synclinal basin under the Pine river and an eroded anticlinal to the south, best reconciles the facts observed. The lithological and topographical characteristics of this quartzite have already been given under A, and need not be repeated.

II. This formation was represented on A by friable sandstone and conglomerate, not observed near this section; the blue and pink *marble* outcropping near centre of Sec. 25, and the marble at the N. W. cor. of Sec. 24, are assumed to belong to one horizon (as shown by map and section), which is supposed to immediately overlie the quartzite. There is no reason to believe that this formation has any great thickness.

III. The "shows" and "signs" of ore to which this number was attached on section A, have developed into certainty on this section, where, near the centre of the N. ½ of Sec. 20, T. 40, R. 30, a considerable outcrop of *iron ore* is seen in the bottom of a small ravine. It is a silicious, red oxide, resembling in its general character the great ore formation of section A. Its continuation eastward is made certain by the magnetic attractions on the south line of Sec. 22, by the iron boulders of N. E. ¼ of Sec. 27, and on the north side of Lake Fumeé, on Sec. 26. Except the slight attractions noted by United States surveyors, at N. E. cor. of Sec. 30, T. 40, R. 29, there is no connecting link, so far as known, between this deposit and the indications of this bed on A. It is not proven that they are identical. Dr. Credner, as will be seen by reference to his paper, believes the ores on the north side of the lakes are the equivalents of those on the south, the two being connected by a synclinal fold.

IV. Crystalline *limestone or marble.* There are immense outcrops of this rock in the S. part of Secs. 34 and 35; large exposures on the S. shore of Lake Antoine; boulders on the W. side of Sec. 30, all in T. 40, R. 30, and a continuation of the boulders in Sec. 25, in the Township west. The apparent thickness is greater than was shown on A., which may be owing to a crumpling or short abrupt folding of this part of the formation; or, it may be due to an actual thickening of the formation to the westward.

Two outcrops referred to, deserve especial mention: that in the N. W. fractional ¼ of Sec. 29, contains beds of a sandy and almost conglomeritic rock, which is associated with thin beds of dark-gray argillaceous limestone. The outcrop on Sec. 35 is the largest marble outcrop in the Menominee region, it being over 1,200 feet wide. As the dip is at a high angle to the S., the perpendicular thickness of the bed cannot be less than 1,000 feet. The S. part of the outcrop shows bands of limestone alternating with thin seams of quartz.

V. The *main iron formation* is marked by an outcrop in the centre of S. E. ¼ of Sec. 25, T. 40, R. 31, and by another which forms the west end of a high ridge on line between Secs. 30 and 31, T. 40, R. 30, the two being connected by a line of magnetic influence.

Attractions also exist near the south ¼ post of Sec. 34, T. 40, R. 30, and in the N. W. ¼ of Sec. 2, T. 39, R. 30, are iron boulders. There is at present (October, 1872) no reason to believe that the ore in Towns 39 and 40, R. 30, is less in quantity, or differs in quality from that already described under the corresponding formation of geological section A.

VI., VII. and VIII. The hypothetical place of these formations on section B, is covered by deep drift—constituting the sandy terraces of the Menominee river. No outcrops of any kind can be seen on this belt of rocks, either in Ranges 30 or 31, except a large exposure of marble observed by Prof. Pumpelly, just south of the ¼ post, between Secs. 32 and 33, which corresponds in strike and dip and in general lithological character with marble formation IV. Reference to the map will show that this rock has no observed equivalent on A, where, if it exists at all, it should be found under Lake Hanbury.

I must confess that the existence of this marble, but lately made known to me, points to the existence of folds in the neighborhood of Lake Antoine, not suggested by my geological sections.

IX., X. The chloritic, hornblendic, and dioritic rocks embraced under these two formations are well exposed at the Great and Little Bequensenec Falls, and at Sand Portage, all in T. 39, R. 30. These falls afford an unsurpassed opportunity to study this series, which was carefully done by Dr. Credner, who made out the following section at the upper fall from north to south :—

a. Crystalline hornblendic rock, consisting of light to dark-green hornblende in crystalline masses, white feldspar, a little chlorite and some quartz.

b. Talcose rock, consisting only of fibrous talc, which forms a kind of soapstone in three heavy beds.

c. Fissile talcose silicious slates, of a reddish color, with small crystals of orthoclase.

d. Soft talcose slates of light green color.

e. Chloritic slates, dark green, with spots and layers of clayish red oxide of iron.

f. Hornblendic rock, dark green, crystalline, coarse-grained to aphanitic, with specks of sulphuret of iron.

By the Little Bequensenec Falls the following series of strata is laid open, from north to south :—

a. Talcose chloritic slates, with a great many segregations of quartz.

b. Hornblendic rocks, with much dark-green chlorite, and many specks of sulphuret of iron and magnetic iron ore, 35 feet.

c. Soft fibrous soapstone in two heavy beds, with some sulphuret of iron, 8 feet.

d. Talcose slates, fissile, with many layers and segregations of white quartz and red limonite.

e. Chloritic slates, 10 feet.

f. Bed of hornblendic crystalline rock, 12 feet.

g. Chloritic slates with seams of iron pyrites, 30 feet.

h. Fibrous talcose slates, reddish, with bands of green color.

i. Chloritic slate.

Geological Section C. (North Belt). On line between Ranges 29 and 30, T. 42.

I. A *quartzite*, which is micaceous at S. ¼ post of Sec. 31, and in south part of Sec. 36, T. 42, R. 29, and ferruginous at the Felch mountain. The lithological character and stratigraphical position of this formation have been fully considered. Although it differs considerably in its character from the equivalent formation of the South belt, there can be little doubt but that it is the same.

North of this quartzite is a considerable breadth of low damp ground, with no outcrops.

II. Crystalline *limestone or marble*, of a quite pure snow-white, to reddish granular variety, outcrops immediately south of the iron on Sec. 31. In the southeast ¼ of Sec. 35, T. 42, R. 30, is an outcrop of marble presenting very distinct bedding planes, which dip to the north. These two outcrops define a range parallel with the quartzite, and probably belong to this bed, II. Another outcrop of marble near the centre of Sec. 35 cannot be reconciled as belonging to this formation, and there is some uncertainty as to whether it lies above or below the iron formation. If below, then it would have the same relative position to the iron as the outcrop first mentioned above. More facts are needed to establish the relations of these marbles. As will be seen by comparing sections C and B, it is assumed that the limestones marked II., on each, are equivalents of this bed.

III. The great *iron-ore formation*, which extends easterly and westerly across Sec. 31, half way across Sec. 36, and probably much farther each way, has already been partially described. This bed is apparently the equivalent of III. of the South belt, but it is certainly more extensive, and, so far as can be seen, contains better ore. If this hypothesis be correct, then the upper and main iron formation of the South belt has no representative in the North belt, unless it be indicated by the slight magnetic attractions already mentioned as having been observed in the north part of Sec. 36. The strongest indication of the continuance of this formation eastward is to be found, so far as known, just six miles due east, in the N. E. ¼ of Sec. 31, T. 42, R. 28, where Prof. Pumpelly observed numerous large angular fragments of specular iron ore, associated with fragments of marble. This deposit should, on this hypothesis, pass just north of the Felch mountain, in its eastward prolongation.* The quartzite near the north ¼ post of Sec. 31, T. 42, R. 28, would, on this hypothesis, be the equivalent of the before mentioned marble in Sec. 36, seven miles west.

IV. Crystalline *limestone or marble*, containing crystals of kyanite, outcrops about 300 steps south of the north ¼ post of S. 36, T. 42, R. 30. Several outcrops of the same rock occur a short distance to the west, and a little south, indicating the probable existence of a large deposit of this rock. Except in the presence of the kyanite crystals, which gives to a weathered surface the rough character heretofore described, this rock has much the character of the marble, with corresponding number of geological sections A and B. Whether these marbles are equivalents is not proven, but it is assumed as being more in accordance with the facts than any other hypothesis.

V. An interesting fact in connection with the limestone outcrops on Sec. 36, just described, is the presence of a very noticeable magnetic attraction on both sides of the marble, or rather associated with it.

Prof. Pumpelly observed south of one of these outcrops of marble "strata made up of layers of quartz, magnetic iron and chlorite, containing garnets, and resembling some of the strata at Republic Mountain, Marquette region." These attractions

---

* The blank space north of and above the iron formation III., on section C, is marked by no outcrops except Potsdam sandstone, which covers the Huronian rocks on Sec. 31, as has been already stated.

are probably due to this rock, which is certainly but a poor representative of the great upper iron bed of the South belt.

VI., VII., VIII. No other rock was observed on this section for several hundred paces; this space may or may not be filled by these formations, which, so far, have only been seen on geological section A. The numbers are introduced here, in order to carry along the hypothesis of structure which will best reconcile and present the observed facts.

IX., X. Just north of the north ¼ post of Sec. 31, T. 42, R. 29, is a large outcrop of gneiss, with thin layers of granite, and adjoining this on the north is the most southerly observed outcrop of the great hornblendic and mica schist series, the geographical extent and general structure of which have been fully considered. Whether this series of schists are the equivalents of beds IX. and X., which occupy the immediate valley of the Menominee, cannot be established. They have the same relative position to the iron ore, marble and quartzite series, and similarity in their lithological character. It must be admitted, however, that the lithological affinities of this series of rocks of the north belt are decidedly Laurentian rather than Huronian. The gneiss and granite outcrop, above described, may be almost regarded as a typical Laurentian rock in its appearance. If future investigations prove them to be Laurentian, a very troublesome structural problem would be presented here, as we would have Laurentian rocks conformably *overlying* beds, unmistakably Huronian. There seem to be fewer difficulties in supposing that the Huronian rocks of the Menominee region embrace lithological families not, so far, found represented in the equivalent series in Marquette region.

An important observation may be made here bearing on the variable thickness of the Huronian series, or else pointing unmistakably to tremendous folds in the rocks of the South iron belt,—it is this: the superficial breadth occupied by formations I. to VIII. inclusive, is nearly four times as great in the South belt as in the North. A portion of this difference may be accounted for by the thinning out of this series to the north; but the folds figured in geological section A, and possibly others not determined, would, I think, account for the greater part of this discrepancy.* There are no evidences of any folds in the corresponding series in the North belt.

---

* See page 169.

A range of marble associated with quartzite, chloritic and talcose rock, and overlaid by a chloritic gneiss, with beds of chloritic schist and gneissoid conglomerate, the whole dipping at a high angle to the south, passes about five miles north of the North belt. These may represent the north side of the trough or basin, of which this iron belt is the south outcrop. No iron has, however, been found, as far as I know, on this range.

Along the Menominee river, where it crosses this broad schistose belt which lies north of the North belt, is a series of north and south dips, observable at the Cedar, Long, and Norway portages, which point unmistakably to intermediate folds in these rocks, whose thickness, therefore, may not be very great.

Nothing remains to be said regarding the Menominee iron region which is of practical importance to the explorer, miner, or capitalist, and which would properly come within the scope of this work, except a statement as to the *quality of the ore*. The quantity has already been described as great, and the chances to mine all that could be desired. The distances by rail from shipping port and grades are most favorable. If the ores are of first quality, this region has a future which will only be surpassed, if it is surpassed, by the Marquette region, now developed to that extent that its ores produce nearly one-fourth of all the iron made in the United States.

Unfortunately at this time the question of quality cannot be fully answered, for the simple reason that up to the date of my last visit, in October, 1872, comparatively little exploring had been done, and iron deposits very seldom expose naturally their best ores; these have to be found by digging. This subject is fully treated in Chap. VII.; but I will repeat here that ninety-nine hundredths, if not nine hundred and ninety-nine thousandths, of all the ore outcropping in the Marquette region (and there is an immense amount of it) is not merchantable, according to the present standard for shipments. Soft hematite ores never outcrop; therefore if pure high grade ores be abundant in the Menominee region, they might not yet have been found from the little work that has been done.

The facts observed by me are as follows, taking the several iron locations in succession:—1st, The *Breen mine* on N. ½ of N. W. ¼ of Sec. 22, T. 39, R. 28, **South belt**. Three kinds of ore occur at this locality, the predominating variety (constituting perhaps four-fifths of all exposed) being a lean, silicious, slaty or flaggy ore, resembling the Iron mountain and Teal lake ores of the Marquette region. It varies in quality from a ferruginous quartz schist, containing but a few per cent. of iron, up to masses as good, if not better, than the second-class or flag ores of the Marquette region, with occasional richer streaks. Careful mining and selecting would produce an ore of this kind that should yield say 45 per cent. in the furnace, but it would be apt to "work hard," from the large amount of silica, and produce a hard iron, suitable, perhaps, for rail-heads. (See Iron Ores, Chap. III.) What percentage of the whole mass would be of this degree of richness, practical mining only can determine; from what could be seen in October, 1872, I should say not exceeding one-third.

The next variety in abundance is a soft, earthy, dark-colored hematite, resembling in its general appearance the Negaunee hematite ore of the Marquette region. A sort of irregular pocket of this ore was found lying in the first described variety, appearing as if it may have been produced by a partial decomposition and disintegration of the flag ore,—that is a secondary form of it. This hematite pocket, so far developed by the shafts and trenches, is of sufficient size to work advantageously, but is divided through the centre by a bar of very silicious ore. Several "shows" of this ore were found in other places, but none were proven to be of workable extent. See Spec. 67, State Coll., App. B, Vol. II.

The third variety of ore is best in quality, but, so far as known, least in quantity. It can be seen near the mouth of a drift on the south side of the ridge next the swamp, where a bed two or three feet thick was passed through, flag ore being found to the north of it. This is a hard, more or less porous, bluish, heavy, red ore, of a hematitic character, and has considerable resemblance to the so-called Jackson "hard hematite." It would undoubtedly work well in the furnace, and would yield not less than 60 per cent. of metallic iron. There are reasons to suppose that there may be a workable bed of this ore on the property; but judging from what is to be seen at the drift above mentioned, it may be under wet ground.

On the whole, it may be said of the Breen location, that the great amount of ferruginous schist there developed, and the tendency shown by it to pass into soft hematite, render it very probable that a considerable quantity of workable ore of this kind may exist. The absence of local magnetic attractions, and of boulders of rich hard ore, leads me to consider it doubtful whether any rich specular and magnetic ores, such as are now produced in the Marquette region, will be found here.

The ore range probably extends east and west, the entire length of the "80," or one-half mile, forming a ridge where the explorations have been made, from 20 to 30 feet high, bounded by a swamp on the south side. The whole iron series dip south, and are underlaid on the north by soft shaly magnesian and argillaceous rocks.

Sections 6, 9, 10 and 11, T. 39, R. 29. The ores on these sections form what appears to be a continuous deposit, and are so much alike in their general character that they can be more commonly and briefly described together. Except a few trenches dug by the Canal Co. on Secs. 9 and 11, and some test-pits sunk this season on Sec. 6, no work had been done on this range at the time of my last visit. Here, as at the Breen, the prevailing variety, in fact the only variety which I saw in quantity, was the silicious flaggy ore already described. The quantity of this ore is enormous, forming as it does the south face, and, perhaps, the great mass of a considerable ridge running west by north. The opportunity for attack by open cuts into the south face of this ridge is unsurpassed. Like the hard ores at the Breen, they vary greatly in richness,—from a quartz schist slightly impregnated with iron up to specimens, and even considerable masses which will yield 50 per cent., and occasionally a specimen that contains 60 per cent. of metallic iron. The prevailing variety, however, is represented by Specimen No. 68, App. B, Vol. II., from Sec. II., which contains from 25 to 45 per cent. of iron.

Dr. Credner reports having found, in "Cut D, on Sec. 11, 28½ feet of good fine-grained, steel-gray iron ore, with here and there a narrow streak of silicious ore, but in such a small proportion as not to spoil the good quality of the mass. The whole series gives a dark-red streak." Specimens designed to represent the average of this deposit gave Dr. C. F. Chandler 52 per cent. of iron. In another place he found a bed "6 feet thick, supposed to be very rich ore." I did not find these trenches (as afterwards appeared), although I designed to see all, and had with me two men, who helped to dig them. Dr. Credner further reports an aggregate of 139 feet in thickness of "workable ore" on Sec. 11, but my own observations lead me to question this, unless the standard of furnace-yield be put considerably lower than at present. It is unwise, however, to predict at this time what thorough explorations may reveal.

The ore on Sec. 9 is very similar to that on 11, but on the whole (so far as can be seen) not so good: the same may be said of that on Sec. 6. Two smaller boulders of rich specular slate ore were found on the latter section, but no large ones. Occasional narrow seams of tolerably rich ore were found, one of them over one foot thick, but nothing that looked like a workable deposit. At the ¼ post between Secs. 9 and 10, north of Lake Hanbury, are to be seen several boulders of a black, porous earthy ore resembling somewhat varieties of the Negaunee manganiferous hematites; the same ore was found in place in a pit near by, and a large boulder of it near the center of S. ½ of N. W. ¼ of Sec. 6, and at other points. A hand specimen gave Mr. Jenney 56.44 per cent. of metallic iron, less than 16 per cent. of insoluble silicious matter, and nearly 1 per cent. of manganese. It is unlike the Breen mine hematite, and, in fact, unlike any Lake Superior ore I have seen. It is not improbable that workable deposits of it may exist, which being soft would not be likely to produce outcrops or boulders. I think it is well worth investigation. I have some reasons for supposing that this ore may be Silurian.

The next exposure of ore west of Sec. 6 on the south range is near the ¼ post between Sections 30 and 31, T. 40, R. 30. This ore is softer and more slaty than those already described, although belonging to the flag ore family. It is apparently more argillaceous, and outcrops conspicuously in several places west of the ¼ post, dipping at a high angle to the *north*, which would necessitate an *overturned dip* in order to harmonize with the hypothetical geological sections given on the map. The exposed bedding-planes are bright and specular, giving the ore the appearance of being richer than it really is. The ore exposed here may yield 45 per cent. in the furnace; see analysis No. 254, Chap. X.

From this locality we are led by a broad belt of very moderate magnetic attractions west by north for half a mile, to the iron ore exposed in the centre of S. E. ¼ of Sec. 25, T. 40, R. 31, where the Canal Company have done some trenching; the exposure here is not great, the ore being in a small ravine on high ground. It is intermediate in character between the flag ores noticed, but most like the last. I followed the attractions about one-eighth of a mile west, to a point where the hill seemed to be capped with Silurian sandstone.

I have now mentioned in order, beginning at the east, all the main exposures of ore in the south range of the South belt, which has already been referred to as the most regular and one of the most extensive deposits of ore in the Lake Superior region; whether it is absolutely continuous for the 16 miles intervening between the extreme exposures, can only be determined by expensive explorations or actual mining.

Passing from the south to the north range of the South belt, we have but one exposure to consider, that near the centre of N. ½ of Sec. 20, T. 40, R. 30. This is in a small ravine, down which, to the south and toward Lake Antoine, a rivulet has its course in wet weather; the water has uncovered a narrow surface of flag ore similar to that seen on the south side of Sec. 30, but less slaty. Iron boulders are strewn along the ravine for over 100 feet. This ore is a red oxide, but holds enough magnetite to give it a moderate magnetic power.

Ten miles northerly across the granite region, from the last mentioned locality on Sec. 20, bring us to the main deposit of ore in the **North belt**—that on Secs. 31 and 36, of T. 42, and Ranges 29 and 30. The great extent of this deposit, and its favorable situation for mining, have already been commented on; it only remains to notice the quality of the ore. It is more granular and massive than the flag ore of the south range, and, as a whole, contains less silica and more metallic iron. The natural exposures of ore in the ledge are greater, no digging or uncovering at all being required to reach a great quantity of the ore. The best ore to be seen outcropping, is just southeast of the centre of Sec. 31: the top of the cliff is here about 100 feet above the low ground at its base on south side; and for about one-third of this height is a ledge of ore, from the foot of which the surface slopes rapidly to the low ground, affording the best possible opportunity for mining. This outcrop was carefully examined for a distance of several hundred feet in length, and from the richest places to be found in it, 29 specimens of ore, of about one pound each, were collected, no two being broken from the same place. The specific gravity of these specimens was approximately determined on the ground, and was found to vary from 3.26 to 4.15, the mean of the 29 specimens being 3.71; this multiplied by 12, according to the empirical rule given under Explorations (Chap. VII.), gives 45 as the average percentage of the whole. An ore which actually analyzes 45 per cent. of metallic iron should yield say 47½ per cent. in the furnace, which is about what I consider this ledge of ore would work, if mined and sorted with ordinary care. Several ounces, chipped from five of the best hand specimens I could find, gave Dr. Wuth, of Pittsburg, 54.81 per cent. of metallic iron (See Analysis No. 98, Chap. X.). Separate analyses of ten hand specimens, selected from same locality by Prof. Pumpelly and Dr. Credner, gave Dr. Chandler from 49 to 64 per cent. of metallic iron, the average being 53.74 per cent. If this higher grade can be found in workable quantities (which is probable), then we should have a 55 per cent. ore, which, considering its granular and semi-porous nature, and the fact of its being a red oxide, would indicate an ore not difficult to reduce, and one which would sell in the present market.

No boulders were observed in this vicinity which would indicate a richer ore than the above of the red oxide variety, and no magnetic attractions were observed which would suggest a workable deposit of magnetic ore, although all the ores of this region are slightly magnetic. As hematite ores do not outcrop, and as no explorations have been directed to finding such ores, nothing can be said regarding them. My impressions are that they will be found on Secs. 31, 32, or 36 of the North belt.

The Felch mountain ore was fully described when considering the lower quartzite. It is totally unlike either of the preceding varieties, and more closely resembles the "mixed ore" which accompanies the rich specular ores of the Marquette region. The laminæ of ore are very rich, analyzing from 63 to 67 per cent. of metallic iron; but the large admixture of quartzite (at least three quarters of the whole) would render it unmerchantable at present. It is by its constitution particularly well adapted to *stamping* and *washing*, and on account of its proximity to several rapids and falls in the Sturgeon river, is well situated to be worked in this way, when the market drives miners to this means of production, as it will sooner or later.

3. Paint River District.

Too little is known about the remote Paint river district, in Towns 42 and 43, Ranges 32 and 33, to enable me to give anything of interest regarding its geological structure. The Huronian rocks are extensively developed there, and contain deposits of hard hematite ore. I had the opportunity to examine only two localities, at the Paint River Falls, Sec. 20, T. 43, R. 32, and on Sec. 13, T. 42, R. 33. The ores are identical, and unlike any in the more easterly part of the Menominee region, in being richer in iron, freer from silica, and in containing more water. (See Analysis 68, Chap. X.)

Explorations now in progress will determine many of the unsettled questions regarding the ores of the Menominee region, especially of the South belt. I regret that I cannot embody their results in this Report, and thus give it a completeness that in the present state of my information is impossible.

CHAPTER VI.

LAKE GOGEBIC AND MONTREAL RIVER IRON RANGE.

An examination of this but little known iron-field was not contemplated in the original plan of the survey. But, having had occasion in the line of my profession to make some explorations there, a few of the general results obtained will be given, with a view of aiding future explorations, and of calling attention to a comparatively unexplored region. The probability of there being early railroad communication through this country, connecting the existing system of roads of the Upper Peninsula with the North Pacific, Minnesota and Wisconsin systems, now radiating from the west end of Lake Superior, attaches additional interest to this most western portion of the Upper Peninsula.

The facts observed and conclusions formed are the joint work of Prof. Raphael Pumpelly and myself, and have, so far as they bear on the stratigraphical relations of the four great systems of rocks, been in substance given to the public, in the American Journal of Science and Arts, Vol. III., June, 1872. Many rock specimens, gathered by us are minutely described by Mr. Julien, in App. A, Vol. II.

The iron range under consideration may be regarded as the eastern prolongation of the Penokie range of Wisconsin, as well as the western extension of the Marquette series, the whole being Huronian. The position of the range is tolerably well defined by magnetic observations and notes on the U. S. land office plats; on these we find mention of iron and magnetic attractions on Secs. 7 and 8, T. 47, N., R. 45, W., as also in Secs. 13 and 14 of the Town west. The belt of Huronian rocks, as made out by us, extends nearly east and west, through the north part of T. 47, Ranges 44, 45, 46 and 47, crossing the Montreal River in Secs. 16 and 21, of the last-named Township. Going east, the range was lost before it reached Lake Gogebic.

The geological boundaries of this range are fortunately of the most unmistakable nature, and render a detailed description of its position unnecessary. (See Map I.)

On the north is the high, broad, irregular ridge, or series of ridges, constituting the South Copper Range, the rocks of which are greenish and brownish, massive and amygdaloidal copper-bearing traps, their bedding being exceedingly obscure, with occasional beds of sandstone and an imperfect conglomerate. The strike of these rocks, so far as it could be made out, was east and west, with a dip to the north at a high angle, thus *conforming* with the Huronian rocks underneath.

Against and over the copper series on the north, abut the horizontally bedded lower Silurian sandstones, which are beautifully exposed on the west branch of the Ontonagon river, in Sec. 23, T. 46, R. 41. These sandstones form the surface rock, and occupy the broad belt between the two copper ranges from the region we are describing to Keweenaw bay, but taper to a point before reaching the Montreal river, in going west.

On the south of the iron-bearing rocks are a series of granites, chloritic gneisses and obscure schists, which, except the latter, are unmistakably Laurentian in their lithological character, and are *non-conformably* overlaid by the Huronian rocks. The general structural relations of the four great systems here enumerated are shown in the accompanying diagram. As the *non-conformability*

Fig. 12. Sketch showing Geological Section—looking west, between Lake Gogebic and Montreal River (in part ideal).

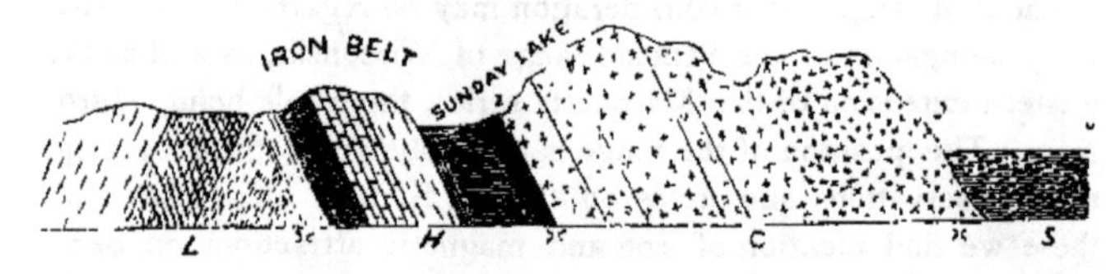

L. Laurentian rocks—gneiss, granite and schists, which are *non-conformably* overlaid by, H. Huronian—Clay slate, ferruginous and jasper schists, flag ores, quartzites and diorites, say 4,000 feet thick, which are *conformably* overlaid by, C. Copper-bearing rocks, chiefly greenish and brownish, massive and amygdaloidal traps, with occasional sandstones and conglomerate layers, which are *non-conformably* overlaid by, S. Lower Silurian sandstone, coarse quartz sandrock.

of the copper-bearing rocks and sandstones is doubted by some geologists, it should perhaps be stated that the actual contact was not seen. But the sandstones were observed lying horizontal, and affording not the slightest evidence of disturbance, within a few miles of highly-tilted copper rocks, which gave every evidence of having been elevated before the deposition of the sandstones. So far as my observation has extended, this rule is general; that is, no Lake Superior sandstone, which is unmistakably lower Silurian, has ever been found in any position other than nearly horizontal; and no rock which was unmistakably of the Copper series has been seen which was not considerably tilted. The fact that certain sandstones belonging to the copper series are very similar, if not lithologically identical with some of the lower Silurian sandstones, has helped to complicate this question. An interesting locality for study in this connection is the west fork of the Ontonagon river, just south of the Forest Copper Mine. I am not sure but that it affords an exception to the rule above stated, as at that point sandstones, *apparently* Silurian, dip south at an angle of 45°.

The best locality in which to study the character of the iron series in the West region, is on Black river and its tributaries, especially on the outlet of Sunday lake, T. 47, Ranges 45 and 46. Here will be found banded ferruginous jaspery schists, chloritic greenstones, brown ferruginous slates, black and gray banded silicious slates, silicious flag ores, several varieties of quartzites and clay slate. The whole series strike east and west, and dip north away from the granites and gneisses and under the copper rocks, at an angle of from 40 to 90°. Several varieties of the Huronian and Laurentian rocks of this vicinity have been examined by Mr. Julien, for descriptions of which see Appendix A, Vol. II. It will be observed from these descriptions that these rocks, although somewhat different from the Huronian series of the Marquette region, are still essentially the same; and I know of no good reason why merchantable ores may not be found amongst them. No ore, however, was found either in place, or in the form of boulders, which would pass for shipping ore in the Marquette region at this time. The absence of strong magnetic attractions renders it improbable that pure magnetic ores will be found here. The most encouraging indications observed pointed towards the existence of soft hematites, which may very likely be found of a quality and in quantity to pay for working. The best "show" observed was in the south ½ of the S. W. ¼, Sec. 18, T. 47, R. 46. It is on the north-easterly side of an east and west ridge, where there is a large exposure of highly ferruginous quartzite in places holding hand-specimens of hematite ore of fair quality. As this kind of ore never outcrops, on account of its soft, earthy character, and as we had no facilities for digging, nothing more definite was determined.

CHAPTER VII.

EXPLORATIONS (*Prospecting for Ore*).

1. HOW FAILURES HAVE OCCURRED, AND HOW TO AVOID THEM.

THE history of the development of a good many of our iron mining enterprises has been somewhat as follows:—The deposit is found, sometimes by accident, but often by systematic explorations made at the expense of corporations, firms, or individuals, by a class of men known as ***explorers***; who are acquainted with woodcraft, are often miners, and who always have some knowledge of structural geology, the different varieties of ore, and the use of the miner's compass. A boulder of ore, red soil in the roots of a fallen tree, the variation of the magnetic needle, the proximity of rocks supposed to belong to the iron range, and often the outcrop of the ore itself, determines where digging shall be commenced.

If the indications are promising, before many marks are made the land is secured, if not already owned or controlled by those interested in the explorations. If government land, it is "entered" at the land office at $1.25 per acre, or $2.50 if within the limits of some railroad grant. If the land is "second-hand," already entered, it may be bought outright, or if the price be regarded as too high, a refusal is often taken with the privilege of exploring.

If the discovery is on the land of some railroad or mining company, it usually cannot be bought. In this case, all trace of the work done is often concealed, secrecy enjoined on all concerned, and the explorer lives in the vain hope that he may sometime have the opportunity to buy the land, an expectation in which he usually dies, as large corporations do not often sell iron deposits for small prices, if at all. Instead of this unwise course, explorers often sell their information to the companies owning the land, which they can usually do at a fair price. Our supposed exploring party having secured the land, begin to dig test-pits and trenches openly and systematically. The solid ledge is usually soon found, which may prove to be some variety of iron ore, perhaps pure, but far more likely a "mixed ore" or lean flag ore, hence not merchantable.

Specimens (which I am sorry to say are apt to be the best that can be found) are sent in as *averages* of the deposit. Experts pronounce them shipping ore, and common talk asserts that So and So have a "good show" for a mine.

Soon the test-pits, trenches and drifts develop a workable width and length of what seems to the explorers to be merchantable ore. "Mixed with a little rock perhaps in places," but this occurs in most mines at the start. Experienced mining men visit the new deposit, examine it carefully, and assert honestly that "it looks better than did the Champion or Barnum locations when they first saw them."

The explorers select what they believe to be strictly *average* specimens of the ore (an impossible thing as will appear), which are sent to some distinguished chemist who reports, perhaps 65 per cent. of metallic iron, and only traces of sulphur and phosphorus, and expresses the opinion that the ore will *work well* in a blast furnace, and is identical with other well-known Lake Superior ores. This report, with the certificates of good practical mining men, and the opinion of some geologist who may have examined the locality, satisfies the owners that they have a workable deposit of "shipping ore."

Next in order, if it has not proceeded simultaneously with the above, is the organization of a company under the general mining law of Michigan,* which prescribes not to exceed 20,000 shares at $25 per share, par value. The property above mentioned is put into the new company at a moderate price; some prominent man of character and means is found to take the presidency of the company, his friends, with others, being "let in" on the "ground floor," and the None-such Iron Co. is organized and at work.

Building up a location is the next thing in order. To this end a contract is usually let to some French Canadian to build a dozen log houses for miners' families, a company's store, barn and shop. For this purpose the contractor lays out *fifteen different lines* on which to put the buildings, being governed in each instance by the ease with which the logs can be got together. In clearing for the foundations it is usual for the Frenchman to find a new deposit of ore

---

* App. I., Vol. II., contains an abstract of the Mining Laws of Michigan.

better than the one first found, to which a part of the mining force is at once transferred, the location of the buildings being changed so as to avoid the fragments which blasting has already begun to throw. The condition of affairs at the new location is at this period about as follows :—houses are going up rapidly, stripping is being pushed to the utmost, several "pairs" of Cornish men are sinking shafts or blasting off the "cap rock" so as to get at the ore. The contract for a first-class wagon road to connect with the State road has been let at $2 per rod, and a party of engineers are at work locating a branch railroad to the mine, and it is confidently predicted that a considerable amount of ore will be shipped from the mine that season.

About this time the president of the company—an old iron man, who has made a fortune by smelting 40 per cent. ores with anthracite coal in Eastern Pennsylvania—and a part of the board of directors visit the mine. One of the directors is an eminent lawyer who helped to "place" the property, another is a stockbroker who had made a fortune in Wall Street, a third is a railroad king, and another a successful whisky distiller. None but the president knew anything of iron before they came into the company. He is of course amazed at the richness of the ore, and tells the captain in charge of the mine truthfully, that he is throwing away as good ore as he ever used in his Pennsylvania furnaces. All collect and examine numerous specimens, which are submitted to the president and captain for their judgment as to richness. Nothing less than 50 per cent. is found, and the average is much higher. The lawyer who has fine muscular sense and a consciousness of its possession, soon discovers that he can judge accurately of the percentage of iron by handling the pieces of ore, and speedily becomes an authority with the broker and distiller. Specimens are hefted which contain 59, 61, 62½, 68, and finally one fine-grained fragment of steely ore, which, after careful manipulation in each hand, it is decided contains 75 per cent. of metallic iron. The captain unhesitatingly admits that to be richer than anything in the Jackson mine. Rock is found in several pits, but the captain explains that it is only greenstone which "caps" the ore, and proves by the magnetic needle which is "dead 90," that the ore is there. Being in a hurry he may not have faced the instrument exactly east and west.

Having spent one half-day in the examination of their property, and becoming satisfied that it is first-class and will prove a profitable investment for themselves and friends, the company leave, having first instructed their superintendent to bend all his energies to getting out ore, without reference to quality, cost, or future condition of the mine—though the whole is not, of course, directly expressed. On their way East, the president perhaps sells a thousand tons or more to some furnace man who is a stockholder in the new company, and telegraphs back to the superintendent to ship it at once.

The foregoing sketch contains the elements on which many Lake Superior iron mining enterprises have been organized, and at the start operated. It is needless to remark that many such undertakings result in utter failure. In the copper region the proportion of failures is far greater, and in oil, gold, and silver enterprises overwhelmingly so. The average human imagination becomes temporarily diseased when stimulated by the chances of possessing hidden mineral wealth. Iron, being the least valuable of the metals, has less of this influence than the others, but is not entirely free from it.

It may interest those who are disposed to identify themselves with Lake Superior iron mining enterprises (and I believe no equal investment has paid better in past time or promises better for the long future) to know the cause of failure in such enterprises. Classifying them carefully, I find that about two thirds of the disastrous enterprises were based on deposits of ore the *quality* of which was not merchantable: they were not rich enough in metallic iron. The extraordinary richness of Lake Superior ore is not generally known. I have reports from 40 furnace stacks in which these ores are smelted, which show that the average furnace yield of 250,000 tons of magnetic and specular ore for 1870 was 65 per cent.

The amount of high grade hard ore is so great that consumers can usually get all they require, and will not buy an inferior grade. For this reason experienced iron men from other regions have often been deceived; they had not a sufficient realization of this question of quality. Marquette ores—which were rich compared with what they were used to—could not be sold on account of their leanness. The soft hematite ores are not considered in this connection.

The remaining third of the failures have come from a lack of *quantity*, the quality of the ore being satisfactory. It follows, therefore, that the question of first importance in a new iron mining enterprise is to know—First, the ***average percentage*** of metallic iron in the deposit. What will the ore, ***mined in the usual way, yield on the average when smelted in the blast furnace?*** Second, approximately or relatively, ***how much is there of it?*** The failure to answer these questions correctly at the start has caused the loss of over one million dollars in the Marquette region during the last ten years, and the business is still going on. Experience is an expensive school, but is always full; no sooner does one class graduate than a new crop of "freshmen" take their places.

I believe it is not impossible nor even difficult to ascertain, at a moderate cost, the average amount of metallic iron, in any given deposit, sufficiently near for all practical purposes, and whether there is enough ore to pay for working.

It is the business of the explorer to find ore deposits and to determine approximately their extent and richness, thereby avoiding such failures as have been described above. This subject will now be considered under the several following heads:—

2. Prospecting and Woodcraft.

As considerable part of the iron exploration work now being carried on in the Lake Superior region involves camping out and a knowledge of woodcraft, some facts regarding this part of the business will not be amiss here, and are the more necessary because very little reliable information on this subject can be found in any book with which I am acquainted. There are no roads through large districts of country, which, in consequence, can only be reached by boats or walking; in either case a considerable part of the labor is ***packing***, which means transporting everything on the backs of men. This mode of transportation costs about $9 per ton per mile at the present time, which is twenty-seven times as much as it costs to move freight on wagon-roads; it is, therefore, important to carry only such articles as are needed. Many an exploration enterprise has practically failed because the chief energies of the party were expended in carrying supplies and material which were not needed, while necessary things were left behind. It is safe to say that two times out of three, even in the case of experienced explorers, supplies do not come out equal. The party will be out of pork and have an abundance of flour, or the converse; will travel in a leaky canoe for the want of a little pitch, or be barefooted because they had no awl; or ragged for want of thread; or suffering for food, where there is plenty of fish and game, because the salt had failed; or have their supplies wet for want of a piece of oilcloth. I have been in all these straits.

Organization of the Party.—Take the ordinary case of searching for mineral or timber, when an explorer and two men constitute the party. As packing is the heavy work, it is indispensable that all hands understand it. An average packer will carry 70 to 80 pounds and his blankets, but loads of 50 to 65 pounds are more common; across portages men often carry 100 pounds, and sometimes a barrel of flour weighing 200 pounds; but the packer who carries 70 pounds and his blankets, 10 to 15 miles per day, on a trail, or 5 to 10 miles through ordinary woods, has earned the $2.25 clear per day, which is the present average wages.

Next to packing, cooking is an indispensable qualification. No man is fit to go in the woods who cannot cook; and many a woodsman, with a frying-pan and two tin pails, will, over his camp-fire prepare a better cooked meal, and in less time, than can be produced in one-third of the kitchens of the country, with all the appliances that belong to modern housekeeping.

An ability to handle a canoe in rapid water is almost as indispensable as the others. Three men with a month's supplies will require a 16-foot canoe, which will weigh, when dry, about 125 pounds, and can easily be carried across a portage by one man; such a canoe will cost, in the Menominee waters, at this time, $15 to $30. The Bad water Indian village is the chief source of supply.

Next to packing, cooking, and canoeing, an ability to travel through the woods, and locate himself, by the United States Land Office plats, or maps made from them, aided by a pocket-compass, is essential. A man who possesses these qualifications is a woodsman, and has a calling which, if he is honest and intelligent, will be profitable in the Lake Superior region for a long time to come. If, in addition to these requirements, he is a judge of timber, and can keep simple accounts, write letters, and locate himself by the "40," then he is fit to lead a party, and become a "pine-looker," or "cruiser." If he add to this, a knowledge of the more common rocks and minerals, and an ability to make rough maps or plans of ground, then he is an explorer. Such men can command from $4 to $6 per day clear, with full time, and often an interest in what they find besides; or if they choose to examine lands (either timber or mineral) on their own account, they can usually sell their "notes" at so much per acre, subject to re-examination; or some one may purchase the land, paying the explorer for his services in an undivided interest in them. Notes of pine lands now sell readily at from 50 to 75 cents per acre.

Supplies.—Pork, flour and tea embrace all that is absolutely essential in the way of supplies, though sugar, beans and dried fruit are usually added; rice, oatmeal or wheat grits are also generally carried, and a little hard bread is convenient, to which a few pounds of cheese may be supplemented. Pickled ham, especially in summer, may take the place of part of the pork, and smoked beef is sometimes used.

The following table of supplies has been prepared with considerable care from actual experience:—

Rations Required for Three Men, One Month.

Rations.	Pounds.	Amount in percentage of the flour.*
Flour, biscuit or crackers, rice, grits or oatmeal, but at least ¾ self-raising flour (equal to 1⅓ lbs. per man, per day)................	125	1.
Extra heavy clear mess pork about ¾; pickled ham, say ¼................................	82	.650
Beans or peas................................	20	.160
Sugar (coffee A).............................	18	.140
Tea (good young Hyson).......................	3	.024
Dried apples.................................	10	.080
Cheese.......................................	4	.032
Salt...	2	.016
Pepper.......................................	¼	.002
Baking powder (Durkee's or Royal), if self-raising flour is not used..........................	2½	.020
Equal to $2\frac{85}{100}$ lbs. per man per day, or total......	266¾ lbs.	

* Supplies purchased in the proportions given in this column should come out even.

Equipment.—A shelter or bake-oven tent is preferable, although a closed A tent is often used in "fly time" (June and July): the former is more cheerful, healthier and warmer, because it lets the fire shine in. The style sketched will hold three men with supplies: it requires 12 yards of cotton drilling, 36 inches wide.

FIG. 13.
Explorers' bake-oven tent.

Two light explorer's axes, weighing with handles 2¾ lbs. for summer use and 5 lbs. for winter, each, are needed; if the exploration is for mineral, the backs or poles should be of steel. For three men a nest of two or three oval tin pails with covers, the largest holding 5 quarts, one frying-pan with socket handle, one 2 or 3 quart tin basin, one large spoon, one butcher or sheath knife, and a tin cup, plate, knife, fork and spoon to each man in the party, is all that is required.

If the party be large, a tin bake-oven will pay; it should be hinged so as to fold up. Canoes have already been mentioned: they are best for most kinds of river and lake service on account of their lightness, which makes them easy to portage, the ease with which they can be repaired with canvas and pitch (or resin and pork fat), and their suitableness for running rapids. But sometimes they

NOTE.—The stools shown in Fig. 13 do not belong to a camp outfit. They were introduced inadvertently by the engraver.

cannot be procured, and in low water are more liable to injury from rocks than are boats; a skiff 2½ fathoms long, pointed at both ends, with flaring sides, and made of ½ inch boards, is a good substitute. Each man in the party should have a pocket compass, water-proof match box, and sheath knife, and there should be at least one leather and one tin map case in each party. Should the exploration be for minerals, a dip compass, and at least one exploring pick ought to be added. A small shovel will pay in such a party, but is seldom carried. A dial compass for use in traveling when there is local attraction, or in discovering the same, is often advantageous. I have found a small horse-shoe magnet and a pocket lens useful. Every party going in the woods should be supplied with the best maps that can be procured (Farmer's are the best I have seen), and always with exact tracings of the U. S. Land Office plats or maps of the Townships they propose visiting; these plats can be obtained at any U. S. Land Office and cost, if they show variations of the needle and geological notes, about $2.25 each at Marquette. The following are the locations of all the U. S. Land Offices in Michigan, with names of officers.

U. S. Land Offices in Michigan.

District.	Office.	Register.	Receiver.
Detroit	Detroit	F. Morley	J. M. Farland.
East Saginaw	East Saginaw	Wm. R. Bates	A. A. Day.
Ionia	Ionia	J. H. Kidd	J. C. Jennings.
Traverse City	Traverse City	Morgan Bates	Perry Hanna.
Marquette	Marquette	A. Campbell	J. M. Wilkinson.

The explorer cannot too carefully study his maps; next to personal examination in the field they are his great original sources of information. The surveys of the Upper Peninsula, as is explained in Chapter I., were made with great care, and embrace topography, timber, soil and geology.

Under sundries which will be found useful in camp, may be mentioned: Soap and towels, thread and needles, buttons, awl, strong twine, some cotton cloth, a file to sharpen axes, a few wrought nails if a boat is used, some extra pairs of moose-skin moccasins (for summer), fish-lines and hooks, extra compass, resin or pitch, blank U. S. plats, and fly-nets or "fly-medicine," or both in "fly-time." A large, stout, water-proof, tin match-box, extra note-book and pencils, paper and envelopes, are desirable. A short, light, single-barreled shot-gun, with bore large enough to chamber buck-shot, may be carried to advantage after the middle of August.

Mode of Working.—Mineral explorations, and especially those for iron, will only be considered under this head. The leading idea is, of course, to make a systematic and exhaustive examination of the surface for the mineral sought: to this end all outcrops of rock of whatever kind, and all boulders must be examined for some "sign" or "show" of mineral. As has been elsewhere remarked, the upturned roots of trees afford one of the best sources of information: the beds of rapid streams, which usually contain boulders and often expose the solid ledge, should be carefully examined. Any indication at all favorable should be followed up by digging. Next in importance to this kind of search is the use of the magnetic needle in discovering local attractions due to iron-ore; it is safe to assume that more than one-half the iron in the Lake Superior iron region is sufficiently magnetic to produce appreciable variations in an ordinary compass; and as magnetic ore will attract the needle at the same distance with equal strength when covered by rock, earth, air or water, this instrument is of great service to the explorer. Its use is fully considered elsewhere, as well as the geological principles applicable to this kind of work.

An explorer should make a careful sketch or map of each section examined, on a scale of 4 inches to 1 mile: on such a scale "a 40" is one inch square. On this should be marked in their proper places all streams, lakes, swamps, hills, etc., and all outcrops, with a name or sign indicating the kind of rock; colored pencils are convenient for delineating the different varieties of rocks. Opposite each such sketch should be a full written description of the rocks and minerals found, as well as notes on timber and soil.

The accompanying sketch (Fig. 14) of Sec. 29, T. 50, R. 30, from the note-book of the late A. M. Brotherton, a perfectly honest and thoroughly competent explorer, will serve as an illustration. To it is appended a map of the same section (Fig. 15), from the U. S. Surveys, which shows, valuable as these surveys are, and reliable, so far as the section lines go, they often are considerably in error in their representations of the interior of sections.

Fig. 14. Sec. 29, T. 50, R. 30. Explorer's Sketch.

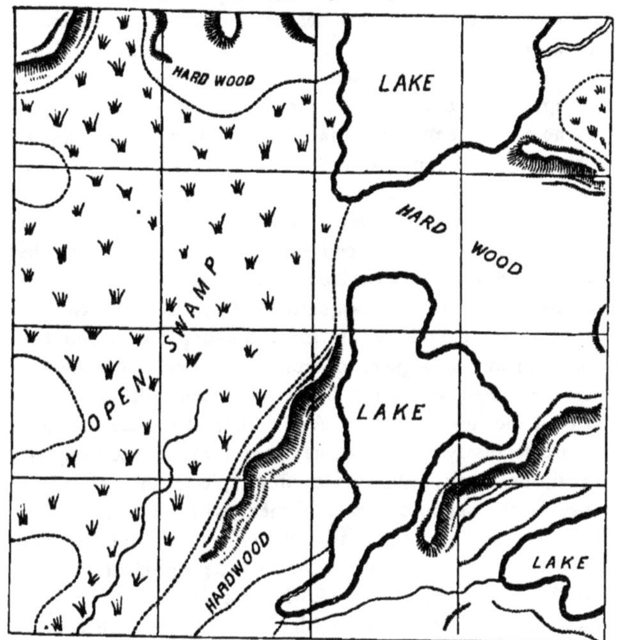

Fig. 15. Same Section, from U. S. Linear Surveys.

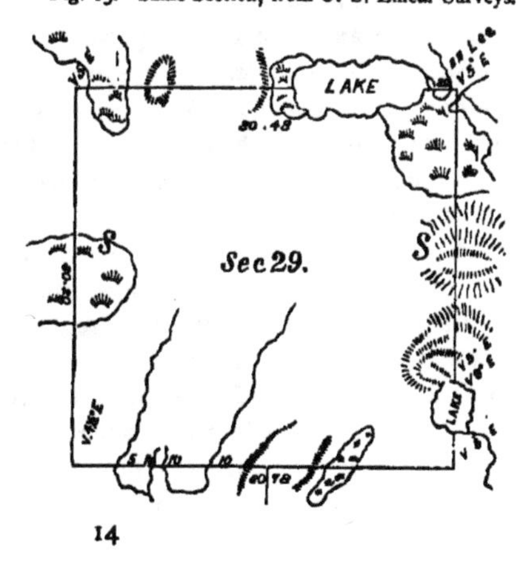

How to Recognize Iron Ores.—As a large majority of the explorers now employed are timber-hunters, they need not necessarily have a knowledge of minerals. I have, however, generally found these men more or less interested in rocks, and often very desirous of knowing how to determine the more common ores, so as to be able to note any they might find. To obtain a good knowledge, the study of a complete collection, or a residence at a mine, is indispensable. A few brief characteristics, only, will here be given, by which explorers may generally recognize iron ores in the woods.—First.—They are considerably *heavier* than any other rocks with which they are associated. Rich, magnetic, and specular ores, like those of the Marquette region, are nearly twice as heavy as the same bulk of the more common rocks, and five-sevenths as heavy as a piece of iron or steel of the same size. The soft hematites are much lighter, but are still appreciably heavier than the heaviest rock. As fine muscular sense and much practice are necessary in this business, the inexperienced explorer is advised, in every instance, to break pieces of rock of the same size as the supposed ore specimen, when, by lifting them together and changing from one hand to the other, the difference in weight will at once be felt if one of the specimens be iron ore. If the explorer is provided with a pair of balances, as is explained hereafter, he may determine, not only as to whether the substance is iron ore or not, but also approximately the percentage of metallic iron.—Second.—As to color, *magnetic* ores are black, and when pounded with the axe give a black powder, which will adhere to the axe or pick. Red *specular* ores are often bright and shining on their weathered surface, almost like polished steel; they give a red powder when pulverized, which does not adhere to the axe. Soft *hematite* ores are reddish and brownish in color, are generally porous, and often soft and earthy, in character; when pulverized they give a brownish and sometimes a yellowish powder, which does not adhere to iron or steel. These characteristics are possessed by none of the rocks of the Marquette region.—Third. Magnetic ores attract the needle of the compass strongly, often causing the north end to point south. Other ores and rocks do not attract it, but a little magnetic ore is often disseminated through rocks, especially greenstone, thereby producing more or less variation of the needle, which may not indicate valuable ores.

The rock which is oftenest mistaken for iron ore is Hornblende, and the related Diorites or Greenstones. These rocks are heavy and dark colored, and often contain enough magnetite to give them some influence on the needle. Many an explorer has carried heavy pieces of this rock many miles through the woods, only to throw them away in disgust on meeting some one who had, perhaps, only so much knowledge of ores, as it is expected these few facts will impart. Some have persisted in their folly, and bought lands on which experienced iron explorers could only find hornblendic rock. This rock differs from the ore, which it most resembles, in being *lighter*, and in giving a *light colored powder*, which does not adhere to iron or steel, as well as in other less important particulars, as may be seen by comparing the two, which should be done.

The text relating to the magnetism of rocks and use of the needle in finding ore might properly have been inserted here as a division under Exploration, of which subject it forms properly a part. But the amount of material which had been prepared on that subject, and other reasons, determined me to place it in a distinct chapter (VIII.), which follows.

3. Digging for Ore.

The exploration work above described is superficial, and will not usually determine whether a certain piece of land contains workable deposits of ore or not. Such examinations are usually made to determine whether lands are worth buying at government price, or as preliminary to a more thorough exploration. When we consider that soft hematite ores never outcrop, and that pure hard ores rarely do, it is evident that something more than looking over the surface is necessary. The excavations of earth and rock required in an exhaustive exploration of a piece of land are mining operations, and will be considered in another chapter. Only a few points will be presented here which bear especially on work of this kind.

This work is simply sinking test-pits and shafts, and opening trenches (costeaning) and drifts to expose the solid ledge. It rarely happens that such work need be prosecuted into the solid ledge. As has been before remarked, if there be pure ore at the locality, i will be almost certain to come to the surface of the ledge somewhere, and will there be found by digging through the earth. This may not always be the case, but it is safe to say that, as a rule, nine-tenths of all the money to be expended in exploring at any given locality, had best be expended in earth excavation.

There is a great deal of vague talk among miners and explorers of the Marquette region about "*cap rock;*" one would get the impression, from much that is said on this subject, that pure ores were always overlaid by rock. The fact is, however, that there are very few workable deposits of ore but what come to the surface, or, at least, connect with those that do. I should distrust any locality where "cap rocks" prevailed to any great extent; our iron-ore deposits are comparatively thin beds, which sit on edge, and come to the surface without wearing any "cap."* There are places, however, where the solid ledge has to be penetrated; when this is necessary, I think it had usually best be done by drilling. By means of hand drills, holes can be sunk 22 feet, and by means of the appliances used in sinking oil-wells to any required depth; an experienced miner will have little difficulty in judging of the material passed through by the drill mud, and if there is any question as to richness, it can easily be settled by an approximate analysis which will be described hereafter. The diamond drill gives the most valuable results, and has been used to some extent in this region, and still more extensively in the Lake Champlain region.

Exploring excavations should always be done by contract; a large amount of "test-pitting" has been done in the Marquette region at seventy-five cents per foot in depth for a 4 × 6 shaft, the miner being paid only for such shafts as were "bottomed," *i. e.*, the solid ledge reached and uncovered, whatever the depth or difficulties. For drifts 3 × 6 which bared the ledge, $1.50 was paid, and for open trenches a price proportionate to depth and width. Good miners can find themselves and make good wages at these prices in much of the ground in the Marquette region. Pits are sometimes sunk 35 feet, but the average depth does not exceed 12 feet. Mr. Colwell sunk 67 feet through sand on Section 24-47-28. Large

* In the Menominee region true "cap rocks" are found in the horizontal sandstones which overlie some of the ore, see page 68.

boulders and water are the difficulties usually encountered; beyond 10 feet a windlass is necessary. A portable forge and mass of iron for an anvil are desirable, but picks can very well be heated in a hard-wood camp-fire and sharpened on a rock.

With regard to the significance of the material passed through, but one remark will be made; mixed drift, that is, large and small boulders, sand, clay, etc., is usually not very deep, 40 feet being the greatest depth I have observed, the average being less than 10 feet. Sand with no boulders is usually deeper and sometimes very deep.

4. Quality and Quantity.—Sampling.—Approximate Analysis.

Up to this point we have considered chiefly the question of finding ore regardless of *quality* and *quantity*. These are, after all, the vital questions, and their importance is rendered still more conspicuous by the statement, that there is at least twenty times as much ore in the Lake Superior region that is worthless from a lack of metallic iron, as there is of merchantable ore, according to the present standard for shipment; and further, it is easy to find specimens of pure ore in almost any body of worthless ore.

To determine approximately the average percentage of metallic iron, proceed as follows:—Open two or more trenches or drifts entirely across such portion of the ore formation as is regarded fit to work. In the region we are considering, the ores usually dip at a high angle, so that the edges of the beds or strata are exposed by such cross cuts; free the solid ledge from all earth and loose material; then, with a heavy hammer, break off small fragments *every two inches across the entire bed*, without reference to whether the pieces are ore or rock. Wash all of those pieces, break them all into fragments of the size of grains of wheat, mix them up thoroughly, send a tea-cupful to a reliable chemist, and his return will be the practical average of metallic iron in the whole bed from which the pieces came.

Of course, in mining, the ore is sorted, so that we should expect to get a somewhat better yield from working the ore, than that found as above, but it is not wise to count much on this. If, after trying, say half a dozen cross cuts in this way, an average yield of fifty per cent. (50%) of metallic iron is not found, the deposit is doubtful; if less than forty per cent. (40%) it is of no value in the present market, should the ore be specular or magnetic. Nineteen times out of twenty, such *mechanical averages*, when honestly taken, would show a yield of less than forty per cent. (40%.)

The plan above described is somewhat expensive and consumes time, which is an important element where one is maintaining an exploring party in the woods. A method which can be used on the ground, and which will give results, according to my experience, within a few per cent. of the above in the case of the silicious or quartzose hard ores (the kind usually found), is the following:—Provide an ordinary swing balance which will sustain at least two pounds, and weights, the smallest of which should not exceed five grains, the whole costing less than $5. Break up numerous hand specimens across the ore deposits as before, wash and dry them. Suspend each in turn by a fine fish-line and weigh it in the air, afterwards weigh it when immersed in water. Divide the weight in air by the difference between the weight in air and the weight in water. The quotient will be the *specific gravity* of the specimen, and will range from 3.17 for very lean ores to 5.13 for very rich compact ores. The specific gravity so obtained, multiplied by thirteen, if the ore be rich (*i.e.*, above 55%), and by twelve, if the ore be lean (*i.e.*, from 40 to 55%), will give the approximate percentage of metallic iron in the specimen.

The mean of a large number of determinations, made with specimens selected promiscuously from the deposit, will give a close approximation to the average percentage of metallic iron in the bed. According to my experience, the error will fall within five per cent., which is nearer the truth than any man can determine by simple inspection. It must be borne in mind that this purely empirical rule applies only to Lake Superior *magnetic and specular* ores, and only to such as contain some form of quartz as gangue, which is true of nearly all. The numbers 12 and 13, given above, as multipliers, were derived from numerous analyses and specific gravity determinations made by Dr. C. F. Chandler, of New York, and J. B. Britton, Esq., of Philadelphia. This plan is not offered as a substitute for chemical analysis, but I believe will often prove useful in the woods, and may sometimes help in deciding whether it is worth while to have an analysis made. As has been before stated, unless the deposit is proven by analysis to contain an average of 50% of metallic iron, if specular or magnetic, and not less than 40%, if soft hematite, it is of doubtful value at the present time.

It would seem as if sufficient experience should enable us to judge of the quality of an ore at sight, or at least enable us to select an average specimen for analysis, without the laborious plan above described; but this is not the case, as is well known to those who have had experience in iron ores. It may be stated as an economic and psychological axiom, *that no man, however honest or skilled, can, on his judgment alone, select an average specimen of ore from a deposit; he will always choose a richer specimen than the average.* This would, of course, be very difficult from the technical stand-point, on account of the delicacy of muscle and skill of sight required; but the greater and insurmountable difficulty is in the human mind. We cannot help feeling that at a new opening there must be somewhere under our feet, or near by, better ore than we can see and the specimen selected is designed to be rather what we suppose, believe or hope the deposit to be, than an average of what we actually see and feel. I have numerous facts under this head, and am able to give an approximate mathematical expression to this form of human hopefulness. In eleven instances the difference between the *average by judgment*, and the *mechanical average* obtained as above described, varied from 6 to 24 per cent., averaging 11; the mechanical average being least in every instance; in each case I had reason to have confidence in the honesty and skill of the parties. It does not seem possible that such errors in average could exist, but they are constantly made, and will continue to be as long as iron ores and human minds are constituted on the present plan.

One of the fallacies which have caused innumerable disappointments in iron mining is the belief, almost universal, that ores grow richer in depth. This may be true of certain ores in some regions, but it is not true of the iron ores here being considered. They are just as good on top as in any part of their extent, and it may be stated as an invariable rule that if there be any good ore in a given deposit which is available for mining, it will somewhere come to the surface, except the earth covering in the Marquette region and the sandstone in the Menominee, which of course have to be removed when found. Hence a sufficient number of earth test pits, trenches and drifts will usually find it, if it exists, without penetrating the rock. I do not mean to say that a deposit of ore may not grow thicker in depth; they often present this feature, and on the other hand sometimes grow thinner, and wedge out entirely. As has been before stated, by far the larger part of the money available for the exploration of any given locality should be spent in earth work.

While it is not difficult to determine with sufficient accuracy for all practical purposes the quality of a deposit of iron ore, as has been above shown, it is often impossible within a reasonable cost, to form so reliable a judgment as to the *quantity*. But a sufficient amount of judicious exploration will usually settle the all-important question as to whether the deposit is large enough to warrant development as a mine, future operations alone determining whether it will prove a great or small one. The method of doing this is obvious; many test-pits and trenches must be dug and drifts made where the earth is deep, the ledge of ore being thus laid bare in as many places as possible. No one engaged in making an exhaustive exploration of an iron-ore property should neglect the advantages of deep drill-holes; these can be sunk 20 feet with the ordinary drills employed at the mines. An inspection of the mud, and especially an analysis of an average of it, will prove of great value.

The annular diamond drill was introduced in 1870, at the Lake Superior mine, and gave very satisfactory results; the core gives almost as good an idea of the nature of the rock passed through as a shaft, and the cost is far less,—about $5 per foot. But being propelled by a steam engine, it is only adapted to work near communications; it cannot be taken into the woods.

In the case of magnetic ores great assistance in determining the extent and position of the bed can be derived from a proper use of the magnetic needle, which subject is considered in the following chapter. Attention will, in this connection, only be directed to one important fact; *worthless ores often attract the needle just as strongly as merchantable ones.* Now, as there are many times more lean magnetic ores than rich, it follows that a variation or dip of the needle may not, probably does not, signify a workable deposit.

CHAPTER VIII.

MAGNETISM OF ROCKS, AND USE OF THE MAGNETIC NEEDLE IN EXPLORING FOR ORE.*

1. Elementary Principles.

A FEW of the elementary principles of the science of magnetism, made use of in the following investigations, will first be given.

Magnetite, or magnetic iron ore, contains, when pure, about 72 per cent. of iron and 28 per cent. of oxygen. The unmixed mineral is black, or blackish in mass and streak, has a specific gravity of 4.9 to 5.2, and hardness of 5.5 to 6.5, which is somewhat less than that of quartz ; its crystals are usually octahedrous, and in the massive state it is often granular, and sometimes friable. Magnetite is one of the most abundant ores of iron in the United States, and, besides occurring in workable masses, is often disseminated through certain rocks, in grains, or in bunches and thin seams or laminæ, thus constituting what will be called "magnetic rocks" in this paper.

Its home is in the oldest rocks:—the primary (azoic, eozoic or archæan), as they have been successively termed. When it occurs in younger rocks, its origin can generally be traced to local metamorphism. The characteristic property of this mineral is its *magnetism*, with reference to which it is sometimes called *lodestone*. When brought near to pieces of iron or steel it often manifests an attraction for them, as it always does for another magnet. It hence causes the magnetic needle to deviate from its normal direction when brought near it. This property does not belong, in any marked extent, to any other mineral, and is the one which we have here chiefly to consider.

A piece of magnetite, broken from its parent bed, and suspended

* A part of this paper was read before the American Philosophical Society, Philadelphia, and published.

by a thread, will take a position, as near as the mode of suspension will permit, corresponding with its original one. If a north and south line be marked on a specimen thus suspended, it would rudely and imperfectly answer the purpose of the magnetic needle; if with this piece of magnetite we rub, in a certain way, a slender bar of hardened steel, it in turn becomes magnetic, and, if properly mounted, will point north and south, and constitute a compass. Mounted in another way, so as to admit of vertical motion, the magnetic needle will, while pointing north, incline downward at an angle of about 76° at Marquette. This "dip," as it is called, increases to the north and decreases to the south.

Two magnetic needles made in this way present these phenomena: their north poles or south poles repel each other, while the north pole of one will attract the south pole of the other, and conversely. The same is, of course, true of two pieces of magnetite, or of a piece of magnetite and a magnetic needle; *opposite poles attract, and similar poles repel.* This property is termed *polarity.* From this it appears that the north magnetic pole of the earth must, in the light of the science of magnetism, be regarded as a south pole, because it attracts the north end of the magnetic needle. The *poles* of any magnet are understood to be those points opposite each other, and near its surface, where the attractive and repulsive power may be supposed to be concentrated. Any magnet, natural or artificial, exerts its influence or sends out its rays in every direction, like a luminous point. The limit of this influence may be designated as the *sphere of its attraction.* A magnetic needle within this sphere, and uninfluenced by other force, would point directly to the centre of the sphere or focus of attraction. The force which holds it in this direction varies inversely as the square of the distance from the centre; hence practically (on account of this rapid diminution of power) we soon get beyond the influence of even a great natural magnet, like a hill of magnetic ore.

All the properties above designated, and numerous others not necessary to our purpose, appertain in general to a mountain of magnetic ore or rock, as well as to the delicate needle of a miniature compass. It is therefore evident that the magnetic needle should assist in determining the position and magnitude of rock formations containing magnetite. It has been extensively used in numerous places in finding iron ore, and to a far less extent, if practically at all, in this country, by field geologists, in determining the geographical extent, and, in part, lithological character of formations containing too little magnetite to give them commercial value, and which have already been designated *magnetic rocks.* The fact that all substances usually encountered in magnetical observations are transparent to the magnetic rays, or permeable by them, enables us to be certain of the existence of magnetic rocks or ores, though they be covered with water, earth, or non-magnetic rocks, to the depth of many feet, or even fathoms. *A given magnetic force affects the needle just as much through one hundred feet of granite as through the same distance of the atmosphere.* Dr. Scoresby gave a fine illustration of this fact, and an important application of the science of magnetism, by measuring, with great precision, 126 feet through solid rock, by observing the deviations in a needle, caused by an artificial magnet.

The earth itself may be regarded as a great magnet, which has the power of inducing this force (all magnets have a similar power) in masses of magnetite, and in all forms of iron and steel. We may suppose the force we have described above, as existing in the magnetic rocks and artificial magnets, to have been derived from the earth. An unmagnetized mass of steel or iron always manifests polarity induced by the earth, the upper or southerly portion being the *south pole,* and the lower or northerly end the *north pole,* in accordance with the law already stated. If the mass of iron or steel be elongated in form and made to stand nearly vertical, or to lie nearly in the plane of the meridian, this force is more manifest. To illustrate :—The upper end of all cast-iron lamp-posts attracts the north end of the needle, and the lower end the south. The magnetism thus induced in the wrought-iron pipes, lining the so-called magnetic wells of Michigan, would probably explain all the phenomena actually observed there. The law is, briefly: *the upper part of every mass* (of whatever form and size) *of iron, steel or magnetite, is a south pole, and the lower part a north pole.* This is, of course, true of magnetic rocks ; hence almost universally the north end of the needle is attracted by such rocks, because it is the south pole of the rock which is uppermost and nearest. South pole or *negative* attractions, which are occasionally observed, come usually from faults or other divisional planes in the rocks ; opposite poles being produced on opposite sides of such breaks which sever the mass; a precisely similar phenomenon can often be observed on opposite sides of the joints in railroad tracks.

From this cause several natural magnets are often encountered in a short distance; and a needle, passing in a few feet from the sphere of the attraction of one of them, will turn round and point toward the pole of a neighboring mass which more strongly attracts it. Hence, in magnetic surveys, we have not the simple focal point first considered to deal with, but often several local centres of attraction, positive and negative, in addition to the directive force of the earth, all influencing the needle at the same time. The recent investigations in the use of "magnetism in testing iron for flaws" would undoubtedly aid in the study of the effect of faults on the magnetism of rocks. See Engineering (London), 1867, p. 550, and 1868, pp. 297 and 440. The magnetism of iron ships should also possess interest in the same connection.

The *direction* which a magnetic needle takes (allowing it to have universal motion), under the circumstances supposed above, and the *power* with which it holds to that direction, must be the mechanical resultant of all the forces acting on it. It cannot point in two directions at the same time, hence stands between, inclining to the greater force. The principle of the parallelogram of forces makes it easy to determine the direction of this resultant, and to measure with mathematical precision the power which urges it. To do this we must know the direction and intensity of all the forces.

As an example, suppose a magnetic needle which, uninfluenced by other force than the earth's attraction, points due north and vibrates 10 times in one minute, to be placed due east from a south pole in a magnetic rock; and that, in this position, the earth's directing force be exactly neutralized by an artificial magnet, placed south of the needle,—it is evident that a needle so situated will point due west, urged by the local force alone, and that its vibrations will be solely due to this force. Suppose, for example, these vibrations to number 20 in one minute, or twice as many as were due to the earth's force. Now remove the artificial magnet; what will be the direction of the needle, and what number of vibrations will it give, urged by the local and cosmical forces?

It is a law of magnetism that the force urging a magnetic needle is proportional to the square of the number of vibrations made in a given time; $10^2 = 100$ and $20^2 = 400$, hence the local force is four times as great as the earth's. Lay off in Fig. 16 the line x N due north, making it equal 100 on some chosen scale: lay off the line x W due west, making it equal by the same scale 400; complete the parallelogram by drawing the lines N y and W y parallel with the first lines. Draw the diagonal x y, it will be the resultant

Fig. 16.

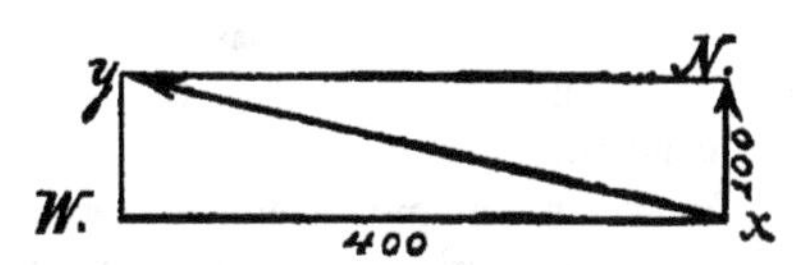

sought. Applying the protractor and scale we find its course to be N. 75° 53 W., and length to be 412.31, the square root of which is 20⅓, which would be the number of vibrations.

Suppose that in another locality the same needle pointed N. 45° E. and vibrated 14¼ times in one minute, what would be the direction and intensity of the local force? In Fig. 17 lay off the line x y N. 45° E., its length equal to the square of the number of vibrations = 200; complete the parallelogram as before. It is evident that the line x E represents the direction and intensity of the local force, which in this case is due east, and has a power just equal to that of the earth. Unfortunately the simple cases here presented

Fig. 17.

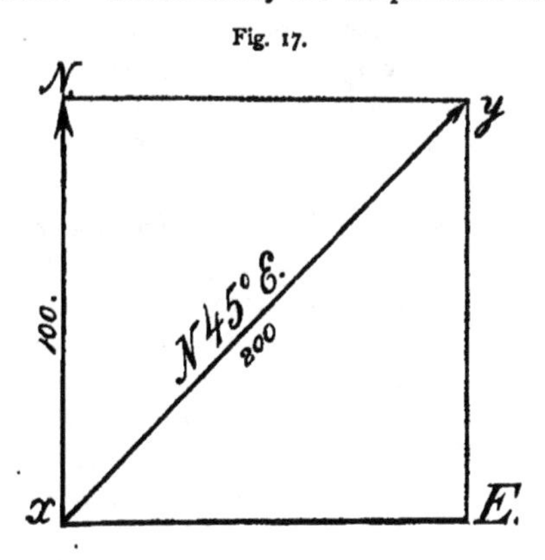

seldom occur,—usually two or more local forces act on the needle at the same time.

In a similar manner any number of forces acting in as many different directions can be resolved. It follows that a magnetic needle, influenced by the earth's force, can never point directly toward a local magnetic pole, but will, with two exceptions which need not be named, always incline to point to the north of it.

It is evident that the degree of magnetism possessed by a needle, while it makes no difference with its direction, will affect the number of vibrations. Take the needle in the last case, and suppose it more highly charged; it will still point N. 45° E., but its vibrations will be increased in number just in proportion to the additional power imparted. Hence, in determining *absolute* terrestrial or local intensity, a standard for comparison is necessary; but this is not required in the work under consideration.

2. Magnetic Instruments—Dip Compass.

As the instruments employed in these observations are quite different from those used in Terrestrial Magnetism, which are described in the works on this science, a brief account of them will be given.

The Dip or Miner's Compass is a circular brass box, a common form being 3¾ inches in diameter, and ¾ inch thick, having a circular glass on each side, which permits a perfect view of the needle. The needle is 2⅞ inches long, weighs 13¾ grains, and is counterpoised so as to stand horizontal where there is no local attraction, the needle being permitted to *swing in a north and south vertical plane*, which is the position in which it is ordinarily used. The axis of the needle is of hard steel, its points resting loosely in conical cavities in agates, fixed in two arms projecting from the sides. Outside is a ring for supporting the instrument when observations are made, so placed that the weight of the suspended instrument brings the zero line of the graduated circle to a horizontal position. Although designed to be used chiefly for determining dips or inclinations of the needle due to local influences, it answers passably well for taking magnetic bearings when laid on its side, and is frequently used in this way in rough work.

As there is usually no means of throwing this needle off its points of support, the wear is great, and the instrument is often out of order. A person going out of the way of shops where repairs can be made, would do well to take two, and then have the means at hand for making ordinary repairs. These compasses generally possess each an individuality of its own, and one must know his instrument before placing much confidence in his results: they will seldom reverse, 30° difference in the two readings being not infrequent. A New Jersey iron explorer informed me that his Dip Compass always indicated 90° when faced west, and the true dip due to local attraction when faced east. He is said to have used one position in buying and the other in selling iron lands very successfully.

My compass was made by Messrs. W. & L. E. Gurley, of Troy, N. Y. I have since seen one made by H. W. Hunter, of N. Y., which promises well. A reliable dip compass is a desideratum.

This is exclusively a hand instrument, and has no support; nearly all the magnetic observations recorded in this paper were made on instruments held in the hand. This may seem rude and unscientific to precise observers of physical phenomena; but it was found by trial that the average error by this mode of observation was less than 3°, which was comparatively small in localities where changing the position of the instrument only a few feet often made 50° difference in the direction of the needle, and deviations of 180° from the normal direction were common. It is not necessary to observe the direction of the wind to the degree to construct a useful theory of storms. Had the accurate instruments and precise methods of terrestrial magnetism been employed, not more than 50 stations could have been occupied with the time at my disposal, while with my rude methods over 1,000 stations were observed at.

The miner's compass above described is now in very general use in the magnetic iron-ore regions of the United States. The object here sought is to endeavor to point out new and perhaps better modes of using that instrument in finding iron ore, and incidentally to ascertain if it has any place in general geological field work. I have long believed that the magnetic needle can be so used as to give more definite information regarding magnetic ores and rocks than has yet been done to my knowledge. I did some rude and incompleted work in this field, at the Ringwood Iron Mines and elsewhere in New Jersey and Southern New York, the results of which are in part published in Prof. Cook's Report on the Geology of New Jersey. The observations of Prof. Cook and Dr. Kitchell on the magnetism of the iron ores of New Jersey, and the use of the magnetic needle in finding them, possess interest; see pp. 532–538 of their report. The map of the Ringwood Iron Mines, accompanying that report, exhibits a part of my own observations above referred to.

The idea of applying Magnetic Science to Geology is not at all new; years ago Bischoff, after citing numerous observations that had been made in various parts of the world by different observers in regard to the influence of mountains on the magnetic needle, concluded as follows: "Assuming that it is magnetic ore alone, either as masses or disseminated through the rocks, to which the magnetic influences are to be ascribed—and in my opinion this is quite unquestionable—it would seem that magnetic observations instituted with the same degree of care as those made by Reich, would be well adapted for the discovery of hidden beds of magnetic iron ore. Such observations might therefore prove eminently serviceable to the iron industry. Certainly it would be requisite first to ascertain whether mountain masses containing only disseminated magnetic iron ore, but extending over a considerable surface, would not produce as great an effect as beds of magnetic iron ore. Sabine's observations do not appear to favor this; but, however this may be, the magnetic needle indicates the presence of magnetic iron ore where it cannot be recognized mineralogically, and demonstrates the very general distribution of this mineral."

My mode of observing was as follows:—To determine "variations" east or west,* the bearings of a standard line were taken as in ordinary surveys. Sometimes a solar compass was used, but oftener a pocket compass. The variations as shown by the miner's compass, termed "dips," were observed on this compass held in the hand generally in the plane of the meridian, hence the instrument would face east and west. Sometimes observations were made with the compass held at right angles with this position; that is, facing north and south. The instrument was always held in the hand and levelled by its own weight.

The *intensity* of the magnetic force for the three positions of the compass above designated, was measured by the number of vibrations† made by the needle in a unit of time, usually taken at ¼ of a

* Declination, or the cosmical deviation of the needle from the true meridian, is not here considered.

† Half-vibrations would be the proper term, as the time from one point of rest to the next was counted and not the complete vibration.

minute. The vibrations varied from 0 to 60 in this time, 6 being the normal for my compass, due to the earth's influence. No attempt was made to eliminate the earth's attraction by neutralizing it with a magnet when the observation was made, or by computation. Of course, when the compass faced north or south, this was partially accomplished, because the earth's attraction would then be nearly in the direction of the axis of the needle. It must be borne in mind that the great amount of friction in this form of compass renders the number of vibrations only a rude approximation to the number which would be indicated by a delicately mounted needle.

The short needle of an ordinary pocket or dip compass, if in good order, will vibrate quickly and for some time where there is no local attraction. This motion is sometimes termed "working," and such normal "working," due simply to the earth's attraction, has often been mistaken by inexperienced persons for an indication of ore.

There is no better instrument for observing variations accurately than Burt's Solar Compass; but it is too heavy for explorers' use. I have found a convenient substitute for rough observations in the Pocket Dial Compass, which, used with a watch indicating local time, is rapid and sufficiently precise. This instrument, or an ordinary portable sundial, can also be used for running lines where there is local attraction; for rough work I have used it instead of the Solar Compass.

I hoped to have made some observations with properly constructed instruments, such as are used in determining the elements of terrestrial magnetism, in order to institute a comparison between accurate results and my own rude work; but the nature of such investigations requires more time than I have thus far had at my disposal. Fortunately Dr. John Locke made complete magnetic observations at several points in the Marquette Iron Region, which are recorded in "Smithsonian Contributions to Knowledge," vol. 3, pp. 25–27. One station was over magnetic rocks in Section 18, Town 47 north, Range 26 west, the geology of which he thus describes: "A loadstone in place broken into sharp angular fragments; here were two poles, 17.67 feet apart, one attracting the north, the other the south pole of the needle." Dr. Locke found the dip to be 42 deg. 53 min., when it should have been about 76 deg. The duration of 500 vibrations was 822 sec., when it should have been about 1,500 sec., and the calculated horizontal intensity was more than four times the normal force computed for that station. If Dr. Locke had occupied 500 stations on that section of land, he would have obtained different results at each, often differing more from each other than the foregoing do from the normal forces.

These observations, like all recorded ones that have come under my notice, have had *terrestrial magnetism* as their chief object; therefore the observers have avoided the very localities which to the geologist and explorer possess the greatest interest—those where local magnetic attractions exist. Dr. Locke calls attention to the importance of magnetic science to the geologist, and gives many interesting isolated facts bearing on the subject, particularly regarding the existence of magnetite in volcanic rocks, where it usually occurs.

Before dismissing the subject of instruments suited to magnetic surveys, I will call attention to a patent mariner's compass made by E. S. Ritchie, Esq., of Boston, in which the needle is entirely supported by a liquid having the same specific gravity, thus giving it universal motion. A needle so mounted and having the earth's attraction neutralized by a magnet, should point directly towards a local magnetic pole when brought within its influence, thus accomplishing with one observation and no calculations what requires at least two with the ordinary compass. For intensity Mr. Ritchie suggested the following mode:—Time the needle from the instant of its being let off at 90 deg. to its passing the resting point. I am of the opinion that a valuable instrument for miners and explorers could be made on Mr. Ritchie's plan.

A modification of the ordinary compass has been made which accomplishes the same thing in part. The agate support is fitted to the needle by a sort of universal joint, which gives the needle a vertical range through half a quadrant in addition to its horizontal motion. The only one I ever saw was made from the design of the late Wm. J. Amsden, Esq., of Scranton, Pa., who made some valuable magnetic surveys.* A pocket compass on a similar idea has lately been patented. A somewhat similar instrument has, I understand, been used for a long time in Sweden and Norway. On the same principle the ordinary surveyor's compass indicates dips rudely. At the west quarter post of Section 7, Town 46

* Messrs. Gurley now make a dip compass which gives the needle limited lateral range.

north, Range 29 west, being on the east side of Republic Mountain, I find marked on the U. S. Survey plat: "End of needle dips ¼ inch, variation 62 deg. west."

C. F. Varley, Esq., the English Electrician, suggested to me that a portable electro-magnetic apparatus could be constructed, with which might be determined the direction and distance to the pole of a magnetic rock by some simple observations and computations. An instrument of this kind would have considerable value in connection with magnetic needles, especially where the magnetic ore or rock was covered with considerable thickness of other material. In 1867 Mr. Varley, with a view to detecting electric currents, if any existed, made some observations both in the copper and iron-bearing rocks of Lake Superior; he found such currents in the mines of native copper, but none in the iron mines. The instruments employed were rude, having been extemporized on the spot. I do not know whether he has published anything on this subject.

Professor Joseph Henry has suggested in a letter that it is "highly probable that the abnormal variations of the magnetic elements in our iron ores are due to *electro-magnetic* action rather than to magnetic."

3. Geological Sketch of the Magnetic Rocks.

In order to make the perusal of this subject to a certain extent independent of the remainder of this report,* a few facts regarding the geological position and lithological character of the magnetic rocks of the Marquette region will here be repeated, the subject having been more fully considered elsewhere.

Rocks of the four oldest geological epochs yet made out on this continent are represented on the Upper Peninsula of Michigan; two belonging to the Azoic, one to the Lower Silurian, and one between these, of questioned age. The equivalency of these with the Canadian series has not been fully established, but the nomenclature of the Canadian geologists will be employed provisionally.

The Laurentian of the Upper Peninsula is like that of Canada in being largely made up of granitic-gneisses, but differs in containing no limestone so far as I have seen, and little, I may say practically

* Many persons have asked for copies of this chapter who do not expect to get the whole Report.

no iron ore, and very little disseminated magnetite. Next above the Laurentian, and resting on it non-conformably, are the Huronian or iron-bearing rocks; these are also called by the Canadian geologists "the lower copper-bearing series." This series comprise several plainly stratified beds of iron ore and ferruginous rock, varying in the percentage of metallic iron from 15 to 67 per cent., interstratified with greenish tough rocks, in which the bedding is obscure, which appear to be more or less altered diorites, together with quartzites (which pass into marble), clay slates, mica schists, and various obscure magnesian schists. The maximum thickness of the whole in the Marquette region is not far from 5,000 feet.

While the great Huronian area of Canada north of Georgian bay bears, so far as I am aware, little or no workable iron, and derives its economic importance from its ores of copper, the Marquette series, supposed to be of the same age, are eminently iron bearing, and have as yet produced no copper. It is doubtful if in the same extent and thickness of rocks, anywhere in the world, there is a larger percentage of iron oxide than in the Marquette series. In the order of relative abundance, so far as made out, the ores are the *flag*, the red *specular* hematites, soft or brown *hematites*, and *magnetites*. These all exist in workable beds, and all as disseminated minerals in rocks usually silicious. The geological distribution of these ores of iron in the Huronian series will be considered in another place. The geographical distribution is less understood; so far there seems to be the greatest concentration of magnetic ores in the Michigamme district of the Marquette region. From this, the relative proportion of magnetite seems to decrease as we go east, north, west and south, although there is a considerable magnetic attraction in the Menominee or southern iron region.*

Next younger than the Huronian are the copper-bearing rocks of Keweenaw peninsula, which extend westward into Wisconsin, the age of which has led to much controversy; good authorities having placed them in different epochs, from the Azoic to the Triassic. Recent observations made by Prof. R. Pumpelly and myself go strongly to confirm the view, if we have not positively demonstrated it, that they are non-conformably overlaid by the Silurian, and are therefore related to the Azoic. The relations of the copper-bearing

* See Appendix H., Vol. II.

rocks to the Huronian are not fully made out. In tracing the dividing line from Bad river in Wisconsin to Lake Gogebic, Michigan, last fall, a distance of sixty miles, we found them nearly, if not precisely conformable, but widely different in lithological character.

With regard to the magnetism of the copper-bearing series, the United States surveyors mark considerable variations at several points on the Land Office plats, due in all probability to disseminated magnetite in the trappean members of the series, although good authorities have ascribed these variations to electric currents. My own observations on the magnetism of these rocks have been limited, but lead me to believe that it is far less in amount and less persistent in character than is usually the case in the Huronian, indicating that the magnetite (to which I ascribe the attractions) is perhaps an accidental rather than essential constituent, and small in amount. Macfarlane found less than one per cent. in one of the Portage lake traps.

The next series of rocks in ascending order are the horizontally-bedded Lower Silurian sandstones, which skirt the south shore of Lake Superior nearly its whole length, called by Foster, Whitney, and Dr. Rominger, Potsdam, and assigned by the Canadian geologists, under the name St. Mary's, to a later period. They have not been proven to be magnetic, although strong magnetic attractions have been observed over this Silurian area, as will be explained hereafter.

To recapitulate, we have: 1. The Laurentian granite and gneiss, practically non-magnetic; 2. The Huronian iron-bearing rocks, often highly magnetic; 3. The copper series, slightly magnetic; and 4th. The Silurian rocks, without magnetism. This classification is intended to apply more particularly to the rocks of the Marquette and Menominee regions proper, embracing the central and southern portions of the Upper Peninsula; and even here, as has been noted above, there are exceptions. This sketch of the Marquette rocks, in the light of the distribution of magnetite, would be incomplete, did I not mention the fact that this mineral is very generally present in the form of fine sand in the drift in the region I am describing. If one moves a magnet about in the sand of a creek it is rarely that *magnetic sand* will not be found adhering. I have never seen it accumulated in quantities that would point towards its being utilized; nor have I ever observed a local variation which I ascribed to the mineral in this form.

We will now return to the Huronian or highly magnetic series, taking up its structure in some detail. About nineteen lithologically distinct beds or strata make up the series; of these, six and probably seven are generally so magnetic as to cause considerable variations in the needle. These beds vary from forty to several hundred feet in thickness, and strike and dip in all directions, and at all angles. The prevailing strike, however, is easterly and westerly, and the dip at high angles, often vertical. These rocks frequently outcrop, when we have no use for the magnetic needle in their study. Again, they are covered by deep drift, where magnetic observations, or workings, can only reveal them.

In order to study the magnetic characteristics of these rocks more minutely than could be done in the field, two hundred and twenty-two specimens, covering all the more common varieties, were collected and are deposited in the cabinet of the University of Michigan; they are fully described under lithology in this Report. Fifty-four, or twenty-four per cent., were found to possess some degree of magnetic power as manifested by their influence on a magnetic needle; each specimen being in turn made to touch each end of a mounted needle. If it had the power to lead it 20 deg. from its normal direction, the specimen was said to be feebly magnetic, and strongly magnetic when the needle followed the specimen round the circle if held about half an inch from it. Of these fifty-four specimens, thirteen were feebly magnetic, twenty-nine magnetic, five decidedly magnetic, and seven strongly magnetic.* None would, however, lift ordinary carpet tacks. Twenty-four, or nearly one-half, possessed polarity in some degree. Thirty were simply magnetic, with no *polarity* that could be detected by the rude means employed: in some instances the specimen would repel the needle at half an inch distance, but would attract it if placed in contact. Such specimens were rated as possessing polarity. All of the strongly magnetic specimens were rich in magnetite and possessed polarity, and it is not improbable that

* Appendix H gives the percentage of material lifted by the magnet in twenty-one specimens of Lake Superior ore, together with the color of the powder.

all would have been found to possess it if tested by more delicate means. Von Cotta, however, speaks of magnetic iron ore which possessed no polarity. The specimens generally attracted the south pole more strongly than the north. When examined, they had been collected about three months. Whether they would have shown more or less magnetic power if tested when freshly broken, I do not know. Dr. Kitchell says that under certain circumstances fragments gain magnetism.

In 1860 I saw a powerful loadstone for its size, in the possession of Professor Trego, of Philadelphia, which he had picked up in New Jersey twenty-two years before. I once collected a number of pieces of loadstone in the Bull Mine, New York, which in the mine would lift small nails; in a few days two-thirds of them had lost this power. This may have been due to the fact that in the mine the nails themselves were made magnetic by induction.

Regarding the *location of the poles* in magnetic rocks, the laws of magnetism would place them near the surface, or next divisional planes or terminations of masses. Observers are generally agreed that iron ore is most magnetic near dykes or volcanic rock. Quoting again from Dr. Kitchell, "Geology of New Jersey," p. 535: "The extent of the magnetic qualities of iron ores depends on their position with respect to the surface; the nearer to the surface the greater will be their magnetic properties. This appears to depend on the action of surface water and atmospheric agents, for it has been frequently observed that ore, when first taken out of a mine at a considerable depth, possessed but slight magnetic properties, but on being exposed to the atmosphere for a few months or years it would increase so much that excellent specimens of loadstone for experimental purposes could be selected therefrom. Seams of ore that contain numerous joints and fissures, through which water and atmospheric agents pass, possess more decided magnetic properties than those which are more compact and free from crevices and fissures." *

These remarks of Dr. Kitchell possess much interest. I have but one fact that bears on this question;—an average sample made up of numerous fragments collected by myself of the Iron Moun-

* If a fact, is this due to the contact of air and water, or is it because the seams necessarily produce small independent magnets.

tain Missouri "surface," or boulder ore, contained only about one-fourth as much magnetite (as measured by the amount lifted with a horse-shoe magnet) as did a specimen of "quarry" (ledge) ore selected at the same time and in the same way.

Classifying the magnetic ores and rocks of the Marquette region economically, the merchantable ores, according to the present standard of richness, would not constitute two per cent. of the whole; the balance being ferruginous quartzites and schists possessing no present value as ores. The merchantable magnetic ores have so far all been found in one formation near the middle of the series, and that is not all pure ore by any means; therefore, when an ore-hunter finds an "*attraction*" in the Lake Superior region, the chances of his having found a mine are not more than one in fifty. Neither the strike nor dip of the formation seems to affect its magnetic power. This depends, so far as my observations throw any light on the question, chiefly on the percentage of magnetite entering into the composition of the rock. Prof. Cook—"Geology of New Jersey," pp. 537-8—says that the magnetism of iron ores was influenced by the "pinch and shoot" structure so prevalent in the iron mines of New York and New Jersey. He points out the analogy between these regular pod-shaped masses—"shoots" of ore,—pitching downward in a northerly direction and an iron bar in the same position; both become magnetic and have polarity.

The "pinch and shoot" structure exists in the magnetic ores of the Marquette region, but is obscure, and in strike and dip there is no parallelism between our rocks and those of New Jersey, as is shown elsewhere. Yet our ores must usually be more strongly magnetic than those of New Jersey; for Prof. Cook says: "It is generally conceded that ore, covered by thirty feet of earth, will attract the needle, and 'large veins' have disturbed it when covered by fifty feet of earth." Now at five and even fifty times these distances horizontally, the needle is often deflected in the Marquette region, and at the Spurr Mountain the needle indicates a dip of 70 degrees at an elevation of 94 feet above the ore.

With regard to the *associations of the various ores* it may be said, that magnetic and specular ores are often found together, as are also the specular and soft hematite ores; but so far the magnetites and hematites have not been found in juxtaposition. If we suppose all our ores to have once been magnetic, and that the red specular was first derived from the magnetite and the hydrated oxide (soft hematites) in turn from it, we have an hypothesis which best explains many facts, and which will be of use to the explorer. As a rule it may be assumed that the hard ores of the Lake Superior region, even although they be rated as red specular, contain a sufficient amount of magnetite to cause some local disturbance in the needle; there are exceptions to this rule, but they are rare. In some instances, especially in the Menominee region, the disturbance is slight, but enough to be noticed by careful observation. It should be noted that the L'Anse Iron Range, so far as known, contains no magnetic ore whatever.

4. Explanation of Magneto-geologic Charts, Plans and Sections.

Having now briefly stated those elementary principles of Magnetism which are involved in our subject, described the instruments employed and their use, and sketched the geology of the rocks whose magnetic forces we are to study, we are fully prepared to examine the results of the observations made, and to draw such conclusions and make such applications as the facts seem to warrant.

It has been found necessary to introduce a few terms which may be new in describing the graphical representations of the phenomena observed. No work to which I could gain access contained expressions such as portions of our work seemed to require. Figures 1 and 2, Republic mountain chart (No. XI. of Atlas), are copied in part from the geological and topographical map of Republic mountain, which see for explanation of geology, relief of ground, and geographical position.

Magnetic observations were made across the entire Huronian series lapping on the Laurentian on each side, along survey lines 26 and 30, which run N. 53° E.; the observations being taken for a considerable part of the distance every 25 feet. The arrows in Fig. 1 indicate the directions which the needle actually pointed under the combined influence of terrestrial and local attraction. The angle between these arrows and the meridian is the *variation* in Azimuth (called simply variation) and ranges, as will be seen, from 0 to 180°. The direction of the arrows, although sometimes irregular, leaves no doubt as to which are the magnetic rocks.

The full significance and value of the common compass in locating magnetic rocks and ores is better shown in Plate V., which represents variations observed at the west end of the Washington mine, embracing the West Cut or Pit No. 10. The stations indicated on the Plate refer to survey lines shown on the map of the mine, No. VIII., to which reference is made for information regarding the geology and topographical features of the locality. A glance at this figure will bring to the mind of all familiar with magnetic experiments, the plumose forms assumed by magnetic sands or iron filings resting on paper and influenced by the magnet. Our figure may be regarded as representing the laboratory experiment greatly magnified. As to the irregularities shown by some of the arrows, it is probable that if the magnetism of ordinary magnets could be studied minutely, as with microscopic needles, that corresponding irregularities would be observed in the directions and polarity of the forces, not unlike those seen on this magnetic plan of the Washington mine. If we admit, as we are forced to do from these facts, that magnetic rocks present phenomena entirely analogous to artificial magnets, then it is not difficult to decide as to the cause of the phenomena exhibited on the sketch before us.

The dotted line is designed to indicate the position of maximum variation, or rather the position of the force which causes the variation. The observations made for *intensity* along this line, indicated by vibrations (six being the normal number), confirm the indications of the horizontal compass. There can be no doubt but that nearly under this line, at no great depth, is a large amount of magnetite; whether free enough from rock to constitute a merchantable ore, explorations only can establish. Since this plan was made, work has been resumed at Pit No. 10, and a tolerably regular bed of ore revealed, having the strike and dip marked on the plan, which coincides closely with what might have been predicted. The relationship of this deposit with the others constituting the mine will be considered elsewhere. This magnetic plan, as well as Fig. 1, Republic mountain chart, shows, that while the variations are governed by a uniform law away from the lines of maxima, within these lines great irregularities of direction exist.*

---

* Since the above was written, I have, by the kindness of Mr. F. Firmstone, of Easton, Pa., been able to inspect some magnetic charts of New Jersey localities, made by the late Mr. Amsden, of Scranton, which are excellent.

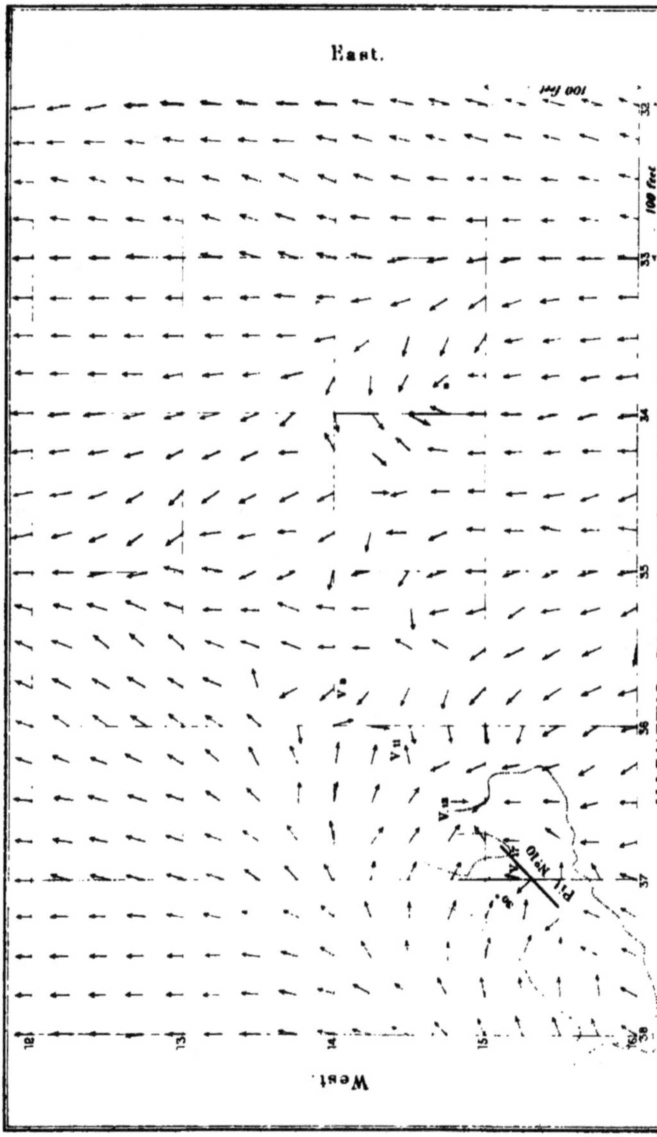

MAGNETIC PLAN — WASHINGTON MINE.

See Map.

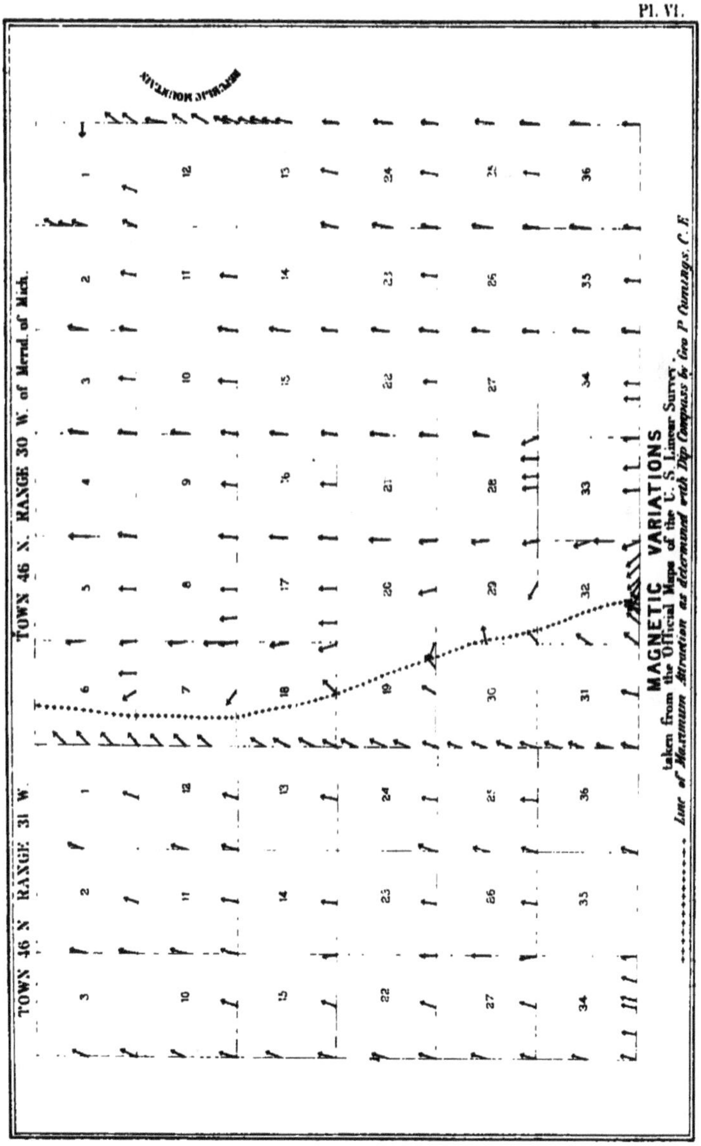

PL. VI.

TOWN 46 N. RANGE 31 W.

TOWN 46 N. RANGE 30 W. of Merid. of Mich.

REPUBLIC MOUNTAIN

MAGNETIC VARIATIONS

taken from the Official Maps of the U. S. Linear Survey.

.... *Line of Maximum Attraction as determined with Dip Compass by Geo P. Cumings, C. E.*

TOWN 40 N. R 30 W. Merid of Mich.

MAGNETIC VARIATIONS

taken from the Official Maps of the U.S Linear Survey.

++++++++ *North and South Ranges of South Belt – Menominee Iron Region*

Passing from Plate V., which represents but a small area, over which the magnetic observations have been very numerous, to a magnetic plan of a large surface, with widely separated observations, we have in Plates VI. and VII., copied from the United States Land Office books, a fine exemplification of the significance of local magnetic variations.

In Plate VI. the magnetic rocks run nearly north and south,—which direction, as has been heretofore stated, produces the maximum variation. It will be seen that the needle is influenced at a distance of nearly, if not quite two miles, and that the variation diminishes rapidly as we depart from the line of maximum attraction. The disturbances recorded on the north-east part of this plan are due to Republic mountain.

Plate VII. represents one of the iron townships in the Menominee region. The variations are scarcely so great, nor do they extend so far as in the other. As the two iron ranges represented run much more nearly east and west in this case, it is interesting to observe the difference in the behavior of the needle. These plans are additional proof of the value of the Linear Surveys to the explorer, a point to which I have often referred.

Figures 3 and 4, Chart XI., Atlas, are *magnetic sections* along lines 26 and 30 of plan. The arrows indicate the direction of the dip-needle vibrating in the plane of the meridian. The normal direction is the horizontal line ; the arrow head indicating north end of needle should therefore normally point to the right hand side of the chart. It will be seen that the dip, like the variation, often attains the maximum of 180°, that is, the north end points south.

The colored curved lines express approximately the *intensity* of the local magnetic force ; their ordinates being the number of vibrations made by the needle in one quarter of a minute, on a vertical scale of eight vibrations to the inch. The *blue* line records the observed vibrations of the *horizontal* needle, the others of the dip-needle. The *black* line refers to the needle vibrating in the plane of the meridian (compass *facing west*). The *red* line refers to the needle vibrating in an east and west plane (compass *facing* south.)

Fig. 2 is a magneto-geological section on the line A—A′ of Fig. 1. The upper curve represents a projection on one plane of the maximum intensities of all the curves of Figs. 3 and 4. The lower curve, Fig. 2, has reference to variations and dips, its ordinates being proportional to the maximum variation in direction of the needle, caused by the magnetic rocks. It is intended as a sort of summary of the facts expressed by all the arrows denoting directions, as the upper curve is a general expression of the intensities. It will be observed that the summits of the lower curve, Fig. 2, which indicates maximum variation, are always northerly from the centre of the magnetic bed. This is as it should be, because the greatest *variation* takes place before we reach the local magnetic pole, when approaching it from the *north*. The intensities, on the other hand, are greatest directly over the magnetic rocks. It should be borne in mind that the intensity of a magnetic force is really proportional to the square of the number of vibrations in a given time; but in these investigations the actual number of vibrations has been used in constructing the sections, as being more convenient.

In addition to the facts observed during this survey, which are recorded on the Republic Mountain Chart, and various figures in this volume, certain others, obtained from the United States Land Office, plats of Towns 46 and 47 north, Ranges 29 and 30 west, will be employed, besides those already given from the same source.

The discussion of the facts in our possession falls conveniently under two heads:—First, Regarding the entire Huronian series as a unit, and the comparison of its magnetism with the Laurentian system. Second, A study of the magnetism of the individual beds of the Huronian or iron-bearing rocks, in detail. Republic mountain and vicinity afford an excellent opportunity for both these investigations.

The Magnetism of the Laurentian System or Granitic Rocks.

The Federal township plats above referred to, cover an area of, say twelve miles in diameter, of which Republic mountain is the centre; at least nine-tenths of this territory is Laurentian. The variations of the needle noted are from two to six degrees east, averaging four and a half degrees, which may be regarded as the *declination* of the needle at the date of the surveys of this locality, due to cosmical causes. From this and similar facts covering the whole Marquette region, we may conclude that this oldest system of all known rocks has here no beds of magnetite, nor does it now contain magnetite as an essential constituent mineral, nor indeed oxide of iron in any form. Prof. Pumpelly and myself found slightly magnetic rocks in the Laurentian south of Lake Gogebic, and the professor mentions in his report to the Portage Lake and Lake Superior Ship Canal Company "a deposit of iron ore in the Laurentian gneiss and hornblendic schist series on Sections 10 and 15, T. 41 N., R. 29 W.," in the Menominee Iron Region, from which I have seen specimens which do not look very promising. One or two other places are mentioned where magnetic beds occur in the Laurentian, but they are exceptional, the rule being as has been stated. But everywhere in the region we are considering, over or near the Huronian Series, the Government surveyors note variations. The approximate boundary between these two systems of the Azoic in some parts of the Upper Peninsula could indeed almost be delineated from their surveys by magnetic variations alone.

Magnetism of the Huronian Series as a Unit—Republic Mountain.

No special observations were made to determine the extreme limit to which its magnetic influence extends. The Federal surveys would make the distance over one mile, and Durocher mentions that he was told in Sweden that "important beds of iron ore produced deviations in the needle up to the distance of nearly two kilometres," or over one mile—*Annales des Mines,* 5 Series, Vol. 8, p. 220. The Federal surveyors note a variation at the north-east corner of Section 7 (See Fig. 1, Republic Mountain Chart) of 25° west, agreeing very nearly with my observations corrected for the change in declination since the survey was made. This corner is at least 600 feet from the nearest Huronian bed, and probably 900 feet from any member of the series containing magnetite. Judging by the direction and intensity of the magnetic force as exhibited by the needle, as we approach the mountain from the north-east (see Figs. 1, 3, and 4), it seems probable that the bed which chiefly produced the effect was No. VI., and still more distant ores. If this be a correct inference, we have the phenomenon of a magnetic needle deflected 25° from its normal direction by a bed of rocks containing not to exceed 33 per cent. of magnetite, distant 1,500 feet horizontally. The facts from the U. S. surveys given above show that the needle is sometimes influenced to a much greater distance.

Passing to the south-west side of the Huronian basin we find the influence exerted by the magnetic rocks to gradually diminish as we recede from their edge, which is believed to be under the Michigamme river. See Fig. 1. Here we find the needle varying 15° at a distance of at least 800 feet from the nearest magnetic rocks.

An inspection of Fig. 1 shows that the variations of the needle are much greater on the north-east than on the south-west side of the mountain, which should evidently be the case from the fact that to the south-west the terrestrial and local forces are more nearly in the same line than on the north-east side; hence in the latter case the mechanical resultant (direction of the needle) would form a greater angle with the direction of the earth's force (magnetic meridian) than in the former.

The question of the distance to which magnetic ore and rocks will attract the needle receives some additional light from the Champion Mine, Plates VIII. to XV. It is evident in this case that the magnetic force of the ore is felt to a distance exceeding 700 feet to the north of the mine. To the south there is less certainty, because of the other magnetic rocks (see sections) which underlie the ore in that direction. It is probable that careful observations would detect the influence of this remarkable deposit of ore through an east and west zone, which in places would attain a breadth of 2,000 feet or more, one-fifth of this area showing a magnetic dip of 90°; but this does not prove the existence of 400 feet of magnetic rocks or ore, by any means, as will be seen below.

At the Spurr Mountain, which is an east-west deposit of highly magnetic ore like the Champion, Mr. Lawton observed just south of the range 23 vibrations in a quarter of a minute; going south the vibrations diminished somewhat regularly, until at 600 feet the needle vibrated but *ten* times in a quarter of a minute. At 300 feet north of the mountain the needle settled indifferently in any direction, owing to the fact that the terrestrial and local forces just balanced each other at that point; further north the vibrations increased somewhat irregularly, owing to the presence of slightly magnetic rocks, until at 1,400 feet *six* vibrations were observed in a quarter of a minute. There must of course be points north and south of all magnetic belts where the vibrations would be equal and normal, but these limits were not reached, the observations proving only that the magnetic belt at Spurr Mountain is over 2,300 feet wide.

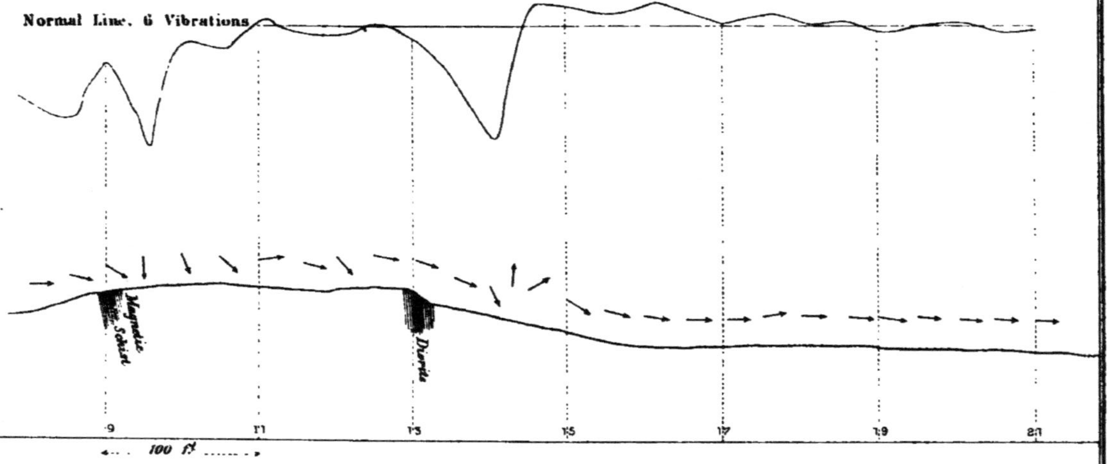

Normal Line, 6 Vibrations

Magnetic Schist

Diorite

9

11

13

15

17

19

21

100 ft

MAGNETIC SECTION – CHAMPION KEYSTONE RANGE.

Survey Line of 42 West.

See Map.

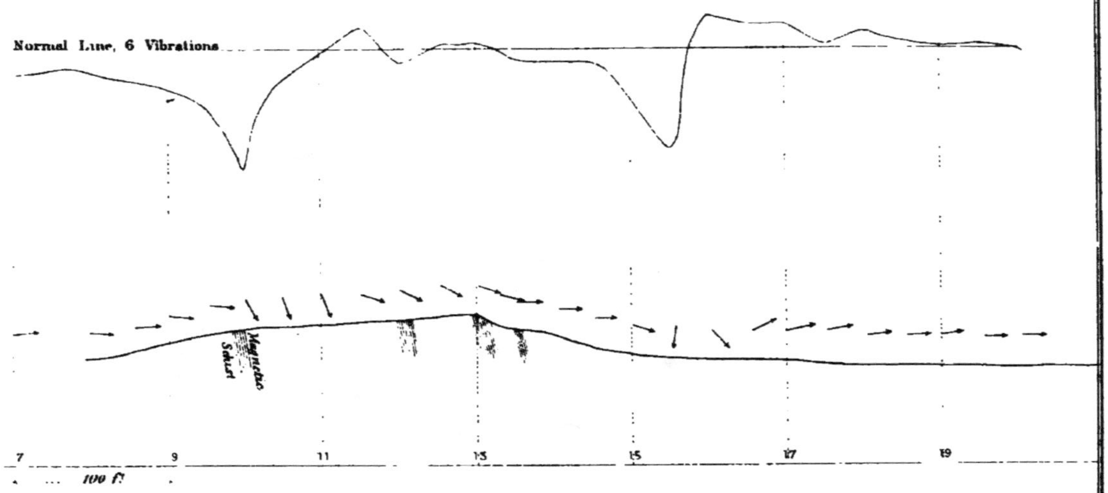

MAGNETIC SECTION - CHAMPION KEYSTONE RANGE.

Survey Line of 50 West.

See Map.

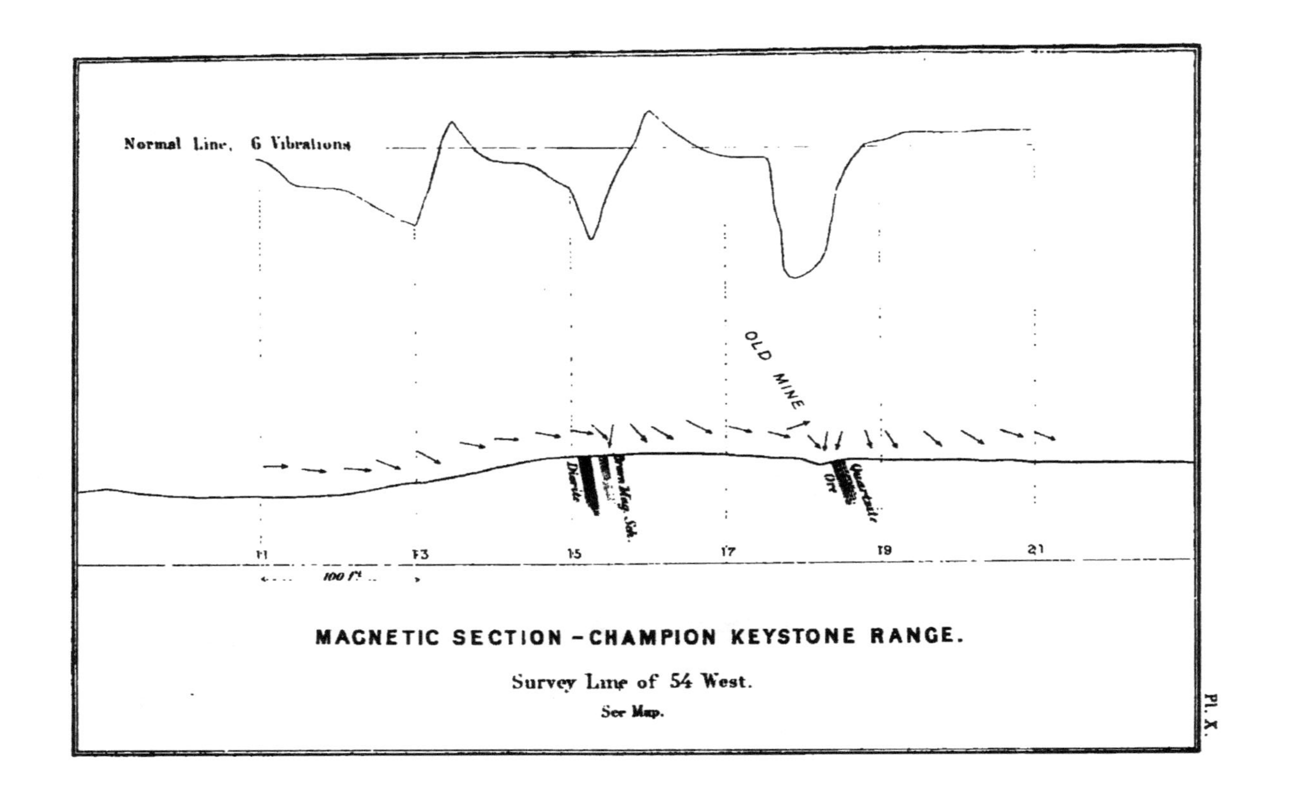

MAGNETIC SECTION – CHAMPION KEYSTONE RANGE.

Survey Line of 54 West.

See Map.

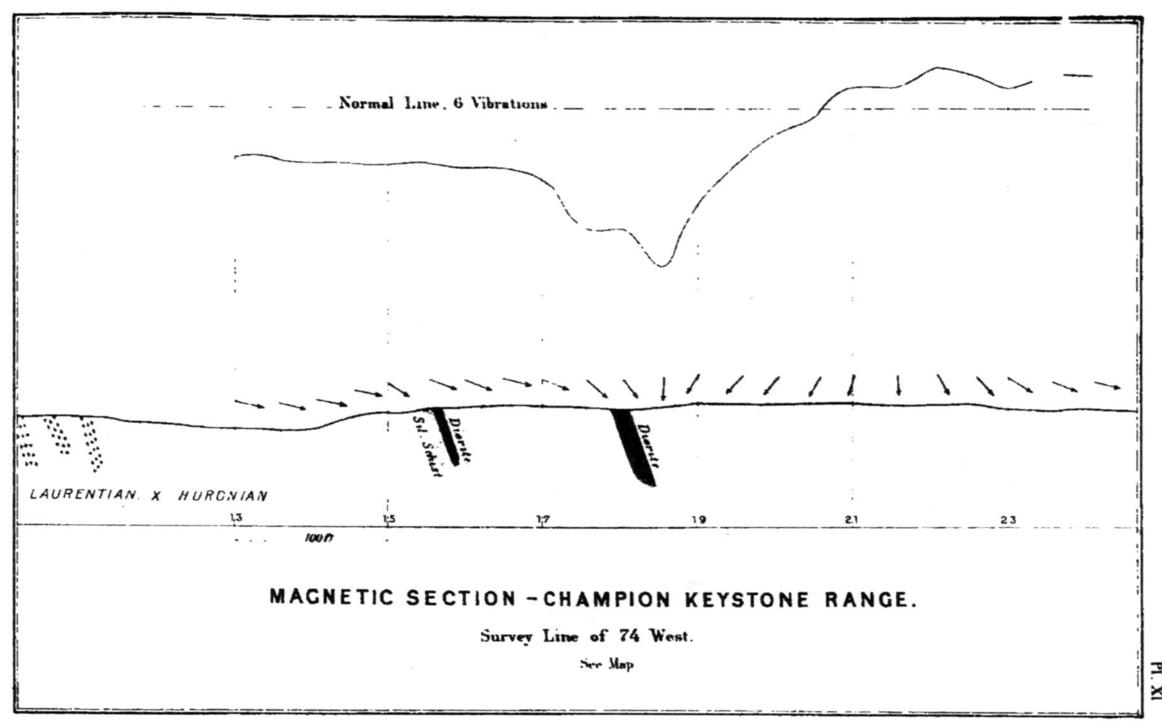

MAGNETIC SECTION - CHAMPION KEYSTONE RANGE.

Survey Line of 74 West.

See Map

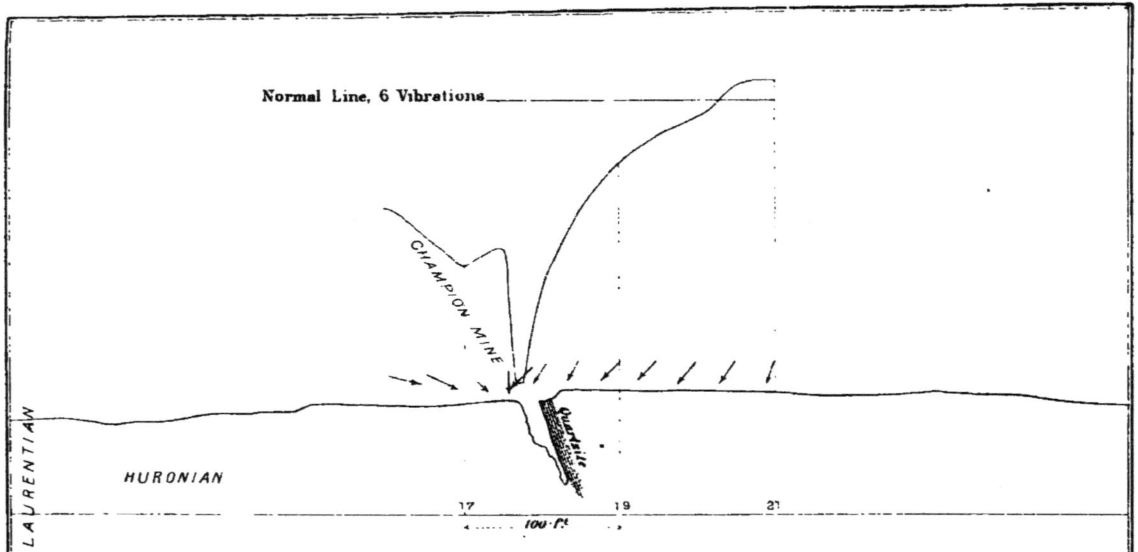

MAGNETIC SECTION – CHAMPION KEYSTONE RANGE.

Survey Line of 76 West.

See Map

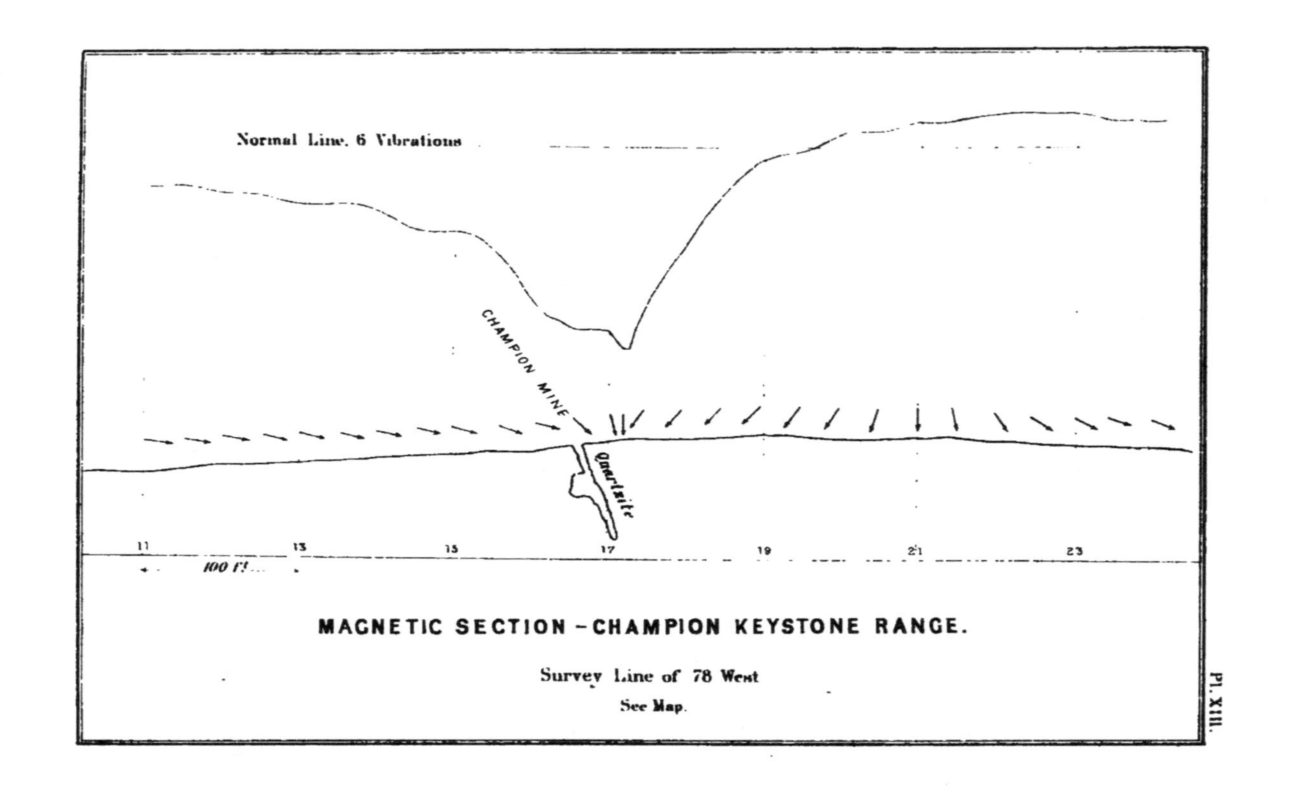

MAGNETIC SECTION – CHAMPION KEYSTONE RANGE.

Survey Line of 78 West

See Map.

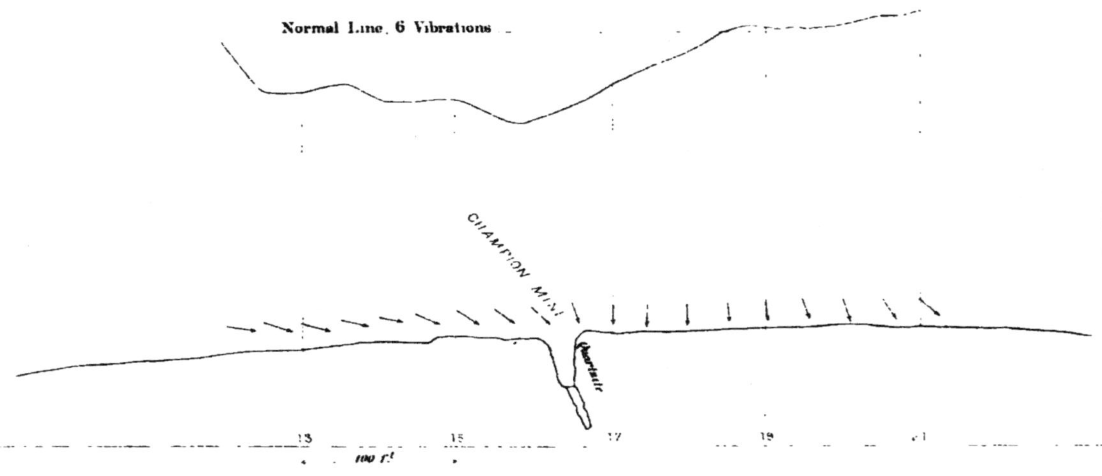

Normal Line, 6 Vibrations

CHAMPION MINE

Quartzite

13 15 17 19 21

100 ft

MAGNETIC SECTION - CHAMPION KEYSTONE RANGE.

Survey Line of 82 West

See Map.

It may be asked why the very silicious magnetic rocks of the Republic Mountain influence the needle at a greater distance than the pure ores of the Champion Mine. It is not at all certain that this is the fact; the limit of the influence has been determined in neither case. The stratigraphical conditions, however, are quite different. The strike of the Republic mountain rocks being north-westerly, is far more favorable for producing variations than is that of the Champion deposit, which is east and west. It is quite evident that a north-south deposit of ore would cause greatest variations (see Plate VI.) and an east-west deposit least. If, in the latter case, we conceive the power to be equally distributed along an east and west mathematical line, there would be produced no variations at all in a horizontal compass. Again, there are four highly magnetic beds at Republic Mountain, while at the Champion there is only one.

Regarding the polarity of the magnetic force: (1) In every instance the north end of the horizontal needle was drawn towards the magnetic rocks; hence, north-easterly of Republic Mountain, the variation was west; and south-westerly, the variation was east. (2) With the dip-needle vibrating in an east and west plane, the north end pointed westerly, or towards the mountain on its north-east side. (3) With the dip-needle vibrating in the plane of the meridian, on the north-east side of the mountain, the south end inclined downward, producing a "negative dip," as shown in Figs. 3 and 4, and this increased as the magnetic rocks were approached until the needle turned entirely over. This apparent *negative* attraction was probably in reality only the effect of an attraction for the north end of the needle, which inclined to the magnetic rocks by the shortest road. Why the north end of the needle moved upward instead of downward (which was apparently just as short a road) as it approached the magnetic rocks over the non-magnetic Laurentian, I can only explain as follows,—which hypothesis may also explain instances other than this where slight negative attractions have been observed over granitic rocks, for example, south and south-east of the Champion mine. My needles were always counterpoised near Negaunee or Marquette, which towns are built on the Huronian. Of course an effort was made to get away from the magnetic members of the series; but this evidently would be impossible if their influence extends to the distance of one half mile.

Magnetic rocks would probably be found throughout the Huronian belt by boring less than 1,000 feet into the earth, owing to the basin-like structure of the series. It is probable, therefore, that my needles were counterpoised under the influence of some ***positive*** magnetic force ; hence, when taken over Laurentian rocks containing no magnetite, they would show "negative" attraction. If this hypothesis is correct, then the negative attraction referred to above is explained.

Regarding the ***intensity of the magnetic force*** exerted by Republic mountain as a whole, but one observation need in this place be made. The vibrations are greater on the south-west than on the north-east side, or exactly the converse of the variations. The Magneto-Geologic Sections of the Champion and Keystone Range (see Plates VIII. to XV.) present the same phenomenon.* As the needle is carried north from the Champion bed, its vibrations rapidly diminish in number until they become ***less*** than the normal number due to the earth's magnetism ; after which, on going still farther, the vibrations will increase until the normal number is reached : but in going south, the diminution is far less rapid, and the number of vibrations never falls below the normal number. The same was observed at the Spurr as is noted on page 226.

The obvious reason is this : when the needle is south of the local force, both it and the terrestrial force act in the same direction, producing a maximum effect ; but when the needle is north of the local force, it can evidently be influenced only by the greater force less the smaller. In the first case the mechanical resultant is the ***sum***, in the other it is the ***difference*** between the two magnetic forces. This readily explains the difference in the slope of the curve of intensity north and south of the magnetic poles, so noticeable in the magnetic sections.

Republic Mountain.

A glance at the directions of the needle as indicated by the arrows in figures 1, 3, and 4 of Chart XI., will impress one with the conviction that there is no direction in azimuth, or inclination which

---

* The survey lines on the Magnetic Sections, Plates VIII. to XV., refer to Map of the Champion Mine, No. VII., which should be examined in connection with them.

the needle does not assume in crossing the series of rocks. The north end of the needle never points north, often east and west, and sometimes south; while in the dip-compass it turns a series of somewhat irregular somersaults, pointing habitually downward, but often towards the zenith. The needle may be said to "box the compass right and left," as we may suppose that feat accomplished by a drunken sailor. A second glance at the arrows will show us that there is much method in the madness of our ge-go-sence;* the needle very generally tends to point toward the blue or red-colored rocks, which contain magnetite, while it is comparatively indifferent to the green, gray, and salmon colored, which contain little or none of this mineral. The particular significance of the variations and dips will be more fully discussed below.

We will leave for the present the consideration of the direction of the magnetic force expressed by the arrows, and return to the subect of the *intensity of the force as expressed by the colored curves* (see page 223). Nothing is more evident on the chart than that these curves indicate with great certainty the position of the magnetic beds over which they are more or less convex, producing summits; and more or less concave or flat over the non-magnetic rocks, pointing literally as a finger in some instances to the location of the magnetic force. Comparing the three curves in figs. 3 and 4, it appears that:—(1) The red line (compass facing south) oftenest rises higher than any other over the magnetic rocks; and sinks lower away from them. It has also fewer changes in direction than the others. (2) The black line (compass facing east-west) falls lower than either of the others over the magnetic rocks. (3) The blue line (compass horizontal) often has an extreme depression, where the others have an extreme elevation.

These, the most obvious generalizations from the curves, are explained by the principles of the mechanics of forces already mentioned.

Fearing there may be some confusion from representing the same element-*intensity* by three curves, I suggest the following conception: Suppose an observer to be provided with a horizontal com-

* A Chippewa word for magnetic needle, signifying "little fish," in allusion to its wiggling motion.

pass having a blue needle and two dip-compasses, one provided with a black and the other with a red needle. Suppose, further, these to be mounted for observing at the same station, but so far apart as not to influence one another; the blue needle moves in a horizontal plane, the red needle in a vertical east and west plane, and the black needle in a vertical north and south plane. Suppose, further, a powerful magnet to be placed (1) directly under or directly over the station, it is evident that only the black and red needles will be influenced. (2) If placed north, the blue and black needles only will be influenced. The directive force in this case would be a maximum; because the magnet's power is added to the earth's, both acting in the same line. (3) If the magnet be placed directly south, the red needle will again be uninfluenced, but the black and blue needles will indicate a minimum of intensity instead of a maximum, for their directive power will be the difference between the force of the magnet and that of the earth. (Places have been observed where the needle gave us no vibrations in any position from this cause. A fine illustration occurs in Fig. 3, Chart XI., Station 24, where there must have been a very strong pole to the south of the station; but this pole is evidently north of Station 24, Fig. 4, where the greatest intensity was observed.) (4) If the magnet be placed east or west of our supposed station, the effect will be the same; the red needle will be most influenced, blue next, and black not at all.

We are now fully prepared to explain the phenomena presented by the colored curves.

(1) Why does the red line usually rise higher over the magnetic rocks, and sink lower away from them, and why does it fluctuate least? When the needle vibrates in an east and west plane, its axis points north,—that is nearly in the line of the directive force of the earth, which it thus partially neutralizes; giving the local forces full power. As these are much stronger than that of the earth at short distances, we should expect the result observed over the magnetic rocks. Away from them, the earth's force being nearly neutralized, we should have the minimum of intensity as is shown by the red line. That the changes in direction in this line are less frequent and less abrupt than the others, indicates, I think, that if the earth's attraction was entirely neutralized and the error of observation reduced to a minimum, the curve derived from the magnetic force resident in the rocks on any particular cross-section might be more

regular than any shown in the chart. It is reasonable to suppose that the red curve has most significance in our investigations. (2) Why do the black and blue lines fall as a rule lowest over the magnetic rocks? Suppose a local force, about equal to the earth's, to exist directly south of a dip-compass placed in the plane of the meridian, or of a horizontal compass; we should evidently have a minimum of intensity, because the terrestrial and local forces would balance each other. The marked exception to this rule over formation XI., Fig. 4, is evidently due to the fact that the magnetic power resident in beds X. and XII. just balance each other, and as the directive power of the earth is neutralized in the case of the red line by the direction in which the needle is held, we have a point of comparative equilibrium. (3) Why does the blue curve sometimes present depressions opposite the summits of the others? This is readily explained by supposing the local force to exist directly under the station; its force would then be entirely neutralized by the centre-pin of the horizontal compass, while having its full effect on the dip-needle in both positions.

5. Diminution of Intensity due to Elevation.

All the observations for intensity above considered were taken at an elevation of about 4 feet from the surface. Sometimes the rocks came to the surface, sometimes there were several feet and perhaps yards of drift between; it is therefore an important practical question to ascertain what effect the elevation of the needle has on the number of its vibrations.

The difficulty of attaining any considerable elevation at which to observe intensity, renders our observations on its rate of diminution due to elevation or vertical distance of little value. The theory of the sphere of attraction and law of decrease of force, as the square of the distance from the centre, has been mentioned; but with several local forces acting on the same point (the case usually presented in nature), the law is greatly modified, the decrease being in a less ratio. This subject possesses especial interest in connection with the determination of the depth at which magnetic rocks, producing a given disturbance, will be found; therefore, the few observations made, unsatisfactory though they are, will be given. At Republic Mountain a staging was erected in the windfall, by means of which eight equi-distant observations were made; the lower one on the magnetic schist, the upper one 14 feet above it. The results were as follows:

Elevation in feet.	Vibrations.		Remarks.
	Facing west.	Facing north.	
0	56	53	On surface of schist.
2	41	41	
4	33	30	
6	27½	30	
8	19	23	
10	15½	24	
12	18	24	
14	12	20	

At another point near the above, and over the same magnetic rock, the following vibrations were observed:

Elevation in feet.	Vibrations.		Remarks.
	Facing west.	Facing north.	
0	60	60	On surface of schist.
3	50	49	
6	36	37	
9	25	26	
12	18½	18	

The observations have all been represented graphically, but as no law was apparent, and as the figures can be easily reproduced, they are not given. The first table gave the most regular curve, but still too angular to attempt the application of a mathematical formula. They do not seem to me to afford a basis for calculation, as to how high the appreciably magnetic influence of these rocks would extend. I have an impression, however, without being able to give any reason, that it would be considerably less than one half mile, which was shown to be the distance to which the influence of the same rocks extended horizontally. I cannot consider it probable that a needle would dip where an earth covering of over 2,000 feet exists, if such a case were possible. At the Champion mine, by the aid of shaft house No. 2, an elevation of 44 feet above the ore was attained, and the following observations made :

Elevation in feet.	Vibrations.			Remarks.
	I.	II.	III.	
0	18½	17½	23	Level of surface of ore in shaft.
18	19	17	17	Surface of ground.
32	16½	16½	16½	Girder of shaft house.
44	15½	..	15	Girder of shaft house.

At other points at the Champion mine, 25, 32, 33 and 40 vibrations were observed, the compass being within 5 feet of the ore. The diminution here is quite regular and nearly as the distance. If the rate continue, the vibrations should reach the normal number (six for the instrument used) at about 150 feet ; but it is highly improbable that this law would hold for the whole height.

The difference between the rate of diminution at the two localities is very marked ; at Republic Mountain an elevation of 12 feet in one instance reduced the vibrations from 60 to 18½, in another 14 feet elevation reduced the number from 56 to 12. At the Champion 44 feet elevation made an average of less than 4 difference in the vibrations. In this comparison the following geological differences must be borne in mind.

The Champion deposit at shaft No. 2 is a heavy bed of nearly pure black oxide running east and west and dipping north at an angle of 68 degrees, and it is the only magnetic rock in the vicinity. The Champion deposit loses its magnetism in going west, specular slate taking the place of the magnetite in that direction. The Republic Mountain bed over which the observations were made (No. X.*) is, on the contrary, a silicious schist, containing not to exceed 33 per cent. of magnetite, (the merchantable ores of Republic Mountain, of which there are large deposits, are in bed No. XIII., and are mostly specular hematites.) This magnetic bed X. is associated with others of a similar character, all striking north-west and south-east and dipping nearly vertical. The specimens of these magnetic schists which were examined possessed marked polarity. The Champion deposit evidently contains far more magnetite within the same sphere of influence than the Republic Mountain.

There is no doubt that variations and dips are a much more delicate and ready means of observing slight magnetic attractions, than vibrations when observed with the hand instruments employed. In one instance at Republic Mountain the dip at 12 feet elevation was 30 degrees, at 9 feet 50 degrees, at 6 feet 70 degrees, at 3 feet 77 degrees, at 0 or on surface of rock 105 degrees. It appears that the magnetic poles of the Champion bed are more deeply seated than those at Republic Mountain, which seem to be at the surface. This may be due to the fact that the upper part of the Champion deposit is mined out. Sets of careful observations made for considerable heights, both for dip and vibrations, would possess great interest, especially if made over beds of ore or rock, the position and character of which were known. In a record of over three thousand magnetic observations made by me in Michigan, Missouri, New York and New Jersey, I have not in more than six instances found the needle in the dip-compass above described to vibrate over 40 times in a quarter of a minute, and in no instance in which this rate was observed was the needle removed more than 5 feet from the magnetic mineral. Of course in the same needle the vibrations will vary with the degree of magnetism that has been imparted to it, and the condition of the instrument in other respects. I have had a rude standard, and when my needle fell below that it was overhauled, so that the numbers are relatively correct. I do not remember to have observed over 15 vibrations in a quarter of a

---

* The Roman numerals refer to the order of the beds of the Huronian series, counting upwards from I. to XIX.

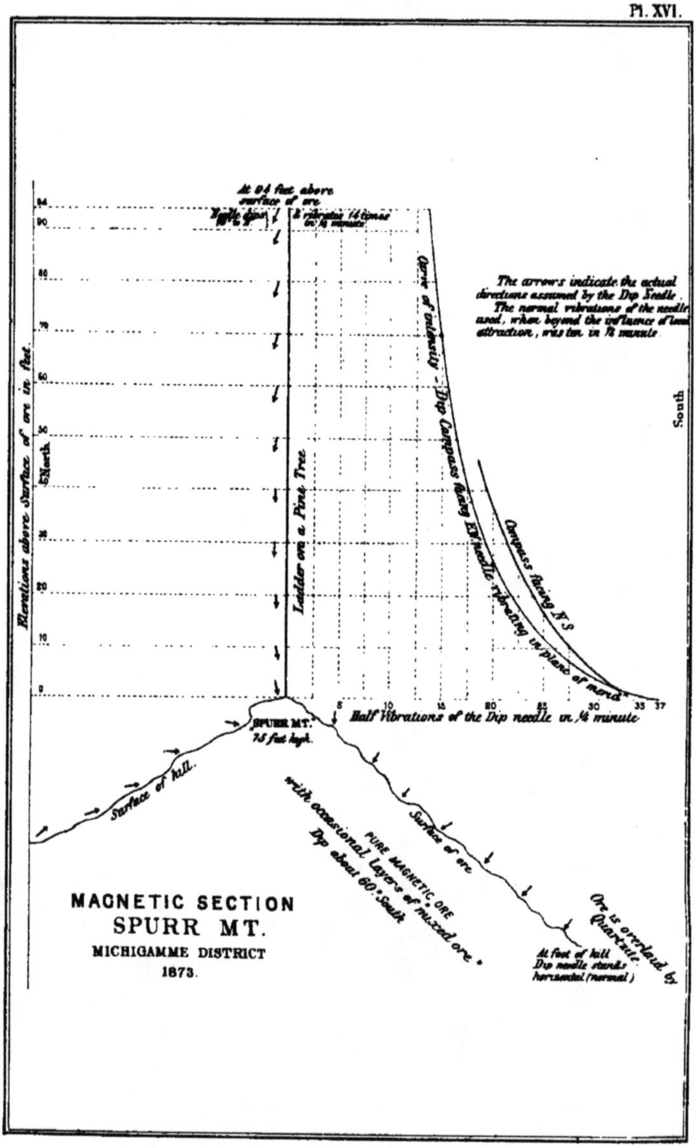
At 94 feet above surface of ore
Needle dips
& vibrates 14 times in ¼ minute
The arrows indicate the actual directions assumed by the Dip Needle.
The normal vibrations of the needle used, when beyond the influence of local attraction, was ten in ½ minute.
Elevations above Surface of ore in feet
North
South
Ladder on a Pine Tree
Curve of intensity – Dip Compass facing E.W. needle vibrating in plane of merid.
Compass facing N.S.
Half Vibrations of the Dip needle in ¼ minute
SPURR MT. 75 feet high.
Surface of hill
Surface of ore
PURE MAGNETIC ORE with occasional layers of mixed ore. Dip about 60° South
Ore is overlaid by Quartzite
At foot of hill Dip needle stands horizontal (normal)
MAGNETIC SECTION
SPURR MT.
MICHIGAMME DISTRICT
1873.

minute, or one per second, at a greater distance than 50 feet from a magnetic bed, and usually this number of vibrations would indicate a distance not exceeding 25 feet in the Marquette Iron Region.

Since the above observations were made and recorded, the development of the Michigamme district has permitted observations to be made at the *Spurr Mountain* which throw much light on the subject of the diminution in dip, and intensity due to elevation. The following table records the observations made,* and Plate XVI. represents the general law of diminution graphically. The observations were carried to an elevation of 94 feet by means of a fortunately situated pine-tree, up which a ladder was constructed. While there are minor irregularities, due wholly or in part to errors of instrument, the presence of nails in the ladder, and personal error, the average curve is remarkably regular, and points, as most of the other facts do, to a far more rapid rate of diminution near the surface than at a considerable elevation.

It is not to be expected that the law of decrease of magnetic force would hold at this locality. Had the local force been concentrated in a focal point directly under the tree, and the force of the earth been neutralized, then we might expect the law to be discernible. Some useful practical rules may be readily drawn from the table and plate under consideration.

1. If, in a locality where magnetic attractions prevail, we find considerable difference in the number of vibrations between the compass, when in contact with the ground, or held six feet above it, we may conclude the ore is very near the surface; if there is but little difference, then the ore is probably deep. 2. The amount of the dip gives but little clue to the depth of the ore. If the Spurr Mountain had been covered by a hundred feet of earth, water, or non-magnetic rock, we would have found at the surface a dip of about 70°, and it is probable, if not certain, that if it were possible to make observations to the north of the mountain, at the same elevation, a greater dip than 70° would be found, due to the changed direction of the local force.

* Mr. C. M. Boss rendered great assistance in these observations.

Observations for diminution of magnetic force in vertical direction—(Needle vibrating in north-south plane)—Spurr Mountain.

Height.	Dip.	Vib. in ¼ min.	Remarks.	Height.	Dip.	Vib. in ¼ min.	Remarks.
0	100°	37	On surface of ore.	25	93°	20½	
2	100°	33		26	93°	20	
4½	100°	30		27	92°	20	
6	100°	28½		28	92°	19½	
7	100°	27		29	92°	19½	
8	100°	27		30	92°	19¼	
9	100°	27½		31¼	91°	19	
10	100°	26		32½	90°	18½	
11	100°	25½	26 facing south.	33½	90°	18½	
12	96°	24½		34½	90°	18	
13	100°	24		35½	88°	18½	
14	96°	24½	24½ facing south.	36½	88°	19	
15	98°	23		37½		18	
16	96°	23½		41	88°	17½	
17	95°	23½		42¼		17	
18	95°	23		45½	86°	17	
19	94°	22½	23 facing south.	48		17	
20	94°	22¼		52	80°	16½	
21	94°	21½		56	80°	16½	
22	93°	22		63	78°	16	15 vib., faced N., dip 86 E.
23	92°	21½		68	80°	15	15 Vib., faced N.
24	93°	20½		79	78°	14½	13 " " "
				92	72°	14½	13 " " S.
				94	70°	14	13½ " " S.

6. What is the significance of a dip of 90° or "dead 90°."

As there is a general impression among those who have made but little use of the dip-needle in exploring for iron ores, that a variation of 90° signifies merchantable ore directly under the feet, it is important to ascertain the exact purport of such great fluctuations in the direction of the needle. For the present we will leave out of the question the unpleasant fact that in 19 cases out of 20, if not 99 out of a 100, the mineral producing the dip would, if found, prove to be only a ferruginous schist or magnetic rock instead of merchantable ore, and consider the case often presented where there is ***a dip of 90° at a place which is not underlaid by magnetic mineral***, or where there is none within several hundred feet. In such cases there are generally two, approximately parallel lines, of 90° dip, one over the ore, where the vibrations are very *quick* (always more than the normal number). The second line (the one we are now considering) will always be found north of the first, and along it the vibrations will be *slow*, always less than the normal number.

A moment's inspection of almost any magnetic section (Plates VIII. to XIV.) will illustrate the fact and suggest the cause. If we hold a dip-compass over a highly magnetic bed the needle will indicate 90°, pointing directly towards it. Moving north, the needle will continue to point towards the ore, that is, be turned backward, thus varying or dipping more than 90° from its normal direction. Continuing north, we soon get so far from the local influence, that its power ceases to entirely overbalance that of the earth, and the needle commences to return to its normal direction. In doing so it must evidently somewhere stand again at 90°, which means simply, that the local force to the south and the earth's force to the north, are so related in intensity, that the resultant is a vertical line. Still going north, the dip grows less and less until the boundary of the local attraction is past, and the needle returns to its terrestrial allegiance. It is evident that no such phenomena can occur to the south of the magnetic bed, for the terrestrial and local influences acting in the same direction, no "dead" points could occur.

This "dead 90°" line, then, instead of proving the immediate presence of ore, proves just the reverse if the phenomena are presented as above, which is the case at the Magnetic, Champion, Spurr, and Michigamme mines, and at one place at the Washington mine. There may be ore under this line, but it will always be deep and have little or no influence in producing the phenomena observed. Rule :—When there is a dip of 90°, and the vibrations *exceed* the normal number, we may conclude that the magnetic mineral is under our feet, or very near us to the south. If, with the same dip, the vibrations are *less* than the normal number, we may conclude that the magnetic bed producing the effect is south of us, and may be at considerable distance. This rule will evidently apply only where there is one strongly magnetic bed not very deep, which is the most common case. If there be several beds, as at Republic Mountain, the application of the principle is more difficult; but in the nature of force, some modification of the phenomena must be presented by all magnetic rocks.

It is worth remarking that the south belt of 90° dip, is more sharply bounded, especially on its south side, and usually narrower, than the north belt.

The lower curve of Fig. 2, Republic Mountain Chart, illustrates what has been said above ; the summits of the curves showing the maximum dips are north of the magnetic beds, while the summits of the curves showing the maximum intensity (see upper curve, same fig.) are over the centres of the magnetic beds.

This subject would be incomplete without considering the case, quite common, where a zone of local attractions has but one line of 90° dip, or to make the case general, but one line of *maximum* dip whether it be 90° or less. It may be said that this last expression, covers the whole question, but with ore-hunters "dead 90" has a peculiar significance, and it is for them that I am writing. This case (one line of 90° dip) is illustrated in some of the Champion mine sections. A few words will explain how it flows out of the first case.

If we follow the two lines of 90° dip to where the earth covering becomes very deep, so that our distance from the magnetic mineral considerably reduces its influence, our two lines would evidently be merged into one, and continuing on to where the earth was still deeper, which has the effect of raising us above the ore deposit, this maximum dip would become less than 90°.

This maxima line would evidently correspond with the south line of 90° dip in the case first supposed, that is to say would lie nearly over the mineral producing it. With great depth of earth covering, it can be proven that it would lie to the north of the magnetic bed.

An inspection and consideration of the facts presented in the Spurr Mountain magnetic section given above, will, I think, convince any one of this without the aid of the rigid mechanical demonstration which the problem admits of.

I have seen large amounts of money unsuccessfully expended in digging for iron ore for want of a knowledge of the simple principles set forth above, hence I have dwelt longer on this point than its importance to the general subject would seem to warrant.

7. Additional Practical Suggestions and Rules.

The facts given above, with others in my possession, enable us to answer provisionally the following practical questions :

I. Can we by means of the magnetic needle determine the order of superposition or succession of beds of the iron-bearing rocks?

Comparing the magnetic sections obtained at the Republic mountain and Champion mines, it is evident that, while there is considerable variation in the details, the salient features agree remarkably, pointing towards the same order and same lithological character in the rocks. A number of other sections made within 10 miles of the above-named localities, across the same belt of rocks, gave the same general result.

It is therefore asserted with much confidence that where a magnetic section similar to these is found in the Michigamme district, a corresponding geological section will be found beneath the surface; and that, as a rule, there will be less difference in the magnetic sections than in the topographical, which we know depends greatly on the underlying rocks. But whoever expects to find many places where so complete sections can be obtained as these localities afford, will be disappointed, for they present rare opportunities for studying the structure and magnetism of the Huronian series.

In places the covering of drift will be so deep as greatly to reduce the intensity, making it exceedingly difficult to observe with ordinary instruments, as was the case at the Cannon location. Again, the lower magnetic rock, beds VI., VIII., and X., are in places far less magnetic, containing sometimes very little magnetite, as is the case south of the Washington mine. In other places the lower magnetic rocks may be entirely wanting, owing probably to a fault, as at the west end of the Champion. On the north shore of Lake Michigamme there is a magnetic bed above XIII. (the ore formation), being therefore younger than any member of the Republic mountain series. In other places XIII. is wanting, and when present it is sometimes highly magnetic, as at the Champion, and again it holds very little magnetite, as at Republic mountain, the pure ore there being mostly specular hematite, as has been elsewhere observed.

With all these uncertainties, however, the results of magnetic surveys cannot but be valuable in the exploration and development of iron properties, and in the solution of all questions of structural geology in regions of magnetic rocks. In such rocks, I believe, their value to the geologist is only second to topographical work, and, considering the cheapness of magnetic surveys, they may often pay best if means be limited.

Detailed magnetic observations, if made with precision, ought to throw light on the lithological character and intricacies of structure of these rocks, and on the nature of the magnetic force resident in them. This could not, however, be undertaken; the work done is more than was contemplated in my instructions and more than was justified by the means at my disposal.

II. Is it possible to determine quality—*i.e.*, the percentage of iron—in a magnetic rock by means of the magnetic needle? In other words, can the needle alone make us sure we have a workable deposit of ore under our feet?

This is the most important practical question connected with this subject, and is the one constantly presented to the miner and explorer. Magnetic observations should always be made in connection with topographical and geological surveys; whether these take such names, and are based on instrumentation, or whether they be such rude work as the explorer is constantly doing, but which are as much topographical and geological as the other, and often quite as valuable. A judgment of the commercial value of a bed of magnetic ore should, of course, be based on all the facts available. *If nothing more was known than what the magnetic needle revealed, I would not venture an opinion as to whether it was merchantable ore or magnetic rock which produced the phenomenon.* In the Marquette Region, as has been before observed, the chances are at least fifty to one that a worthless ferruginous rock is the cause of any observed attraction. But this case never occurs; we always know something more than the needle reveals. One of the most important uses of the needle, and one for which it can within certain limits be depended on, is in tracing magnetic beds in the direction of their strike *until some outcrop*, which may give us the information sought, is found. I have in this way traced magnetic beds for many miles both in the Marquette Region and in New York and New Jersey.

Preparatory to the examination of any particular range of ore, the exp'orer should thoroughly study up, with his own instrument, the phenomena presented at some exposed or developed part of the range he is exploring. This will give him data relating to variations, dips, and vibrations, which can be used where the rocks are covered and unknown. By means of the quickness of the vibrations, or of the rapidity with which they decrease as the compass is elevated, he may judge approximately of the depth of the drift, and so of other phenomena.

III. Does the magnetic needle afford the means of determining the absolute thickness of a bed of magnetic ore or rock?

My observations do not permit an affirmative answer to this question, especially if there be much earth covering. A study of all the magnetic sections which have come under my observation, indicates that, while in some instances the *comparative width* is plainly shown, the boundaries between the magnetic and non-magnetic rocks are not generally brought out sufficiently to warrant a definite expression as to thickness. We should expect this, because the magnetic influence is centred in the poles of the masses, and towards such foci the needle tends to point.

IV. Can we by means of the magnetic needle ascertain the direction and depth of a local magnetic pole? In other words, can we determine the thickness of rock or earth covering which overlies a given magnetic rock?

Often I think we can, with much precision, locate a point in the surface over the pole and determine its depth, by making what may be called a *magnetic triangulation*. Proceed thus: Remote from any magnetic rocks, neutralize, by means of a bar magnet, the earth's influence on the needle of a solar compass. The needle will then stand indifferently in all directions, and will not vibrate. Record carefully the distance and position of the neutralizing magnet; the compass is then ready for use. Set it up near the magnetic pole to be determined, and fix the magnet in exactly the same relative position it had before. The earth's directive power on the needle will again be neutralized, and the needle will point as nearly towards the local pole as its mode of mounting will permit; mark the line indicated by the needle on the ground; remove the compass to one, or, better, two other positions, and repeat the operation. If there is no other local force to interfere, the three lines must intersect in one point, which will be directly over the pole whose position is sought. By using a dip-compass in a similar manner, it is evident that the data to determine the depth, by the simple solution of a triangle, would be obtained. The fact that several local poles often influence the needle at each station renders this operation difficult in practice;—we should endeavor to find a place where but one strong pole exists.

A magnetic needle having universal motion, like Mr. Ritchie's, would evidently determine both position and depth at the same time; but a solar compass would have to be used to fix the position of the artificial magnet used in neutralizing the earth's force, unless it be fixed by an observation on the North Star, or by a meridian line brought in from a non-magnetic area.

V. When considering the magnetism of the rocks of the four great geological epochs represented on the Upper Peninsula of Michigan, I observed that considerable magnetic variations were noted by the Federal surveyors, over rocks of Silurian age, which had never been observed to be in themselves magnetic. In some instances these variations had been observed over a limestone, supposed to be Trenton, and at a distance of 75 miles from the nearest Huronian, or other (known to be) magnetic rocks.

This phenomenon may be due either: 1. To the presence of magnetite in such rocks, due to local metamorphism or other cause. 2. To accumulations of magnetic sand in the drift; or, 3. To the underlying Huronian rocks, which may be supposed to exert their influence up through the overlying Silurian.

Without having made a study of any of these localities, I incline decidedly to the latter hypothesis, as accounting for the known facts better than either of the others.

Should this prove true (and I hope to settle it at some future time) it may lead to a novel and interesting application of the science of magnetism to some important questions in geology—the determination of the thickness of sedimentary rocks by *magnetic triangulation* in places where it would otherwise be difficult to arrive at such thickness. It might also enable us to work out the structure and distribution, in a rough way, of these oldest rocks which underlie great Silurian areas, which would in no other practicable way be possible, thus throwing light on the nature of the rocky bottom of the ancient seas.

On the same principle we can, of course, trace magnetic iron belts under water. I have in many instances made very satisfactory magnetic observations from a canoe in the inland lakes of the Upper Peninsula. The bottom of Lake Superior may be thus partially mapped. Silt and sand will make no difference with the needle; it looks through everything but iron.

I have endeavored in the above to set forth plainly just what has been done in this comparatively new field, to give the results obtained, and to call attention to those principles which underlie the use of the magnetic needle in exploration for iron ores. The time and means at my disposal were meagre, my instruments imperfect, and I had no precedent to follow. I am persuaded that the subject is worth the attention of the explorer, miner, geologist, and physicist.

There has been a good deal written bearing on the subject of the Magnetism of Rocks, my references having very much increased of late. I had proposed to examine these authorities before writing this paper, but unfortunately the best libraries of Michigan do not contain any of the works referred to, and not being able to have abstracts made in Eastern libraries, I have derived no benefit from these authorities.* Could I have examined the results of the magnetic observations which must have been made in the great iron regions of Sweden, Norway, and Russia, I should probably have found my meagre results anticipated, and this article might not have been written. I am confident, however, that the Huronian rocks of Michigan have never been magnetically studied, and it may be that the methods that have been used in Europe are not such as would commend themselves to Lake Superior explorers, miners, and surveyors, who require cheap, light, and simple instruments that admit of rapid use.

The State of Michigan, or those interested in her Iron Regions, may at some future time see fit to have this subject thoroughly investigated. To that future investigator I commend my notes, trusting that he may find in them a reconnoissance of his rich field of labor.

---

* Gilbert's Annalen (German) contains several papers. See volumes 3, 4, 5, 16, 26, 28, 32, 35, 44, 52, 53, and 75.

•

CHAPTER IX.

METHOD AND COST OF MINING SPECULAR AND MAGNETIC ORES.*

THE iron ores of the Marquette region are mostly extracted in open excavations; hence the process is more nearly allied to quarrying. Several attempts at underground work have been made, which have not, on the whole, been successful. The Edwards mine has been almost entirely wrought by candle-light. The slate ore pit No. 1 of the New England mine was worked in the same way, as is also the Pioneer furnace pit of the Jackson mine.

The Champion mine was opened systematically for underground work, with two levels, sixty feet apart, and three shafts at distances apart along the bed of about 200 feet; but this idea has been so far modified that one-third of the ore of this mine is now extracted by daylight. The Cleveland mine has recently commenced to mine considerable ore underground.

Several other mines have, from time to time, worked underground stopes, but so far only temporarily; if such stopes could not be opened out to daylight, they have usually been abandoned. In brief, it may be said that no considerable amount of ore has as yet (1870) been mined underground in this region, and of that so mined very little has been taken out at a profit, and I may add that it seems to be the belief of the most experienced mining men that this state of things will hold for some time to come, for reasons which will appear.

Nearly the same remarks may be applied to the mines of the Iron Mountain region, Missouri, the ores of which are very similar in character to those of Marquette. Some of the New York and New Jersey magnetic deposits are also wrought open, but this is the exception, underground mining being there the rule.

---

* Two papers on this subject read before the American Institute of Mining Engineers and the American Society of Civil Engineers, and published, are embodied in this chapter.

The following brief sketch of the geological structure of the Marquette iron deposits will indicate some advantages of the method of mining employed; the subject being more fully considered in the chapters on the geology of the Marquette and Menominee regions, and illustrated in maps Nos. III. to X. of Atlas. See also Plate VIII., representing Edwards mine.* The iron-bearing or Huronian series of rocks are stratified beds, the principal ore formation being overlaid by a quartzite, XIV., and underlaid by a diorite, or greenstone, XI. This ore formation is made up, first, of pure ore; second, of "mixed ore" (*i. e.*, banded jasper and ore); and third, a soft, greenish schistose, or slaty rock (magnesian), which occurs in lens-shaped beds which alternate with ore, thus often dividing the formation into two or more beds of ore, separated by rock. Usually the beds of both ore and rock thin out as they are followed in the direction of the strike from a centre of maximum thickness, producing irregular lentiform masses. Since their original deposition, if we may assume they were laid down under water, the whole series, including the iron beds, have been bent, folded and corrugated into irregular troughs, basins and domes, which often present at the surface their upturned edges of pure ore, standing nearly vertical. A cross-section, finely illustrating this structure, can be seen on the west of the great south-west opening of the Lake Superior mine. It is locally known as the "Big W," which letter is plainly suggested by the sharp folding of rock and ore. See Fig. 7 and View IV.

The fact that, as a rule, the richest ore is found near the upper part of the formation, and the most jaspery part near the base, has led to the separation of this formation into two beds, Nos. XII. and XIII.

This structure, involving sudden changes in the amount and direction of the dip, from horizontal to vertical, would evidently necessitate, in the case of underground work, constant changes in the plan of attacking the ore, as well as in the mode of supporting the roof.

The magnetic iron deposits in the Eastern States may also be regarded as true beds, but are far more regular in strike and dip, extending downward at a high angle to an undetermined depth, and appearing more like veins. If folds exist, they are much deeper and more regular than in the deposits under consideration. The

---

* Many copies of this chapter will be distributed separately, rendering this geological *résumé* necessary.

Marquette ore deposits are often very thick, 50 feet being not infrequent, which makes ordinary timbering difficult, if not practically impossible; while the eastern deposits, so far as my observations have extended, are seldom over 20 feet, and average considerably less than that thickness.

The "pinch and shoot" structure, suggesting what are termed "chimneys" and "courses of ores" in some metalliferous mines and which is very apparent in the New York and New Jersey mines (practically dividing the ore into pod-shaped masses, the axes of which "pitch" in the planes of stratification in a direction quite different from the dip), can at this time be best observed in the Marquette region at the Edwards mine, Plate XIX. and Map No. VIII. Atlas. The intervening barren streaks where the hanging and foot-walls come near together, and which therefore divide the "shoots," form excellent supports to the overlying rocks and give the mine great security, as all who have worked deposits having this structure will testify.

The soft schist mentioned as occasionally bedded with the Marquette ores, often constitutes the hanging wall in parts of the mine, but does not possess the requisite strength to make a good roof. It is impossible to support such rock with occasional timbers or pillars, for it will scale off between the supports, demoralizing the men, if not actually endangering their lives. Even when the works reach the solid quartzite XIV., which, as has been stated, is the true hanging wall-rock of the ore formation, it is sometimes not safe, particularly near the surface. These facts make open workings a practical necessity at the start, and the great economy of breaking ore from high stopes with heavy charges of powder induces a continuation of the method, even when the rock covering has attained a thickness of many yards, and underground work would seem to be advisable. It is, indeed, hard to say what thickness of solid rock a Marquette mine-superintendent would hesitate to remove if it covered a large deposit of ore. Forty feet of earth and nearly as many of quartzite (as hard as granite) have been "stripped," and the thickness of rock is daily growing greater as the beds of ore are followed in depth.

It may be said, and I do not know but that it is a canon of mining, that all mines, which sooner or later have to be wrought underground, should be systematically opened as mines at the start, but this is not Marquette practice; and I have undertaken to describe, and, so far as I am justified, defend the methods there employed. It would be difficult to convince our people that, having a large deposit of pure ore before them of unknown form and size, covered often by but little earth, and backed by perhaps a small amount of money in the company's treasury, it is best to incur the delay and cost incident to sinking and drifting to open ground already opened by nature and ready to win. Wrought as open quarries, several of our mines have paid their way from the start, while, had they been opened on a regular system of mining, they would have required an investment of $50,000 in plant and improvements before shipments could have begun, and at least one year's time. Such facts settle such questions with American capitalists; and with the uncertainties which attend the opening of new mines in new districts, the high rate of interest in this country, and uncertainty of tariff legislation regarding iron, it may be a question whether this hand to mouth—quick return—let the future take care of itself—view of the question, is not in a certain degree defensible.

The appearance of our mines is anything but pleasing. They consist of several (sometimes of ten or more) irregular elongated pits, often very large and generally more or less connected, having usually an easterly and westerly trend imposed by the strike of the rocks. Everywhere are great piles of waste earth and rock, which are often in the way of the miner, and which in some instances have been handled over three times.

There are two principal advantages in open works. First, the preparatory work is all reduced to the simplest and safest kind of pick and shovel, hammer and drill, horse and cart business; such as can be let to the common run of mine contractors. On the other hand, underground mining involves sinking, drifting, timbering and elaborate machinery, all of which require skilled labor and large investments. In an isolated cold country like Marquette, the *quality* of the labor demanded is an important consideration. The second advantage, already mentioned, is the great economy in cost of drilling and explosives which high stopes in open works permit. These elements of cost are important items in all mining where hard ores are encountered. It is believed that they have been reduced to a minimum in the Marquette iron mines, where holes two inches in diameter are sometimes sunk 22 feet, and 15 feet is common. Such holes are not fired directly with the blasting charge, but are "shook" several times first, that is, fired with small charges which produce cracks and cavities about the bottom of the hole; when these are large enough to contain a sufficient amount of powder, the lifting charge is put in and the great mass thrown down. Twenty kegs of powder, of 25 pounds each, are sometimes fired at once, and from five to ten kegs is not an uncommon charge for a stope hole. By this method 5,000 tons of material have, in some instances, been removed at one blast, and one-third of that amount is quite common at some of the mines. In this way the entire cost for labor of drilling and explosives has been reduced, for a single blast, to less than three cents per ton. But the average cost is of course much greater, being at some of the mines 50 cents for all the drilling and powder consumed in the mine; about one-third of this is for block-holing the large masses thrown down by the stope holes, which are often so large that they have in turn to be broken by powder. The cost of powder and fuse for the hard ore mines, it is believed, does not exceed ten cents per ton. In some of the New York and New Jersey mines, which are worked underground, I am informed that these items cost much more. In the Persberg mines, Sweden, the drilling and explosives cost 65 cents per ton of ore in 1870.

It may be inferred, from the above description, that Marquette iron mining does not differ essentially from ordinary rock excavation on public works, being work that may be let by the cubic yard or ton. Until quite recently this has been very near the truth, the difference being in the skill and care required in separating the ore and rock which are often mixed together in the deposit. But these palmy days are rapidly passing for most of the mines now worked. An increase of water and greater cost of handling incident to increased depth, and, what is still more costly, the increase in thickness of the rock covering, will soon require, in fact does now (1870) really require, more expensive plants, different methods, and more skill.

The transition from the present system of quarrying to the future method of underground mining, which will have to be made in the Marquette region, will be a critical period, and will possess great interest, as affording a solution of a mining problem such as may not yet have been presented anywhere. Attempts at its solution have already been made, but, as has been remarked, very little ore has as yet been extracted at a profit by candle-light. To recapitulate, the system adopted will have to meet the case, 1st, of beds of ore varying, often abruptly, in thickness from 0 to 50 feet; 2d, of beds varying in dip from nearly vertical to horizontal, and passing by a curve of small radius from one inclination to another; 3d, of beds varying in character of hanging wall from a solid quartzite, which will stand with ordinary supports, to a soft schist, which can only be kept in place by a continuous support, or by actual filling in—"remblais." Again, the axes of the folds are not horizontal, but sometimes "pitch" at angles of 30 degs. or more in the direction of the strike, producing a fourth troublesome feature. See Map IX. Now, when we consider that the dressed ore is expected to yield 65 per cent. in the furnace, and is seldom worth on the average over $4 or $5 per gross ton on the cars at the mine, including royalty, the general character of the problem will be understood.

In New Jersey, with perfect regularity in the dip, better hanging walls, thickness within the limits of easy timbering, cheaper fuel and labor, and material which breaks easier than that of Lake Superior, the ores of several well-known mines, I am told, cost fully this amount.

Steam machinery for hoisting and pumping, which has cost from six to not less than fifty thousand dollars, has been erected at most of the Marquette mines, as shown by the table at the end of this chapter. In 1870, however, not much more than one-half of the entire ore product of the region was handled by steam, and much less than this proportion of all the material, the balance being done by horses, the use of which, however, is decreasing.

From these facts it may be inferred, that while the cost of breaking ore may have been reduced to a minimum by the system of mining employed, not so much can be said in favor of the methods of handling the ore from the miner's hands to the cars. The expensive horse and cart, swing derrick and whim, are in too general use, and the roads over which the loads are hauled are often not above criticism as to grades and surface. The causes which have led to this extensive use of horses are considered in another place.

The local staff of a Lake Superior Iron Mining Company usually consists of the *agent*, who is often secretary or treasurer of the company, and whose duty it is to take general charge of the company's business, except selling the ore, which is commonly done by a special agent in Cleveland, who may or may not be an officer of the company. This agent supervises the accounts, makes the payments, attends to shipping the ore and to ordering supplies, and often assists in selling ore. One man sometimes represents more than one company in this capacity. A majority of the agents reside in Marquette. The *superintendent*, who by custom has the title of captain, always resides at the mine, directs the work, and is in the main responsible for it. On him as much or more than on any other officer of the company does the success of its operations depend.

The offices of agent and superintendent are sometimes united in the same man. Large mines have a *chief clerk*, who is practically assistant superintendent. Next in order of rank are the foremen, master mechanics, and time-keepers. For names and addresses of agents, superintendents and managing officers, see Statistical Table, Plate XII. of Atlas.

The *organization* of the force of two large mines in the summer of 1870 is shown below. The first mine (I.) shipped the greatest amount of ore, and the second (II.) did most of its dead work in the winter, the aggregate shipments for the two, for that year, being 300,000 tons.

	I.		II.	
Contractors engaged in stripping, sinking, etc.	77		7	
Company account, men, laborers and mechanics on miscellaneous work	65		32	
Total employed in *dead work*		142		39
Contractors breaking ore	117		114	
Company account, men breaking ore			25	
Total at *mining* proper		117		139
Carpenters and wagon-makers	6		6	
Blacksmiths and helpers	17		10	
Total mechanics employed in *repairing*		23		16
Drivers and stable-men	20		12	
Engineers and firemen	11		8	
Loading ore from stock-pile	18			
Total *handling* ore		49		20
Superintendent and clerks	3		3	
Foreman, blaster and watchman	6		7	
Total *staff* at mine		9		10
Total force employed		340		224

This force was employed during the period of shipments, hence of greatest activity; after the close of navigation, in November, it would probably be reduced 25 per cent. Less than one-half of the men employed have families, many single men going "outside" in the fall and returning in the spring.

One large mine, the best managed in the region, expended in 1872, 51,000 days' work all told, of which 48 per cent. was by contractors, and 52 per cent. by the day or on company's account: it produced about 2¼ tons of ore for each day's work.

The *wages* of the men employed in and about the mines, in 1869 and 1870, were about as follows: Common labor was nominally $1.80 per day for most of the time, but by far the largest part of the mining work was done under contracts. Contractors made, clear of costs, from $60 to $77 per month as high and low averages; $70 is probably near the mean of the whole. It was not uncommon for a "pair" (two or more men working jointly) to make $100 per month each, and again the earnings will fall so low as barely to pay board; but such are extreme cases. Leaving out the staff of the mine and the contractors, the wages of all others, mechanics, engineers, firemen, drivers, but mostly common laborers, averaged in 1869 and 1870 about $2.12 per man per day. Mechanics received from $2.50 to $4.00. In 1872 the wages of men and contract prices were from 25 to 50 per cent. above the figures here given.

The *nationality* at three mines, which employed an aggregate of over 600 men, was in 1870 as follows, expressed in percentages:

Irish	31
English (Cornishmen)	27
Swedes	18
Canadians (French)	5
Americans	5
Germans	4
Norwegians, Danes, and Scotch	10
	100

The relative proportion of the Irish element is decreasing; a few years since nearly all the men employed at some mines being of this nationality. The percentage of Cornishmen is increasing, owing largely to a want of work in the copper region. These men are skilled miners, and do a large part of the sinking and drifting. Swedes are rapidly gaining in numbers, many of them having been miners in their own country.

The exodus of Swedes to the United States apparently threatens to depopulate that country. There can be little doubt but that a more genial climate and better food will improve the lower class, from whom the emigrants come. Statistics of the population of the Upper Peninsula are given in App. G, Vol. II.

The unit of measure and comparison in the following table is the *gross ton of merchantable ore*. The ore is the object of the miner's efforts, and the tons sold measure his business. The items of cost in all that follows express the expenditure per ton of ore mined, prepared for market, and loaded on the cars. In instituting a comparison between these figures and those obtained by the civil engineer on public works excavations, where the cubic yard of vacant space is the ordinary unit of work accomplished, it must be borne in mind that the labor incident to sledging up and sorting out the ore from the rock considerably enhances the cost of mining.

In order to more intelligently follow the methods of working the Marquette mines, we must classify the various items of cost under appropriate heads, and assume some absolute cost per gross ton, as near the actual fact as possible, as a basis of comparison of these items with each other, and with other mining regions.

No discussion of the question which leaves out the *cost*, would possess much practical interest; but all who have undertaken to obtain such facts for publication, know the difficulty, and will not place implicit reliance on the accuracy of what follows. $2.64 per gross ton will be assumed as the entire cost of mining the hard ore, and delivering it in the cars ready for shipment (in 1870); but this sum does not include interest on capital, expense of selling, royalty or mine rent, nor depreciation of the mining property. The cost of mining the soft hematite ores is considerably less, and the methods much simpler.

Royalties or mine rents have not become settled; there are not many leased mines; one of the best of its kind (the New York) pays but 20 cents per ton for first-class specular ore. In other instances 75 cents is paid for a lean hematite. Time and experience will settle these prices on an equitable basis. See Atlas, Tables XII., XIII.

Before dismissing the subject of royalty or mine rent, which is not again noticed in the following discussion, I will make a few remarks. Marquette mines, as has been stated, are generally owned and worked by the same parties, hence royalty does not enter directly as an item of cost, but it exists in substance, and may be called ***depreciation of the mine***, an item in the cost of ore often not sufficiently considered. One of the best organized and successfully operated iron companies in eastern Pennsylvania place this item at fifty cents per ton of ore. That is to say, every ton of ore sent from a New Jersey mine (which they own) is charged with fifty cents over and above its cost, as shown by the mine accounts, and a like sum is credited to the capital stock account, or to a sinking fund. This fifty cents stands for the original cost of the ore in the ground, and is all the more real, that it was paid in advance in the price of property and improvements. Any mining company which fails to recognize this principle is doomed some day to serious disappointment. Whoever has had experience with charcoal blast furnaces, which so rapidly sink their capital by the consumption of timber, will be fully alive to the importance of this matter. It is a delusion to suppose that our mines will not eventually be *exhausted*; iron ores do not grow; a ton shipped from a mine is gone forever, and the property has one ton less remaining, and is therefore worth less money. Continued shipments will eventually exhaust any and all deposits. Abandoned pits, in which no ore can be found, now exist at all of our mines, and in this class are some that two years ago were the best. The Andover mine, New Jersey, once presented as good opportunity to break ore as any pit now worked in the Marquette region; but about 150,000 tons aggregate product exhausted the mine, and to-day the owners do not know where to find a ton of merchantable ore on the property. I do not wish to be understood as predicting the exhaustion of the whole region; I think Marquette will produce iron as long as that article is wanted. New deposits of rich ore will be found, and leaner ones, which now have no value, will be worked, and the old deposits will be followed deeper; but this implies new mines, the building up of new locations, new railroads, new men and more capital. What I wish to say is, that unless present holders of average Lake Superior iron mining stocks are receiving fair interest on their investments, and in addition are being paid back the capital they have invested at the rate of, say, 50 to 75 cents per ton of ore sold, they are not doing a good business.

Therefore the $2.64 assumed in the following table should be increased by this royalty, making it $3.14. Commission for selling, interest and exchange, insurance and expenses of the general office of the company (including salaries), will increase this sum to at least $3.50, which will more truly represent the actual cost per gross ton of ore on cars and sold. This, from the amount assumed before as selling price, leaves from 50 cents to $1.50 per ton for interest on all fixed capital invested; in an exceptional condition of the market, like 1872 and 1873, the margins are of course larger.

There may be no better place than in this connection, to speak of another fruitful source of the disappointments which are sometimes experienced by stockholders. I refer to those delusive "*permanent improvement accounts*," better named permanent disappointment accounts, which are too often kept open, and in which are too frequently placed awkward sums which should properly go to running expenses, and be paid for by the pig-iron, ore, lumber, or whatever is produced. After the necessary real estate is bought, the mining or manufacturing plant built, and the business of production actually commenced, the improvement account should be closed forever. Some kinds of business, in some places, under some managements, may permit an opposite course, but the above is the only safe rule. If in any particular year an extraordinary expenditure is made which is not likely to be repeated, a part of it may properly be held in some open account, in order that it may be distributed over more than one year's product. But this is a different thing from piling up a permanent account under the delusion that the property is enhancing in value.

There are few kinds of business in which there is more danger from this cause than in iron mining, for not only is an iron-ore property depreciating from the exhaustion of the ore, but at any time it may be still more depreciated by unfriendly tariff legislation, for which the iron-master must be prepared.

TABLE *showing the Approximate Cost of Mining the Specular and Magnetic Ores of Lake Superior, made in 1870.*

General heads under which cost of mining is classified.		Elements of cost, not including royalty or depreciation.	APPROXIMATE COST OF EACH ITEM. In per cent. of the whole. Items.	Totals.	Based on a total cost of $2.64 per ton. Items.	Totals.	Amounts. Labor.	Supplies.
I. Dead work (preparation).		1. Explorations	00.6	28.1	.015	.742	Eighty per cent. .620	Twenty per cent. .122
		2. Sinking shafts	01.5		.040			
		3. Drifts and tunnels	06.1		.160			
		4. Roads	00.6		.017			
		5. Stripping earth and rock	13.2		.350			
		6. Miscellaneous work and minor improvements*	06.1		.160			
II. Mining proper (labor).	Drilling.	1. Ledge holes (in stope)	04.2	39.8	.110	1.050	1.050	....
		2. Block holes (in fragments)	04.9		.130			
	Other work.	3. Sledging, sorting, and loading	13.3		.350			
		4. Handling rock	09.5		.250			
		5. Miscellaneous work	07.9		.210			
III. Mining materials and implements ("mine costs").	Explosives.	1. Powder and fuse	03.6	11.9	.095	.313	.103	.210
		2. Nitro-glycerine	...†		...†			
	Tools.	3. Steel (drills)	00.7		.018			
		4. Tools other than drills	01.6		.043			
	Repairs.	5. Blacksmiths' supplies	01.8		.047			
		6. Blacksmiths' labor	04.2		.110			
IV. Handling ore from miners' hands to cars, and pumping.	By horses.	1. Teaming, labor of drivers and stablemen	05.7	15.6	.150	.413	.272	.141
		2. Forage	04.2		.110			
		3. Carts, sleds, harness, etc.	00.2		.006			
	By men.	4. Loading ore from stock pile	01.3		.035			
	By steam.	5. Labor, supplies, and repairs	04.2		.112			
V. Management and general expenses.		1. Salaries and office expenses. 2. Tax of all kinds.	04.6	04.6	.122	.122	.062	.060
			100.0	100.0	2.64	2.64	2.107	0.533

* Does not include exceptional permanent improvements.

† No reliable figures obtained.

In order to institute a comparison between American open-excavation mining and the systematic underground work of Sweden, I append the following table, for which I am indebted to Prof. Richard Akerman, of Stockholm :—

COST OF MINING ORE IN PERSBERG MINES, SWEDEN, 1870.

In currency.

General heads under which cost of mining is classified.	Elements or items of cost, not including royalty or depreciation.	In percentage of whole.		Based on a total cost of $2.20 p. ton.	
		Items.	Totals.	Items.	Totals.
I. In the mine	1. Boring	22.73	38.66	.50	.85
	2. Powder	5.82		.13	
	3. Priming reed	0.84		.02	
	4. Clay	0.50		.01	
	5. Candles, augers, and sledges	2.33		.05	
	6. Charcoal	0.44		.01	
	7. Auger whetting	3.20		.07	
	8. Shooters' fees	2.80		.06	
II. Water drawing (or pumping)	1. Water drawing	3.50	3.50	.07	.07
III. Bringing up the mountain (hoisting rock and ore)	1. Putting into the ton	4.68	11.17	.10	.26
	2. Receiving	1.11		.03	
	3. Down freight	1.10		.03	
	4. Hoisting	2.74		.06	
	5. Oil and lines	0.28		.01	
	6. Mine tubs and ladders	1.26		.03	
IV. Dressing	1. Dressing	8.12	8.12	.18	.18
V. Picking and washing	1. Picking and washing	5.65	5.65	.12	.12
VI. Buildings	1. Buildings	16.45	16.45	.36	.36
VII. General expenses	1. General expenses	16.45	16.45	.36	.36
		100.00	100.00	2.20	2.20

Professor Akerman furnished also these explanations :—

a. Our drill holes are about one inch in diameter and cost 7½ to 12 cents currency, per foot, when boring downwards, and twice as much when boring upwards.

b. Powder costs 11½ cents, dynamite 43 cents, and ammonium powder 40½ cents per Swedish pound (the Swedish lb. equals .93 of the English).

c. The reason why blasting with us is more expensive than with you, must partly depend upon stronger mountain ground and partly upon the small diameter of our augers.

d. "Dressing" on the Persberg table is to be understood as sledging and sorting.

e. "Picking and washing" is a kind of after-sorting by hand of the smaller pieces (of which about a third of the ore consists), got partly by blasting and partly by the first sorting.

f. "Buildings" include timbering in the mines and all buildings made for pumping and hoisting.

g. "General expenses" include some benefits for the laborers, such as domiciles, potatoes, gardens, expenses for schools, medicine, administration, etc., etc.

h. "Down freight" is the cost for bringing down the ore a short distance from the mines to the lake-shore, where it is sold.

i. Water power is used at Persberg both for pumping and hoisting.

j. Our miners receive from 48 to 75 cents per day, besides what I above called benefits.

k. The mining costs at Persberg are among the highest in Sweden.

The titles of the several heads under which mining costs may be divided, and the number of the items, depend on the object sought: the classification employed in the Marquette table, seemed best adapted to the presentation of the facts in hand. It will be observed that the form of the Swedish table differs materially and is of course better adapted to underground work, and to a more careful and laborious selection of ore.

I believe that considerable advantage would accrue to many of the Marquette mines, if the accounts were so kept that cost sheets similar to the foregoing could be prepared from time to time. It is well known that the cost of mining varies greatly in the different mines, some costing twice as much as others. This difference is often largely owing to natural causes, but sometimes it is, in part at least, in the management. There is no better way, in fact there is no other way, of stopping "leaks" of this sort, than by first finding where they are.

A comparison of such cost sheets from different mines, for the same time, or from the same mine for different periods, would indicate at once to which items the excessive cost belongs, and thereby direct the attention of the management to the leak. I therefore venture the opinion, that a carefully prepared cost sheet is one of the first steps in attempting to reduce the cost of ore.

In the detailed description of methods which follows, the items will be taken up in the order of the table.*

I. Dead Work.

This general head embraces all the work and costs incident to getting ready to mine the ore, and is subdivided into—1. Explorations (embracing only such searches for ore as are in progress from year to year about the mine). 2. Sinking shafts. 3. Drifts and tunnels. 4. Roads for wagons. 5. Stripping earth and rock, or uncovering the ore. 6. Miscellaneous work and minor improvements. The entire expenditure for dead work is 74 cents per ton of ore produced, which equals 28 per cent. of the whole cost.

1. **Explorations.**—More or less digging of test-pits, sinking shafts, drifting, trenching, and sinking drill holes is constantly in progress at most of the mines. My facts indicate that this work varies in amount from one-half to three cents per ton at the producing mines, being of course greatest at the new locations. It is not carried on systematically, being pushed when there is an increased demand for ore, or some old pit shows signs of failing, and again entirely discontinued. The price paid for pits 4 feet by 6 feet, and not over 10 feet deep, is from 30 to 60 cents per foot, depending on the ground; when so deep as to require a windlass, 50 to 75 cents and up to $1.25, if the shaft reach the depth of 30 feet and is wet. Drifting in firm earth will cost about the same per foot, depending

* For detailed descriptions of all the mine workings as they were at the close of the season of 1872, see "Appendix to A. P. Swineford's History of the Lake Superior Iron Region," being a review of its mines and furnaces for 1872, published by the Marquette Mining Journal.

on the depth below the surface and nature of the earth. Drill holes sunk by hand, material 15 feet deep, will cost from 75 cents to $1.00, and if deeper, considerably more per foot. There seems to be no reason why more use should not be made of the drill in this work. By means of a simple spring pole, such as was used in early days in the oil region, holes could be easily sunk 100 feet, which is as deep as it is usually necessary to go at this time. An experienced miner will judge very accurately of the ground passed through by the mud, and if there was any doubt, chemical analysis would determine the nature of the material; the mud furnishing a strictly average specimen, so desirable in an analysis for practical purposes. As has been mentioned, the *annular diamond drill* was introduced last season (in 1869) at the Lake Superior mine with success. A hole 130 feet deep was sunk at a cost of about $5 per foot; the core produced furnished very satisfactory knowledge of the substance passed through. The drill did not perform as well at the Washington mine, where several holes were sunk, the deepest 96 feet. In two instances the annular diamond bit got fast in an oblique seam and two were lost; not counting loss of diamonds, the work cost about $1.50 per foot: whether larger bits, a different setting of the diamonds, or more experience would overcome this difficulty, I do not know. It is a matter of great importance, and is worth thoroughly working out. As the subject of exploration for ore has been fully considered in another chapter, it is not necessary to treat it farther here.

2. **Sinking Shafts.**—This work, which forms so large an item of cost in some underground mines, varies in the Marquette Region, so far as I have ascertained, from 1½ to 5½ cents per ton of ore. Our open and comparatively shallow workings do not call for many shafts or winzes; the deepest shaft in the region is now (1870) not over 200 feet. The prices for this work range from a mean of $22.50 to $31.50 per foot in depth, depending on the hardness of the ground. In some mines, extreme prices range from $15.00 to $40.00, and even more if the shaft be very wet. Miners are often permitted to select the size most advantageous to themselves, which may be four feet by six; but eight by twelve feet is more common. The material is generally hoisted with the ordinary hand windlass, but sometimes with a horse-whip or whim, the miner having to deliver the stuff at the mouth of the shaft. From 10 to 15 per cent. of the

price received by the miner for sinking has to be expended in *mine costs; i.e.*, powder, fuse, candles, steel, tools, etc. No charge is made against him for smith's work. Sometimes the contract is let at so much per foot of shaft and so much per ton of ore, which gives the miner an interest in separating ore from rock.

3. **Drifting and Tunnelling.**—This element of cost varied more widely than any other, and might have been divided into two: (1) Drifts designed to open ground for stoping; and (2) Tunnels or adits for drainage and transportation of ore, the latter being of the nature of a permanent improvement. But on the principle that permanent improvement accounts are often permanent disappointment accounts, and to be avoided, and considering the fact that this kind of work is actually going on year by year, and must do so as long as the mine is worked, it does not seem wise to separate it from the current cost of getting ore. Ordinary 4 × 7 drifts cost, in hard ore, from an average of $22.50 to $24.50 per foot, the miners delivering the material behind them, and paying their own costs, as in the case of shafts.

Tunnels large enough to admit railroad cars and small locomotives cost from $30.00 to $50.00 per foot. The Washington tunnel, now over 1,100 feet long, and timbered a considerable part of the way, cost an average of about $40.00, not including rails. The timbered portion is twelve feet wide at the bottom, ten feet at the top, and ten feet high in the clear. No machinery has yet been brought to bear on either sinking shafts or drifting; the labor required is more than one-half expended in drilling holes for blasting. The subject of drilling is fully considered under its proper head.

4. **Making Wagon-Roads.**—The great amount of team-work employed about the mines requires a complete system of roads for summer and winter use. These are sometimes expensive on account of rock-cuts, costing, in some instances, as high as four cents per ton of ore in the early stages of work.

5. **Stripping Earth and Rock,** or uncovering the ore. This constitutes on the average nearly one-half of the dead-work, and is one of the largest single items in the whole cost of mining. So far as my inquiries extended I found it to vary from 20 to 52 cents per ton of ore. This cost is necessarily increasing at all of the mines worked as open cuts. It is simple rock and earth-work, the material being removed on wagons, carts, or sleds, drawn by horses.

The advantages of light railroads and small locomotives do not seem to have commended themselves for this work. There would, of course, be considerable danger of destroying tracks from blasting, and it often happens that not much work has to be done in one place; still there is no doubt but that a large saving would be effected by substituting steam for horses in portions of this work, as will be more fully considered hereafter.

The aggregate amount of material which has been handled in stripping is very great. Thirty and even forty feet of earth have been removed, and nearly as great a depth of rock; but this is the experience in open workings everywhere. I have seen twenty-one feet of earth and soft, shaly rock stripped from a nearly horizontal bed of 44 per cent. Clinton ore in Western New York, which did not average over thirty inches thick. In South-eastern Kentucky I found the rule among the miners of sub-carboniferous ores to be, that it would pay to remove a foot of earth for the sake of an inch of ore, which does not differ widely from the Western New York practice. In both of these instances the stripping was nearly the entire cost of mining, and labor was much lower than in the Marquette region. The usual contract price for removing ordinary earth (sand, clay, and boulders mixed together) is fifty cents per cubic yard, the digging costing about one-half, and the hauling one-half. Hauls vary from 100 to 800 feet. The highest price paid for excavating any considerable quantity of rock in open cuts, which has come to my notice, was $3.00 per cubic yard, equal to $24.00 per fathom, or about $1.00 per ton. This was a very hard jasper rock, containing but little ore. Large quantities of rock have been excavated and hauled over 500 feet at the Lake Superior mine for $2.50 per yard. The soft greenish schist, so common at all the mines, can be moved for from $1.00 to $1.40 per yard, including hauling. When a good face can be obtained on the overlying quartzite, which is likely to constitute the greater part of the rock to be moved in future, it should be broken down and loaded on wagons for from $1.50 to $2.00 per cubic yard.

The amount of money which it will pay to expend in stripping of course depends chiefly on the quantity of ore uncovered. If we assume fifty cents to be the maximum expenditure per ton of ore for this work (this amount has been greatly exceeded), the problem of what thickness of rock may be stripped admits of an easy theoretical solution. One cubic yard of solid ore (allowing for wastage on account of associated rock) may be considered to yield three tons of merchantable ore, which, at the allowance above assumed, would give us $1.50 to be expended per square yard in stripping a bed of ore only one yard thick. Hence in this case it would pay to remove nine feet in thickness of earth, or about three feet in thickness of rock. But suppose we have a bed of ore twenty-four feet in vertical thickness, which is a more common case, what amount of earth or rock would it pay to remove under the assumed limit of expenditure? Twenty-four feet of ore will yield twenty-four tons per square yard of surface, which, at fifty cents per ton, gives $12.00 available for stripping per square yard. This sum would remove twenty-four feet thickness of solid rock; or a foot in thickness of rock may be stripped for every foot in thickness of ore uncovered, at a cost of fifty cents per ton of ore. The same expenditure will remove three times this thickness of earth.

An important and often neglected question connected with this subject is, *where to deposit waste*, that it may be out of the way of future mining operations. Some material has been already handled twice in the Marquette region, and I know of a mine in Southern New York where the same earth was three times handled before it was finally permitted to rest. In a new region, like Marquette, where comparatively little thorough exploring has been done, it is often difficult to decide where waste piles will be out of the way for all future time. If a drill hole were put down for fifty feet in rock, and no ore found, it would be safe to say, that if ore existed under that spot, it would have to be mined under ground; hence, that so far as future stripping was concerned, a waste pile placed there would be out of the way. A very common practice in under-ground work, in some mining regions, is *to fill up the worked-out places with the waste*, and this can undoubtedly be done to advantage in some instances in open works, although it has not as yet been practised in the Marquette region. The trouble is to find out when a pit is exhausted—it is so common to break through a thin layer of rock and find a bed of workable ore behind it. But there are parts of most mines where the foot-wall has unquestionably been reached, and if any doubt exists, a few deep drill-holes will settle the point. When this is the case, and the foot-wall has a sufficiently gentle slope to permit of its holding materials deposited on it, it will, I think, be often found advantageous to use it to support a waste pile.

For the sake of illustration, take the New York and Cleveland Mine workings, which are adjacent. In this instance the slope of the foot-wall is so steep that it would probably be necessary to cut in it a rude step on which to rest a rough retaining wall, which could be built of blocks of quartzite swung across from the hanging-wall by means of a derrick. The triangular space thus formed would hold all the waste rock for a long time to come, and would afford a minimum haul. It might not answer to deposit earth in such positions, as heavy rains would be likely to wash it into the pits. The dip of the foot-wall in this, as well as in most cases, will, I think, become flatter in depth, so that a better opportunity will be afforded for a second similar waste receptacle at greater depth, if one should be required.

6. **Miscellaneous Dead Work.**—Under this head are included several items which were not of sufficient importance to require separate treatment. Improvements such as dwellings, shops, fences, tracks, trestle-works, pockets, docks, whims, skip-ways, pumping-fixtures, etc., etc., occurring from year to year, are embraced here. These items are in part embraced under " Building " in the Swedish table. This head was originally also designed to cover those exceptional expensive improvements which are of occasional occurrence only, and the cost of which might properly be distributed over several years' product. Additional facts, however, lead me to believe that the amount given (16 cents per ton) is too small. The expensive pumping and winding plants now being erected, and which will continue to be built for a long time to come, increase the cost of the ore materially unless we charge them to permanent improvement accounts, which is not altogether a safe course, as has been already pointed out.

II. Mining Proper, or Breaking Ore.

This general head embraces all the labor incident to blasting the materials down from the solid ledge, breaking it up into fragments that may be easily handled, the separation of the ore from the rock by hand and loading. The average cost of this is $1.05 per ton of ore produced, which equals forty per cent.

of the whole. The character of this work will be sufficiently well understood from the table and the following explanation :—

1. **Ledge or Stope Holes.**—The drilling or rock-boring is now (1870) entirely done by hand. The steel used for drills is 1¼ inch octagon, with a bit 2 inches, making a hole nearly 2¼ inches in diameter. Drills vary in length up to 24 feet. English steel is used at some mines, but a majority use American steel, and the most experienced men who have employed both, inform me that the drill steel made by Hussey & Wells and Parke Bros., Pittsburgh, answers as well as the best imported steel, and much better than the average. The drill is turned by one man sitting and struck by two standing, with eight-pound hammers, at the rate of about thirty-six blows per minute each. In this way from nine to eleven feet of hole are sunk per day, the men working usually on contract. The price of stope holes ranges from 60 to 80 cents per foot in depth, the mean being not far from 75 cents ; no mine costs have to be paid out of this price. When there is a large proportion of block holes, which admit of the use of smaller steel, the whole drilling of a pit is often let at from 60 to 65 cents. Very deep holes, say from fifteen to twenty-two feet, are sometimes sunk with still larger bits, which about doubles the cost. In these cases two men are required to turn the drill and three to strike.

The cost of drilling ledge-holes per ton of ore, varies from a mere trifle in the case where one twenty-two foot hole throws down 4,000 tons, as has been done, to a very large item on low stopes with perhaps tight, hard ground. From 3 cents to 25 cents per ton may be regarded as extreme averages, although 35 and even 48 cents have been reached, for short periods, under very unfavorable circumstances. The price given in the table (11 cents) approximates to the average for hard ores ; this number divided into 75 cents, the average cost of drilling per foot, gives, say 7, which should represent the number of tons of ore broken per foot of stope-hole drilled. The data obtained directly under this head confirm this amount, which is also equivalent to about two cubic yards per foot of hole.

The depth of stope-holes varies from two to twenty-two feet, the short ones being employed in "taking up bottom," that is, in squaring the stope so as to give the best chance for the deep holes. The average of 1,500 holes of all kinds in one part of the

Washington mine was four feet nine inches, but the stopes which furnished this result were below average height. It is believed that nine or ten feet would be nearer the average for deep holes, and say three and a half feet for the short ones.

2. **Block-Holes.**—The masses of rock and ore loosened by the heavy blasts already described, are often so large that they have in turn to be broken with explosives, which operation is termed block-holing. The amount of this work varies from almost nothing in some pits and in certain mines, to four-fifths of all the drilling required in others, the maximum being reached on high stopes of hard, tough ore. Over two hundred block-holes have been employed to one stope-hole in the Cleveland Mine, one hole being required to every two to four tons of ore. Block-holes sometimes produce fragments so large as to require block-holing in turn, before they are made small enough to be mastered by the sledge. These holes vary in depth from eight to twenty-four inches, the mean ranging near one foot. With nitro-glycerine the holes need not be so deep as for powder. One inch octagon steel is often used in this work, making a hole nearly 1¼ inches in diameter. The drilling is performed as in the case of stope-holes, but usually only one man strikes.

In the same ground, the same drill-gang will sink more than twice the number of feet of block-hole in a day with small steel, than of stope-hole with large steel,—ranging from twenty-four to twenty-seven feet. In open mines of strictly hard ore, this work costs more than stope-holes, and is set down in the table at 13 cents per ton. This amount added to the 11 cents given as the cost of stope-holes per ton, equals 24 cents for the total cost of the labor of *drilling* required under breaking ore:—this would also equal about 70 cents per cubic yard, which would pay for one foot of two-inch drill-hole. But this is by no means the whole; the work of sinking and drifting, which is set down as aggregating 20 cents, is more than half drilling; and a part of the cost of rock-stripping is also for this work. I estimate that 40 *cents per ton of ore* is not far from the actual price paid for this kind of labor in the hard-ore mines, equal to fifteen per cent. of the whole cost. On this estimate, not less than $300,000 were paid out for drilling in 1870. This work, from the favorable circumstances under which much of it is done in open excavations, no scaffolding being required, is by far the most purely mechanical labor performed about the mines. While the absolute cost of this item of drilling is very large, and can undoubtedly be reduced by the use of the *power-drill*, it is, as compared with some other mines and regions, small. Our open cuts or quarries afford far better facilities for blasting than under-ground mines. In one Southern New York mine the drilling cost, in 1870, $1.25 per ton of ore, or forty per cent. of the whole cost of mining; in a large magnetic mine in New Jersey, it cost from 60 to 80 cents per ton of ore. In the Persberg mines, Sweden, when the ore cost, in 1870, $2.20 currency per ton, the drilling was 40 cents per ton, equal to twenty-three per cent. of the whole cost, being considerably more than ours, absolutely and relatively. When we consider that the average of wages in Sweden is not far from 65 cents per day, or say one-fourth of what is paid Lake Superior miners, it would seem as if Sweden would be a good field for a power-drill.

The facts relating to drilling have been given in much detail in the hope that inventors and owners of rock-drilling machines may become acquainted with the wants of the Marquette region in this regard. I have had my attention called to several of these machines, but have not had opportunity to make such investigation of their respective merits as would justify an opinion. I have no hesitation in saying that a machine which would do the work required at a less cost than it is now done (75 cents per foot) would find ready sale, and every facility would be afforded for experiments.

I need not here remark that a power-drill, adapted to Marquette iron mines, must be portable, as it would have to be shifted every few hours; and I should say that two men, or at most three, should be able to handle it on a ragged rock surface. Again, it must be capable of being set up anywhere, to accomplish which, I think that movable tripod, telescopic legs, like those with which engineers' instruments are often supplied, would be convenient.*

3. **Sledging, Sorting, and Loading.**—In considering this item, it must be borne in mind that the ore and rock have not only to be broken so that they can be removed, but must be made so fine as to

---

* Since the above was written the Burleigh Drill has been tried at several mines with varied success. My facts are quite insufficient to enable me to form a judgment as to its fitness to do the required work, or to know whether it has had a fair trial.

be easily separated, and so that the pieces can be fed into a Blake crusher. This work requires more muscle and as much skill and care as any other done at the mine. Eighteen to twenty-three pound sledges are employed, and the difference in results, between the experienced miner who strikes the lump of ore the right blow in the right place, with this immense hand hammer, and the tyro, is very great. Contracts for sledging and loading, which sometimes include a little block-holing and short tramming, have been let at prices varying from 20 to 50 cents per ton. The loading usually costs not to exceed 10 or 12 cents, the balance being chiefly sledging. There is a wide difference in the texture of ore, some kinds requiring five times as much sledging as others. On the whole, Marquette ores break with much greater difficulty than those of the Eastern magnetic mines. With poorer ground worked and the market more in favor of buyers (which makes them more exacting on quality), the cost of this element will be increased.

Drops, similar to those used at foundries to break old castings, have been employed to break very hard lumps of ore, but the expense of getting the lumps of ore to them has caused this plan to be abandoned. In the copper region powerful steam hammers have been used for a similar purpose, but the same objection as that given above would apply to their introduction at the iron mines. It must be borne in mind that a lump of iron-ore is not worth more than about one-hundredth part as much as a lump of copper of the same weight, and therefore will not bear as much handling.

A steam miner who can walk up to the lump of ore and sledge it to pieces where it lies is what is wanted. Nitro-glycerine or duallin breaks the material finer, producing by its explosion more of a smashing effect than powder, and thereby requiring less sledging. There is no doubt, as is elsewhere stated, about the advantage of employing these new explosives in block-holing.

4. **Handling Rock.**—In addition to the rock which overlies the ore, considered under stripping, at most of the mines more or less rock is found mixed with the ore through the mines, which has to be removed during the process of mining. The proportion varies from none up to one-half of the whole, and often for short periods more than this; the average at this time is believed to be twenty per cent. The 25 cents placed against this item in the table is intended

to cover the cost of sorting out and handling this rock under average circumstances. This cost will be increased as poorer grades of stuff are worked.

5. **Miscellaneous Work.**—The 21 cents opposite this item in the table is no more than sufficient to pay for foremen, repairs of tracks and roads, wheeling, tramming, blaster, sometimes hand-pumping, and such securing of the workings as may be necessary, etc.

III. Mining Materials and Implements, embracing "Mine Costs."

This general head is subdivided in the table into Explosives, Tools, and Repairs, which are in turn itemized, as will appear below. The expense incurred here is 31⅓ cents per ton of ore produced, equal to about twelve per cent. of the whole cost.

1, 2. **Explosives.**—Powder and fuse and nitro-glycerine. The present (1870) is an unfortunate time to collect statistics regarding the cost of explosives, for the reason that nitro-glycerine is to a certain extent on trial, and most of the mines employ both it and powder in the same pits, making it difficult to separate the results. The place of the new explosive cannot be said to be wholly fixed in our mines. It is more powerful than powder, bulk for bulk, or weight for weight; can be used in wet as well or better than in dry ground, which is very important in some places; it has so far proved no more dangerous than powder, and its fumes have not been found objectionable. As has been stated, the fragments resulting from its use are usually smaller, hence require less sledging, and, it being more powerful than powder, less drilling is needed.

In the case of wet holes intended for sand-blasting, nitro-glycerine can often be used in small charges to produce cracks which carry off the water and thus prepare the way for the powder. Overhanging loose rock can often be advantageously brought down by a flat cartridge of glycerine.

In short holes, 3 to 6 feet, glycerine will sometimes break two or three times as much ground as powder, thus making the saving on the drilling more than balance the extra cost of the explosive.

The quantity of glycerine used per hole, of course, varies with its depth and other circumstances, and is at the Washington and Republic Mines, according to Captain Peter Pascoe, as follows :—

Depth of hole.	Glycerine.
3 feet	¾ lbs.
4 "	1½ "
5 "	2¼ "
6 "	3½ "
8 "	5 "
10 "	7 "
12 "	10 "
14 "	14 "
16 "	18 "
18 "	21 "
20 "	24 "

There can be no doubt but that the use of this explosive hastens work. Sinking and drifting can be more speedily done with it than without.

Whether it is suited to breaking the great masses from the solid ledge remains to be seen. Certainly it cannot be used to fill the cracks produced by shaking, where heavy sand blasts are required; and it is doubtful whether drill-holes large enough to contain the requisite amount of the blasting oil can be profitably employed; two or more holes could be used, but this would greatly increase the cost of drilling. It certainly costs *more* per ton of ore mined than powder, but how far this greater cost is balanced by other advantages experience must determine. It is significant that in 1870, being the next year after its introduction, over $40,000 worth was sold in the Marquette region at $1.50 per pound. In 1872 about 40,000 pounds were used, the price being $1.25 per pound. The Painsville Ohio Co. erected (1871) a factory near Negaunee. Duallin and giant powder have recently been introduced.

The figures given in the table, and in what follows, refer exclusively to powder, the nitro-glycerine element having been eliminated as far as was possible. Fuse costs about ½ cent per ton, leaving 9 cents per ton for powder, which, according to the data obtained, varied from 7 to 10 cents. The price of powder ranged from $3.75 to $4.50 per keg of 25 pounds. Therefore an average of 45 tons of ore should have been broken with one keg of powder, or about ½ pound of powder to one ton of ore. This, it must be remembered, does not express the actual work of the powder, on account of the amount of rock moved in addition to the ore—in one instance 23,000 weighed tons of material required 320 kegs of powder, or 72 tons per keg. In another instance 31 kegs threw down 3,500 tons (approximate) of quartzite, or 113 tons per keg. One mine, which produced over 100,000 tons of ore in 1869, consumed for all purposes one keg of powder to every 43 tons of ore produced. The waste material in this case did not amount to over 20 per cent., hence about 52 tons, or, say, 18 cubic yards of material, were moved per keg of powder. The consumption of explosives per ton of ore must increase as the mines grow deeper, either by the greater amount required to remove the rock covering, or by the less favorable opportunity afforded for blasting, if the ore be won underground.

In one group of New Jersey mines, the powder and fuse in 1870 cost 18 cents per ton; in another mine in Southern New York, 14½ cents; in Sweden, at the Persberg mines, 15 cents. All of which figures considerably exceed those reached in Marquette, which is proof of the economy in explosives from working iron mines as open quarries as long as possible.

3. **Steel.**—The use of steel drills has already been described, and reference made to the brands in use. My data, which are far from complete, under this head, indicate that the cost of steel per ton of ore ranges from ¾ to $3\frac{3}{10}$ cents, averaging perhaps $1\frac{8}{10}$ cents; the price of steel being 20 cents per pound. This would give about 11 tons of ore, or about 3 cubic yards per pound of steel consumed, which is less than the data obtained direct on this point seemed to indicate.

It is the practice of some mines to charge the ore contractors 2 per cent. on their contracts for wear of steel, which agrees nearly with the above. At other mines the steel is weighed at the end of each month, and the contractor charged with the shortage, whatever it be.

4. **Tools, other than Drills.**—Cost about $4\frac{3}{10}$ cents per ton of ore.

The Ames No. 2 D-handled, square, and round-pointed, strap-backed, solid steel shovel is the favorite.

Washoe picks, Nos. 5 and 6, and Powell, same numbers, both

railroad (25 inches long), and pole (19 inches long) are extensively used. Certain mines make their own picks after a fashion of their own.

Solid steel crow-bars, both single and double-pointed, are used.

Solid cast-steel sledges, both American and chrome, weighing from 16 to 18 pounds, and often 25 lbs., are extensively used.

Solid cast-steel striking-hammers, 8 to 9 pounds, and in some instances 11 pounds, are employed.

5. **Blacksmiths' Supplies.**—This item is largely made up of coal and iron, steel being embraced under another head. Charcoal was formerly used exclusively for working steel; but mineral coal is now employed with good results at most mines. The table shows this item to be a trifle less than five cents per ton of ore.

6. **Blacksmiths' Labor.**—This is largely sharpening drills. The number dulled per day by a gang of three drillers will average about 75, in hard ore. One blacksmith and helper will sharpen about 275 drills per day of ten hours. The 11 cents marked opposite this item embraces all the blacksmiths' work done in and about the mine, for whatever purpose. Therefore strictly, it should have been divided, part going to dead work.

IV. Handling Ore from Miners' Hands to Cars, and Pumping.

Pumping, which has heretofore been a small item in the Marquette region, cannot well be separated from hoisting ore, as the same machinery does both. This item, in the case of some New Jersey magnetic mines, costs 75 cents per ton of ore: at the Persberg mines, Sweden, it costs but 7 cents. The entire cost under this head, in the Marquette region, including hoisting and pumping, is 41 cents per ton of ore produced, which equals 15½ per cent. of the whole. This work is done in part by horses, part by men, and part by steam.

1, 2, 3. **The Work of Horses in Handling Ore.**—The team work employed at the Marquette mines, apart from the stripping, amounts, according to my inquiries, which have been quite full on this point, to 10 per cent. of the whole cost of mining, or say 27 cents per ton of ore, the drivers' wages being the largest item. This cost is obtained by dividing the total expenditure for teaming, by the total number of tons of ore produced. If it was figured only on the ore actually handled by the horses, it would be much greater. If to this were added the cost of the team-work employed in stripping, the total would not be less than 30 cents per ton of ore, or, say $250,000 on the product of 1870, a sum sufficient in itself to supply all the mines in the region with all the additional steam-hoisting and pumping machinery and small locomotives required to do the work now done by horses, and at a very much less yearly cost. We may verify this almost incredible estimate in another way. The total number of horses employed at all the mines in 1870, including hired teams, was about 364, or an average of 30 to each mine, varying from 9 to 74. The best data I can get indicate that to work a lot of horses for one year, including wages of drivers, stable-men, smiths' work, forage, repairs of vehicles, and depreciation, in the years 1869 and 1870, cost an average of $650 per horse. The wages of hired teams, including drivers, for the same period, was $6 per day. At this rate, 364 horses would have cost nearly $240,000, a sum sufficiently near the other to confirm the general truth of the estimate.

These figures surely justify the prediction, that if there ever comes a period when our mines do not pay, it may be due largely to horses. In this age of steam, has a business any just right to prosper which employs horses to do work that can be more cheaply done by machinery? The average number of tons of ore handled per horse employed in and about the mines for all work in 1870 was 2,350, ranging from 1,150 to 5,300 tons. In considering these facts it must be borne in mind that the mines in question are not by any means without steam power. Twelve engines, varying in power from say 10 to 50 horse, were at work. To prove that this item of cost is unusually large in the Marquette region, I will give a few facts regarding the employment of live stock at mines, which have come under my notice elsewhere. While the cases cited do not present all circumstances like the Marquette mines, they are sufficiently near to afford interesting comparisons.

The Cornwall Ore Bank Co., Penn., shipped from their one immense deposit, in 1870, over 174,000 tons, employing no horses in the work. The ore was all handled by one locomotive, the cars being loaded by wheelbarrows. No pumping is required in this mine, and the facilities for reaching the ore with cars are unusually good. The ore is quite soft, so that the blasting does not endanger the tracks.

The Iron mountain mine, Missouri, shipped in 1870 more ore than any one mine in the Marquette region. It employed during the winter 68, and during the summer a somewhat less number of horses, mules, and oxen. One animal moved about twelve tons per day, or 3,600 tons per year; but more than three-fourths of this stock was employed in getting "surface ore," a feature which does not exist in Marquette mining. The bluff (quarried) ore moved per horse employed was more than five times the above amount. No steam-engine or locomotive was in use at the mine.

At the Caledonia and Keene mines, St. Lawrence County, New York, in 1869, three horses handled 27,500 tons of ore and waste, the average haul being over 700 feet, all up grade, in places steep. This gives over 9,000 tons per head; steam was not employed for handling material at either mine.

The Sterling mine, Orange County, New York, shipped in 1869 40,000 tons of ore, which was handled under circumstances quite similar to those encountered in the Marquette region, by two horses and one small stationary engine, which gives 20,000 tons per animal employed. The system of tramways and sidings at this mine is very complete.

Passing from American to Swedish mines, which are far deeper, and in which there is a larger percentage of rock mixed with ore, we find that in the Persberg mines, in 1870 (see table), the total cost for handling ore and water drawing was 14[3] per cent. of the whole cost, or 33 cents per ton of ore; and this amount included the handling of all the rock and other waste material which in our table is embraced under *Dead-work*. If we take out of dead-work 10 cents for handling this waste and add it to the amount found above, we have 51 cents as total cost of handling Lake Superior ores, equal to twenty per cent. of the whole cost, or about fifty per cent. greater than in the Swedish mines, but there water was exclusively used.

It is not difficult to understand how *horses** have come to play so important a part at our mines.

---

* It should be noted that oxen have been in use for some time at the Lake Superior mine, but, so far as I am informed, at no other.

The first operation in opening a new mine is, usually, to strip off the earth and rock covering, which can be best accomplished with the horse and cart. On the ore face thus exposed, mining is begun, the ore being hauled to the cars (often not brought very near to the pit), and such rock as is mixed with the ore is sorted out and hauled in another direction. It is very convenient and economical to back a cart directly to the miners' hands, and this was done until it came to be regarded as *the way* to get out ore. There was certainly no better way at the start in many cases; but when horses come to be used on hauls of over 500 feet and up grades, in places as steep as 1 in 10, the operation costing 25 to 30 cents per ton, it may be worth while to ask if such ore had not better be left in the ground until machinery propelled by steam can be brought to bear on it. Another cause which conspired to prolong this expensive mode, was the great demand for ore during the war and the consequent high prices. Mine superintendents were given no time to plan nor make improvements looking to future economy. Mine owners did not then want surveys, nor machinery, nor tunnels, nor anything that had reference to the future; they only wanted ore, nor did they care much what it cost, nor what the quality was (so consumers say): it was ore, ore, ore! Wherever three men could be set at work, a cart was backed up to them and shipments began from a new pit.

On short hauls, smooth roads, and light grades, horses can be used to advantage, and will continue to be so used, especially where there is more or less uncertainty as to the quantity of ore in the pit worked, which is often the case. But where there is a large mass of ore, rock, or earth to be moved under any other circumstances, it will usually pay to bring steam-power to bear upon it. Portable, or easily-to-be-moved railroads, and small locomotives for long hauls are in much favor at this time, and would have the advantage of utilizing existing wagon-roads. But the first step in many cases is undoubtedly to lay horse railways on the present roads. As is shown above in the remarks on the use of horses in certain New York iron mines, one animal can move from ten to twenty thousand tons on such roads in one year. If the horses at our Marquette mines can be made to perform one third this amount of work, the present cost of hauling will be reduced fifty per cent.

Portable hoisting-engines are extensively used in New Jersey and

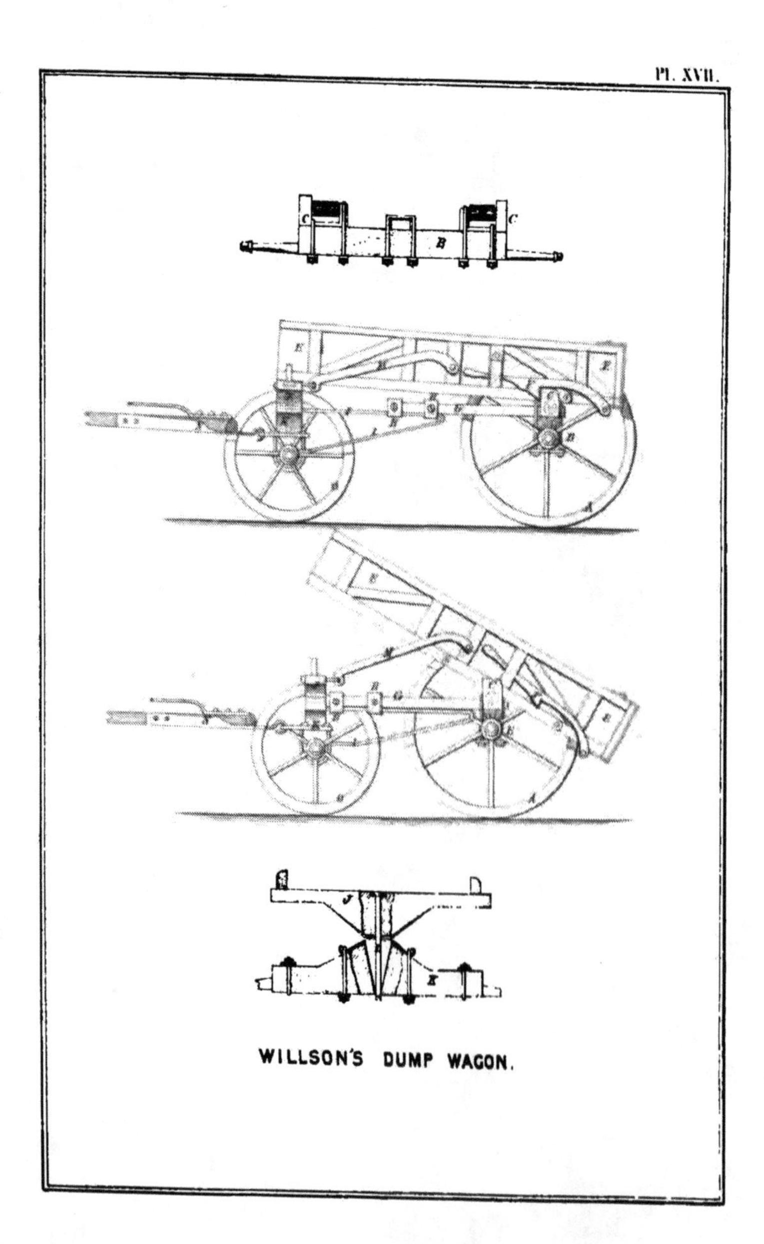

WILLSON'S DUMP WAGON.

Pennsylvania; they can be set up quickly just where wanted, and handle material rapidly and with great economy. A thorough system of under-ground communications which would bring all or most of the material to the main hoisting-shaft is always to be aimed at, as in this way the dead lift may be made by steam. At present, owing to the continued pressure for ore, it is not uncommon to see ore and rock carted up-hill, over abominable roads, from pits which in a few months, perhaps, will or could be reached by drifts along which the ore could be cheaply trammed to a steam hoisting-shaft.

As may be supposed, this extensive use of draught animals has led to great perfection in the carts, wagons, and sleds. A dump-sled for winter use, contrived by Captain Merry, of the Jackson mine, is a perfect vehicle of its kind. I am unable to give drawings of but one, known as Daniel Willson's Patent Dump Wagon, of which over 50 are in use in the region. See Plate XVII.

While harnessed to the cart or wagon is the favorite mode of using the horse, it is by no means the only way. Some pits in the course of mining became too deep for cart roads; these were in many instances worked by *swing derricks*, horses being the power employed; the long booms of these derricks made it possible to drop the bucket in different parts of a wide pit. This method is, however, very expensive, as the following figures will show. The total lift from bottom of pit to bottom of cart was in one case 79 feet; the cost being as follows:—

2 men filling	$4 00
1 man to land	2 00
2 derrick horses and driver	5 25
	$11 25

This sum paid for hoisting 45 tons in 10 hours, is equal to 25 cents per ton. In one case, where the hoist was 55 feet, the cost was 16 cents per ton.

In another case, with the ordinary two-bucket horse-whim, the cost of hoisting 65 feet, and landing, was 6 cents per ton; this did not include filling the buckets. In another case the ore was hoisted 40 feet, and landed for 5 cents per ton, not including the filling. Estimating the filling at 10 cents, these facts show that it costs in the cases cited an average of 1 cent to lift one ton of ore 7 feet, including the landing or dumping, which employs one man.

Without attempting to fully solve the important problem of the best mode of handling the material at Marquette mines, for that is beyond the scope of this report, I would suggest the following general policy as being safe for the mines to pursue :—

Let all large pits now worked, where a considerable amount of horse labor is required, be suspended until some form of steam machinery can be brought to bear on them. There are, of course, exceptions to this rule : for instance, where the other costs are unusually light, more money may be expended in handling the ore, as is often the case with the soft hematites ; but the principle is, I think, correct. It would not be difficult to find many instances of this kind ; for example, a given pit is worked, the ore being moved by horses, at a profit say of 50 cents per ton, which if left for one year could be reached by some tunnel or other improvement which would permit the same ore to be taken out at a profit of $1.00 per ton ; it would certainly pay to wait in such instances. In these cases it will usually be found that the superintendent has been persuaded into promising that his mine can be made to produce a certain amount of ore which may have been already sold, his attention being thereby fixed on a large product, rather than cheap mining. This subject will be considered more fully below. I will here only ask, if it is not better policy for a mine to net say $50,000 on 50,000 tons of ore, than to make the same sum on 100,000 tons. If the mines were inexhaustible it might not make much difference, but as it is, it may make all the difference there is between a profitable business and an unprofitable one in the end. It must be borne in mind, that while the ore business has been on the whole profitable, there are large mines that have been producing ore for years that have never returned a dollar to their stockholders.

Among the mining appliances which have been brought to great perfection in the Marquette region, are the various forms of pockets and shoots for transferring the ore, first, from the mine cars, buckets, and carts to the railroad cars, and second, from these to the vessel.

The magnificent ore docks at Marquette, Escanaba, and L'Anse belong to the latter class, and are undoubtedly the best of the kind

Description.
a. Ore pockets from which it slides into cars or carts.
b. Door. b'. Door open.
c. Rollers, for mine hoisting.
d. Pulleys with rope to open the Door.
e. Pulleys with chains to shut the Door.
f. Toothed wheel, with
g. Lever and
h. Catch, to turn the chainpulleys in shutting the Door.
i. Catch, for toothed wheel after shutting the Door.
k. Summit of track, coming out of the mine and going on by horse.
Scale ½ inch - 1 foot.
ORE POCKET - CLEVELAND MINE.
Vertical Elevation.
Cross Section.
Scale ¼ inch - 1 foot.

in the United States if not in the world. They are described and illustrated in Chapter I., and in Appendix F. of Vol. II.

Of the first class there are numerous varieties, from the simple log crib built up alongside and above the track, into which the ore is dumped from elevated railways, and from the sloping bottom of which it is "shot" through holes closed by rods into cars at a cost of not over $3\frac{1}{2}$ cents per ton, to the more expensive and perfect contrivance employed at the Cleveland mine, which is shown in Plate XVIII.

The mine car in this case passes over the centre of the pocket, which dumps its ore in turn into a car or cart below, by an ingeniously arranged door which is shown on an enlarged scale.

4. **Loading Ore from Stock Pile.**—During the winter no shipments are made from the mines, hence the product has to be piled up. It is the policy of some mines, and I think it is the best, to do most of their dead work in the winter, hence to stock but little ore; others maintain nearly the same rate of production in proportion to the force employed, winter and summer. Stocked ore has to be loaded in cars by hand, which is always contract work and costs from 9 to 12 cents per ton, the mean being, say 11 cents, including all costs connected with it. This amount, distributed over the whole product for the year, was found to average for the cases inquired into, $3\frac{5}{10}$ cents per ton.

5. **Machinery for Pumping and Hoisting.**—Notwithstanding the great cost of the work of horses, a large amount of machinery, as has already been remarked, is now in use, as the following statements will prove:—

The introduction of machinery has so far seemed to make but little relative diminution in the number of horses employed, because of the greater amount of waste material which has to be moved in the later years. The amount given in the table, opposite this item, $11\frac{2}{10}$ cents, is designed to be an approximation to the cost of running the machinery of such mines as have plants distributed over the entire product of those mines. I estimate that less than one-half of the product of such mines was handled by machinery in 1870. The actual cost of moving the ore so handled, including the *pumping*, varied from 14 to 21 cents, the mean, as shown by my data, being about 18 cents. This cost is made up of wages of engineers and firemen, say fifteen per cent.; fillers, landers, and surface tramming, sixty per cent.; fuel, repairs of machinery and supplies, say twenty-five per cent. This covers the cost from miners' hands to cars or stock pile.

While this sum is materially less than the cost of the same work by horses, it is much greater than in the Copper region of Lake Superior, where this work is brought to great perfection. Some of the appliances employed in the Copper region cannot be used at iron mines on account of the greater irregularity of the deposits. But time will introduce many economies which will reduce this item below the figures given. It must be borne in mind, in comparing the cost of steam machinery with horses, that in the case of the engines all the pumping is included, while the horses handle only the ore and rock. Making this correction, it is safe to say that it costs at least four times as much to handle the same material by horses as by machinery.

The following description of recently erected plants will give a good idea of the machinery now in use at the iron mines, it being essentially such as is employed at the copper mines.

The *Macomber mine machinery* consists of one steam-engine with cylinder 18 x 24 inches, with bed cast solid in one piece. Valve is of the kind known as the H valve, and is worked by link motion; steam pipe 4 inches in diameter; exhaust pipe 6 inches in diameter; engine supplied with the Judson governor. Pump for feeding boiler is worked from cross-head; also an auxiliary for fire protection, etc. Main shaft is 5 inches in diameter, of hammered iron, and 16 feet long. One boiler 48 inch shell, 26 feet long, with two 18-inch flues. Smoke-stack is 40 feet high and 24 inches in diameter. The winding drums are 4 feet in diameter, and of sufficient capacity to contain 525 feet of 1¼ inch wire rope. They are worked by a friction movement, thrown in and out of gear by means of eccentrics with lever attachments. The brakes are known as band-brakes, which clamp the entire surface of the drum, 5 inches in width, and are of sufficient power to hold a loaded skip at any point in case of accident. They are worked by levers with hand or foot, as may be desired. The drums make about 13½ revolutions per minute, the engine making 80, which gives the skip a speed of a trifle less than 3 feet per second. The skips are of heavy boiler iron, each having four 12-inch wheels. The capacity of each is 35 cubic feet, equal to about 2½ tons of ore. The pump is 10 inches in diameter by 6 feet stroke, capable of discharging 660 gallons of

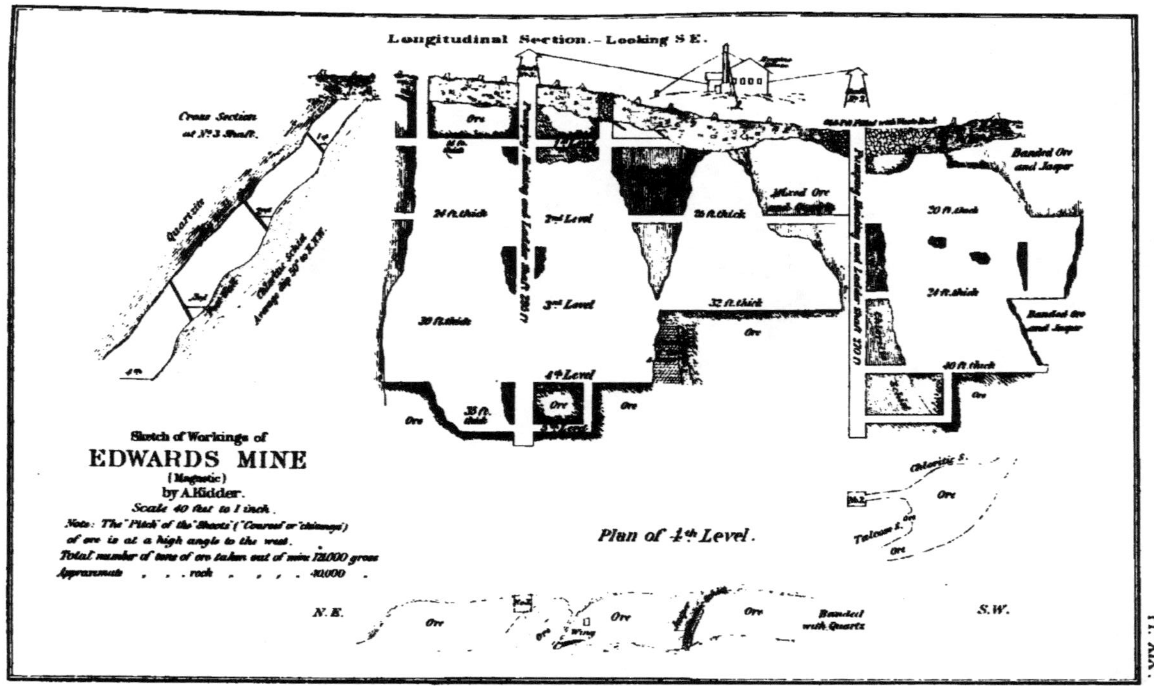
Longitudinal Section.- Looking S.E.
Cross Section at No. 3 Shaft.
Quartzite
Chloritic Schist Average dip 50° to S.E.
Pumping, Hoisting and Ladder Shaft 230 ft
Pumping, Hoisting and Ladder Shaft 270 ft
Ore
24 ft. thick
2nd Level
3rd Level
4th Level
30 ft. thick
26 ft. thick
32 ft. thick
Mixed Ore
Banded Ore and Jasper
30 ft. thick
24 ft. thick
40 ft. thick
Plan of 4th Level.
Chloritic S.
Talcose S.
N.E.
S.W.
Banded with Quartz
Wing
Sketch of Workings of
EDWARDS MINE
(Magnetic)
by A. Kidder.
Scale 40 feet to 1 inch.

water per minute. It is worked from a slotted crank arm, on end of main drum shaft, which admits of lengthening or shortening the stroke at pleasure. The pump is double acting, with single valve on a new plan. It is furnished with rods, travellers, connections, balance bobs, etc. This machinery was furnished complete in all its parts, and set up at the mine in working order for pumping and hoisting by the Iron Bay Foundry, Marquette, Mich., 1872.

The *Barnum mine plant* consists of one horizontal high pressure steam-engine of 20 inches diameter of cylinder and 30 inches stroke; steam furnished by two tubular boilers, each 48 inches in diameter and 14 feet long, and each containing 50 tubes, three inches in diameter. Maximum power of this engine is 120 horse, but is working at present at one-third its capacity. There are two winding drums, each 5 feet in diameter; speed of engine about 60 revolutions per minute, and of drums about 12. Drums are attached to main shaft by cone-gears, which are operated by steam cylinders and levers; screw-levers control the brakes and drums during the descent of the skip.

Engine is connected to the drum-shaft by spur-gearing in the proportion of one to five; speed of skip in shaft, about 3 feet per second; load of ore, 5,000 pounds; weight of skip, which is self-dumping, is 2,400 pounds, making the total load 7,400 pounds. Actual power employed, about 47 horse; engine also draws water with a 6-inch Cornish pump. Total weight of this machinery about 42 tons, and total cost about $10,000. Built at the Michigan Iron Foundry, Detroit, in 1869.

The foregoing described plants, together with those given in the subjoined tabular statement (pages 280 and 281), embrace over three-fourths of all the machinery employed in hoisting and pumping in the entire region.

V. Management and General Expenses.

This covers only such expenses as are incurred in the mining region, and not salaries of officers above the superintendent, nor the cost of selling the ore.

1, 2.—**Salaries, Office Expenses, and Taxes.**—This element of cost constitutes less than 5 per cent. of the whole cost of the ore,

DESCRIPTION OF STATIONARY ENGINES, WITH THEIR

Name of Mine.	Number or name of Pit or Shaft.	Size of Cylinder, Length and Diameter.	No. of Cylinders.	Number, size, and kind of Boiler.	Average working-pressure.	Nominal horse-power.	Kind of work, as pumping, hoisting, etc.	Height to which ore is lifted in feet.	Average number of tons hoisted in 24 hours.
Jackson ..	Pit No. 4.	13"x30", one 40 horse, Root's pat'nt trunk engine	2	Steam supplied for this double and single engine from two of Root's patent boilers, 50 horse-power each, connected together.	70 lbs.	140	Hoisting.	125 feet.	120
	Pit No. 6.	13"x 30", one 40 horse Root's pat'nt trunk engine	1	do. do. do.	70 lbs.	40	Hoisting and pumping.	From 80 ft. to 125 ft.	200
	Pit No. 7.	8"x 12"	2	One boiler, 42" diameter x 12 feet long, tubular, 40–3 in. flues.	70 lbs.	20	Hoisting and pumping.	50 feet.	50
	Pit No. 5.	5" x 8"	2	Tubular boiler, 40 in. diam. x 13 ft. long, 40–3 in. tubes.	70 lbs.	8	Hoisting.	50 feet.	40 tons of ore, rock and water.
	Machine shop.	8 x 16 inches.	1	Tubular boiler, 60 in. diam. x 25 ft. long, 121–2 in. tubes.			Running machinery in shop.		
Champion	4 shafts now worked by main engine. 2 new shafts now being sunk.	One horizontal engine, 14 inches bore, 20 in. stroke of piston.	1	Two return flue boilers, 42 inches diam., 28 ft. long, 2 flues in each, 16 inches diameter. One locomotive boiler, 28 inches diam. of shell, 26 x 30 in. fire-box, 36–2 in. flues, at 1st level of No. 3 shaft.	65 lbs. to the sq. in.	60 on hoisting drums	One 6-inch plunge pump. One No. 7 Earle pump, at 3d level of No. 3 shaft; elevating water to surface; supplied with steam from boiler at 1st level.	180 feet.	400 from 4 shafts.
Edwards.	Nos. 2 & 3.	24 x 36 inches.	1	Two. 5 ft. diam., 27 ft. long, with return flues each.	70 lbs.	150	Pumping and hoisting.	300 feet.	200
Lake Angeline.......	2 Pits.	16 x 24 inches.	1	One. 42 in. shell, 20 ft. long, with 2–14 in. flues.	60 lbs.	60	Pumping and hoisting.	75 feet.	150
Washington.	At No. 7 opening, known as No. 1 & 2, skip roads.	16 x 24 inches.	1	One boiler, 2 flues, 24 ft. 6 in. length, 44 in. diameter.	90 lbs.	50	Hoisting.	No. 1 skip, 130 feet. No. 2 skip, 55 feet.	44 35
Lake Superior ..	Main shaft.	20 x 30 inches.	1	Two boilers, 3 flues, 43 x 6 feet.	60 lbs.		Hoisting and pumping.	160 feet.	350
	Hematite.	12 x 20 inches.	1	One boiler, 2 flues, 3½ x 24 feet.	65 lbs.		Hoisting and pumping.	130 feet.	100
	Portable engine & boiler—"Sect. 16."	10 x 18 inches.	1	One boiler, flue.	30 lbs.		Hoisting and pumping.	60 feet.	
	Sect. 21.	10 x 12 inches.	1	One boiler, upright flue.	35 lbs.		Hoisting.	60 feet.	

WORK, AT SIX MARQUETTE MINES, JANUARY, 1873.

Kind of Skip and its load.	Diameter of Barrel-pump in inches.	Kind of Pumps.	Revolutions of Engines per minute.	Hours per day that Pump is worked.	Shaft.		Kind and quantity of fuel used in 24 hours.	Remarks.
					Vertical.	Inclined.		
Two 5 ft. drums, with 4 wheel, self-dumping. Skip-car 2½ tons.			75			2	One cord of wood per day; don't run at night.	There are also two (2) 12-horse power locomotives, which are used for distributing cars in the tunnels during the shipping season.
Two 3 ft. drums, one hoisting skip-car 2½ tons, 4 hoisting patent dump buckets 1 ton each.	8	Cornish jack-head.	100	10	1	1		
Inside dump-car 3 tons.	8	Cornish jack-head.	100	12		1	One cord of wood per day; don't run at night.	Also four (4) steam pumps, which are used in various parts of the mine, viz.: 1 No. 9 Earle steam pump; 1 No. 8 Knowles steam pump; 1 No. G Cameron steam pump; 1 Worthington duplex pump; also one 8 (eight) inch double-acting bucket pump.
Bucket 1 ton.			100	10	1		¾ cord per day.	
			80				One cord per day.	
Wrought-iron skips, 42 inches long, 30 in width and depth. Hold 3,000 lbs. of ore.	6	Plunge-pump, 6 in. diameter of cylinder and 6 in. column, elevating the water to the surface 180 feet.	20	22		incl'd	Mixed wood, four cords in 24 hours.	Makers—Hodge & Christie, Detroit, Mich. This one engine does all the work of this mine.
Cornish skip 1½ tons.	Two 6 in. One 7 in. One 8 in.	Two 6-in. draw-lifts from 5th to 4th levels, at Nos. 2 and 3 shafts. One 8-in. draw-lift, at No. 2 shaft, from 5th level to 2d. One 7-inch plunger-pole, from 4th level at No. 3 shaft, taking also No. 2 water to surface.	30	20		incl'd	Wood, six cords.	Makers—Hodge & Christie, Detroit. See plan of mine—Plate XIX.
Cornish skip 1½ tons.		One 10-inch double-acting pump.	30	10		incl'd	Wood, 2½ cords. Coal, ¾ ton.	Maker—D. H. Merritt, Marquette.
Iron self-dumper about 1 ton.		Earle. Nos. 4, 6, and 7.	see catalogue.	3	vert'l		Coal, hard & soft wood; about 1,500 lbs. coal in 24 h.; 4 cords wood in do.	Furnishing steam for Burleigh Drill Compressor, 3 Earle pumps (2 No. 4 & 1 No. 7), besides to hoisting engi's.
Iron skip 3 tons.	10	Plunger.	about 30	10		incl'd	Six cords wood.	Makers—Wash'n Iron Works, N'burgh, N. Y.
Iron car 3 tons.	8	Bucket plunger.	about 60	14		incl'd	Three cords wood.	
Iron skip 2 tons.	6	Bucket plunger.	about 40	10		incl'd	Three cords wood.	
Iron skip 2 tons.			about 80			incl'd	Three cords wood.	

amounting to about 12 cents per ton. I am happy to note here a much better showing than in the Persberg mines, Sweden, where this item, in 1870, cost 16½ per cent. of the whole, or 36 cents per ton of ore; nearly three times its cost with us. I presume the excess of this item in Sweden may be largely due to heavier taxes, and smaller production.

CHAPTER X.

CHEMICAL COMPOSITION OF ORES.—ANALYSES.

THIS chapter contains the results of over one hundred and fifty analyses, more or less complete, of iron ores from the Upper Peninsula of Michigan, mostly from the Marquette region, together with five analyses of pig-iron produced from these ores ; and several analyses of ores from other parts of the U. S., which are largely used with Lake Superior ores as mixtures. In order to bring out the variations in quality of the ores, and to obtain *reliable practical averages*, seldom less than two and in one instance eight samples were analyzed from the same mine.

By far the largest portion of the samples, the analyses of which appear in this Report, were selected by myself with a view to obtaining a fair and *safe average* of the ore sampled, one that would be borne out and confirmed by practically working the same ore in the furnace. I am well aware, from extended observation and practical experience, that a large majority of the published analyses of iron ores, not only have no practical value, but are positively detrimental to the best interests of the iron trade, representing as they so often do the ores to be richer in iron than they actually are, simply because the samples analyzed were not honestly or skilfully collected. Even the most skilful and conscientious men, if they err at all in collecting a sample from a new iron location, are almost sure to err on the side of finding too much, rather than too little iron. The chemist is often wrongly blamed for these false results. My experience with many analysts leads me to believe that they are, as a rule, thoroughly honest and painstaking men, who return correct results for the *samples sent them ;* the trouble is with the samplers. This point receives further consideration under Explorations, Chapter VII.

In earnestly endeavoring to avoid this rock on which so many mining engineers and geologists have wrecked their reputations, I may in some instances have gone to the opposite extreme and collected samples which were below the average richness—at least I am quite persuaded that I shall be charged with this—hence venture this explanation in advance of the charge. If such mistakes are found, I can only say myself and not the analysts are to blame, and I stand ready to make such corrections as lie in my power.

My *method of sampling* is as follows:—1st. To obtain an average of a producing mine; I found that the immense stock piles accumulated at Cleveland, Ohio, at the end of the shipping season, afforded excellent opportunities for sampling. The stock piles at the mines or a large number of loaded cars were often resorted to, and in many instances it was thought best to go into the mine and take the samples from the solid ledge or the loose ore as it was being taken out. In either case an ordinary shot bag, holding 4 *or* 5 *pounds of ore, was filled with small fragments, varying from the size of a pea to that of a walnut, of all kinds of ore, from all parts of the pile, together with the rock, if any, which was found mixed with the ore.* Some of these fragments were picked up and some were broken from larger pieces; the dust and mud over the ore made it often impossible to distinguish whether the pieces taken were ore or rock. These samples were all pulverized and thoroughly mixed, and from this the specimens were taken for the chemist, the same being forwarded by mail in small numbered tin tubes; and in each instance a pound or more of the pulverized ore was retained for future reference. The reserved portions are now in my safe in Marquette, from which samples will be furnished to any who may desire. 2d. To obtain an average sample from a new locality or from exploration pits is more difficult and unsatisfactory. This subject is fully treated under Explorations, Chapter VII.

With all this care my results varied, in extreme cases, from 10 per cent. below to 5 per cent. above the true average, but the common variation was not more than three per cent. Two or three of the extreme results, known to be wrong, are omitted from the tables. The name of the sampler is in every case given when known, and the circumstances of its collection are briefly stated in the notes. The samples collected by E. R. Taylor, of Cleveland, were, at my request, taken in accordance with the rules above given.

The surname of the chemists and date at which analysis was

made, as near as could be ascertained, are given under the result in every instance except one. The number of analyses made, with names in full and address of these gentlemen, are as follows :—

	No. Made.
Professor Oscar D. Allen, New Haven, Conn....	17
Professor Geo. J. Brush, New Haven, Conn.....	1
J. Blodgett Britton, Philadelphia, Pa............	56
A. A. Blair, St. Louis, Mo....................	2
Dr. C. F. Chandler, School of Mines, N. Y......	8
Dr. C. F. Chandler and F. A. Cairns, School of Mines, N. Y...........................	12
Chandler and Schweitzer.......................	1
F. H. Emmerton, Chicago, Ill..................	1
F. B. Jenney, Marquette, Mich.................	8
Prof. Geo. W. Maynard, New York............	5
Maynard and Wendel.........................	3
Ed. R. Taylor, Cleveland, Ohio................	14
Dr. A. Wendel, Troy, N. Y....................	20
Dr. Otto Wuth, Pittsburgh, Pa................	30
Samuel Peters...............................	1
T. G. Wormley.............................	4

The metallic iron was usually determined by but one chemist, as the chances of difference on this element are small. Phosphorus determinations are more difficult, and considerable differences in the amount of this element found in the same sample by different chemists, will be observed. For this reason duplicates were often sent to two and sometimes to three ; the results being given as returned by them. If any one supposes the differences to be due to errors in samples, which is improbable, I will gladly furnish duplicates for re-examination. The specific gravities of powder were mostly determined by Mr. Jenney, and not by the chemists over whose names they are sometimes placed.

The subjoined table contains an approximate general summary of the results, exhibiting the average composition of the four classes of ore now produced by the following mines :—

I. *Red Specular Ores.* Barnum, Cleveland, Jackson, Lake Superior, New York, Republic, and Kloman.

II. *Black Magnetic and Slate Ores.* Champion, Edwards, Michigan, Spurr, and Washington.

III. *Soft Hematites.* Foster, Lake Superior, Lake Angeline, Taylor, Macomber, New England, Shenango, S. C. Smith, and Winthrop.

IV. *Flag Ore.* Cascade.

Table No. XIII. of Atlas contains a somewhat similar summary so far as metallic iron and phosphorus are concerned. More facts are incorporated in this table, which has slightly changed the averages.

	I.	II.	III.	IV.
Protoxide of Iron		19.639		
Sesqui- or Peroxide of Iron	90.52	67.761	75.75	70.98
Oxide of Manganese	Trace.	0.13	0.80	Trace.
Alumina	1.39	2.13	1.536	2.01
Lime	0.70	0.68	0.36	0.45
Magnesia	0.42	0.69	0.294	0.20
Sulphur	0.05	0.132	0.110	0.03
Phosphoric Acid	0.258	0.199	0.185	0.13
Silicic Acid, Silica, or Insoluble Silicious Matter	5.892	7.828	14.035	25.12
Water, Combined			3.94	
" Uncombined			1.18	
" Total	0.77	0.811		1.08
Volatile Matter			1.81	
	100.000	100.000	100.000	100.00
Metallic Iron	62.915	62.930	52.649	49.332
Phosphorus	0.111	0.085	0.078	0.053
Sulphur	0.05	0.132	0.110	0.03
Metallic Manganese	Trace.	0.091	0.56	Trace.
Specific Gravity	4.74	4.59	3.88	4.09

A glance at this table shows us that, except the soft hematite III., which contains about 5 per cent. of water, all the ores are essentially and chiefly composed of oxide of iron and silica or insoluble silicious matter. The other elements, viz., oxide of manganese, alumina, lime, magnesia, sulphur, phosphoric acid, and water amount in the aggregate to only about 5 per cent. in the I., II., and IV. classes. So constant is this ratio that a valuable determination of iron in a hard ore, and one sufficiently accurate for practical purposes, can be made by ascertaining the percentage of insoluble silicious matter, adding 5 to it and subtracting the sum from 100. The result is the iron oxide, which, multiplied by .70 for red, and .72 for black oxides, gives the metallic iron.

Regarding the percentage of metallic iron, consumers of Lake

Superior ores will at once note that their furnace books very often show a higher yield than 62.9 per cent., which is given in the table as the average percentage for first-class ores. This may not have been the case in exceptional years, like 1872, when the consumption so crowded the production that mines had not the time nor skilled labor to make such selection as they usually make. But that furnaces running on first-class ores usually make a better yield than that given, is shown by "Table of Metallurgical Qualities of certain Lake Superior Ores by Consumers," Plate No. XIII. of Atlas, where various consumers credited these ores, in 1870, with an average of *over sixty-four per cent. of iron*, as shown by their furnace-books. This discrepancy is easily accounted for; the chemist's result is in *pure metallic iron*, the furnace man's is in *pig iron*, which contains several per cent. of carbon and silicon, and other substances,—see subjoined analyses. Therefore the chemist should always find *less* iron than is shown by the furnace accounts if he has an *average* sample of the ore. Just what this difference is depends on the grade of iron made, on the waste in the slag, and other things: good authorities have placed it at 2½ per cent.

Passing to a more detailed examination of the facts recorded in the table, we find, in descending order,—oxide of *manganese* has a maximum of nearly one per cent. in the hematite, and is nothing in the specular and flag ores. If the hematite was subdivided into manganiferous and non-manganiferous varieties, as suggested under Lithology, Chapter III., then one variety would contain only a minute quantity of manganese, while the other would reach an average of, say 3 per cent. of the oxide. The presence of manganese adds to the value of an ore, especially for making steel. *Alumina* reaches a maximum of over 2 per cent. in the magnetite ores, and is least in the specular ores. The earthy character of the hematites would lead one to expect more of this element in that class. *Lime* and *magnesia* aggregate a trifle over one per cent. in the high grade ore, and less than this amount in those of low grade. *Sulphur* is relatively most abundant in the magnetites; but, so far as I know, the minute quantity found has never been objected to by consumers of the ore. The quantity of *phosphoric acid* and phosphorus is of such moment in connection with the wants of the Bessemer steel manufacture, now rapidly developing in the West, that this subject will receive especial attention hereafter.

The distribution and relations of the *silicious matter* have been mentioned ;—it has its maximum in the flag ores where it reaches one-fourth of the whole weight, and is least in the rich speculars, which contain only about 6 per cent. on the average.

The total *water* in the hard ores is only about 1 per cent. In the soft hematites it rises to an average of over 5 per cent., and, as will be seen in the subjoined analyses, increases in a few instances to about twice this amount, the greater part of which is combined with the limonite, which largely makes up the soft ore. An appreciable amount of *volatile* matter, supposed to be mostly carbonaceous, occurs only in the hematite ores. The specific gravities given will be observed to have a very significant relation to the amount of iron, which subject is considered fully in Chapter III.

Phosphorus in Lake Superior Ores.

Pig-iron intended for the use of *steel* makers must be remarkably free from phosphorus, *one-tenth of one per cent.*, according to some authorities, being the maximum amount allowable for many purposes. As it has been found impossible, up to this time, to eliminate this element from the metal either in the blast furnace or in any of the various processes for making steel, it is indispensable, in steel manufacture, that we start with an ore comparatively free from it; and for the best bar iron, only a very small amount of phosphorus is admissible,—its effect being to produce cold shortness.

It is a safe practical maxim of iron metallurgy that all the phosphorus contained in the coal, limestone, and ore charged into a blast furnace will be found in the resulting pig-iron, and that the conversion of such pig-iron into steel will increase the phosphorus just in the ratio in which the metal is wasted in the process. It is therefore very evident, if say one-tenth of one per cent. only is admissible in steel, not only our ores but fuel and flux must be very free from phosphorus at the start. In considering the facts regarding this element here given, it must be constantly borne in mind that a rich ore may contain more phosphorus than a lean ore, and yet produce a pig-iron containing less phosphorus than the other, because *less of the rich ore is required* to make a ton of iron.

To illustrate: an ore yielding 66⅔ per cent. in the furnace, and containing .06 of phosphorus, will produce a pig containing .09 of phosphorus; while an ore containing but 50 per cent. of iron and .05 of phosphorus will produce a pig containing .10 of phosphorus; therefore the amount of iron in the ore must be always considered in comparing the amounts of phosphorus. Applying this rule to the facts given in the foregoing table, we shall find that the apparent greater freedom of the hematite and flag ores from phosphorus is nearly balanced by their comparative poverty in iron.

The distribution of phosphorus among the Lake Superior ores, so far as my facts go, follows no obvious law; it seems to have little, if any, relation to the kind of ore. Some of the hematite ores are among the lowest and others among the richest in this element, and so of the specular and magnetic ores.

A rule, to which there are, however, several exceptions, seems to be that the ores poor in iron and rich in silica, contain least phosphorus; but the analyses of the Republic mountain ore show more iron and less silica than in any other, and that it is also very low in phosphorus. The table of analyses, in Plate No. XIII. of Atlas, presents most of the facts in a compact form; but as this subject is of peculiar interest at this time in connection with the Bessemer steel manufacture, I venture to incorporate a second tabular statement here, in which the mines are arranged in order of the quantity of phosphorus, beginning with the lowest. No mine is included from which less than two samples have been analyzed. The deposits and mines marked with a * are new, and not sufficiently developed to enable me to say that an average sample of the ore was obtained.

Mine.	Kind of Ore.	Phosphorus.	Iron.
Lake Angeline.......	Jaspery Specular.....	0.031	53.83
Winthrop............	Soft Hematite........	0.037	54.63
Republic*...........	Specular and Magnetic	0.040	66.51
Michigamme*.........	Magnetic............	0.041	64.388
Silas C. Smith.......	Hematite............	0.047	49.70
Cascade.............	Flag................	0.053	49.332
Menominee Iron reg'n*	Specular & Hematite.	0.054	48.209
Edwards.............	2d Class Magnetic....	0.055	49.190
Macomber...........	Hematite............	0.058	54.92
Cascade.............	Flag and Specular....	0.061	51.253

Mine.	Kind of Ore.	Phosphorus.	Iron.
Jackson	Specular	0.066	63.715
Magnetic*	Magnetic	0.067	54.72
Edwards	Do.	0.067	61.60
Shenango	Hematite	0 070	56.315
Champion	Magnetic and Slate	0.072	63.55
Negaunee *	Manganifs. Soft Hem'e	0.074	44.29
Lake Angeline	Hematite	0.079	50.70
New England	Soft Hematite	0.080	48.24
Kloman*	Specular	0.089	63.55
Foster	Hematite	0.094	52.27
Spurr Mountain*	Magnetic	0.104	63.81
Lake Superior	Specular	0.104	62.11
Taylor (L'Anse)*	Hematite	0.107	52.88
Jackson	Hematite and Jaspery.	0.124	57.155
Cleveland	Specular	0.126	61.092
Lake Superior	Hematite	0.130	54.19
Saginaw*	Specular and Hematite	0.132	52.40
Barnum	Specular	0.134	61.69
Washington	Magnetic	0.141	61.305
New York	Specular	0.224	61.74

It has been stated that an inspection of the first table did not warrant us in asserting that either of the four classes of ore represented could be easily recognized as being comparatively free from phosphorus; so an examination of the above presentation of the facts forces us to the conclusion that the distribution is not geographical; for we here see widely-separated mines containing the same amount of phosphorus, whilst contiguous mines vary widely. In fact, in different parts of the same mine there is found a wide difference in the quantity of this noxious element; *e. g.:* The New York mine results show more than twice as much phosphorus in the ore from pit No. 1 as from pit No. 2; and the Lake Superior ore appears to contain less phosphorus than the Barnum, although they belong to one deposit. A part of this difference is undoubtedly due to errors in sampling and errors in the analysis; but the number of samples analyzed, the care taken in collecting them, and the reputation of the chemists, leave but little doubt that the relative and absolute average amounts of phosphorus in the ores from the developed mines are nearly expressed in the foregoing table.

At the suggestion of Mr. A. L. Holley, I selected, with much care, an average sample of the rock which occurs in the hard ores, more or less of which goes into the furnace, and had it analyzed; the result was less than the average amount of phosphorus. This fact, in connection with the low amount found in the second class and flag ores, leads me to believe that no care in selecting and sorting ore will diminish the quantity of phosphorus.

By way of verifying the amount of phosphorus in Lake Superior ores, here given, there are presented in the following table five analyses of pig-iron made from them with charcoal, and a flux containing no appreciable amount of phosphorus. They may, therefore, be said to indicate very accurately the amount of phosphorus in the ores, which, as will be seen, averages about the maximum amount given above as admissible in steel.

	1	2	3	4	5	Average.
Magnesia			0.47			
Silicic Acid, or Silica		1.16	1.83	3.21	2.91	2.28
Silicon	2.245					
Graphitic Carbon	2.88	3.72	3.35		3.61	3.39
Combined "	.80	0.30	0.00		.05	.38
Metallic Iron	93.201		93.49			93.34
Phosphorus	.138	0.104	0.082	0.126	.092	0.108
Sulphur	.011	0.045	trace.		.04	0.032
Metallic Manganese	.174					.174

No. 1 was chipped from many pigs of No. 1 gray foundry iron made at the Pioneer furnace Negaunee, of Jackson ore. Analysis by Dr. C. F. Chandler. No. 2 is a pig-iron made from assorted Lake Superior ores at the Appleton Furnace, Wisconsin. Analysis by Mr. Morrell. No. 3 is also a specimen of Appleton iron. Analysis by Dr. Wuth. No. 4 is No. 1 gray foundry iron made by the Jackson Iron Co. at Fayette, Michigan, of Jackson ore with charcoal, and is extensively used in the manufacture of Bessemer steel. Analysis by Mr. Morrell. No. 5 is a specimen of pig made by the Michigan Iron Co. in Marquette County, of a mixture of specular, magnetic, and hematite ores. Analysis by Mr. Morrell. The analysis of Pioneer pig was at the expense of the Survey; the others were furnished by Mr. Holley. It was proposed to carry this work much further, but the limited means would not permit.

For contributions in money, and valuable suggestions and encouragement in obtaining the results set forth in this chapter, I am under especial obligation to John Fritz, of Bethlehem, Pa., and S. P. Ely, of Marquette; A. Pardee, Daniel J. Morrell, A. B. Meeker, and W. H. Barnum also contributed liberally towards paying for the chemical work, which cost nearly $2,000.

The physical and mineralogical character of the following ores is given under Lithology, in Chapter III. For commercial statistics, and, incompletely, the metallurgical qualities, see Plates XII. and XIII. of Atlas.

(*The mines are arranged alphabetically. The upper line gives the number of the sample.*)

BARNUM MINE—*Specular Ore.*

	58.	14.*	14.*	262.*	262.*	277.*	277.*	*Averages.*
Sesqui- or Peroxide of Iron	93.40	86.71						
Oxide of Manganese	......	trace						
Alumina	0.33	1.92						
Lime	0.30	0.64						
Magnesia	0.15	0.53						
Sulphur	min'te trace.	0.04						
Phosphoric Acid	0.23	0.25	0.189	0.31	0.288	......	0.649	
Silicic Acid, or Silica	4.80	9.43						
Water, Total	0.29	0.46						
	99.50	99.98						
Metallic Iron	65.38	60.79	......	64.30	......	56.31	......	61.69
Phosphorus	0.10	0.11	0.083	.133	0.123	0.149	0.278	.134
Sulphur	......	0.04						
Specific Gravity	......	4.58						
	Chemist, Allen. Sept. 2, 1871. *Sampler,* Brooks.	*Chemists,* Chandler & Cairns. Mar. 4, 1872. *Sampler,* Brooks.	*Chemist,* Wuth. *Sampler,*	*Chemist,* Taylor. *Sampler,* Taylor.	*Chemist,* Wendel. *Sampler,*	*Chemist,* Britton. Dec. 31, 1872. *Sampler,* Brooks.	*Chemist,* Wendel. Feb. 6, 1873. *Sampler,*	

NOTES.—58. From three stock piles at mine. 14. Large stock pile at Cleveland. 262. Stock pile at Cleveland. 277. All parts of mine.

CLEVELAND MINE—*Specular Ore.*

	36.	36.	36.	260.	260.	271.	271.	272.	272.	273.	*Averages.*
Sesqui- or Peroxide of Iron	88.50										
Oxide of Manganese	trace.										
Alumina	1.84										
Lime	0.89										
Magnesia	0.75										
Sulphur	0.01										
Phosphoric Acid	0.46	0.229	......	0.178	0.14	0.343	......	0.218			
Silicic Acid, or Silica	6.40										
Water, Total	1.23										
	100.08										
Metallic Iron	61.95	......	......	......	62.10	......	56.590	......	......	63.730	61.092
Phosphorus	0.20	0.100	0.187	0.076	.061	0.147	0.168	0.093	0.154	0.111	0.119
Sulphur	0.01										
Metallic Manganese	trace.										
Specific Gravity	4.64	......	......	......	......	4.42	......	4.37	......	4.93	
	Chemists, Chandler & Cairns. Mar. 9, 1872. *Sampler,* Brooks.	*Chemist,* Wuth. *Sampler,*	*Chemist,* Britton. *Sampler,*	*Chemist,* Wendel. *Sampler,* Taylor.	*Chemist,* Taylor. *Sampler,*	*Chemist,* Wendel. Feb. 6, 1873. *Sampler,* Brooks.	*Chemist,* Britton. Dec. 31, 1872. *Sampler,*	*Chemist,* Wendel. Feb. 6, 1873. *Sampler,* Brooks.	*Chemist,* Britton. Dec. 31, 1872. *Sampler,*	*Chemist,* Britton. Dec. 31, 1872. *Sampler,*	

NOTES.—36. Large stock pile in Cleveland. 260. Stock pile in Cleveland. 271. Laurie Genth's Pit, No. 3. 272. Swede's pit. 273. School House opening. The last three were from mine.

* The occurrence of the same number more than once in this line, signifies that duplicates of the same sample were sent to different chemists.

CHAMPION MINE—*Magnetic and Slate Ore.*

	38.	38.	38.	227.	227.	228.	228.	*Averages.*
Protoxide of Iron	17.87							
Sesqui- or Peroxide of Iron	74.93							
Oxide of Manganese	0.05							
Alumina	1.15							
Lime	0.52							
Magnesia	0.92							
Sulphur	0.12							
Phosphoric Acid	0.28	0.116	......	0.021	0.337	0.161	0.316	
Silicic Acid, or Silica	3.70							
Water, Total	0.52							
	100.06							
Metallic Iron	66.04	......	......	......	57.97	......	66.65	63.55
Phosphorus	0.12	0.051	0.048	0.009	0.143	0.070	0.136	.084
Sulphur	0.12							
Metallic Manganese	0.03							
Specific Gravity	4.75	......	4.43	......	......	4.87		
	Chemists, Chandler & Cairns. Mar. 9, 1872. *Sampler,* Brooks.	*Chemist,* Wuth. *Sampler,*	*Chemist,* Britton. *Sampler,*	*Chemist,* Wuth. *Sampler,* Brooks.	*Chemist,* Wendel, Feb. 6, 1873. *Sampler,*	*Chemist,* Wuth. *Sampler,* Brooks.	*Chemist,* Wendel, Feb. 6, 1873. *Sampler,*	

NOTES.—38. Large stock pile in Cleveland, all varieties. 227. "Slate ore," Shaft No. 4. 228. "Black ore," Shafts Nos. 1, 2, and 3. The two last from mine.

CASCADE MINES—*Flag Ore.*

	17.	22.	22.	257.	257.	258.	258.	15.	*Averages.*
Sesqui- or Peroxide of Iron	71.98	83.70	......	......	......	......	......	66.20	
Protoxide of Manganese	0.01	trace.							
Alumina	0.68	3.34							
Lime	0.16	0.75							
Magnesia	0.06	0.34							
Sulphur	0.04	0.03							
Phosphoric Acid	0.07	0.24	0.248	......	......	......	......	.14	
Silicic Acid, or Silica	......	10.67							
Insoluble Silicious Matter	25.26	......	......	......	......	......	......	31.02	
Water, Total	1.03	0.87	......	......	......	......	......	1.29	
Alkalies, undetermined and lost	0.71								
	100.00	99.94							
Metallic Iron	50.49	58.59	......	46.120	......	45.010	......	46.450	49.332
Phosphorus	0.03	0.10	0.108	.042	0.043	0.027	0.036	0.060	.053
Sulphur	0.04	0.03							
Specific Gravity	......	4.43	......	3.95	......	4.01			
	Chemist, Britton, Aug. 18, 1870. *Sampler,* Brooks.	*Chemists,* Chandler & Cairns, Mar. 4, 1872. *Sampler,* Brooks, J'y, '72	*Chemist,* Wuth, 1872. *Sampler,*	*Chemist,* Britton, 1872. *Sampler,* Brooks.	*Chemist,* Allen, 1872. *Sampler,*	*Chemist,* Britton. *Sampler,* Brooks.	*Chemist,* Allen. *Sampler,*	*Chemist,* Britton, Aug., 1870. *Sampler,* Brooks.	

NOTES.—17. Selected bird's-eye slate ore. Exploration pit. 22. The richest pieces from a small stock pile in Cleveland. 257. Emma mine. 258. Bagley mine; bird's-eye slate ore. The two last were obtained from the mine workings. 15. Old opening, north face ridge, S.W. corner. Sect. 29.

CASCADE MINES—*Flag and Specular Ore.*

	259.	259.	266.	266.	256.	256.	*Averages.*
Phosphoric Acid	....	....	0.16	0.096			
Metallic Iron	50.820	....	44.00	....	58.940	....	51.253
Phosphorus	0.078	0.073	0.069	0.041	0.055	0.055	0.061
Specific Gravity	4.13	....	....	....	4.44		
	Chemist, Britton. 1872. *Sampler*, Brooks.	*Chemist*, Allen. 1872. *Sampler*,	*Chemist*, Taylor. *Sampler*, Taylor.	*Chemist*, Wendel. *Sampler*,	*Chemist*, Britton. 1872. *Samplers*, J. Fritz & A. Pardee.	*Chemist*, Allen. 1872. *Sampler*,	

NOTES.—259. Saw-Mill opening, west of stream. 256. West End Mine (specular ore). Stock pile at Mine. 266. Stock pile at Cleveland.

CANADIAN ORES—*Magnetic.*

	222.	217.	216.	220, *b*.	220, *a*.	*Averages.*
Bisulphide of Iron	....	....	....	....	2.19	
Sesqui- or Peroxide of Iron	....	....	....	78.03		
Proto-sesquioxide of Iron	92.19	....	....	....	84.38	
Alumina	.68	....	....	1.17	2.86	
Lime	.28	....	....	....	0.74	
Magnesia	.83	....	....	....	5.61	
Sulphur	.78					
Phosphoric Acid	.14	0.21	....	0.077	0.087	
Silicic Acid, or Silica	3.55	....	....	6.10	4.13	
Water, Total (moisture)	.48					
Carbonate of Lime	....	....	....	13.71	....	
Carbonate of Magnesia	....	....	....	0.91	....	
Oxygen with the Sulphur and loss	1.07					
	100.00			99.997	99.997	
Metallic Iron	66.86	51.40	45.20	54.00	60.00	55.49
Phosphorus	.06	0.092	.037	.033	0.037	0.052
	Chemist, Britton. Nov. 17, 1870. *Sampler*, Brooks.	*Chemist*, E.R. Taylor. Jan. 2, 1873. *Sampler*, Taylor.	*Chemist*, Taylor. Jan. 2, 1873. *Sampler*, Taylor.	*Chemist*, Wuth. *Sampler*,	*Chemist*, Wuth. *Sampler*,	

NOTES.—222. Analysis furnished by Redington and Adams, Cleveland. 217. Stock pile at Cleveland. 216. Stock pile at Cleveland. 220. Analyses furnished by Dr. Wuth. *a*. Magnetic ore after roasting; *b*. Red hematite. 222 and 217 are Forsyth ore. 216 and 220 are Marmora ore.

EDWARDS MINE—*Magnetic Ore.*

	41.	41.	41.	199.	*Averages.*
Protoxide of Iron	21.60	....	....	9.98	
Besqui- or Peroxide of Iron	55.80	....	....	85.41	
Oxide of Manganese	0.10				
Alumina	4.34				
Lime	0.77				
Magnesia	0.84				
Sulphur	0.16				
Sulphuric Acid	....	....	....	.03	
Phosphoric Acid	0.12	0.288	....	.07	
Silicic Acid, or Silica	15.41				
Insoluble Silicious Matter	....	....	....	2.43	
Water, Total	0.81				
	99.95				
Metallic Iron	55.75	....	....	67.45	61.60
Phosphorus	0.05	0.125	0.137	.030	.067
Sulphur	0.16				
Metallic Manganese	0.06				
Specific Gravity	4.24				
	Chemists, Chandler & Cairns. March 9, 1872. *Sampler*, Brooks.	*Chemist*, Wuth. *Sampler*,	*Chemist*, Britton. *Sampler*,	*Chemist*, Taylor. Jan., 1873. *Sampler*, Unknown.	

NOTE.—41. Large stock pile in Cleveland.

EDWARDS MINE—*Second-class Magnetic Ore.*

	265.	265.	286.	*Averages.*
Phosphoric Acid	0.10	0.136		
Metallic Iron	48.80	....	49.580	49.190
Phosphorus	0.043	0.058	0.061	.055
	Chemist, Taylor. *Sampler*, Taylor.	*Chemist*, Wendel. *Sampler*,	*Chemist*, Britton. Jan. 9, 1873. *Sampler*, Brooks.	

NOTES.—265. Stock pile at Cleveland. 286. From mine.

FOSTER MINE—*Hematite Ore.*

	49.	87.	270.	270.	26.	*Averages.*
Sesqui- or Peroxide of Iron	74.69	....	....	....	79.49	
Oxide of Manganese	.42	....	....	....	0.25	
Alumina	.50	....	....	....	1.19	
Lime	.37	....	....	....	0.27	
Magnesia	.63	....	....	....	0.33	
Sulphuric Acid	....	....	....	....	0.17	
Phosphoric Acid	.18	....	0.33	0.226	0.19	
Silicic Acid, or Silica	16.44	....	....	....	9.28	
Insoluble Silicious Matter	....	20.68				
Water, Combined	....	6.12				
Water, Total	7.16	....	....	....	8.74	
	100.39	....	....	....	99.91	
Metallic Iron	52.28	49.78	51.40	....	55.64	52.27
Phosphorus	.080	....	0.144	0.097	0.083	.094
Sulphur	....	....	....	....	0.068	
	Chemist, Brush. July 5, 1871. *Sampler*, Brooks.	*Chemist*, Britton. Nov. 4, 1871. *Sampler*, Brooks.	*Chemist*, Taylor. *Sampler*, Taylor.	*Chemist*, Wendel. *Sampler*,	*Chemist*, Chandler. May 14, 1866. *Sampler*, Brooks.	

NOTES.—49. Stock pile at Pioneer Furnace, Negaunee, Mich. 87. From mine, numerous fragments. 270. Stock pile at Cleveland. 26. From mine when first opened.

JACKSON MINE—*Specular Ore.*

	24.	24.	24.	51.	230.	230.	230.	*Averages.*
Sesqui- or Peroxide of Iron	93.75							
Oxide of Manganese	trace.	......	......	0.60				
Alumina	0.73							
Lime	0.61							
Magnesia	0.23							
Sulphur	0.03	......	......	0.18				
Phosphoric Acid	0.32	0.127	......	0.10	0.144			
Silicic Acid, or Silica	3.27	......	......	1.45				
Water, Total	1.09							
Alumina, Lime, Magnesia, Water, etc.	......	......	......	1.67				
	100.03							
Metallic Iron	65.62	......	......	......	......	61.810	......	63.715
Phosphorus	0.14	0.055	0.069	0.04	0.063	0.078	0.073	.066
Sulphur	0.03	......	......	0.18				
Specific Gravity	......	......	......	......	4.95			
	Chemists, Chandler & Cairns. Mar. 9, 1872. *Sampler*, Brooks.	*Chemist*, Wuth. *Sampler*, Brooks.	*Chemist*, Britton. Dec. 20, 1872. *Sampler*, Brooks.	*Chemist*, Chandler. July 13, 1871. *Sampler*, Brooks.	*Chemist*, Wuth. *Sampler*, Brooks.	*Chemist*, Britton. Feb. 18, 1873. *Sampler*,	*Chemist*, Allen. *Sampler*,	

NOTES.—24. Large stock pile in Cleveland. 51. Stock pile at Pioneer Furnace, Negaunee, Mich. 230. Slate ore. West end of mine.

JACKSON MINE—*Hematite and Jaspery Ores.*

	231.	231.	231.	229.	229.	229.	*Averages.*
Phosphoric Acid	0.316	0.523	......	......	0.338	0.054	
Metallic Iron	......	59.30	54.530	56.590	58.20	......	57.155
Phosphorus	0.138	0.224	0.154	0.061	0.144	0.023	.124
Specific Gravity	4.20	......	......	4.59			
	Chemist, Wuth. 1872. *Sampler,* Brooks.	*Chemist,* Wendel. Feb. 6, 1873. *Sampler,*	*Chemist,* Britton. Feb. 18, 1873. *Sampler,*	*Chemist,* Britton. Feb. 18, 1873. *Sampler,* Brooks.	*Chemist,* Wendel. Feb. 6, 1873. *Sampler,*	*Chemist,* Wuth. *Sampler,*	

NOTES.—231. Hematite ore—west part of mine. 229. Old Pioneer opening—Jaspery ore. The Hematite and Specular ores occur together in this mine.

KLOMAN MINE—*Specular Ore.*

	235.	225.	*Averages.*
Metallic Iron	63.55	....	63.55
Phosphorus	0.097	0.081	.089
Specific Gravity	4.90		
	Chemist, Britton. *Sampler,* Brooks.	*Chemist,* Allen. *Sampler,*	

NOTE.—235. Fragments broken from outcrop, before work began.

LAKE SUPERIOR MINE—*Specular Ore.*

	37.	37.	37.	261.	261.	44.	274.	*Averages.*
Sesqui- or Peroxide of Iron	86.70							
Oxide of Manganese	trace.							
Alumina	1.64							
Lime	0.57							
Magnesia	0.24							
Sulphur	0.02							
Phosphoric Acid	0.14	0.075	......	0.24	0.239			
Silicic Acid, or Silica	9.82							
Water Total	0.61							
	99.74							
Metallic Iron	60.69	......	......	63.50	......	64.37	59.89	62.11
Phosphorus	0.06	0.033	0.046	0.103	0.102	0.10	.065	.078
Sulphur	0.02							
Specific Gravity	4.55						4.69	
	Chemists, Chandler & Cairns, Mar. 9, 1872. *Sampler,* Brooks.	*Chemist,* Wuth. *Sampler,*	*Chemist,* Britton. *Sampler,*	*Chemist,* Taylor. *Sampler,* Taylor.	*Chemist,* Wendel. *Sampler,*	*Chemist,* Britton. *Sampler,* John Fritz.	*Chemist,* Britton, Dec., 1872. *Sampler,* Brooks.	

NOTES.—37. Large Stock pile in Cleveland. 261. Stock pile at Cleveland. 44. Stock pile at Bethlehem Furnace. 274. Lower bed. Pit No. 1. Pennsylvania mine.

LAKE SUPERIOR MINE—*Hematite.*

	10.	10.	10.	269.	269.	276.	276.	87.	*Averages.*
Sesqui- or Peroxide of Iron	79.80								
Oxide of Manganese	0.10								
Alumina	2.05								
Lime	0.45								
Magnesia	0.53								
Sulphur	0.03								
Phosphoric Acid	0.30	0.104	.....	0.24	0.237	.....	0.668		
Insoluble Silicious Matter	.....	.....	.....	.....	.....	.....	.....	15.42	
Silicic Acid, or Silica	12.52	.....	.....	.....	.....	.....	.....	.....	
Water, Combined	4.11	.....	.....	.....	.....	.....	.....	4.66	
" Uncombined	0.14								
	100.03								
Metallic Iron	55.86	.....	.....	52.00	.....	.....	.....	55.00	54.28
Phosphorus	0.13	0.045	0.066	0.103	0.101	0.131	0.286	.....	.130
Sulphur	0.03								
Metallic Manganese	0.07								
Specific Gravity	4.12								
	Chemists, Chandler & Cairns, March 4, 1872. *Sampler,* Brooks.	*Chemist,* Wuth. *Sampler,*	*Chemist,* Britton. *Sampler,*	*Chemist,* Taylor. *Sampler,* Taylor.	*Chemist,* Wendel. *Sampler,*	*Chemist,* Britton. Dec. 31, 1872. *Sampler,* Brooks.	*Chemist,* Wendel. Feb. 6, 1873. *Sampler,*	*Chemist,* Britton. Nov. 4, 1871. *Sampler,* Brooks.	

NOTES.—10. Large Stock pile at Cleveland. 269. Stock pile at Cleveland. 276. Hematite workings of mine.

LAKE ANGELINE—*Specular Ore (Jaspery).*

	21.	21.	267.	267.	34.	*Averages.*
Sesqui- or Peroxide of Iron	72.00	....	....	....	85.43	
Oxide of Manganese	trace.					
Alumina	0.92	....	....	....	1.89	
Lime	0.33	....	....	....	0.24	
Magnesia	0.34	....	....	....	0.13	
Sulphur	0.02	....	....	....	none.	
Phosphoric Acid	0.08	0.101	0.04	0.083	none.	
Silicic Acid, or Silica	25.09	....	....	....	12.31	
Water, Combined	1.09					
" Uncombined	0.12					
	99.99	....	....	....	100.00	
Metallic Iron	50.40	....	52.00	....	59.08	53.83
Phosphorus	0.03	0.044	0.017	0.036	none.	.031
Sulphur	0.02				.	
Specific Gravity	3.97					
	Chemists, Chandler & Cairns, March 4, 1872. *Sampler,* Brooks.	*Chemist,* Wuth. 1872. *Sampler,*	*Chemist,* Taylor. 1872. *Sampler,* Taylor.	*Chemist,* Wendel. 1872. *Sampler,*	*Chemist,* Wuth. Dec. 29, 1865. *Sampler,* Unknown.	

NOTES.—21. Stock pile in Cleveland. 267. Stock pile in Cleveland.

LAKE ANGELINE MINE—*Hematite.*

	268.	268.	280.	*Averages.*
Phosphoric Acid	0.09	0.160		
Metallic Iron	51.40	....	50.000	50.70
Phosphorus	.038	0.070	0.104	.079
	Chemist, Taylor. 1872. *Sampler,* Taylor.	*Chemist,* Wendel. *Sampler,*	*Chemist,* Britton. Dec. 31, 1872. *Sampler,* Brooks.	

NOTES.—268. Stock pile at Cleveland. 280. Stock pile at mine.

MICHIGAMME MINE—*Magnetic Ore.*

	1.	197.	225.	225.	225.	*Averages.*
Protoxide of Iron	....	29.109				
Sesqui- or Peroxide of Iron	....	61.631				
Protoxide of Manganese	1.01	traces.				
Alumina	2.12	2.120				
Lime	.12	1.070				
Sulphur	....	0.002				
Sulphuric Acid	....	0.008				
Phosphoric Acid	.05	0.057	0.067	....	0.392	
Silicic Acid, or Silica	3.06	3.280				
Water, Total	.57	1.497				
Organic or Carbonaceous Matter	....	0.340				
Titanic Acid	....	0.032				
Copper and Carbonic Acid	....	none.				
		99.146				
Metallic Iron	....	65.767	....	....	63.01	64.388
Phosphorus	0.027	0.024	0.029	0.019	0.168	.041
Sulphur	....	0.005				
Specific Gravity	....	....	4.61			
	Chemist, Britton. Sept. 21, 1870. *Sampler,* Brooks.	*Chemist,* Ralph Crooker, Boston. *Sampler,* Ralph Crooker.	*Chemist,* Wuth. 1872. *Sampler,* Brooks.	*Chemist,* Britton. 1872. *Sampler,* Brooks.	*Chemist,* Wendel. Feb. 6, 1873. *Sampler,* Brooks.	

NOTES.—1. Drill mud from 3 holes. 197. Numerous fragments from Exploration pits. 225. Taken at mine, fragments after blasting. All were taken before mine was opened.

MACOMBER MINE—*Hematite.*

	35.	35.	87.	*Averages.*
Sesqui- or Peroxide of Iron	76.80			
Oxide of Manganese	2.06			
Alkalies (by difference)	3.47			
Sulphur	0.14			
Phosphoric Acid	0.15	0.130		
Silicic Acid, or Silica	14.64			
Insoluble Silicious Matter	....	....	14.51	
Water, Combined	....	....	2.23	
Water, Total	2.74			
	100.00			
Metallic Iron	53.76	....	56.08	54.92
Phosphorus	0.06	0.057	....	0.058
Sulphur	0.14			
Metallic Manganese	1.51			
	Chemist, Chandler. Oct. 6, 1871. *Sampler*, Brooks.	*Chemist*, Wuth. *Sampler*,	*Chemist*, Britton. Nov. 4, 1871. *Sampler*, Brooks.	

NOTES.—35. From two trains of 16 cars each, one month apart. 87. From Mine. Numerous fragments. This mine belongs to the Negaunee hematite group, and contains considerable manganese.

MAGNETIC MINE—*Magnetic Flag Ore.*

	69.	54.	232.	232.	*Averages.*
Proto-sesquioxide of Iron	78.35	78.42			
Oxide of Manganese	0.10	trace.			
Alumina	trace.	.43			
Lime	0.69	.19			
Magnesia	0.21	.17			
Sulphur	0.58	none.			
Phosphoric Acid	0.151	.13			
Insoluble Silicious Matter	19.64	19.44			
Soluble Silica	....	.41			
Water, Total	....	0.42			
Undetermined and Loss	0.279	0.39			
	100.000	100.00			
Metallic Iron	55.16	56.78	52.22	....	54.72
Phosphorus	0.066	0.057	0.087	0.071	.067
Sulphur	....	none.			
Specific Gravity	....	....	4.30		
	Chemist, Jenney. 1872. *Sampler*, Brooks.	*Chemist*, Britton. Nov. 21, 1870. *Sampler*, Brooks.	*Chemist*, Britton. 1872. *Sampler*, Brooks.	*Chemist*, Allen. Jan. 1, 1873. *Sampler*,	

NOTES.—69. From small stock pile at mine. 54. From layers of rich ore banded with rock. From outcrop. 232. Small stock pile at mine.

MENOMINEE IRON REGION—*Specular Ores and Hematites.*

	95.	98.	102.	246.	74.	254.	68.	68.
Sesqui- or Peroxide of Iron	47.27	78.30			80.63		81.35	
Oxide of Manganese					3.075		1.32	
Alumina							trace.	
Lime		0.53					0.41	
Magnesia		0.17					0.237	
Sulphur							0.14	
Phosphoric Acid		0.044						0.260
Silicic Acid, or Silica		19.52					12.043	
Insoluble Silicious Matter	50.22				15.54			
Water, Total							3.498	
		98.564			99.245			
Metallic Iron	33.09	54.81	53.742	37.720	56.44	44.720	56.944	
Phosphorus		0.019		0.053		0.033		0.113
Metallic Manganese					0.735		0.313	
Specific Gravity				3.45		3.83		
	Chemist, Jenney. 1872. *Sampler*, Brooks.	*Chemist*, Wuth. Sept. 19, 1871. *Sampler*, Brooks.	*Chemist*, Chandler. *Sampler*, R. Pumpelly.	*Chemist*, Britton. *Sampler*, Brooks.	*Chemist*, Jenney. 1872. *Sampler*, Brooks.	*Chemist*, Britton. *Sampler*, Brooks.	*Chemist*, Jenney. 1872. *Sampler*, Brooks.	*Chemist*, Wuth. *Sampler*,

NOTES.—95. Average of prevailing variety of lean ore, Sect. 31, T. 42, R. 29. 98. Average of five of the richest pieces found, S. 31, T. 42, R. 29. 102. Average of 10 analyses for P. S. and L. S. Ship Canal Co., Sect. 31, T. 42, R. 29. 246. Same as 95. 74. Boulders at west ¼ post, Sect. 10. T. 39, R. 29. 68. From outcrop in swamp, Sect. 13, T. 42, R. 23. 254. Slate ore south ¼ post, Sect. 30, T. 40, R. 30.

MISSOURI—IRON MOUNTAIN MINE—*Specular Ore.*

	127.	127.	127.	128.	128.	128.	*Averages.*
Sesqui- or Peroxide of Iron	93.57			95.42			
Proto-sesquioxide of Iron	0.76			0.86			
Alumina	0.08			0.06			
Lime	0.46			0.32			
Magnesia	0.23			0.21			
Sulphur			0.008			.012	
Phosphoric Acid	0.035		0.112	0.036		0.067	
Silicic Acid, or Silica	4.75			3.02			
Metallic Manganese	0.12			0.07			
	100.005			99.996			
Metallic Iron	66.049			67.416			66.732
Phosphorus	0.016	0.043	0.049	0.016	0.025	.029	.029
Sulphur			0.008			.012	.010
Metallic Manganese	0.12			0.07			.095
Specific Gravity	4.944			5.002			
	Chemist, Wuth. April, 1872. *Sampler*, Brooks.	*Chemist*, Allen. May, 1872. *Sampler*, Brooks.	*Chemist*, A. A. Blair. 1872. *Sampler*, Brooks.	*Chemist*, Wuth. April, 1872. *Sampler*, Brooks.	*Chemist*, Allen. *Sampler*,	*Chemist*, A. A. Blair. *Sampler*,	

NOTES.—127. "Quarry Ore." Chippings from all parts of the pit and Stock piles. 128. "Surface Ore" (Boulders). Chippings and pebbles from all the diggings and Stock piles.

NEW YORK MINE—*Specular Ore.*

	20.	20.	20.	237.	237.	238.	238.	*Averages.*
Sesqui- or Peroxide of Iron	90.00							
Oxide of Manganese	trace.							
Alumina	1.87							
Lime	1.20							
Magnesia	0.60							
Sulphur	0.03							
Phosphoric Acid	0.57	0.428						
Silicic Acid, or Silica	4.72							
Water, Total	0.98							
	99.97							
Metallic Iron	63.00	......	......	62.13	......	60.10	......	61.74
Phosphorus	0.22	0.187	0.204	.1385	0.151	0.326	0.326	.225
Sulphur	0.03							
Specific Gravity	4.64	......	......	4.88	......	4.63		
	Chemists, Chandler & Cairns. March 4, 1872. *Sampler*, Brooks.	*Chemist*, Wuth. *Sampler*,	*Chemist*, Britton. *Sampler*,	*Chemist*, Britton. September, 1872. *Sampler*, Brooks.	*Chemist*, Allen. *Sampler*,	*Chemist*, Britton. March, 1873. *Sampler*, Brooks.	*Chemist*, Allen. *Sampler*,	

NOTES.—20. Large Stock pile at Cleveland—all varieties. 237. Great South Opening—Pit No. 1. 238. Beardsley's Pit—No. 2. The two last from mine.

NEW ENGLAND MINE—*Soft Hematite.*

	87.	239.	*Averages.*
Sulphur	....	None.	
Insoluble Silicious Matter	25.66	23.30	
Water Combined	1.42		
Volatile Matter (a little organic and water)	....	2.69	
Metallic Iron	49.64	46.84	48.24
Phosphorus	....	0.08	.08
Sulphur	....	none.	
Metallic Manganese	....	0.18	
Specific Gravity	....	3.79	
	Chemist, Britton. Nov. 4, 1872. *Sampler*, Brooks.	*Chemist*, Britton. December 27, 1872. *Sampler*, Brooks.	

NOTES.—87. From mine, numerous fragments. 239. From cars and stock pile at mine. First-class specular ore was formerly mined here, but is not at present.

NEGAUNEE HEMATITES—*Manganiferous Soft Hematite.*

	243.	243.	243.	11.	11.	108.	116.	*Averages.*
Sesqui- or Peroxide of Iron	......	......	......	65.40	......	65.48		
Oxide of Manganese	......	......	......	6.71	......	1.54		
Alumina	......	......	......	1.46				
Lime	......	......	......	0.45				
Magnesia	......	......	......	0.66				
Sulphur	......	......	......	0.04				
Phosphoric Acid	......	......	......	0.16	0.171			
Silicic Acid, or Silica	......	......	......	22.67	......	29.25		
Water Combined	......	......	......	1.88				
" Uncombined	......	......	......	0.58				
				100.01				
Metallic Iron	......	35.00	......	45.78	......	45.83	50.58	44.29
Phosphorus	0.067	0.065	0.099	0.07	0.074	......	......	.074
Sulphur	......	......	......	0.04				
Metallic Manganese	0.42	......	......	4.67	......	1.03	......	2.04
Specific Gravity	3.47	......	......	3.83				
	Chemist, Jenney. 1872. *Sampler,* Brooks.	*Chemist,* Britton. 1872. *Sampler,*	*Chemist,* Allen. 1872. *Sampler,*	*Chemists,* Chandler & Cairns. Mar. 4, 1872. *Sampler,* Brooks.	*Chemist,* Wuth. 1872. *Sampler,*	*Chemist,* Chandler. June 29, 1872. *Sampler,* Brooks.	*Chemist,* Jenney. July 13, 1872. *Sampler,* Brooks.	

NOTES.—243. From exploration pits. 11. Small stock pile at Cleveland. 108. Average of three analyses of ore from exploration pits. 116. Dark brown chalky ore. All from Sects. 6, 7, and 8, T. 47, R. 26.

NEW YORK STATE ORES (ST. LAWRENCE & WAYNE CO.)—*Hematites.*

	203.	206.	205.	204.	215.	209.
Protoxide of Iron	....	....	12.49	12.72		
Sesqui- or Peroxide of Iron	75.30	77.24	56.54	57.93	63.31	
Protoxide of Manganese	0.15	....	trace.	0.07		
Alumina	1.69	0.45	0.69	4.54		
Lime	7.04	1.60	8.23	2.32	6.03	
Magnesia	0.38	0.23	2.13	0.85		
Sulphur	0.03	0.05	none.	0.07		
Phosphoric Acid	trace.	trace.	0.36	0.16	....	1.49
Silicic Acid, or Silica	10.12	....	4.28	10.97		
Insoluble Silicious Matter	....	12.93				
Water, Total	....	2.107	....	0.62		
Carbonic Acid	5.41	....	15.01	9.75		
	100.12	94.607	99.73	100.00		
Metallic Iron	52.71	54.07	49.30	50.23	44.31	41.80
Phosphorus	....	....	0.16	0.07	0.43	0.64
	Chemists, Maynard & Wendel. Mar., 1871. *Sampler,* Geo. W. Maynard.	*Chemist,* Jenney. 1872. *Sampler,* Brooks. Jan. 1871.	*Chemists,* Maynard & Wendel. *Sampler,* Geo. W. Maynard.	*Chemists,* Maynard & Wendel. *Sampler,* Geo. W. Maynard.	*Chemist,* Chandler. *Sampler,* Unknown.	*Chemist,* Taylor. Jan. 2, 1873. *Sampler,* Geo. R. Tuttle.

NOTES.—203. Sampled for John A. Griswold & Co., at mine. 204. Do. do. 205. Do. do. 206. From small stock pile at Cleveland. 215. Analysis furnished by H. B. Tuttle. 203 and 206 are from Keene Mine. 204 and 205 are from the Caledonia Mine, both owned by Rossie Iron Works. 209 and 215 are Wayne Co. ore.

NEW YORK (LAKE CHAMPLAIN REGION)—*Magnetic.*

	288.	289.	290.	291.	292.
Protoxide of Iron	26.69	25.35	23.29	8.87	19.05
Sesqui- or Peroxide of Iron	59.84	56.19	50.13	69.99	42.97
Protoxide of Manganese	.55	0.12	0.38	0.38	
Alumina	1.87	3.56	4.22	3.67	3.47
Lime	....	0.82	1.28	1.90	1.19
Magnesia	....	....	0.85	traces.	0.09
Sulphur	.20	....	....	0.24	
Phosphoric Acid	1.94	trace.	trace.	0.07	trace.
Silicic Acid, or Silica	3.45	12.34	20.02	14.60	32.94
Water, Total	....	0.47			
Carbonate of Lime	6.02				
	100.56	98.85	100.17	99.72	99.71
Metallic Iron	62.61	59.02	53.21	55.91	44.98
	Chemist, Geo. W. Maynard. *Sampler*, Geo. W. Maynard.	*Chemist*, Geo. W. Maynard. *Sampler*, Geo. W. Maynard.	*Chemist*, Geo. W. Maynard. *Sampler*, Geo. W. Maynard.	*Chemist*, Geo. W. Maynard. *Sampler*, Geo. W. Maynard.	*Chemist*, Geo. W. Maynard. *Sampler*, Geo. W. Maynard.

NOTES.—288. Wetherby, Sherman & Co., and Port Huron Iron Ore Co., No. 21. 289. New Bed; Wetherby, Sherman & Co. 290. Hammond, Crown Point. 291. Indian; Ferrona ore; Hassey, Wells & Co. 292. Fisher; Port Henry Iron Ore Co.

OHIO IRON ORES—*Black Band and Kidney.*

	293.	294.	295.	296.	297.	298.	*Averages.*
Protoxide of Iron	26.82	23.02					
Sesqui- or Peroxide of Iron	8.94	8.79	75.00	7.60	75.00	12.34	
Oxide of Manganese	1.00	1.70	1.65	1.35	1.85	1.70	
Alumina	trace.	0.70	0.60	2.60	0.60	0.50	
Lime	1.05	1.70	2.80	....	5.94	....	
Magnesia	0.97	0.88	1.48	{ carbo. 6.50 }	3.64	{ carbo. 5.33 }	
Sulphur	0.18	0.11	trace.	0.18	0.12	trace.	
Phosphoric Acid	trace.	0.492	0.773	0.863	1.26		
Silicic Acid, or Silica	11.84	26.22	17.02	8.96	8.46	11.94	
Water, Combined	.....	......	......	....	2.28	.78	
Water, Total	......	......	0.25				
Volatile Matter	30.50	21.10					
Carbonic Acid	18.30	15.00					
Lime Phosphate	......	......	......	....	......	1.74	
Lime Carbonate	......	......	......	7.35	......	8.59	
Iron Carbonate	......	......	......	64.17	......	56.23	
	99.60	99.712	99.573	99.573	99.15	99.15	
Metallic Iron	27.12	24.06	52.50	36.31	52.50	35.88	
Phosphorus	.....	0.216	0.34	0.379	0.554	0.797	
Specific Gravity	2.494	2.321	3.411	3.434	4.076	2.539	
	Chemist, T. G. Wormley. *Sampler*.	*Chemist*, T. G. Wormley. *Sampler*,	*Chemist*, T. G. Wormley. *Sampler*,	*Chemist*, T. G. Wormley. *Sampler*,	*Chemist*. *Sampler*.	*Chemist*. *Sampler*.	

NOTES.—293. Black Band, Mineral Ridge, Mahoning Co., O. 294. Black Band, Tuscarawas Coal and Iron Co., Tuscarawas Co., O. Raw. 295. Black Band, Tuscarawas Coal and Iron Co., Tuscarawas Co., O. Calcined. 296. "Shell" or "Kidney Ore," Tuscarawas Coal and Iron Co., Tuscarawas Co., O. Raw. 297. "Shell" or "Kidney Ore," Tuscarawas Coal and Iron Co., Tuscarawas Co., O. Calcined. 298. Nodular Ore, Washingtonville Co., Columbiana Co., O.

For further analyses of Ohio Iron Ores, consult *Geological Survey of Ohio*, 1870, pp. 47, 48, 49, 219, 223.

REPUBLIC MINE—*Specular and Magnetic.*

	233.	233.	234.	234.	Averages.
Metallic Iron	67.21	......	65.81	......	66.51
Phosphorus	0.03	0.025	0.061	0.045	.040
Specific Gravity	5.19	......	5.07	......	......
	Chemist, Britton. 1872. *Sampler,* Brooks.	*Chemist,* Allen. 1872. *Sampler,*	*Chemist,* Britton. 1872. *Sampler,* Brooks.	*Chemist,* Allen. 1872. *Sampler,*	

NOTES.—233. Specular ore. First stock pile at opening of mine. 234. Magnetic ore. First stock pile at opening of mine.

SAGINAW MINE—*Specular and Hematite.*

	281.	282.	Averages.
Metallic Iron	50.820	53.980	52.40
Phosphorus	0.184	0.080	.132
	Chemist, Britton. Jan. 9, 1873. *Sampler,* Brooks.	*Chemist,* Britton. Jan. 9, 1873. *Sampler,* Brooks.	

NOTES.—281. Small stock pile (first mined) at mine. 282. Ditto. Both samples are soft hematite. By oversight no sample of the specular ore, which is first-class, was collected.

SHENANGO MINE—*hematite.*

	242.	242.	78.	Averages.
Sesqui- or Peroxide of Iron	....	....	82.13	
Oxide of Manganese	....	....	.15	
Alumina	....	....	2.32	
Lime	....	....	.41	
Magnesia	....	....	.08	
Phosphoric Acid	....	....	.186	
Silicic Acid, or Silica	....	....	14.46	
Water, Combined	....	....	.26	
			99.996	
Metallic Iron	55.140	....	57.49	56.315
Phosphorus	.049	0.071	0.081	.070
Specific Gravity	3.60			
	Chemist, Britton. 1872. *Sampler,* Brooks.	*Chemist,* Allen. 1872. *Sampler,*	*Chemist,* Wuth. 1872. *Samplers,* Davock, Glidden & Co.	

NOTES.—242. From small stock pile at mine. 78. "Taken from under snow, with no possible selection."—Letter from Davock, Glidden & Co.

SILAS C. SMITH MINE—*Hematite.*

	70.	70.	87.	*Averages.*
Sesqui- or Peroxide of Iron	71.70			
Oxide of Manganese	0.10			
Alkalies	2.03			
Sulphur	0.27			
Phosphoric Acid	0.09	0.127		
Silicic Acid, or Silica	23.38			
Insoluble Silicious Matter	....	....	23.79	
Water, Total	2.43	....	2.43	
	100.00			
Metallic Iron	50.19	....	49.21	49.70
Phosphorus	0.04	0.055	....	.047
Sulphur	0.27			
	Chemist, Chandler. Sept. 7, 1871. *Sampler,* Brooks.	*Chemist,* Wuth. *Sampler,*	*Chemist,* Britton. Nov. 4, 1871. *Sampler,* Brooks.	

NOTES.—70. From small stock pile at mine when first opened. 87. From mine when first opened, numerous fragments.

SPURR MOUNTAIN MINE—*Magnetic Ore.*

	2.	226.	226.	226.	97.	*Averages.*
Proto-sesquioxide of Iron	89.21	....	....	....	92.36	
Oxide of Manganese	traces.	....	....	....	0.15	
Alumina	2.67	....	....	....	1.66	
Lime	.67	....	....	....	0.73	
Magnesia	0.19	....	....	....	0.75	
Sulphur	0.35					
Phosphoric Acid	trace.	0.259	....	....	0.221	
Silicic Acid, or Silica	6.28	....	....	....	4.31	
	99.37				100.181	
Metallic Iron	64.60	....	....	59.96	66.87	63.81
Phosphorus	....	0.113	0.112	....	0.096	.104
Sulphur	0.35					
Specific Gravity	....	4.62				
	Chemist, Chandler. Nov. 1868. *Sampler,* Brooks.	*Chemist,* Wuth. 1872. *Sampler,* Brooks.	*Chemist,* Britton. 1872. *Sampler,*	*Chemist,* Wendel. Feb. 6, 1873. *Sampler,*	*Chemist,* Wuth. Sept. 14, 1872. *Samplers,* Morgan & Herne.	

NOTES.—2. Numerous fragments broken from outcrops of ore. 226. Fragments broken from outcrop 97 Numerous fragments broken from outcrop. All before mine was opened.

TAYLOR MINE—L'ANSE RANGE—*Hematite.*

	88.	81.	89.	89.	87.	*Averages.*
Sesqui- or Peroxide of Iron	82.664	....	62.25			
Oxide of Manganese	0.894	....	1.87			
Alumina	....	....	3.028			
Lime	0.312	....	0.31			
Magnesia	0.226	....	0.26			
Sulphur	0.090	....	trace.			
Phosphoric Acid	0.236	....	0.31			
Silicic Acid, or Silica	6.180					
Insoluble Silicious Matter	....	10.75	21.30	....	5.29	
Water, Combined	....	....	....	....	8.70	
" Total	9.438	9.41	8.10			
	100.040		97.428			
Metallic Iron	57.86	52.00	43.576	44.78	57.51	52.88
Phosphorus	0.102	....	0.13	0.097	....	.107
Sulphur	0.090					
Metallic Manganese	0.62	....	0.45	....	....	.53
	Chemist, Chandler & Schweitzer. *Sampler,* Brooks.	*Chemist,* Britton. 1871. *Sampler,* Brooks.	*Chemist,* Jenney. 1872. *Sampler,* Brooks.	*Chemist,* Britton. Feb. 18, 1873. *Sampler,*	*Chemist,* Britton. *Sampler,* Brooks.	

NOTES.—88. From shaft 20 feet in ore. 81. From three trenches across ore deposit. 89. From all pits, shafts and trenches showing ore. 87. From mine, numerous fragments. All before mine was opened.

WASHINGTON MINE—*Magnetic Ore.*

	39.	39.	39.	264.	264.	284.	285.	*Averages.*
Proto-sesquioxide of Iron	91.06							
Oxide of Manganese	0.23							
Alumina	0.85							
Lime	0.92							
Magnesia	0.77							
Sulphur	0.03							
Phosphoric Acid	0.25	0.406	......	0.21	0.170			
Silicic Acid, or Silica	5.13							
Water, Total	0.66							
	99.90							
Metallic Iron	65.94	......	......	67.20	......	57.280	54.800	61.305
Phosphorus	0.11	0.177	0.149	0.09	0.073	0.195	0.146	.141
Sulphur	0.03							
Metallic Manganese	0.14							
Specific Gravity	4.66							
	Chemist, Chandler & Cairns. March, 1872. *Sampler,* Brooks.	*Chemist,* Wuth. 1872. *Sampler,*	*Chemist,* Britton. 1872. *Sampler,*	*Chemist,* Taylor. 1872. *Sampler,* Taylor.	*Chemist,* Wendel. 1872. *Sampler,*	*Chemist,* Britton. Jan. 9, 1873. *Sampler,* Brooks.	*Chemist,* Britton. Jan. 9, 1873. *Sampler,* Brooks.	

NOTES.—39. Large stock pile in Cleveland. 264. Stock pile at Cleveland. 284. Shafts Nos. 1 and 4 at mine. 285. Shafts Nos. 2 and 6 at mine.

WINTHROP MINE—*Soft Hematite.*

	240.	240.	287.	*Averages.*
Sesqui- or Peroxide of Iron	....	....	84.66	
Protoxide of Manganese	....	....	1.41	
Alumina	....	....	0.13	
Lime	....	....	0.40	
Magnesia	....	....	0.007	
Sulphur	....	....	0.02	
Phosphoric Acid	....	....	0 084	
Silicic Acid, or Silica	....	....	12.70	
Insoluble Silicious Matter	24.34			
Water, Total	....	....	0.71	
Volatile Matter	0.93			
			100.191	
Metallic Iron	50.00	....	59.26	54.63
Phosphorus	0.03	0.045	0.037	.037
Sulphur	none.			
Metallic Manganese	0.53			
Specific Gravity	4.03			
	Chemist, Britton. Dec. 26, 1872. *Sampler,* Brooks.	*Chemist,* Allen. *Sampler,*	*Chemist,* Fred. H. Emmerton. Feb. 13, 1873. *Sampler,* J. T. Torrence.	

NOTES.—240. From all parts of mine. 287. Stock pile at a Chicago furnace.

WISCONSIN IRON ORES—*Iron Ridge* (*Hematite*).

	298.	*Averages.*
Pure Metallic Iron	56.44	
Oxygen with the Iron	24.91	
Protoxide of Manganese	.47	
Alumina	2.30	
Lime	2.28	
Magnesia	.99	
Sulphuric Acid	.15	
Phosphoric Acid	1.10	
Insoluble Silicious Matter	3.57	
Soluble Silica	.75	
Water and Carb. Acid	6.30	
	99.26	
Metallic Iron	56.44	
Phosphorus	.48	
Sulphur	.06	
	Chemist, J. B. Britton. *Sampler,* J. J. Hagerman.	

NOTE.—298. The ore is a fossil ore, in grains about the size and shape of flax-seed; Dr. Topham calls it Oolitic ore. There are several old analyses showing no phosphorus, but they are not reliable.

INDEX.

PART II

PPER DISTRICT

BY

RAPHAEL PUMPELLY.

ASSISTANTS:

ARVINE, M.E., L. G. EMERSON, C.E., AND S. B. LADD.

ATLAS PLATES

REFERRED TO IN PART II

TABLE OF CONTENTS.

To the Honorable Board of Geological Survey of the State of Michigan:

Gentlemen:—I have the honor to submit herewith a Report on the work done by the Geological Survey of the Copper District of the State, during the seasons of 1870 and 1872.

Yours very respectfully,

RAPHAEL PUMPELLY.

New York, *May* 1, 1873.

PREFACE.

THE work of the Geological Survey of the Copper District in 1870 and 1872 was confined chiefly to the construction of as perfect a series of cross-sections as circumstances would permit, in the Portage Lake, Calumet, and Eagle River districts. This seemed to be the most desirable method of expending the very limited annual appropriation of two thousand dollars. Any attempt to cover a wider field would have resulted in a simple reconnoissance of the district, without adding essentially to the previously existing information. To obtain the cross-sections which accompany this report, it was necessary to make an accurate triangulation of the country examined, without which it would have been impossible to represent in their relative positions, in a very uneven country, the outcroppings of the very numerous thin beds of nearly similar rocks, possessing varying thicknesses, an almost constantly changing course, and a high and varying degree of inclination. During 1870, the work was confined to the Portage Lake district, the survey of which was platted on a large working-map to the scale of 300 feet to one inch $= \frac{1}{3600}$. The cross-sections are on the scale of 90 feet to one inch. This scale appeared to be necessary in order to represent many beds of from two to six feet only in thickness, and especially to admit of adopting the same scale for both vertical and horizontal distances, without which it would not be possible to correlate the different sections as they are now exhibited. The cross-sections are accompanied by an accurate copy of the large working-map reduced to the scale of $\frac{1}{12,000}$, on which the various cross-section lines are delineated. This map represents only the work of the Geological Survey during the summers of 1870 and 1872, except the line of the northern shore of Portage Lake, which was taken from the United States Lake Survey Charts, and some results compiled from railroad surveys.

Both the maps and the cross-sections represent, with a few rare exceptions, only actual observations of rocks *in situ*. The Survey is indebted to Mr. A. B. Wood, Mr. R. J. Wood, and Mr. Mabbs (Agent of the Isle Royale Mine), for notes of measurements made by them in trenches dug under their supervision. These old explorations have been remeasured by the Survey, and where the rocks were not covered by the fallen *débris* they are described in the accompanying text. In the work of 1870 I was ably assisted by Mr. A. R. Marvine, M.E., and Mr. L. G. Emerson, C.E.

During the early part of the summer of 1872, Mr. Marvine was employed, in connection with Mr. Emerson, in completing some unfinished geological work on the north side of Portage Lake.

During the rest of the season, Mr. Marvine, assisted by Mr. S. B. Ladd, was engaged in making a cross-section in the Eagle River district. The results of this survey are embodied in the map and sheets of cross-section of the Eagle River district, which, like those of the Portage Lake district, are founded upon a careful trigonometrical survey, in which every pains was taken to insure extreme accuracy.

For comparison with the southern part of the Eagle River section I made a detailed geological section underground in the Central Mine.

Careful measurements were made by Mr. Marvine between the conglomerate beds at various points along the Range, in order to correlate stratigraphically the rocks of Portage Lake with those of Keweenaw County. The results of these measurements are given on the plate of "Grouped Sections," together with reduced sections of the Portage Lake and Eagle River districts, in order that the geological relations between the various sections may be more readily compared. The subject, which presents many difficulties, is ably discussed by Mr. Marvine in a separate chapter.

The chief aim of this report is to present to owners of mining properties a number of accurately determined geological landmarks, as guides in exploring for any given bed, and to establish an accurate base from which the Geological Survey may be prosecuted in Houghton and Keweenaw Counties on the one side, and toward Ontonagon on the other.

The chief portion of the present report is, therefore, to be found in the plates in the Atlas. I had intended to confine the text simply to the limits of the description of the cross-sections; but as many interesting observations were made which throw light on the age of the copper-bearing rocks, and on the distribution and manner of occurrence of the copper, I have thought they might very properly be added to the text.

The Survey is indebted to Mr. Pictrie, agent of the Central Mine, Mr. J. Chassells, Mr. Bolton, C.E., and Col. J. H. Foster, in many ways for aiding in its progress.

R. P.

CHAPTER I.

ON THE AGE OF THE COPPER-BEARING ROCKS.

KEWEENAW POINT, from its beginning at the Montreal River to where its extreme point is beaten by the storms at the middle of the Great Lake, is formed by two, stratigraphically, and (for the greater part) lithologically, distinct formations. These are separated by a line dividing the peninsula longitudinally. On the eastern side of this line are beds of sandstone, sloping gently to the *south-east;* on the western side is an immense development of alternating "trappean" rocks and conglomerates, dipping to the north-west at an angle of 55° to 60° at Portage Lake, and of 23° to 33° at other points.

As the question of the relative age of these two formations has given rise to considerable discussion, I will give, as briefly as possible, the principal facts in the case, and certain observations made by Major T. B. Brooks and myself, which bear directly upon the subject. Both formations have been referred by Foster and Whitney to the Potsdam, by Sir William Logan to the Chazy, while Mr. Bell, of the Canadian Corps, considers the Cupriferous series to be Triassic, agreeing herein with Jackson and with the view afterwards abandoned by Owen.

The principal facts on the south shore of Lake Superior are as follows: A series of red sandstone and shales, lying everywhere nearly horizontally, borders the Michigan shore between the Sault St. Mary and Bête Gris Bay on Keweenaw Point. From the former place to west of Grand Island, this sandstone is overlaid on the south by other Silurian rocks, and between Grand Island and Marquette the whole series sweeps around to the south-west, on its way to form the western, as it had hitherto formed the northern rim of the great Michigan basin. Where this south-westerly bend begins, the outcrop-line of the sandstone divides, and from Marquette westward we find, with short interruptions, the sandstone beds flanking the northern foot of the Huron Mountains, and dipping gently, 5° to 15°, toward the trough of Lake Superior.

In this part of its course, where it may be said to belong to the Lake Superior basin proper, it forms a marginal band along the lake shore, varying in breadth from a few rods to one or two miles. But west of the Huron Islands it widens with the south-westerly curving of the topographical axis of the Huron mountains, and fills with its horizontal strata the broad trough lying between these hills and the range of copper-bearing rocks of Keweenaw Point. In this depression there still remain one or two hills formed by remnants of the younger Trenton limestone. The trough, partly occupied by the waters of Keweenaw Bay, has for its western slope the beds of this Lower Silurian sandstone, which rise, at what seems to be the original angle of deposition, from the waters of the bay, to form the broad belt of nearly level sandstone country, which makes up the eastern half of Keweenaw Point.

At the western edge of this belt, its nearly horizontal strata abut against the steep face of a wall formed by the upturned edges of beds of the Cupriferous series of melaphyr and conglomerate, which dip away from the sandstone, at angles of 40° to 60°, according to geographical position. This sharply defined and often nearly vertical plane of contact, having been seen by the earlier geologists at several points along a distance of many miles, and having been found to be often occupied by a thick bed of chloritic fluccan, which was looked upon as the product of faulting motion, was considered as a dislocation.

This idea seemed to gain corroboration in the fact that, on the western side of Keweenaw Point, sandstones bearing considerable resemblance to those of the eastern horizontal beds occur, apparently conformably overlying the Cupriferous series.* Both sandstones came to be considered as identical in age, and as forming the upper member of the group.

There are many circumstances which make it difficult for us to accept this conclusion. One obstacle lies in the enormous amount of dislocation required, for instance, at Portage Lake, where the strata of the Cupriferous series, with an actual thickness of several

* It is not yet known whether these sandstones on the west side of Keweenaw Point are upper members of the Cupriferous series or belong to the Lower Silurian.

miles, dip away from the supposed *longitudinal* fault at an angle of about 60°.

Again, there are at least two patches of sandstone lying on the upturned melaphyr beds near Houghton, though it was not easy to prove that they were not brought thither by glacial action. Mr. Alexander Agassiz informed me that he has found in the horizontal sandstones near this so-called "fault," abundant pebbles of the melaphyr and conglomerate of the Cupriferous series, a fact which I found abundantly confirmed on the spot.

Sir William Logan hints at a similar doubt as to the proximate equivalence in age of these two series of rocks.* But the most decided facts were gathered by Major Brooks and myself, during a reconnoissance of the country between Bad River in Wisconsin, and the middle branch of the Ontonagon, east of Lake Gogebic † in Michigan. Our route was chiefly confined to the surface of the upper member of the Michigan Azoic, which we have provisionally considered to be the equivalent of the Huronian.

From Penokie Gap, on Bad River, to near Lake Gogebic—a distance of nearly sixty miles—the quartzites and schists of this formation are tilted at high angles, and form a belt one-fourth to one-half mile in width, bordered on the south by Laurentian gneiss and schists. On the north it is everywhere overlaid by the bedded melaphyr (containing interstratified sandstones) of the Cupriferous series. These form ridges and peaks which rise 200 to 300 feet above the surface of the Huronian belt.

These ridges, forming the "South Mineral Range," unite at their western end with the Mineral Range proper, which forms really through its whole length the back-bone of the tongue of land known as Keweenaw Point. Between these two ranges lies the south-western part of the Silurian trough, which has been mentioned before as extending inland from Keweenaw Bay.

Here, as there, it is filled with the horizontally stratified Silurian sandstone, forming a generally level country. For a distance of nearly thirty miles, between the Montreal River in Town 47 and Lake Gogebic, we found the Cupriferous series apparently conform-

* Geol. of Canada, page 85. And again in Geol. Survey of Canada. Rep. of Progr., 1866–69, pp. 472–475. In the last-mentioned place, he protests strongly against the idea that the copper-bearing rocks of Lake Superior are Triassic.

† Wrongly written Agogebic on many maps.

ing in strike and dip with the Huronian schists, and both uniformly dipping to the north at angles of 50° to 70°.* But in approaching Lake Gogebic from the west, we find that erosion of Silurian or pre-Silurian age has made a deep indentation entirely across the Cupriferous series and the Huronian, and into the Laurentian, so that at a short distance west of the lake these rocks end in steep and high declivities, at the base of which lies the level country of the Silurian sandstone, in which is cut the basin of the lake. From this point eastward, this ancient erosion had made great inroads upon the continuity of the Cupriferous and older rocks before the deposition of the Silurian sandstone. The melaphyr ridges are broken into knobs, or are wanting, and no Huronian was found as far as the Ontonagon River, seven miles away, and the limit of our observations.

On this river, in the centre of the north-west quarter of Sec. 13, Town 46, Range 41, the Silurian sandstone was found exposed in cliffs 50 to 60 feet high. The strata are horizontal, or at most have a barely perceptible tendency to a northerly dip. About 150 steps from the base of this cliff, there are outcrops of Laurentian schists whose bedding trends north-east toward the cliff of horizontal sandstone, and dips 45° to 60° south-east. The nearest observed outcrop of the Cupriferous series is in the south-east corner of Sec. 5, about four miles distant. It is a characteristic amygdaloidal melaphyr, whose bedding planes strike nearly east and west, and dip 50° to north. In general terms, the conclusions we are drawn to are these :

I. The Cupriferous series was formed before the tilting of the Huronian beds upon which it rests conformably, and consequently before the elevation of the great Azoic area,† whose existence during the Potsdam period predetermined the Silurian basins of Michigan and Lake Superior.

II. After the elevation of these rocks, and after they had assumed their essential lithological characteristics, came the deposition of the sandstone, and its accompanying shales, as products of the

* We observed several dips in the Huronian of 25° to 40°, while in the overlying Cupriferous series none lower than 50° were found. While this may point toward non-conformability, the greater dip of the overlying beds would make it probable that the lower dips were of a local character and due to minor undulations in the Huronian.

† Islands of Laurentian gneiss, etc., existed in the Huronian sea over parts of this area.

erosion of these older rocks, and containing fossils which show them to belong to the Lower Silurian, though it is still uncertain whether they should be referred to the Potsdam, Calciferous or Chazy. The question would still seem to be an open one. whether the Cupriferous series is not more nearly related in point of time to the Huronian than to the Silurian.

Our observations have detected a lack of conformability between the Laurentian and Huronian, at several points on the Upper Peninsula. On the other hand, in the region we have been discussing, which is the only one where the Huronian and Cupriferous are seen in contact, there seems to be a well-marked concordance between these two. There is abundant evidence on the Upper Peninsula, that the Silurian sandstone was not deposited until after the Huronian beds had assumed both their present structural position and lithological characteristics, and after they had undergone an enormous amount of erosion.

Some of the most salient topographical features of the Upper Menominee had been sculptured to the depth of two hundred feet or more before that time, and were afterward buried and wholly obliterated by the Lower Silurian deposits, and have been partially restored by the subsequent erosion to which that valley now owes its features. We now find ridges, consisting of the nearly vertical beds of Huronian quartzite and iron ores, capped with the horizontal sandstone, of which last, patches still remain in place on the end and side declivities. Where the sandstone was deposited at the base of these cliffs, we find it consisting largely of a breccia of the debris of quartzite and iron ore, identical in character with these substances in the unbroken ledge. It would probably be perfectly safe to apply the same remark to the Cupriferous series. Its members were formed, as we have seen in the previous pages, before the elevation of the Huronian rocks. The deposition of the Silurian rocks bordering on Lake Superior, should seem to have taken place during the progress of a gradual depression, which caused the coast line of that part of the Silurian sea to be represented by the bold cliffs of the interior of the Azoic land. In the eastern declivity of the mineral range of Keweenaw Point, we may see, then, one of these shore cliffs instead of the exposed side of a gigantic fault.

It is probably to this process of deposition of the Silurian sandstone over an area, which, after having undergone an enormous amount of erosion, was being gradually submerged, that we owe the absence of outcrops of the Cupriferous series beneath the sandstone at so many points on the south shore. It would be difficult indeed to account for their total absence at L'Anse, for instance, by supposing them to have thinned out, when at a distance of eighteen miles they have a thickness measured by miles, a thickness they exhibit wherever they are known, at points hundreds of miles apart on the north and south shores.

CHAPTER II.

ON THE LITHOLOGY OF THE COPPER-BEARING ROCKS OF THE PORTAGE LAKE DISTRICT.

In the immediate neighborhood of Portage Lake, the strata composing the "Mineral Range" have a uniform trend of north 35° east, and a nearly equally regular dip of 55° to 60° to west-north-west.

The series consists of beds of melaphyr, varying in thickness from twenty feet to more than one hundred feet, the demarkation being frequently defined by the amygdaloidal or epidotic character of the upper portion of each bed.

At intervals, varying from a few yards to several thousand feet, beds of conglomerate occur intercalated in the series.

This is the general character of the country near Portage Lake for a distance of about three miles, measured west-north-west across the formation.

Still farther west-north-west the rocks are little known, but seem to consist chiefly of sandstones and conglomerates.

The trappean rocks of Portage Lake occur uniformly in beds varying from a few feet to one hundred feet, or more, in thickness. Frequently an appearance of subordinate bedding is observable, arising, perhaps, from the existence of joints lying parallel to the plane of stratification, which divide the rock into plates a few inches to several feet thick.

The texture of the many varieties varies from compact and sometimes porphyritic, through fine-grained subcrystalline or earthy, to coarse-grained and distinctly crystalline. In individual beds, the texture is usually found to undergo a more or less gradual change from compact or granular at the bottom, to a vesicular or amygdaloidal condition, in the neighborhood of the hanging wall.

Green, of various shades, is the dominating color, and next to this brown and dirty red. Light and dark green, mottled or speckled with brown; dirty brownish-green; reddish-gray; and dark green, almost black, are the usual colors.

Even in the fresh state these rocks may be easily scratched with a knife, but they are exceedingly tough under the hammer, and the force which crushes a fragment often leaves the powder very firmly compacted.

The fracture is generally uneven, or hackly, to imperfectly conchoidal, but in the freshest, and especially in the compact varieties, it is often highly conchoidal. They have an earthy odor often even without having been breathed upon.

Some varieties yield a thick beard of a magnetic iron ore to the magnet, while others contain very little of this mineral.

The ingredients which are visible under the glass, and which seem to be common to all varieties, are a light ***green triclinic feldspar***, apparently labradorite and chloritic mineral of different shades of green, while the magnet reveals a very variable percentage of a magnetic iron; and in some of the coarser-grained varieties small jet black crystals, apparently of augite or hornblende, are occasionally visible. The accessory minerals observed, many or all of which are probably products of the alteration of the above constituents, are:

A brick-red foliaceous mineral resembling rubellan, occurring as very minute specks in some fine-grained varieties; it lends a soft rusty-brown appearance to the weathered surface, and speckles the interior with red.

Specular-iron in minute flakes, seemingly more frequent in the coarser-grained varieties.

Calcite in seams, and more frequently in grains and amygdules, especially in the amygdaloidal portion of the beds.

Epidote rarely crystallized; most common in the amygdaloidal varieties, but frequently in seams and impregnations, and nearly always associated with quartz.

Quartz which occurs in amygdules and seams, and also as an indurating medium near the hanging wall of many beds.

Prehnite in amygdules and seams, mostly confined to the amygdaloidal portion of the beds.

A chlorite-like mineral, soft, compact, amorphous, greenish-black, sometimes altered to brick-red, occurring in grains from pin-head to walnut size.

A yellowish-green soft earthy mineral, probably a green earth.

Laumontite and leonhardite in seams and amygdules.

Analcite in amygdules.

Orthoclase in small crystals and massive; in amygdaloidal cavities.

Native copper sometimes in fine impregnations in the fine-grained rock, also in thin sheets in jointing cracks, but chiefly in the amygdules, masses, sheets and impregnations which form the metalliferous deposits in the amygdaloids, where it is occasionally associated with native silver.

Native silver.

Datholite massive in the amygdaloidal portion of some beds, and also in small aggregations of microscopical crystals in the same positions.

We have fortunately several recent analyses of different and typical varieties of these rocks, made by Mr. Thomas Macfarlane.*

Of one of the coarser-grained varieties which forms very thick beds several hundred feet west of the Quincy "vein," Mr. Macfarlane says: "It is distinctly of a compound nature, but all its constituent minerals are not large enough to be accurately determined. Conspicuous among them is a dark green chloritic mineral, the grains of which vary from the smallest size to one-fourth of an inch in diameter. In the latter case they are irregularly shaped, with rounded angles, but they are never quite round or amygdaloidal [?]. They frequently consist in the centre of dark green laminæ. The mineral is very soft, and has a light greenish-gray streak. It fuses readily before the blowpipe to a black magnetic glass, and it would seem to be the preponderating element in the rock. The other constituents are in very fine grains, and consist of a reddish gray feldspathic mineral, with distinct cleavage planes, and another closely resembling it, in light greenish-gray particles, but whether of a feldspathic, pyroxenic or hornblendic nature could not be determined.

"The prevailing color of the rock is dark grayish-green. Hydrochloric acid produces no effervescence with it, even when in the state of fine powder. Its specific gravity is 2.83, and the magnet attracts a very small quantity of magnetite from its powder. The color of the powder when fine is light greenish-gray.

"When ignited it loses 3.09 per cent. of its weight, and changes to a light brown color. When digested with nitric acid, and afterwards with a weak solution of caustic potash (to remove free silica), it experiences, including the loss by ignition, a loss of 46.36 per cent. This consists of

* Canadian Geological Survey, Report of Progress, 1863–1866, p. 149.

Silica	14.73
Alumina	7.17
Peroxide of iron	14.87
Lime	4.47
Magnesia	2.03
Water	3.09
	46.36

"In the undecomposed residue light-red and dark-colored particles are discernible. On digesting it with hydrochloric acid, and subsequently with a weak solution of potash, it sustains a further loss of 10.6 per cent., which consists of

Silica	3.48
Alumina	3.03
Peroxide of iron	1.98
Lime	1.76
Magnesia	.35
	10.60

"The undecomposed residue was still found to consist of a light-red and a dark-colored constituent. The latter was the heavier, and an approximate separation was accomplished by washing. The dark-colored particles, which could not, however, be freed wholly from the light-colored feldspathic constituent, fused readily to a dark-brown glass. To judge from its gravity and fusibility, it would not appear unreasonable to regard it as either pyroxene or hornblende. In quantity it did not, however, exceed one-eighth of the feldspar. The latter fused easily before the blowpipe to a colorless glass, tinging the flame strongly yellow. It would therefore seem to be of the nature of labradorite, although it is only slightly decomposed by hydrochloric acid. Since, according to Girard, neither labradorite, pyroxene nor magnetite are decomposable by nitric acid, it may reasonably be concluded that the constituents removed by the nitric acid are those of the chloritic mineral. On treating the rock previous to ignition, much of the iron is removed as protoxide.

"Although some peroxide is also possibly present, I have calculated the whole of the iron as protoxide, and have, moreover, added the difference of the weight between it and the iron estimated as peroxide to the loss sustained by ignition, and put it down as water. In this way the composition of the chloritic mineral, calculated to 100 parts, would be

Silica	31.78
Alumina	15.47
Protoxide of iron	28.87
Lime	9.64
Magnesia	4.37
Water	9.87
	100.00

"In these figures the quantity of iron is much greater and that of magnesia much less than in ordinary chlorite. In its composition, and in being easily decomposed by acids, the mineral most closely resembles the ferruginous chlorite of Delesse (the delessite of Naumann), but differs from it in containing a considerable amount of lime, and in being readily fused before the blowpipe. Assuming, nevertheless, that the chloritic constituent is delessite, and that one-half of the iron removed by hydrochloric acid belongs to the magnetite, then the rock would be composed mineralogically of

Delessite	46.36
Labradorite	47.43
Pyroxene or hornblende	5.26
Magnetite	0.95
	100.00"

By the same method of analysis, Mr. Macfarlane found the rock underlying the copper-bearing bed of the Quincy mine to consist of

Delessite in amygdules and grains	38.00
Labradorite	62.00
	100.00

This rock is distinctly amygdaloidal. "The matrix is fine-grained, but it is crystalline, and is seen to consist of different constituents. Its color is dark reddish-gray." Its cavities, rarely the size of a pea, are filled with what seems to be the same chloritic mineral which occurs as a constituent of the rock above described.

Mr. Macfarlane also examined the rock which overlies the Albany and Boston conglomerate at the Albany and Boston mine. "It is a fine-grained mixture of dark-green delessite, greenish-gray feldspar, and reddish-brown mica, some of the laminæ of the latter showing ruby-red reflections. Its specific gravity is 2.81, and the smallest trace only of its powder is attracted by the magnet." He considers the mineralogical composition of this rock to be

Delessite	40.00
Mica	20.00
Labradorite	40.00
	100.00

The rocks to which the above given analyses refer are representatives of the three predominating types of the melaphyr of Portage Lake. Mr. Macfarlane's results agree very closely with my own observations on several hundred specimens, aided by the blowpipe and examination of external characteristics.

Everything goes to show that the normal, essential constituents of these rocks are, in their present condition, a triclinic feldspar, probably labradorite, and a ferruginous chlorite closely allied to delessite. This composition places these rocks among the typical melaphyrs, the greater specific gravity of the Portage Lake varieties being accounted for by the fact that the sp. gr. of delessite is 2.89, while that of ordinary chlorite ranges from 2.65 to 2.78.

Although the name melaphyr is an unfortunate one, having been first used to designate an entirely different rock, and having been successively applied to others of very various characters, it is now the only distinctive name for the class we have under consideration. All the trap rocks and associated amygdaloids of Portage Lake are varieties of melaphyr.

But I do not doubt that any one who will carefully study the melaphyrs of Portage Lake, and compare them with their equivalents in Keweenaw County, will feel convinced that the melaphyr owes its distinctive character to a process of metamorphism, in which the chlorite resulted, largely or wholly, from the alteration of hornblende or pyroxene. In the more distinctly crystallized traps of Keweenaw County, the pseudomorphic occurrence of chlorite after the hornblende or pyroxene constituent of the trap, may be traced through all the stages to a complete replacement of the latter by chlorite.

The principal varieties of melaphyr on Portage Lake are :—

1. *Coarse-grained;* in which the crystals of feldspar and grains of delessite are more or less distinct. The color is greenish-gray. It contains generally grains of magnetite and small tabular crystals of specular iron.

2. *Fine-grained;* the constituents, light-green or reddish triclinic feldspar and dark-green delessite, are sometimes distinguishable, but more generally they are not so. The usual color is grayish-green, but it sometimes is speckled with brown, through the presence of small flakes of rubellan ; or mixed green and brown, from the oxide of iron produced in the decomposition of some of the constituents. As a rule, the greater the amount of rubellan the less there seems to be of magnetite. In some instances, especially in some of the beds east of the Isle Royale copper-bearing bed, the rock is fine-grained and subcrystalline, brilliant black-green, sometimes purplish ; slightly shimmering ; easily scratched with the knife ; contains considerable magnetite, small pieces of rock adhering to the magnet. It weathers rusty gray.

3. *Melaphyr-porphyry;* dark-green, often nearly black ; compact, with perfect conchoidal fracture ; very hard ; contains minute crystals of triclinic feldspar.

AMYGDALOIDS.

The amygdaloids are merely varieties of the melaphyr. On Portage Lake they always form the upper or hanging-wall portion of beds of trap, into which they pass by a more or less gradual transition.

It is rare that one finds a bed of trap which does not contain, here and there, scattered segregations of secondary minerals, especially delessite, but often calcite, laumontite, quartz, or chalcedony, or prehnite, occupying cavities which are often well defined and spherical or ovoidal, but sometimes wholly irregular in shape, and without definite walls. These enclosures usually become more frequent in ascending from the foot-wall of a bed toward the hanging wall. The plane of demarkation between the amygdaloidal upper portion of a bed and the overlying rock is always well defined. Where they are sufficiently numerous to impress a distinctive character upon the rock, while at the same time the matrix retains the essential features, in regard to color and texture of the parent trap, I have designated the variety

AMYGDALOIDAL MELAPHYR.

All the varieties of melaphyr on Portage Lake are subject to this modification, but there is a considerable variation among different beds in regard to the nature of the minerals in the amygdaloidal cavities. In all the varieties, amygdules of delessite, or calcite, or quartz coated with delessite, or again spots of epidote, occur here and there in the body of the rock. In some beds the rock is characterized throughout by the presence of laumontite in small amygdules and minute seams.

In the belt occupying 1,000 feet or more on either side of the Isle Royale copper-bearing bed, many of the beds assume towards the top amygdules of delessite and of a green flinty mineral, resembling chrysoprase, coated with delessite. These are gradually succeeded nearer the top by ovoidal, lenticular, or irregular amygdules, from the size of a bean to several inches in diameter, of prehnite, greenish-white, or tinged with pink generally amorphous, but often with a radiating structure, and sometimes slightly impregnated with native copper.

The portion of the bed nearest the hanging wall is often highly amygdaloidal, while the matrix has at the same time a different degree of hardness, texture and color, and often a different mineralogical constitution from the parent trap. These varieties form the

AMYGDALOIDS PROPER.

The amygdaloids are the most highly altered form of the melaphyr, and present themselves under a variety of characters in different beds and in different parts of the same bed. The colors of the matrix are different shades of brown or red, and of green, or of these mixed; its texture varies from fine-grained or sometimes subcrystalline to compact; and its hardness ranges from that of limestone to that of quartz.

Two quite different kinds of amygdaloid occur on Portage Lake, both separately, and intimately associated in the same bed, and are easily distinguished by their different colors, the one being brown and the other green.

The brown, which exhibits the amygdaloidal character in its highest development, has a chocolate-brown to dirty red matrix, which generally is easily scratched with the knife, but is sometimes indurated and hard; it has a fine-grained to subcrystalline texture, and now and then contains minute reddish crystals of feldspar, and fuses easily to a dark-green and somewhat magnetic glass.

The amygdules in this variety are more generally spherical, but often somewhat irregular and connected, and more rarely long-cylindrical, and then usually perpendicular to the plane of bedding. The contents of these cavities, for they are very rarely empty, are laumontite, leonhardite, calcite, quartz, a soft green mineral, apparently green-earth, delessite (more rarely), native copper, epidote, prehnite, analcite, orthoclase. In places one, in others another, of these predominates; generally several are associated.

The green variety is a very fine-grained to compact light grayish-green rock. It is generally very hard, striking fire under the steel. Its constituents are very largely free silica, and a green mineral which has been generally taken for epidote, but which is so minutely disseminated as to render it difficult of determination. Small pieces of the rock fuse easily on the edges to a dark enamel which gelatinizes with acids. These beds are called epidote "veins," and they are probably, in many instances, at least, an intimate mixture of quartz and epidote, though in otherwise nearly similar beds the green mineral is soft, and is probably either a green-earth or a chlorite. The cavities in this variety are often less regularly defined in shape than in the brown amygdaloid. The enclosed minerals are quartz, epidote, calcite, delessite, prehnite, laumontite, green-earth, analcite, native copper, orthoclase. These two varieties of amygdaloids often occur together without any well defined lines of separation, the bed being made up of irregular masses of the two rocks. In places, however, the brown amygdaloid forms a band one to two feet thick on the hanging wall, with a rather abrupt transition into the green amygdaloid underlying it; I have never observed the reverse.

Some beds have an exceedingly mixed character; the amygdaloidal portions are associated with massive segregations of calcite, quartz and epidote, and are traversed by seams and irregular veins of these minerals; this structure is especially noticeable in the beds worked for copper. A somewhat similar structure occurs in other beds on a smaller scale, giving to them a brecciated or even a conglomerate-like appearance, which seems, however, to be due to purely metamorphic action; the best example of this is in the "Ancient Pit" bed, on the Shelden and Columbian property.

CONGLOMERATES.

The conglomerates of Portage Lake differ from each other but little, if at all, in lithological characteristics. The pebbles vary from the size of a pea to one foot or more in diameter, being coarser in some beds than in others. The different beds vary in thickness from mere seams to several hundred feet, and the same bed often varies greatly in width.

The pebbles, in most of the beds on Portage Lake, consist almost exclusively of varieties of non-quartziferous felsitic porphyry; two kinds predominate; one of these has a chocolate-brown to liver-brown, subcrystalline to compact almost vitreous matrix containing very scattered minute crystals of triclinic feldspar of the same color as the base. The other and rarer variety, also non-quartziferous, has a chocolate-brown, compact to minutely crystalline matrix, in which lie crystals, 1/8 to 1/2 inch long, of a flesh-colored triclinic feldspar.

In some beds there appear pebbles of a flesh-red rock, composed almost entirely of granular feldspar, containing small specks of a black undetermined mineral. In some instances the feldspar is wholly triclinic, in others the twin-striation is frequently absent. This variety of pebble is altogether absent in some beds, at least where they are opened, while in others they predominate, as in the

Albany and Boston Conglomerate. Pebbles of compact melaphyr, and of melaphyr amygdaloid also occur, but are quite subordinate in number to those already enumerated.

The normal form of cement is a fine-grained sandstone, composed apparently of the same material as the pebbles. Often the cement is very subordinate in volume, the pebbles touching each other. Frequently, however, the reverse is the case, and often, the sandstone forms layers from less than an inch to many feet in thickness.

The original character of the cement is often entirely lost; the interstices between the pebbles are sometimes, though rarely, empty; in places, the sand is associated with oxide of iron, chlorite, a white talc-like mineral, and carbonate of lime, or it is entirely replaced by calcite, chlorite, epidote or even native copper.

It is a remarkable fact that while all the conglomerate beds near Portage Lake are free from pebbles of quartz-porphyry, those in the neighborhood of Calumet are characterized by pebbles rich in grains of quartz. This abrupt change takes place about six miles northeast from the lake.

Different horizons of the Portage Lake series of rocks are marked by certain distinguishing lithological characteristics, which, without in any instance being peculiar to a given horizon, still serve to mark decidedly those parts of the series where they are, respectively, most frequent.

Thus, to begin toward the eastern part of the field, from the neighborhood of "Mabbs' vein" to within, say, 1,000 feet east of the Isle Royale "vein," there is a tendency, among the different traps, to a compact or fine-grained texture with a dark-green, almost black color, sometimes slightly mottled, especially on the weathered surface. The fracture is brilliant, and the trap contains enough magnetite to cause small bits of the rock to adhere to the magnet.

From this region till 1,500 feet or more west of the Isle Royale copper-bearing bed, the upper portions of very many of the beds have the amygdaloidal cavities filled with a light-greenish white or pale pink prehnite, which sometimes, for a width of 2 to 6 feet, form from 10 to 40 per cent. of the rock, and lend it a very characteristic spotted appearance.

During the next 2,000 feet or more, the traps have frequent seams 3 to 20 inches thick, consisting of distinctly individualized triclinic feldspar, delessite, prehnite and specular iron ; these occur both parallel to the plane of bedding and oblique to it. The traps through a portion of this distance are frequently impregnated with epidote, as is also the cement of the conglomerate beds.

On the "Dacotah" property we come to a belt of the formation in which many beds have a tendency to a coarse-grained, crystalline texture, and in some, the character is highly developed, giving the rock, at a distance, almost the appearance of a chloritic granite. Still farther west, on the "Southside" property, the brown amygdaloids often present a scoriaceous appearance which is quite characteristic.

Some, at least, of these features are traceable for miles in the longitudinal extension of the zones in which they occur. Thus, the prehnite amygdaloid of the Isle Royale series is found in the northeast extension of this zone, near where the road to Eagle River crosses the line between Townships 55 and 56 north, or about 7 miles from Portage Lake.

The coarse-grained melaphyr of the "Dacotah" is found extensively developed in the extension of the same zone on the South-Pewabic, Quincy and St. Mary's properties. The brown amygdaloids of the "Southside" reappear with their peculiar scoriaceous structure in the South-Pewabic and Hancock beds, and in the trenches on the St. Mary's, and are unquestionably the equivalents of the "Ash-bed" rocks of Keweenaw County, which they resemble.

CHAPTER III.

PARAGENESIS OF THE MINERALS ASSOCIATED WITH COPPER.

(*With a Comparative Table.*)

No. 1. CAPEN VEIN.—This is apparently a true fissure vein. It occurs in a compact and very tough melaphyr, which is exceedingly chloritic near the vein. All the joints within a distance of several yards from the vein are covered with a coating $\frac{1}{10}$ to ½ inch thick, of dark-green and bluish-green chlorite, having a combined fibrous and foliated structure oblique to the joint surfaces. The melaphyr is rich in magnetite. Sheet copper was found in mining, but not in paying quantity.

1. *Laumontite*, in thin seams.

2. *Prehnite*, in seams which cut through those of laumontite, also between symmetrically arranged bands of laumontite.

3. *Chlorite*, as destroyer and replacer of prehnite, and as lining of cavities in the latter.

4. *Analcite*, in clear crystals on the prehnite and chlorite.

5. *Calcite*.

No. 2. HURON MINE.—1. *Laumontite*, in thin crystalline bands on the sides of a cavity; the free ends of the opposed crystals nearly meet.

2. *Prehnite*, filling the space between the bands of laumontite.

No. 3.* COPPER FALLS MINE.—Fissure vein. 1. (?) *Natrolite*. 2. *Laumontite*. 3. *Analcite*.

No. 4. PHŒNIX MINE.—(Fissure vein.)

1. *Datholite*, on rock.—2. *Calcite*, scalenohedrons on the datholite.—3. *Feldspar*, crystals on both the datholite and calcite.—4. *Apophyllite*, in very clear crystals on datholite, calcite, and feldspar.

---

* Taken from a list given by Hilary Bauerman, Quart. Journ. Geol. Society, Nov., 1866.

No. 5.* BAY STATE MINE.—1. *Prehnite.* 2. *Quartz.* 3. *Copper.* 4. (?) *Laumontite.*

No. 6.* PHŒNIX MINE.—Fissure vein. 1. *Laumontite.* 2. *Quartz.* 3. "*Green-Earth.*"

No. 7.* BAY STATE MINE.—Fissure vein. 1. *Quartz.* 2. *Apophyllite.* 3. *Calcite.*

No. 8.* BOHEMIAN MINE.—1. *Analcite.* 2. *Copper.* 3. *Orthoclase.*

No. 9. AMYGDALOID MINE.—Fissure vein. 1. *Prehnite*, in its characteristic reniform shape.

2. *Quartz*, in small crystals on the prehnite.

3. *Analcite* crystals, covering the quartz.

4. *Orthoclase* crystals, on the analcite and quartz.

No. 10. BAY STATE MINE.—Fissure veins. On the soft brown gangue: 1. *Analcite*, lining part of a vugg. The crystals are ¼ inch in diameter, often white and transparent, but very much fractured. Near the contact with the rock they are often reddened internally and much altered, and then surmounted by the next member.

2. *Orthoclase*, in the usual minute crystals, some of which are scattered over the altered analcites.

No. 11. AMYGDALOID MINE.—Fissure vein. 1. *Prehnite*, in the characteristic reniform shape, forming the body of the specimen; fresh-looking on the free surface, but on the under broken side somewhat porous, with earthy fracture, and then rather intimately associated with datolite and a soft green (chloritic ?) mineral.

2. *Copper*, in films traversing the prehnite, and moulded to the reniform surface. While the under surface of the copper bears the impression of the prehnite, the upper surface, now free, bears that of some mineral that is gone; threads of copper rising from this free surface ¼ inch are crystallized at the tips, where they stood above the mineral that has disappeared.

3. Minute grains of a hard yellowish-white mineral, sprinkled like meal over the prehnite and copper; under the microscope appear to consist of sheaf-like clusters of minute rhomboidal plates; fuses with difficulty.

4. *Datolite*, in microscopic crystals on No. 3; others, one line in diameter, rosy, with suspended flakes of copper, lie upon the prehnite.

No. 12. AMYGDALOID MINE.—(Fissure vein.) On the gangue —here chloritic—lie, 1. *Calcite*, imbedded between the gangue and No. 2.

2. *Prehnite*, forming the greater part of the specimen, and having a tolerably fresh lustre.

3. *Copper*, in grains, flakes and threads conforming to the radiating cleavage planes of the prehnite.

4. *Datolite;* compact amorphous, white, translucent mass, covering the prehnite with a layer of which $\frac{3}{4}$ inch thickness still remains. The copper threads do not penetrate it.

No. 13. PEWABIC COPPER-BEARING BED.—This specimen—about $2\frac{1}{2}$ inches by $3\frac{1}{2}$ by $\frac{3}{4}$—is evidently from the interior of a druse, to whose wall it was attached by only a small part of its surface. The body of the specimen is copper, very cavernous, much of it pseudomorphous after laumontite. The copper is very thickly bestrewn with small green crystals of quartz—prisms terminated at both ends—which are, however, *older* than the copper. On the sides and around the edges of the specimen there are beautifully modified scalenohedrons of calcite. The successions are :

1. *The rock or mineral* to which the laumontite was originally attached, and which has disappeared.

2. *Laumontite* or leonhardite ; has also disappeared ; the prisms were $\frac{1}{8}$ to $\frac{1}{2}$ inch long, terminated at one end with a hemidome.

3. *A mineral*, now gone, which must have been present to support the quartz crystals (see Quartz). It may perhaps have been the alteration-product of the laumontite or an enclosing mineral.*

4. *Quartz*, in prisms $\frac{1}{20}$ to $_{8}$ inch long, often terminated at both ends. They occur on parts of nearly every one of the pseudomorphs after laumontite ; the copper is moulded to them, giving casts even of the striæ on the prisms, and they frequently pass entirely through the pseudomorphs after laumontite, so that the two ends of a quartz prism frequently just appear on opposite sides of the pseudomorph and transmit light. In some instances, the quartz crystals are so numerous as to touch each other, but they are often wholly isolated, and supported only by the copper which is younger.

The quartz crystals contain minute, brilliant particles of copper, wholly isolated, in the interior.

---

* Laumontite crystals often occur enveloped in younger calcite, except at the base.

5. *Copper*, in the form of laumontite, preserving often the sharpness of the angles and smoothness of the faces of the original mineral, when seen by the naked eye; under the glass the surface is less even. The pseudomorphs are not solid copper, as will appear in describing other specimens.

6. *Chlorite?* a soft, light-green mineral in minute hemispherical forms, with radiating structure, scattered over the quartz and copper. Wherever this mineral lies upon the quartz crystals, these are more or less penetrated by it, and some of them are eaten through and through to such an extent that the crystalline form is no longer recognizable.

7. *2d Copper and Calcite;* Calcite crystals—scalenohedrons—⅛ to ¾ of an inch long, lie on the sides and around the edges of the specimen. These crystals, in forming, adapted themselves with partially entering faces to the rough surface of the preëxisting quartz and copper. Some of the calcite crystals contain brilliant isolated films of copper, which must have been formed at least after the calcite had begun to crystallize, and is therefore younger than the copper previously mentioned.

8. *Datolite*, in exceedingly minute crystals, lying on both the chlorite and calcite; they are less than $\frac{1}{200}$ inch in diameter, but the datolite form is distinctly visible under the microscope; they fuse easily with the characteristic green flame.

No. 14. PEWABIC COPPER-BEARING BED.—1. *The rock* or mineral to which the laumontite was originally attached, which has disappeared, copper now forming the support of all the members.

2. *Laumontite*, of which only the form now remains.

3. *A mineral*, now gone, which seems necessarily to have been present to support the isolated crystals of quartz.

4. *Quartz*, in minute prisms, containing brilliant particles of copper.

5. ? *Calcite*, represented only by impressions in the copper. This calcite may, perhaps, be older than some of the foregoing members.

6. *Copper*, now forming the body of the specimen. It is very cavernous, and besides forming in places pseudomorphs after laumontite, it is the support of every member of the series. That it is younger than the quartz crystals is shown by the fact that

on removing these, perfect casts of them are visible in the copper. The copper also contains impressions of calcite crystals (see above).

7. *Chlorite?* the same mineral as the 6th member of No. 13, and occurring in the same manner.

8. *Calcite;* a few small scalenohedrons planted on the copper in the impressions of the older calcite = 5th above.

No. 14 (*a*). *Copper after laumontite*, from the PEWABIC COPPER-BEARING BED.

The upper face of this specimen is part of a partially filled cavity in a cupriferous and highly altered amygdaloid; the lower or broken face is a portion of the altered amygdaloid itself. The general appearance of the specimen at first glance is that of a drusy cavity, nearly filled, except in the middle, with broken crystals of calcite, whose interiors contain many thin plates and threads of native copper.

The amygdaloid is a soft compact brown and green rock with earthy fracture—an altered amygdaloidal melaphyr. The small amygdules near the wall of the larger cavity are of calcite.

The pseudomorphs of copper after laumontite are prisms $\frac{1}{8}$ to $\frac{1}{2}$ inch long and about $\frac{1}{8}$ of an inch square, and are terminated with a hemidome; they are each attached by one end to the wall of the cavity and project out toward the interior. The angles are often sharp, though in some instances the junction of two faces of a prism presents something of the appearance of a copper cast made in a mould whose two halves fit only imperfectly together. Sometimes, under a strong glass, the joining is clearly imperfect, and the pseudomorph has the effect of a prism built with four badly-soldered plates of metal.

Minute prisms of quartz (colored green by the chlorite-like mineral mentioned in specimens No. 13 and 14) project from the interior of the pseudomorphs, through the copper, to $\frac{1}{100}$ of an inch above the surface.

In one place I cut to the depth of $\frac{1}{16}$ of an inch in solid copper; but a cross fracture in another prism showed that the copper was, there, a mere superficial film, while the interior was occupied by a confused and rather porous mass of quartz prisms, copper, and the green chlorite-like mineral mentioned above. It is these quartz crystals whose ends just pierce the copper coating. In some instances, a prism of quartz terminated at both ends passes entirely through the pseudomorph and appears on both sides, and allows the light to pass through. After removing a quartz crystal from the copper, a perfect cast, even to the horizontal prism-striæ, is found in the copper.

The copper-surface of the pseudomorphs seems nearly smooth to the naked eye, but under a strong glass it appears not only often perforated with holes, but it often shows flakes of copper rising on edge to a height of $\frac{1}{100}$ of an inch above the face.

These pseudomorphs before the breaking of the specimen were imbedded in the interior of scalenohedrons of calcite, except at the attached ends of the prisms. At the contact planes between the calcite and the pseudomorphs, the former seems to adapt itself fully to all the irregularities of surface of the latter.

On the bottom of the specimen the calcite amygdules exhibit marked signs of change to datolite. The transparent crystals become gradually opaque, with a pearly lustre on the cleavage planes, and a little farther away this condition merges almost insensibly into a lustreless white mass composed of an aggregation of exceedingly minute crystals, which exhibit the datolite form under the microscope, and fuse easily with the characteristic green flame before the blowpipe. The same change is visible, in places, on the crystals of calcite enveloping the pseudomorphs after laumontite.

The relative ages here appear to be, 1, *The amygdaloid*, though probably not in its present condition; 2, *Laumontite;* 3, *Quartz;* 4, *Copper*, *chlorite-like mineral;* 5, *Calcite;* 6, *Datolite.*

No. 14 (*b*). *Another specimen* from the same locality exhibits—besides pseudomorphs of copper after laumontite—pseudomorphs of quartz after laumontite. In these last, the ends of the pseudomorphs are broken off, leaving only the prism. The faces are formed by a tolerably even mass of quartz, on the outer surface of which a crystalline form appears here and there under the glass. The interior of the prism is not a compact mass of quartz, but is merely filled with quartz prisms projecting from all sides toward the middle, and containing minute brilliant and isolated particles of copper.

Near these are the copper pseudomorphs; they are mere hollow shells, scarcely as thick as paper; the angles are sharp and the faces tolerably smooth, but often pierced with holes. The hemidome of one of these is studded with the ends of minute quartz prisms, which occupy the interior of that part of the pseudomorph and project through the copper shell.

In this specimen, also, some of the pseudomorphs are imbedded in scalenohedrons of calcite, which sparkle with brilliant particles of copper swimming in the transparent crystals.

In these remarkable pseudomorphs, the quartz is undoubtedly the oldest existing intruder, while the copper, which far more generally preserves the crystalline form of the laumontite, seems to be pseudomorphous in, at least, the second degree of removal. Yet in nearly all instances the older quartz is present, occupying a part of the space originally belonging to the laumontite crystal; and very often these quartz crystals are wholly separate from each other and supported only by the younger copper. Something which is now gone must have existed to perform this office of support before the copper was deposited.

No. 15. "RAGGED AMYGDALOID," ALBANY AND BOSTON M. —On the rock lie: 1, *Prehnite*—2, *Orthoclase*, in minute crystals on the prehnite.

No. 16. "EPIDOTE LODE," ST. MARY'S.—In a cavity in the quartz-epidote rock, which forms a frequent feature of this bed, lie: 1, *Prehnite* crystals, disposed as a reniform lining of the cavity—2, *Quartz*, in transparent prisms on the prehnite—3, *Analcite*, crystal ⅛ inch in diameter, slightly opaque and somewhat cavernous internally, planted on the quartz—4, *Orthoclase* crystals planted on the prehnite, quartz, and analcite.

The prehnite is partially altered, containing cavities lined or filled with a soft green mineral, chlorite or green-earth. There is also a greenish-yellow chlorite-like mineral which incrusts and has eaten away the surface of the quartz crystals.

No. 17. AMYGDALOID ON THE KEARSARGE LOCATION.—On the rock lie: 1, *Prehnite*—2, *Quartz* on the prehnite.

No. 18. HURON MINE.—On the rock lie: 1, *Analcite*, in a continuous band ½ inch thick, crystallized on the inner face; it is reddish and perhaps much altered, though still hard—2, *Calcite* incrusting the analcite crystals, and occupying cavities in their interior.

No. 19. ALBANY AND BOSTON AMYGDALOID.—The rock of this bed is a wholly irregular mixture of hard light-green amygdaloid, and soft-brown amygdaloid, in which the vesicular form is frequently lost, from the fact that the cavities containing secondary minerals have extended and become merged together, forming a confused patch-and-vein structure: 1, *Prehnite*, amorphous and altered to a slightly cavernous appearance on the surface—2, *Quartz* in prisms—3, *Orthoclase* in minute crystals chiefly on the altered prehnite, with which its formation is probably connected, and also on the quartz.

No. 20. SAME BED.—On the amygdaloid, which contains quartz amygdules, lie: 1, *Prehnite* penetrated with strings and films of copper—2, *Quartz* in prisms; *Chlorite-like mineral* in hemispherical forms, with radiating structure; *Orthoclase* in minute crystals; all these lie separately on the prehnite—3, *Calcite* covering all the above-mentioned members.

No. 21. SAME BED.—1, *Quartz* in prisms—2, *Chlorite-like mineral* in hemispherical forms, with radiating structure; wherever it is in contact with the quartz it has pitted it and eaten into it—3, *Calcite*.

No. 22. SAME BED.—On the amygdaloid lie: 1, *Prehnite* crystals in reniforn masses—2, *Quartz*, in prisms on the prehnite crystals—3, *Orthoclase*; *Calcite*; the orthoclase is in minute crystals on the prehnite and quartz.

No. 23. SAME BED.—The amygdaloid on which the following succession occurs consists of quartz and chlorite, and is wholly altered, so much so that the quartz which now composes a large part of it is evidently of the same age as that which follows the prehnite: 1, *Prehnite*, in crystalline reniform masses—2, *Quartz* prisms—3, *Copper* in cubes, with octahedral modifications, planted on, and moulded to, the quartz crystals.

No. 24. SAME BED.—1, *Prehnite*—2, *Quartz*—3, *Calcite*.

No. 25. SAME BED.—*Analcite* in pellucid crystals—2, *Calcite*; *Orthoclase*; in this specimen the analcite appears to have incrusted some mineral which has disappeared, and the feldspar crystals occur in the cavity thus formed as well as on the outer surface of the analcite.

No. 26. SAME BED.—1, *Prehnite*—2, *Quartz*—3, *Copper*, in threads often moulded to the quartz—4, *Orthoclase* in minute crystals planted on the prehnite, quartz and copper.

No. 27. SAME BED.—1, *Prehnite*, penetrated with copper threads—2, *Quartz* in prisms—3, *Chlorite-like mineral* mentioned in Nos. 20 and 21; here also it has eaten into the faces of the quartz crystals—4, *Analcite* crystals, much fractured and eaten away, and sometimes quite hollow.

No. 28. SAME BED.—1, *Prehnite* in places cavernous—2, *Quartz* in prisms in the cavities in the altered prehnite—3, *Orthoclase* crystals planted on the quartz.

No. 29. SAME BED.—1, *Prehnite*—2, *Copper* traversing the prehnite in the form of threads, etc., ending in crystals which adapt themselves to the crystalline surface of the prehnite.

No. 30. SAME BED.—1, *Prehnite*—2, *Analcite*—3, *Copper* in flakes on the analcite—4, *Orthoclase; chlorite-like mineral.*

No. 31. SAME BED.—*Quartz* in prisms—2, *Orthoclase* crystals, planted on the quartz.

No. 32. SAME BED.—1, *Prehnite*—2, *Copper* in crystals whose under surfaces are moulded to the crystalline surface of the prehnite.

No. 33. HURON MINE.—On the amygdaloid containing smaller amygdules of delessite and quartz, lie: 1, *Laumontite*, a crystalline layer with projecting crystals—2, *Calcite* crystallized upon and wholly enveloping the laumontite crystals.

No. 34. WESTERNMOST ADIT ON THE "SOUTHSIDE."—1, *Analcite*, in opaque crystals $\frac{1}{8}$ to $\frac{1}{2}$ inch in diameter—2, *Orthoclase* crystals planted on the analcite. (The rock containing this is chocolate-brown, and filled with small amygdules of 1st, Laumontite, 2d, Calcite).

No. 35. "RAGGED AMYGDALOID." ST. MARY'S.—This is a soft brown amygdaloid with brown streak, in which the cavities have assumed the most irregular shapes, and merge into each other in a manner which gives to the rock a highly brecciated appearance. The cavities are generally partially open at their wider points; and the minerals occupying them are chiefly the following, often accompanied by a white clay: On the rock lie, 1, *Analcite*—2, *Orthoclase* crystals on the analcite—3, *Calcite* over both the foregoing members.

No. 36. SAME BED.—On the rock lies *Calcite*. *Orthoclase* crystals on the calcite.

No. 37. SAME BED. On the rock are scattered: 1, *Analcite* crystals—2, *Calcite* on the analcite.

No. 38. SAME BED.—1, *Analcite* in large crystals; much altered—2, *Orthoclase* crystals planted upon the outer surface of (and in cavities in) the analcite.

No. 39. PEWABIC COPPER-BEARING BED.—On the amygdaloidal rock lie: 1, *Calcite; Copper*—2, *Datolite* in a granular mass incrusting the calcite crystals.

No. 40. SAME BED?—1, *Calcite* in scalenohedrons—2, *Datolite*, a granular mass of microscopical crystals. Here the datolite has partially displaced the calcite; only the points of the crystals of the latter were exposed, all the rest being imbedded in the younger datolite. These free-standing points remain perfect in glance and form, while wherever the calcite crystals are in contact with the datolite, their surfaces are roughened and perceptibly eaten into. The calcite crystals rest upon a granular mass of the same variety of datolite, which is perhaps the result of a displacement of calcite.

No. 41. "EVERGREEN BLUFF."—1, *Quartz* prism—2, *Orthoclase* in minute crystals—3, *Calcite* in simple and twin scalenohedrons.

No. 42. "AMYGDALOID" MINE.—(Fissure vein). 1, *Copper*—2, *Compact datolite*.

No. 43. SAME VEIN.—1, *Prehnite* in its characteristic form—2, *Copper* conforming to the radiating cleavage-structure of the prehnite—3, *Datolite*, compact.

No. 44. LOCALITY UNKNOWN.—1, *Prehnite* in its characteristic form—2, *Quartz* in prisms on the prehnite—3, *Analcite* crystals on the quartz—4, *Orthoclase* crystals on the analcite.

No. 45. WESTERNMOST ADIT. "SOUTHSIDE."—Small amygdules consist of: 1, *Laumontite*—2, *Quartz* surrounded by the laumontite.

No. 46. MICHIGAN MINE.—(Fissure vein). On the veinstone, which is here a greenish-gray, hardened clay-like material with flakes of chlorite and copper, and which becomes brown and soft near the contact with No. 1, lies: 1, *Datolite* in a uniformly distributed layer $\frac{1}{10}$ of a line to 2 lines thick, with the free surface highly crystallized. The crystals are transparent and rose-colored from the presence of minute particles of copper. The datolite appears quite fresh, and the copper seems to be confined to it—2, *Calcite*, four small slightly yellowish semi-transparent rhombohedrons, modified with steeper scalenohedron faces, lie upon the datolite—3, *Orthoclase*, yellowish crystals, $\frac{1}{10}$ of an inch long, are scattered over the surface of the specimen, some lying upon the calcite and some upon the datolite.

No. 47. MANY LOCALITIES.—1, *Prehnite*—2, *Delessite*. The prehnite which occurs as the solid filling of amygdaloidal cavities in the upper part of many beds, is subject to alteration to chlorite. It is very common to see the prehnite soft and green to a slight depth from the outer surface of the amygdule, without any line of separation between this portion and the hard centre; and in the interior the prehnite often passes gradually into spots and flakes of chlorite. In these amygdules the prehnite is characterized by a radiating structure starting from a single centre. It is along these planes of radiation that the change begins. Every possible gradation is observable. The resulting product is sometimes a mass of foliated chlorite, but more generally, it is an amygdule of compact chlorite, exhibiting in its fracture the same radiating structure as the prehnite.

No. 48. SHELDEN AND COLUMBIAN LOCATION.—1, *Prehnite*, which is the general filling of the cavities in the upper part of the amygdaloid of this locality—2, *Feldspar*, red. It is quite an exceptional occurrence in this neighborhood, and it is intimately associated with the prehnite in a manner that makes it seem to be pseudomorphous after it.

Crystals of epidote and of quartz occur on this feldspar, but the specimen gives no insight into their relation, as regards age, either to each other, or to the feldspar, as a secondary product.

No. 49. SOUTH PEWABIC MINE.—In this bed, a frequent form of the rock is a compact amygdaloid, of which 50 or 60 per cent. of the volume consists of amygdules from the size of a pin-head to $\frac{1}{3}$ inch in diameter. The matrix in a specimen before me is brown, and too hard to be scratched with a knife. The amygdules are sometimes of calcite, but more generally contain, 1, *Quartz*, clear and filling the cavity—2, *Chlorite; copper*. The chlorite (apparently delessite) appears to displace the quartz; in some amygdules, it merely penetrates the fissures of the quartz, giving to this a green color; in others, nothing remains but a cavernous mass of quartz and chlorite usually well charged with copper; indeed the copper occurs here only with this chlorite.

The series to which this bed belongs is represented on the "Southside," and again on the St. Mary's, by amygdaloids which resemble this one in all essential particulars, except that the amygdules are there filled chiefly with laumontite, and are free from copper and nearly free from chlorite.

No. 50. "OSSIPEE AMYGDALOID."—The rock is compact, spotted green and brown, and contains small *separate* amygdules of prehnite, quartz, epidote, calcite, chlorite. The chlorite appears as a destroyer of prehnite, quartz and calcite. A larger cavity shows the following succession: 1, *Prehnite;* a lining $\frac{1}{8}$ to $\frac{1}{3}$ inch thick, much altered and in places changed to chlorite—2, *Orthoclase,* minute crystal—3, *Epidote* on the feldspar.

No. 51. HURON MINE.—In places, the Isle Royale copper-bearing bed consists largely of a light grayish-green, fine-grained rock of epidote and chlorite indurated with quartz; small, irregular cavities in this contain: 1, *Epidote;* crystalline lining $\frac{1}{4}$ inch thick—2, *Quartz* filling the interior.

No. 52. SHELDEN AND COLUMBIAN MINE.—In cavities, in a brown and green amygdaloid, lie: 1, *Quartz* in well-shaped prisms—2, *Calcite; Quartz.* This second quartz is in small and very much distorted crystals, which are often partially imbedded in the calcite, and are also often planted on the older quartz, from which they can be easily removed without fracture.

No. 53. HURON MINE.—1, *Quartz* with more or less crystalline structure—2, *Copper* moulded on the quartz and filling cracks and interstices in it.

No. 54. RAGGED AMYGDALOID. ST. MARY'S.—In the rock (see No. 35), some of the smaller cavities contain: 1, *Orthoclase* as a thin crystalline lining—2, *Calcite* filling the interior—3, *Chlorite-like mineral* penetrating and apparently replacing the calcite.

In a larger cavity occur the following:

No. 55.—1, *Analcite* crystals $\frac{1}{4}$ inch to 2 inches in diameter, much reddened—2, *Orthoclase;* small crystals on the analcite—3, *Clay?* a soft white mineral, apparently the result of continued decomposition of the analcite under conditions unfavorable for the formation of new silicates as feldspar.

No. 56. "ANCIENT PIT" BED. DOUGLASS LOCATION.—1, *Epidote* forming a crystalline lining of a cavity—2, *Quartz* filling the interior.

No. 57. SULPHURET VEIN (Fissure vein). HURON LOCATION.—

The vein (6 to 8 inches wide) consists of the following: 1, *Ankerite?* (crystalline, massive, white on fresh surface, but rusty brown on exposed fractures) forming the member nearest the wall on each side—2, *Quartz*, in two symmetrical comby bands on the dolomite, and in thin seams in the dolomite connected by cross-seams with the quartz-comb—3, *Chalcocite*, black—bluish-black *with distinct cleavage*. It resembles the pseudomorphous chalcocite of the Lac la Belle mine. Bornite occurs sprinkled through the chalcocite in minute specks; in places it predominates, sometimes to the exclusion of the latter.

These sulphurets form the central member, and bunches of them are often surrounded by the older members, giving the "cockade" structure to the vein.

No. 58. UNKNOWN.—A specimen (in Mr. W. H. Stevens' collection at Copper Harbor) consisting chiefly of prehnite and containing beautiful cast-impressions of calcite.

1st. *Calcite* scalenohedrons 1½ inches long, represented now only by perfect casts.

2d. *Prehnite* massive and botryoidal containing the casts of calcite.

3d. *Quartz* upon the upper surface of the prehnite.

No. 59. SCHOOLCRAFT MINE.—(See cut on next page). Small amygdules in the brown amygdaloid immediately overlying the conglomerate bed. This rock is very rich in copper surrounding amygdules and filling their cracks.

1st. *Copper* forming outer shell of amygdule.

2d. *Delessite*.

No. 60. IBID.—(See cut on next page). In same specimen with No. 59.

1st. *Copper* forming the outer shell of a large amygdule.

2d. *Red feldspar*, forming the next inner layer.

3d. Calcite filling the interior.

No. 61. PHŒNIX MINE.—(Fissure vein). A specimen of purple and green vein rock, 5 inches by 4 inches, is covered on one side by

1st. *Datholite?* crystals closely studded over the whole surface.

NOTE.—It is in another portion of this vein that the arseniurets, whitneyite and domeykite are found.

2d. *Green-earth or chlorite?* forming a thin film over much of the datholite.

3d. *Analcite* in numerous $\frac{1}{8}$ inch semi-limpid crystals, often very cavernous.

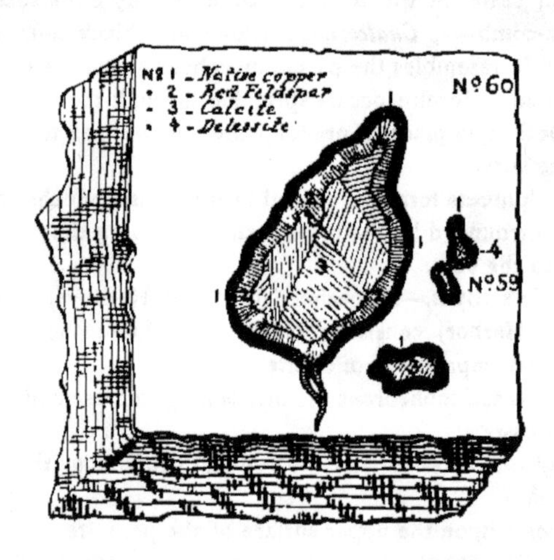

4th. *Feldspar* in small limpid crystals scattered profusely over the datholite and green-earth, but rarely sitting on the analcite.

5th. *Calcite* in scalenohedrons sitting on the green-earth and feldspar.

6th. *Apophyllite* in countless $\frac{1}{16}$ inch crystals perfectly transparent, and sitting on 2, 3, 4 and 5; where it occurs on calcite the crystals are pure and colorless; elsewhere they are colored red, possibly through particles of oxide of copper.

I have attempted to bring the foregoing observations into the following tabular form for greater convenience of comparison. The table is unavoidably imperfect, owing to the necessary difficulties which arise in attempting to compare the successions of different localities. The detailed observations, however, will serve for a check upon the imperfections of the schedule.

COPPER AND CALCITE.—In many of the instances in which calcite crystals are found enclosing copper, it is difficult and often impossible to distinguish as to the relative ages of the two. But

specimens in my collection offer conclusive proof that each of the following cases occur:

I.—*The copper was present before the calcite began to form, and became enclosed in the growing crystal.*

In this case the copper and its associated minerals generally form the basis on which the calcite rests, and the crystals of the latter exhibit entering faces wherever the surface of the crystal is in contact with the copper; it should seem to indicate an effort at those points to crystallize free from the foreign substance, by forming separate individuals. But on the finished crystal the traces of this tendency are visible, generally, only in the comparatively very small entering faces at the contact with the copper.

In this way calcite crystals, formed in a cavernous mass of copper, are intersected internally by a perfect network of thin plates of the metal, and yet preserve their cleavage unaffected; but wherever the copper comes in contact with the surface of the crystal, the small entering faces are present.

II.—*The crystal of calcite was partly formed, then became incrusted with copper, and was finished by a new growth of calcite over the metallic film.*

A most remarkable instance of this case is that of a crystal about 2 inches long—a steep scalenohedron—with a basal termination of about 1 square inch surface. At this stage of its growth it was covered, over nearly the whole surface, with a thin coating of copper. The basal termination on scalenohedrons of calcite is as rare on Lake Superior as elsewhere, and in the few instances where I have seen it, it lacks the polish which indicates perfect growth. The tendency to complete the point of the scalenohedron is well shown on this specimen; over the partially copper-coated basal plane there are scattered a large number of perfectly pointed scalenohedrons—two or three of these are ⅓ to ½ inch high—and others are scattered over the side-faces. All of these younger crystals are arranged in perfect uniformity with the plan of the underlying, older individual.

Those portions of the surface on which the copper-coating is perfect have no younger calcite crystals; these occur where the metallic film is thinnest and more or less perforated.

The copper is not confined absolutely to the surface of the crystal on which it lies; it penetrates to a slight distance along the cleavage-planes, and the result is an exceedingly delicate reticulation on its under surface. The calcites which are planted on the copper contain brilliant particles of the metals swimming, if one may use the word, in the interior of the crystals; and these are so disposed as to lead to the idea that, throughout the growth of the younger crystals, they had to contend with the continued deposition of the metal. Thus one of the new scalenohedrons, after growing to the height of $\frac{1}{4}$ inch, was, like the underlying one, also ended with a basal termination, on which again smaller new and well-pointed individuals were built up.

III.—*The copper has entered the calcite crystal since its growth was finished.*

A specimen, in my collection, illustrates this remarkably well. It is a cleavage-rhombohedron of opaque calcite, traversed by intersecting sheets of copper, which are wholly independent of the cleavage planes. On detaching the copper from the calcite, the surface of the latter appears rough; it is a fracture oblique to the cleavage, and the face of the fracture is formed by countless corners, or solid angles, of minute cleavage-rhombohedrons, as is fully proved by the reflection of the light. The copper-sheets, which are about $\frac{1}{40}$ inch thick, reproduce this very completely.

Another very remarkable specimen is from the cement of the Albany and Boston conglomerate. It is about 1 inch in diameter, and consists of opaque white calcite. The continuity of the cleavage shows it to be a single individual, though it passes on the edges without any sharp demarkation into the common cement of the conglomerate. This calcite is traversed by continuous sheets of copper $\frac{1}{300}$ to $\frac{1}{40}$ inch thick, which are perfectly straight. These sheets are parallel to several planes (nearly all of which are independent of the cleavage), and intersect each other. In each of the sets thus formed the sheets are perfectly parallel, and are separated by plates of calcite, which are, in places, as thin as the copper itself. Where three such sets intersect each other, the resulting solid appears composed of concentrically arranged laminæ of copper and calcite. In some parts of the specimen, the copper predominated over the calcite. Wherever the faces of the copper laminæ are exposed, they are marked with a delicate, reticulated tracery, indicating the lines of intersection of the sheet with the cleavage planes of the crystal. The cement in the vicinity of the calcite is impregnated with copper; in places it is almost wholly replaced by the metal in the fine granular condition called "brick copper," and into this the laminæ of metal extend, without break, from the calcite. This specimen is really a pseudomorph of copper after calcite.

Copper and Silver.—It is a well-known fact that these two metals occur in the metallic state, in the Lake Superior deposits, in the most intimate contact with each other, and yet without being mutually alloyed. Even at the contact they are not absolutely joined together, for after rolling out a piece of copper containing spots of silver, the two metals become more or less separated, and may often be readily detached from each other.

In all the instances that have come under my observation the silver appears to be younger than the associated copper, and seems to have been directly precipitated by copper, and to have replaced the latter.

Chalcocite, Bornite, Whitneyite, Domeykite.—Two fissure-veins are known in the neighborhood of Portage Lake which carry these ores. They have been examined only very superficially; but it is a remarkable fact that the amygdaloids traversed by these veins contain only native copper. One of the fissure-veins, bearing both sulphides and arsenides of copper, enters the Grand Portage cupriferous amygdaloid bed, which bears only native copper, and remains in it with a changed direction for a short distance. The gangue of these veins is quartz, calcite, and a carbonate of lime containing some iron and magnesia—ankerite?

The only other instance I have observed of the occurrence of copper in combination with sulphur, is in the fissure-vein of the Mendota property, near Lac la Belle. This vein appears to traverse the entire trappean series from Agate Harbor on the north to Lac la Belle.

Wherever this vein has been opened or uncovered, along the greater part of its course, north of the Mendota property, only *native* copper has been found; but when it enters a bed of conglomerate on the north flank of Mt. Bohemia, the little copper it contains is combined with sulphur in a very pure chalcocite. Where the vein passes from the conglomerate into the underlying amygdaloid, a fine deposit of chalcocite with calcite was found to have been formed, for a short distance, on both sides of the vein, between the two beds.

Still farther south the vein enters a mass of syenite, consisting of a pink triclinic feldspar, some hornblende and much chlorite, as an alteration-product of hornblende, and containing frequent impregnations of chalcopyrite, bornite, and, more rarely, chalcocite. In this syenite the vein and its many feeders carry bornite and considerable quantities of a bluish sulphuret of copper, in sheets ⅓ to ¾ inch thick, which has a very crystalline structure and exhibits octahedral cleavage.*

Near the contact between the syenite and trap, the latter rock is impregnated with magnetite, specular iron, chalcopyrite, and bornite. Excepting the syenite, wherever copper is found in the traps and amygdaloids on the Mendota property, it is in the metallic state. The occurrence of the sulphides and arsenides of copper in this isolated manner and in fissure-veins traversing rocks more or less impregnated with metallic copper, seems to show a diversity of origin for the sulphur and arsenic on the one hand, and the copper on the other. It does not seem unreasonable to suppose the copper to have entered the vein-fissure from the adjoining rocks in solution, as carbonate, sulphate, or silicate, and to have been then precipitated by sulphuretted hydrogen and arseniuretted hydrogen respectively. Or the copper may have been deposited in these as in other veins, in the metallic state, and have been subsequently changed by the same gases. In the case of the pseudomorphous chalcocite, where the Mendota vein traverses syenite, cuprite must have been formed by the oxidation of chalcocite or of native copper, and the oxide must have been subsequently decomposed by sulphuretted hydrogen.

The Huronian formation, which probably underlies all this region, contains in its upper members large amounts of carbonaceous matter in the form of graphite; the gases may have originated

* Prof. Cooke, after a casual examination of this mineral, suggests that it is probably a pseudomorph of chalcocite after cuprite.

in a reduction of sulphates and arseniates by the carbon of these beds.

REPLACEMENT OF PORPHYRY MATRIX BY CHLORITE AND COPPER.

Among the pebbles in the Calumet conglomerate there is a variety of quartz porphyry, with a brown, compact, almost jaspery matrix, which only glazes slightly before the blowpipe. In this paste there are numerous grains of dark quartz $\frac{1}{20}$ to $\frac{1}{4}$ inch in diameter, and often more frequent crystals of flesh-red feldspar, apparently orthoclase,—$\frac{1}{10}$ to $\frac{7}{10}$ inch in length.

It not rarely happens, that in these flesh-red crystals there appear dirty green portions exhibiting the twin-striation of a triclinic variety. The feldspar is hard and brilliant, but is nevertheless no longer intact; under the glass the crystals appear cavernous, 10 per cent. or more of the substance being gone. This is the character of this porphyry in the freshest pebbles.

I have before me a pebble 4 inches in diameter, broken through the middle. It was the same variety of porphyry I have just described—the same brown matrix, with the same grains of quartz, and the same large crystals of orthoclase, often enclosing crystals of triclinic feldspar. But this pebble carries on its face the history of an extreme change. In the interior, where it is freshest, the matrix, still of the same brown color, has become so soft as to be easily scratched with the point of a needle. The quartz grains are highly fissured, and the surfaces of the fissures are covered with a soft, light-green magnesian mineral. The feldspar, although it still resists the point of the steel needle, has generally lost its glance, and has an almost earthy fracture; it is lighter colored, and tends to spotted dirty-red and white. In places, specks of chlorite are visible in the holes in the altered feldspar, and the cleavage planes often glisten with flakes of copper. As we go farther from the middle of the specimen toward the original surface of the pebble, the matrix becomes much softer, though still with brown color and brown streak, and then changes to a soft, green, chloritic mineral, which whitens before the blowpipe, and fuses on the edges to a gray glass. A little farther from the centre there is no longer a trace of the porphyry matrix: it is altered wholly to chlorite. The feldspar crystals are somewhat more altered here than they are in the middle of the pebble, but the quartz grains seem to have been in part replaced by chlorite. The change to chlorite is accompanied throughout by the presence of a large amount of copper. While in the interior of the pebble, the flakes of copper are confined to the cleavage planes of the feldspar, and the porphyry matrix exhibits scarcely a trace of the metal, the chlorite which has replaced the matrix contains in different parts of the specimen from 10 to 60 per cent., by weight, of copper.

In another pebble of the same porphyry, not only is the original matrix gone, but the usurping chlorite has been almost, if not wholly, replaced by copper; and we have as the remarkable result a quartz-porphyry, whose crystals of feldspar and grains of quartz lie in a matrix of metallic copper. There is still a very small amount of chlorite present, but it seems to have come from the change of the feldspar crystals and quartz grains.

In other pebbles of the same quartz-porphyry, containing, perhaps, less quartz, the alteration seems to have taken a somewhat different direction, or at least the result before us is different. In the interior of the pebble, the matrix is of a darker and dirtier brown than in the previous cases, which may be due to the presence of manganese in the alteration-product. Going from the middle, the brown color changes rather abruptly to a dirty greenish-gray; the material also becomes softer, but it is earthy, with an earthy odor, and gritty to the touch. The change seems here to be in the direction of kaolinization.

The entire pebble is permeated with minute shining threads and plates of carbonate of lime. The lighter-colored portion contains considerable copper, while nearer the surface of the pebble it is largely replaced by that metal. Pebbles showing the various alterations described above are by no means rare. Many of them, from 1 inch to 1 foot in diameter, are found every day.

A very interesting occurrence of copper and silver is visible in specimens, in my collection, from the abandoned Suffolk mine in the "South Range," south of Eagle River.

There is here an extensive development of feldspathic porphyry (without visible free quartz). In this rock a vein carrying sulphuret of copper was once worked. The specimens referred to were found loose in a gorge near the vein, but are porphyry. They have a dark purplish brown compact matrix, with uneven to semi-conchoidal fracture, and contain numerous amber-colored to flesh-colored crystals of triclinic feldspar, averaging ¼ inch in diameter by ⅛ inch thick, and occupying about ⅛ of the surface area of fracture.

The rock contains, disseminated through the matrix, small particles of black sulphuret of copper. The feldspar crystals often contain minute flakes of native copper, and in some instances flakes of native silver occur in the same manner in the cleavage planes of the feldspar. The chain of changes was, perhaps, initiated by the forming of chloritic substance in the feldspar crystals, the ferrous oxide then reduced the copper from solution, and the copper precipitated the silver.

CONCLUSIONS.

We may be permitted to draw a few conclusions from the facts brought out in the observations thrown together in the foregoing pages.

I. The *Chlorite* of the melaphyr, and consequently the distinctive character of that rock, is due to the alteration of hornblende or pyroxene. This seems to have been the first step toward the production of melaphyr proper. *Laumontite*, which we find alike in the beds containing the least and in those containing the most chlorite, and occurring both diffused and concentrated in seams, appears to have been formed either contemporaneously with the chlorite, or as the next step in the process.

In the fissure-veins of Keweenaw County, laumontite is most abundant near and under the "Greenstone," as in the north end of the Central Mine. Here the great body of overlying rock is one in which the hornblende has undergone but little change.

The next step appears to have been the individualization, in amygdaloidal cavities, of *non-alkaline silicates*, viz., *laumontite*, *prehnite*, *epidote* respectively, according as the conditions favored the formation of one or the other of these.

In the fissure-veins of Keweenaw County, prehnite is the most abundant silicate found in depth, *i.e.*, below the 80 or 90 fathom level. The alkaline silicates are found chiefly in the upper levels.

Following the non-alkaline silicates came the individualization of *quartz* in these cavities.

Perhaps we may be warranted in considering these minerals, together with the lime of the calcite that more rarely occurs in this portion of the series, as chiefly due to the decomposition of the pyroxenic ingredient of the rock.

So far as we may infer from the tabulated results, the concentration of *copper* in the amygdaloidal cavities does not appear to have begun till after the formation of the quartz.

In this part of the series falls also the formation of a chloritic or green-earth mineral, which in some manner has displaced prehnite, quartz, calcite, and with which copper, when present, appears to stand in intimate relation. Subsequently to this came the individualization of the alkaline silicates, viz., analcite, apophyllite, orthoclase. Here also seems to belong the formation of datolite.

The alkaline silicates represent the period of decomposition of the labradorite ingredient of the original rock, and when they occur in the mass of the rock (as distinguished from veins), it is only where the alteration of the rock has proceeded so far that the amygdaloidal form has merged into the brecciated through the enlargement and union of the cavities.

In the fissure-veins of Keweenaw County, the alkaline silicates, as before stated, abound in the upper levels, and are rare in depth; in other words, they are abundant in that zone of the veins which lies between walls of those portions of the beds of melaphyr in which we should look for the most advanced stages of alteration in the components of the melaphyr, supposing such alteration to be due to the action of *descending* solutions.

The fact that calcite occurs at almost every step in the paragenetic series, and forms one of the most common of the secondary minerals, is proof that carbonic acid was very generally present throughout the whole period of metamorphism; it was probably the chief mediating agent in the processes, without being sufficiently abundant to prevent the formation of silicates.

II. The change of pyroxene to chlorite, as illustrated on an immense scale in the formation of the melaphyr, and the displacement of feldspar and quartz—quartz-porphyry—by chlorite, as exhibited in pebbles of the conglomerate, point to an extremely important line of investigation for the chemical geologist. The alteration of the pebbles appears to have followed two different directions according to the ruling conditions, viz., either toward chlorite or toward kaolinization; and as the result of the latter process is impregnated with calcite, while the result of the former is free from carbonates, it would seem that the direction was determined by the presence or relative freedom from free carbonic acid. The deposition of calcite, if formed from the acid carbonate, would set free sufficient carbonic acid to prevent the formation of silicates of iron and magnesia.

III. *Copper*, wherever we can detect it with the eye, has already gone through a partial concentration. The presence of this metal in minute quantity in the sandstones of Lake Superior, is made evident by the stains of carbonate which form on the cliffs of the "Pictured Rocks."

It has also been found in the form of thin sheets of native copper occupying thin vertical fissures in the cliffs of Lower Silurian sandstone on the lake shore north of the Huron mountains. It is found here and there in the less amygdaloidal melaphyr in minute specks and impregnations, or even in a more concentrated form, as thin sheets occupying the joint-cracks.

These occurrences increase in frequency in proportion as the rock is more amygdaloidal; in other words, the copper is more concentrated in those portions of the beds where the chemical change has been greatest. Where the rock has not passed beyond the strictly amygdaloidal stage, the copper occurs in the amygdules, traversing these in flakes, or coating them in a film of greater or less thickness, to such an extent as to form from ¼ per cent. to 3 per cent., by weight, of the rock over considerable areas. Finally, in those beds where the metamorphism has proceeded to such an extent as to wholly replace large portions of the amygdaloid by secondary minerals, epidote, calcite, quartz, chlorite, laumontite, etc., there the copper occurs in masses of many pounds, and sometimes of several tons weight, and in forms equalled in their irregularity only by those of the masses of secondary minerals accompanying the metal.

In each and all of these positions we find that the deposition of the copper took place subsequently to the decomposition and removal of a portion of the rocks, and subsequently to the deposition of laumontite, epidote, prehnite, and quartz, where these accompany it

In all this we have direct evidence of the movement of some salt of copper in wet solution, and the concentration of the metal by accumulating deposition in places where the precipitating agent existed.

In the fissure-veins of Keweenaw County, the widening of the vein is frequently due to "splicing," *i.e.*, a portion of the wall rock became detached and split by countless small cracks having a general parallelism to the main vein. Thus, instead of forming a solid "horse of ground," it consists of myriads of small and large lenses of often wholly decomposed rock, surrounded by films and seams of chlorite, laumontite, calcite, and clay. These places are often the home of large "masses" of copper, which then also have a spliced structure. Where the masses occur in a gangue of calcite, they are not spliced, but have a solid texture and the most irregular shapes. These facts point, perhaps, toward the formation of "masses" by replacement.

According to Mr. Pietrie, the superintendent of the Central mine (fissure-vein), where a "horse" occurs in the vein, the regular vein filling follows the foot-wall side of the "horse," and the younger fissure on the hanging-wall side is filled with calcspar and mass-copper. The foot-wall branch contains (like the regular filling elsewhere) prehnite, one of the earlier-formed minerals.

Except in the melaphyrs of Lake Superior, the copper so widely diffused in the Palæozoic and older stratified rocks, exists either in the various sulphurets, or as oxidation products of these. Indeed, we cannot well suppose the copper to have been deposited in submarine formations in any other condition than as sulphuret. Nor can we suppose it to have taken any other form permanently, so long as unoxidized organic matter remained in the beds. An oxidation of the sulphuret would be followed by reduction of the resulting sulphate to new sulphurets around the organic remains. In this way we may suppose the simplest and most common form of concentrated deposits—the impregnations—to have originated, as well as the farther enrichment of particular beds or zones—*fahlbands*—which may represent strata which were originally richer in organic substances, or which may have retained these longer than the other beds.

The trappean series of Keweenaw Point differ from all the other cupriferous rocks of the Northwest in lithological constitution, and in having the copper in the metallic state. It is still an open question whether the trap which formed the parent rock of the melaphyr was an eruptive or a purely metamorphic rock. If it was eruptive, it was spread over the bottom of the sea in beds of great regularity, and with intervals which were occupied by the deposition of the beds of conglomerate and sandstones.

It should seem probable that the copper in the melaphyrs was derived by concentration from the whole thickness of the sedimentary members of the group, including the thousands of feet of sandstones, conglomerates, and shales which overlie the melaphyrs, and including melaphyrs also—and especially, if these are purely metamorphic.

Among the most interesting questions connected with the occurrence of the copper are those touching its condition previous to concentration during the amygdaloidal stage of metamorphism, the chemical combination by which this concentration was effected, and the character of the precipitating agent.

The great persistency of metallic sulphurets through the usual processes of metamorphism, and the almost universal association of sulphur with copper in crystalline rocks, renders it perhaps probable that this was here also the combination in which the metal was diffused, or rather, very partially concentrated. Traces of sulphur detected by Mr. Hochstetter in the melaphyr contiguous to the Hecla conglomerate point also in this direction, considering that the only acids generally present in the melaphyrs are silicic and carbonic acids, and if we add sulphuric acid as an oxidation product of the sulphurets, our choice of the form of solution, by which the final concentration was effected, should seem to be limited to silicates, carbonates, and sulphates of copper. Probably all of these combinations took part in the process, but while we may consider the translocation of the copper to have been initiated by the sulphate, this salt must have been so soon decomposed by the abundant acid carbonate of lime,* as well as by the alkaline silicates, that we cannot readily suppose the sulphate † to have

---

* A coating of gypsum covering very thin sheets of copper from the jointing-cracks of the melaphyr contiguous to the Hecla conglomerate, may be due to this decomposition, followed by the reduction of the copper.

† Compare Bischof, Chem. u. Phys. Geol., I., p. 52, and III., p. 716.

generally effected the final concentration of large deposits. It is more probable that this was accomplished by the more permanent solutions of carbonate and silicate of copper respectively, as the circumstances favored. The position of the metallic copper in the paragenetic series, shows it to have been deposited after the non-alkaline silicates, and before the formation of the alkaline silicates, *i.e.*, after those minerals which resulted from the decomposition of the pyroxenic constituent of the rock, and before those which were formed by the destruction of the feldspar. Now this is what we should expect if we suppose the pyroxenic rock to have been altered to its present condition under the co-operation of water carrying carbonic acid and some free oxygen, because the oxygen must have been employed in oxidizing the carbonate of iron resulting from the decomposition of the pyroxene ;* the oxidation of the sulphuret of copper could not, therefore, take place until the pyroxene had so far disappeared as to leave a relative excess of oxygen as compared with the amount of ferrous salts exposed to a higher oxidation. Throughout its deposits the copper exhibits a decidedly intimate connection with delessite, epidote, and green-earth silicates, containing a considerable percentage of peroxide of iron as a more or less essential constituent; while among the other silicates, viz., analcite, laumontite, datolite, prehnite, only the last named, which alone seems subject to a considerable replacement of its alumina by ferric oxide, is especially favored by copper. This association is so invariable and so intimate that one is forced to the conclusion that there exists a close genetic relation between the metallic state of the copper and the ferric condition of the iron oxide in the associated silicates; that the higher oxidation of the iron was effected through the reduction of the oxide of copper and at the expense of the oxygen of the latter.

As regards the green-earth and that variety of chlorite or delessite which is intimately associated with the copper, they either immediately follow the copper in point of age or are contemporaneous with it, and they may be looked upon as having been formed under the influence of this reduction. Where copper is associated with prehnite it is invariably younger than the latter, a fact which would

* The result of this oxidation is seen in the brick-red color of the amygdaloids and in the brown color and spots of many of the melaphyr beds.

seem at the first glance to oppose the supposition that there is any relation between the peroxide of iron in the zeolite and the deposition of the copper. But we have seen that prehnite undergoes a change to delessite; we find these pseudomorphs in every stage of the process from the first green discoloration on the cleavage planes to the amygdule of delessite with prehnite structure. Now, may we not consider the presence of iron in prehnite generally to be due to a beginning change, and the deposition of native copper in the Lake Superior prehnites to be partially or wholly correlated with the higher oxidation of the iron?

In at least very many instances, if not in all, the deposition of the copper has been a result of a process of displacement of pre-existing minerals. In some rare instances the metal retains the form of its more or less remote predecessor, as in the pseudomorphs after some mineral (clay?) after laumontite.

Nowhere is this displacement more apparent than in the cupriferous conglomerates. In these, the cement is the home of the metal, and in some places, as in portions of the Hecla and Calumet mines, it is wholly replaced by it; copper forming 20 to 50 per cent., by weight, of the rock. In these instances, either chlorite or epidote is associated with the copper as minerals formed since the deposition of the conglomerate, while calcite very frequently replaces the cement in barren portions of the bed.

The cement of the conglomerates is of the same materials as the pebbles in a more comminuted form. The displacement of the whole mass of quartz porphyry in large pebbles by chlorite and copper described above, is probably an illustration of the manner in which the cement was displaced on a more extended scale.

The absence of the ores of the baser metals—lead, zinc, nickel, etc., from the deposits of the trappean series, while they are present in the less metamorphosed rocks of the Quebec group in other localities—may be due to the greater intensity of the chemical action to which the melaphyrs have been subjected; an intensity which may be measured by the extent to which the process of concentration has been carried. Concentration is a process of removal, relatively speaking, and concentrated deposits are accumulated masses of material arrested in the drainage channels of rock masses by the action of competent forces; if the arresting cause is absent from a given region, the removal will continue to another where it is present. If causes exist which are able to arrest one class of the substances in the passing solution, and are powerless as regards another class, then a separation will occur between the two classes.

Now, copper and silver belong to a class distinct from the baser metals, in that, by reason of their smaller affinity for oxygen, they are more readily reduced to the metallic state, the condition of greatest permanence in presence of the usual reagents to which they are exposed. If the arresting cause of these metals was, as we have supposed, their reduction by protoxide of iron, it is a cause which would have been powerless as regards the salts of the baser metals, and we may suppose these to have continued in solution till they reached some region where they were arrested by the presence of organic matter, or of sulphuretted hydrogen, etc.

CHAPTER IV.

CORRELATION OF THE ROCKS OF HOUGHTON AND KEWEENAW COUNTIES.

BY A. R. MARVINE.

(See table at the end of this chapter.)

UPON the Plate of Grouped Sections are gathered the various sections that have been made by the Geological Survey in the years 1870 and 1872, that their relations may be more readily studied. They are placed down the Plate in the order in which they occur, going from south to north. All the sections but I. and II. are taken very nearly across the formation. I. and II. of the Portage Lake District, however, contain many points scattered along the formation, as well as across it, as may be seen by reference to the map; the two being connected by the Roman numerals.

In Section I., the exposures have been projected upon the plane of section from the various points at which they occur along such lines of strike, as near as may be, that the formation has at them. In Section II. (which stands in a relation to Section I. to be hereafter explained) the "Albany and Boston" conglomerate, No. 15, has been assumed as a straight line between the Pewabic and Rhode Island mines, and the remaining beds plotted off from it at right angles, at the distances which they were found to have at those points where they happened to be exposed. Under these circumstances, both in Sections I. and II., points exposed directly across the formation from one another, would appear upon the sections as having the same geological relations to one another that they have along the line at which they actually occur. Exposures not lying directly across the formation from one another would also carry with them to the section their true relations, provided that the intervening beds had neither thickened nor thinned in going along the formation from one point to the other. But if thickening or thinning has taken place, two points so exposed would be projected upon the section, either farther apart or nearer together than they would if either point had been exposed directly across the formation from the other; while they might or might not appear as they would, supposing the beds prolonged till they actually intersected the plane of section.

Thus Sections I. and II. show the relations the beds have at the various points along the range at which they happen to be exposed. Were it an object to form a true section across the range at some one point, the rates of thickening or thinning of the beds would have to be first ascertained and a correction—depending on those rates and the distances of the exposures from the section—applied to the various beds, as shown upon the printed sections, thus reducing them to their proper places upon the ideal section.*

The differences in dip which occur along the mineral range are apt to lead to very deceptive results in comparing the distances between beds as shown upon the surface. Two beds, whose dip angle is growing less as we follow them along the formation, will appear to gradually separate from one another, while the actual distance between them, measured perpendicular to the plane of bedding, remains the same or may even diminish. Thus, the horizontal distance between the Allouez and first conglomerate, south at the Central mine, is more than four times the horizontal distance between the Albany and Boston and Houghton conglomerates as exposed on the Mesnard property. The dip at the former, however, is twice as flat as at the latter, thus accounting for most of the difference in distance; while if allowance be made for the rate at which we know the intervening beds to be actually widening at the Mesnard, the remaining discrepancy is entirely accounted for.

In order, then, to impartially compare different portions of the

* It is with the object of avoiding the publication of a number of such sections, while at the same time affording data for forming them when necessary, that the table which is to follow has been prepared. Having some known bed to measure from, this table, in connection with the accompanying sections, enables the positions of beds, covered upon the property in question, but known elsewhere, to be ascertained with considerable accuracy.

mineral range, the perpendicular distances between the beds only must be regarded.*

The number and lithological similarity of the amygdaloids, together with the fact that they are prone to wholly disappear, as may be readily seen from the sections, render them unfit for the purpose of tracing the formation. The great similarity among the melaphyrs renders them also doubtful guides; though in both amygdaloids and melaphyrs there are certain broad features obtaining over more or less wide belts, which are serviceable in correlating, in a general way, different parts of the range. Such are the coarse-grained melaphyr belt of the Dacotah, and the belt containing scoriaceous beds to the west.

The relative infrequency of the conglomerates, notwithstanding their lithological similarity, renders them, therefore, the only reliable stratigraphical guides of the formation.

The measurements between the conglomerates, to which the survey paid much attention, have therefore been discussed, and the results gathered into the accompanying table, (opposite page 60) a description of which follows :—

The right-hand column contains the names of the various properties along the range on or opposite to which the various measurements were made. These, in going down the page, read from south to north. The strip across the paper opposite each is supposed to represent a narrow strip running directly across the formation. Column two contains the dip of the formation, as near as

---

* The perpendicular distance between two beds equals their horizontal distance multiplied by the sine of the dip angle. A measurement made between two beds that are exposed at different heights does not give the horizontal distance between them. With strata dipping to the north-west, if the north-west point is the higher the measurement will be shorter than it should be,—if lower, longer than it should be. The correction to be added or subtracted, as the case may be, to a measurement itself reduced to the horizontal, in order to give the horizontal distance between the beds, equals the difference in height of the observed points divided by the tangent of the dip angle. The following table gives the natural sines and tangents of the dips of the principal mining locations :—

Dip angle.	56°	55	54	53½	53	52¾	52	45	43	39¼	31	30	26
Sine......	.82904	.81915	.80902	.80386	.79864	.79600	.78801	.70711	.68200	.63271	.51504	.50000	.43837
Tangent..	1.48256	1.42815	1.37638	1.35142	1.32704	1.31507	1.27994	1.00000	.93252	.81703	.60086	.57735	.48773

could be ascertained, on the property it is opposite. Column three contains the approximate distances, in feet, along the formation of the various strips from the southernmost one; and subtraction of the numbers opposite any two measurements enables one to obtain the distance along the formation between them. The conglomerates are numbered uniformly with the grouped sections. The names of several near Portage Lake are placed along the top of the table, of several in Keweenaw County, along the bottom.

The Albany and Boston and the Allouez conglomerates being represented by a straight line (No. 15), the various other conglomerates are plotted off from it, and from one another, to a scale of 1,600 feet to one inch, the distances from foot-wall to foot-wall* perpendicular to the formation being taken. This part of the table, then, may be taken to represent the appearance the outcrops of the conglomerates would present, on the supposition that the mineral range, instead of being tilted into its present position, had been tipped up, without faulting, until the strata at all points stood vertical, and then straightened out until the Albany and Boston and the Allouez conglomerates formed one straight line.†

The horizontal measurements between the conglomerates are placed opposite the points in column 1, at which they were made, and between the lines representing the conglomerates, their direction and limits being indicated by the fine dotted lines. Under each of these, in a parenthesis, is placed the resulting perpendicular distance between the beds. It is those numbers which are drawn to a scale of 1,600 feet to an inch. Numbers in parentheses which stand alone are not derived from horizontal measurements, but are obtained indirectly either by subtracting adjacent parenthetical numbers, or by applying a correction to direct or diagonal measurements on account of thickening or thinning of the formation. Placed between those numbers in parentheses, which represent the perpendicular distances between the same two conglomerates at different points of the range, are numbers in brackets. These represent the number of feet of thickening or thinning, per each

---

* Except No. 14, in which the hanging-wall is taken, the foot-wall not being observed.

† The distances across the table being drawn to scale, it may be cut in two, just above the Kearsarge, when, by moving the lower part back or forth upon the upper, any other apposition of the conglomerates than that here adopted may be made.

thousand feet along the formation between the measurements, that has taken place in the beds lying between the conglomerates in question. They are obtained by subtracting the perpendicular distances between two conglomerates at different points, and dividing the difference by the number of thousand feet lying between the points, as obtained from column 3. Thus, take conglomerates 14 and 15, between the Mesnard and Albany and Boston, when measurements between them were made. At the former place the two are 411 feet apart, and having a dip of about 53°, the perpendicular distance between them would be (328) feet. At the latter, these distances are 460 and (362) feet respectively. The total thickening is, therefore, 34 feet, which has taken place in a distance along the range (column 3) of 12,800 feet, or at the rate of [2.7] feet per thousand. Thickening toward the north is indicated by the simple rate; thinning toward the north, by the minus sign,—thus [−1.6].

Rates that are derived from measurements made directly across the formation are the most reliable. Having these we may correct diagonal measurements, or calculate the distances between the same beds at other points, and from these obtain rates of thickening of beds not directly measured. Thus, at the Pewabic we have a measurement (3,166) between 9 and 15, but no direct one elsewhere. At the Montezuma, however, we have from 9 to 12, and at the Dacotah from 12–14 (1467), which, corrected for the rate [3] to the Montezuma, becomes 1,480. Reducing the distance between 14–15 at the Mesnard for the rate [2.7] gives (297) feet at the Montezuma, or, in all, from 9 to 15 at the Montezuma (3,151) feet. Comparing this with (3,166) at the Pewabic, gives a rate of widening, per 1,000 feet, of [2.8] feet for the total space between conglomerates 9 and 15. If we add the rates of widening given in the table for the spaces included between 10 and 15, we have for this whole space [−1.8] feet which would leave [4.6] for the space 9–10. It will be observed that between the Montezuma and the Kearsarge, a rate of [4.1] was obtained for this same space. It may be objected that in the above calculation the rate [2.7] which 14–15 has between the Albany and Boston and the Mesnard was assumed to continue to the Montezuma.

The projected distance in Sec. I. between 14 and 15, corrected, for a reason not readily explained, for the rate [6.4] between 15 and 17 gives [285] at the Montezuma, and this number, compared with the distance at the Mesnard, gives the same rate of widening as from there to the Albany and Boston, or, indeed, from the latter point to the central. But be this as it may, the assumption may be avoided by regarding the region between 9 and 14 only, and considering that the rate [2.7] holds good only so far as the Pewabic. Here 14 and 15 would therefore be (310) feet apart, leaving (2,856) feet between 14 and 9. Proceeding as before, we find that between the Montezuma and the Pewabic the zone 9–14 is widening at the rate of [0.3] per thousand, leaving [4.8] for the zone 9–10, or [0.2] larger than before. These examples, taken at random, serve to show the manner of obtaining rates not entered in the table, for the purpose of ascertaining the approximate positions of unexposed conglomerates, and, with the help of the sections, of intermediate beds, should there be reason to suppose them continuous. It must be observed that numbers thus derived from the table are ***thicknesses***, or distances perpendicular to the plane of bedding. To reduce them to the horizontal, they must be divided by the sine of the dip on the property on which a bed is sought (p. 3), and this number should be increased if the ground slopes downward to the west, or decreased to the east.

General Conclusions.—Regarding, first, the region lying between the South Pewabic and the Albany and Boston, we see that there is a very general thickening of the beds in going north along the formation.* In only one zone does the reverse take place, viz., between conglomerates 11 and 12. The large rate here obtained [−10] may be somewhat in excess owing to some error. It could not be varied much, however, and shows that there is a very decided narrowing going on in this belt. Eastward of No. 9 the thickening toward the north is small throughout, while west of it, excepting the narrowing zone, it is much greater, being especially marked west of 14. The rate [5.9] between 15 and 16, derived from a measurement probably not reduced to the horizontal, must be inaccurate; but it serves to show that most of the total thickening [6.4] between 15 and 16 is taking place in the zone 15–16.

* Upon the table, a sudden bend or widening in the lines representing the conglomerates does not indicate a similar change in the formation, for distances up and down the page are not drawn to any scale, so that some parts of the Range are much crowded together compared with others. Two lines that are separating indicate a thickening, the actual amount of which, however, must be obtained from the numbers between them.

Identity of the Allouez and Albany and Boston Conglomerates.—It will be observed that in the table the Albany and Boston conglomerate has been assumed to be the same as the Allouez. We may seek the proof of this in three separate groups of facts—geographical, stratigraphical, and lithological.

1. *Geographical.*—The extension of the line between the Albany and Boston and Rhode Island shafts northward would intersect the line of the Calumet conglomerate near the line between the Calumet and Hecla properties, at an angle of about 5¾ degrees, and it has been supposed that by curving in a peculiar way they might be the same bed. The extension of the Calumet base, however, passes over 2,400 feet east of the Rhode Island, and not far from where the North Star conglomerate (No. 12) should be; while the extension of the line of strike of the Allouez conglomerate from the Allouez mine would intersect the Albany and Boston conglomerate not far from the Rhode Island. Might not a curving throw either of these two together? * The Albany and Boston conglomerate, between the Pewabic and Albany and Boston mines, is bending eastward toward the Calumet, partly in virtue of the flattening of the dip which is here taking place. This effect is much exaggerated upon the map by the depression of the Albany and Boston Creek, which throws the outcrop here too far to the westward. Reduced to Lake Livet, the line of the bed still shows considerable eastward curvature, which, however, notably diminishes at the Albany and Boston, and is hardly apparent at the Rhode Island, where there seems to be almost a slight reversal of the curve. It is a line bearing eastward of this which fails to intersect the Calumet till reaching that mine. A bend at the Rhode Island, similar to that which takes place between the South Pewabic and Portage Lake, or at the Pewabic mine, would throw the Albany and Boston conglomerate west even of the Allouez. The Calumet conglomerate, in going south, also swings westward toward the Albany and Boston. Here, fortunately, we have data which give an approximate idea of the amount

---

* The Calumet base line was extended by Mr. L. G. Emerson southward to the Rhode Island. Mr. A. B. Wood prolonged the Albany and Boston strike, as a picket line, northward toward the Calumet, while the Allouez was prolonged, as a transit line, from the Allouez mine to Calumet by Mr. Harry Beaseley, formerly county surveyor. A line nearly parallel with, but about 900 feet west of, the latter was afterwards run by Mr. L. G. Emerson, southward, I believe, to the Rhode Island.

of curving. Explorations on the Ossipee and Sawabic show that at about one-fourth of the distance toward the Rhode Island, the Calumet has swerved 60 feet westward from the extended Calumet Base Line, not taking difference in height into account, which might or might not increase this distance. Supposing that this rate continued, not directly as the distance from Calumet, which would give a departure from the extended line at the Rhode Island of 300 feet, but as the *square* of the distance, a most liberal estimate, and which would give a departure of 960 feet, still leaving the bed over 1,400 feet east of the Albany and Boston. Upon previous considerations, it has been seen to be between 1,600 and 1,700 feet east of the Albany and Boston. So far as curvature goes, these two cannot be identical, while placing the Calumet where curvature would probably carry it, leaves the Allouez the same as the Albany and Boston.

At the Calumet mine, the dip, according to Mr. G. D. Bolton, is 39° 15′. Both toward the north and south it is increasing, becoming, according to Mr. E. J. Hulburt, 52° at the Albany and Boston, and, according to Mr. G. D. Bolton, 46° at the Allouez, while the survey obtained 43° in several observations on the Kearsarge. It is partly in virtue of this steepening in the dip that the Calumet curves toward the Albany and Boston as we have seen. But if this curvature is small, smaller than is due to the mere steepening of the dip, it shows a bending in the opposite direction of the whole formation, in which case the widening upon the surface between outcrops, also due to decreasing dip, would increase the curvature of beds to the east, while it would diminish it to the west, and at a point far enough off in this direction actually reversing it. This is what takes place. The Allouez conglomerate, as exposed at Calumet, notwithstanding the flattening dip and difference in height between the two, is exposed 120 feet *west* of the extended Allouez base. If, with a flattening dip, this bed has swerved from the line, with a steepening dip it would again approach it, and, with the thinning of the strata, at the Rhode Island could not be far from the Albany and Boston conglomerate. The Pewabic West conglomerate (No. 16) is too far west to be it, and between the latter and the Albany and Boston the explorations on the St. Mary's show that no conglomerate exists. It must either be the Albany and Boston, or have vanished. It is hardly possible that a dislocation of the strata could have affected this reasoning. The glacial, or preceding eroding agencies, which have so strongly modified the surface of this region, have selected those weak points formed by all known faults of any magnitude upon which to leave their deepest impress. No such surface evidence of any fault exists between Calumet and the Rhode Island.

2d. *Stratigraphical.*—*a*. Number of beds. It will be seen from the table that, considering the Albany and Boston the same as the Allouez, only four, out of fifteen conglomerates considered, are not found in both districts, and these would all pass, supposing them persistent beds, through regions as yet not uncovered from the "drift," unless it be the Calumet. This could easily escape notice on the south shore of Portage Lake under the covered portions of the Dacotah and Huron Properties;* but on the north shore of the lake, certain now-filled-up explorations on the Pewabic should have passed near it. The Survey has been unable to obtain any definite information of these, except that the deep "drift" upon the surface was not always penetrated, and the bed may have been thus not uncovered.

b. Thickening of beds. It is unfortunate that more exposures of the various beds are not known in Keweenaw County, in order that rates of thickening could here be obtained to compare with those of the Portage Lake District. Between Calumet and Kearsarge thickening continues, and also between the Central and the Delaware, though the latter is approximate. The rates of thickening, however, obtained in the Portage Lake District are singularly in harmony with those obtained by comparing Portage Lake distances with those north, and which are placed in the table between the Albany and Boston and Calumet properties. We have already referred to the widening in the zone 14–15 between the Albany and Boston and Central as being the same as that from the Albany and Boston southward, and also to the similarity of rates between 9 and 10. The rate [3–3] between Nos. 7 and 10 is small when the thickness between these beds is considered,—though larger than the rate that exists between these beds from the Huron to the Douglass (which may be obtained by taking the rate before derived [4.6] between 9 and 10 and deriving the distance 6–9 at the Huron, to

* Fragments of a conglomerate resembling the Calumet have been seen near the road following up the valley of "Huron Creek."

compare with (2,756) at the Douglass). The distance between 6 and 9 on the Kearsarge is probably also somewhat in error, being scaled from a small map made from measurements by Capt. Newcomb.

Though No. 12 has not been opened to the north, the narrowing zone between it and 11 still continues through. The absence of 12 makes it impossible to directly ascertain the amount of narrowing, but we may approximate to it. Regarding the zone 11–14, we may obtain the rate of widening in it in two ways: first, 14 not being exposed on the Kearsarge, we may calculate its distance from 15 by the rate [2.7] between the two. This gives (462) leaving (2,034) between 14 and 11, which may be compared with the distance between them at the Montezuma:—or, second, we may obtain the rate in the zone 11–15 between the Montezuma and the Kearsarge, and subtract the rate due to the zone 14–15 [2.7]. These two methods give [−1.6] and [−1.7] feet per thousand, respectively, as the rate of narrowing that has taken place between 14 and 11 from the Montezuma to the Kearsarge. This is made up of a widening between 12 and 14, and a narrowing between 11 and 12. At the Kearsarge the rate between 13 and 14 would be about [3.8]—[2.7] or [1.1], leaving [−2.7] between 11 and 14. Supposing the [3] between 12 and 14 to the south continues through, it would leave [1.9] for the zone 12–13 or [−4.6] for the zone 11-12, giving, as at the south, a very decided narrowing going on in this one belt. In the Eagle River District, the "slide" lying some sixty feet above the so-called "Ash-bed," has unmistakable patches of sandstone in it, and is the only bed corresponding with No. 17 in the Portage Lake District. It gives a rate of [7.2] between Eagle River and St. Mary's,—the rate from St. Mary's to the South Pewabic being [6.4]. No. 16 would be represented at Eagle River by two thin sandstone seams, giving a rate of [4.7] between 15 and 16 from St. Mary's to Eagle River. These seams widen to true sandstones at Copper Falls, six miles farther north-east.

It appears, then, that this particular apposition of the conglomerates, gives rates of widening from the Portage Lake District northward differing but very little from those obtained from actual measurements between known beds in that district,—so little, in fact, that it seems absolutely impossible that it could be owing to mere chance.

3. *Lithological.*—I am unprepared to enter into a detailed correlation of the lithological characters of individual beds, and can only speak of broader distinctions extending over wider areas. Among the conglomerates themselves too much similarity prevails on any one line across the formation to give any conclusive evidence. At Portage Lake, all are much alike. At Calumet, all have gained in addition to their components at Portage Lake, a greater or less percent. of pebbles, of grains of quartz in the pebbles. The representative of the "greenstone" series of Keweenaw County would be the "coarse-grained melaphyrs" occurring between the Albany and Boston and Pewabic west conglomerates in the Portage Lake District. In some beds of the latter, as for instance a belt exposed in a ravine just east of the Hancock Mine, hornblende crystals are abundant, and they closely resemble some of the upper strata of the greenstone series as exposed near Eagle River. There is a decided dissimilarity, however, between the beds lying immediately above conglomerate 15, in Sections II. and V. This can be no argument against the correlation, however, as the change has taken place before the space to be spanned is reached. At the Allouez mine, the lower greenstone beds have lost much of their character, while at the Kearsarge and at Calumet the beds just above No. 15 are more like the beds above it at the St. Mary's than at the Allouez. West of 16, however, the accordance in the general characters of the formations north and south is very striking.

The lithology of the formation lying at any distance south of the greenstone in Keweenaw County, is too little known to compare with equivalents in Houghton County.

Thus, three separate lines of evidence converge to establish the fact that the Albany and Boston and the Allouez conglomerates are one and the same bed.

1. Geographically, they are either identical or else one disappears toward the south, the other probably to the north.

2. Stratigraphically, 11 out of 15 conglomerates have equivalents in both regions, and in such a way that the rates of thickening observed in the Portage Lake District are so nearly reproduced as to render it most improbable that it can be due to chance.

If the alternative of reason 1 is accepted, but few beds would have equivalents, and there would be no general harmony in rates of thickening.

3. Lithologically, conditions, so far as known, are satisfied.

It will be observed, that under these circumstances, the "ash-bed" of Keweenaw County becomes the equivalent of the Hancock, or South Pewabic bed of Houghton County, a correlation that has been suggested by several, and arising, no doubt, from their general lithological resemblance. It is an exception to the general rule, however, of non-persistence of individual amygdaloid beds, and is not a proof of the general reliability of the correlation of individual beds based on individual lithological characters. The family resemblance, as it were, often existing between the beds of certain zones, while it may enable one to say that a distant bed belongs to the zone in question, renders it impossible for any one to prove, on mere lithological evidence, that it is the equivalent of any one certain bed of that zone. In its minerals and peculiar structure, as well as in the accompanying trap, the South Pewabic amygdaloid and the "ash-bed" are much alike. Yet these facts, judging from the interchange of characters sometimes occurring between neighboring beds in other parts of the range, as well as the thinning out of beds and replacement by others, would not prove them to be identical; a fact, the probability of which is made much greater by their occupying the same horizon beneath the same conglomerate.

The Pewabic Lode, which, judged by lithological characters, has been found by different persons at such a variety of points in Keweenaw County, probably does not pass far beyond the Kearsarge; while the bed on which the Concord and Douglass mines are situated, is some distance east of either the Isle Royale or Grand Portage beds. The amygdaloid over conglomerate 9 on the Kearsarge, which is opened on the Huel Vivien, and which is thought by some to resemble the Douglass or Isle Royale bed, must be some 3,000 feet west of them.

The Calumet conglomerate furnishes an instance of a bed, of which there is perhaps a more general desire in the Copper region to trace than any other bed. While, as will appear in the sequel, there is reason to suppose that it is persistent, there is a possibility that it may continue as a mere seam, and no reason to suppose that it should carry with it the remarkable richness which it has at the Calumet and Hecla mines. The short distance which it has been traced renders prediction as to its place probably more uncertain than of almost any other bed. The only two measurements to it are near together, so that any error in either would strongly affect the rate of widening, and there are no minor facts which can serve to indicate in which way the rate may err. However, its place upon various properties has been calculated, with such data as the table gives, with the following results, which are probably correct within 50 to 150 feet, either one way or the other, depending on the distance of the point from Calumet.

At South Pewabic, 1,390 feet east of conglomerate No. 15 (Albany and Boston) or 3,070 feet (about) east of the South Pewabic amygdaloid.

Huron and Montezuma, between 750 and 850 feet west of No. 12.

On the Huron, 12 is exposed near the Slaughter House, and is the third conglomerate west of Huron Finishing Mills.

On the Montezuma, 12 is exposed on the Lake Side Road, west of Houghton. (See map.)

Pewabic, 1,500 feet east of the Albany and Boston conglomerate.
Mesnard, 1,560 " " " "
St. Mary's, 1,590 " " " "
Albany and Boston, 1,640 " " " "

Phoenix Mine, 3,290 (about) south of the "slide" under the "greenstone."

Central (dip 26?) 2,245 (about) south of second conglomerate south of the greenstone.

These are approximately the horizontal distances between the beds, and, with a flat dip and much difference in height, should be amended accordingly.

General Conclusions.—There are certain general conclusions which result from the facts presented in the table, and which may be of interest, concerning the original structure of the range. The very uniform thinning of the zones included between the conglomerates toward the south-west, excepting always the zone 11–12, shows that when first laid down in a nearly horizontal position, whatever may have been their mode of origin, the *source* from which the material forming them came, was toward the north-east. I believe that near the extremity of Keweenaw Point the formation is said to be again thinning. If so, it would place this source somewhere abreast of the centre of the Point, and probably to the south or south-east, rather than north of it.

The conglomerates do not seem subject to the general thinning toward the south characteristic of the intermediate beds. In fact, their thickness seems to follow no well-defined law, but to be wholly irregular, some seeming to thin or thicken in one direction as well as another. This fact, together with a pretty general uniform coarseness, would seem to show that the direction of the source from which they were derived was different from that of the melaphyrs, and that the source must have been a shore line not far from parallel to what is now Keweenaw Point. Upon this shore, opposite and north of Calumet, a quartz-porphyry must have predominated among the rocks, and which did not exist further south. This shore line, however, was variable. Conglomerate 9, on the Pewabic, is of the normal Portage Lake type, while at the Kearsarge it is 50 feet wide, and is composed of two parts, a lower of dark gray, more or less fine shaly sandstone, some layers approaching slate in character, an upper of a red, exceedingly fine arenaceous non-shaly sandstone. This portion of the bed must have been further from the shore than that near Portage Lake. The reverse takes place in the first broad conglomerate skirting the north-west border of the range—a heavy conglomerate all along the upper point, near Portage Lake it is composed of sandstones and fine arenaceous shales, which must have been a deep-water deposit compared with the conglomerate.

The conglomerate beds of Keweenaw Point have been generally considered as mere local deposits, rapidly fading out in either direction. The table would seem to show, on the contrary, that for conglomerates they are unusually persistent, and that while a bed may thin out and lose its character as a conglomerate, it may still exist even as a mere seam. Thus the Allouez conglomerate has a thickness of from 15 to 20 feet at the Central and Allouez mines, while near the Phœnix mine, between them, it does not appear, being replaced by what is locally known as the "slide," a layer of soft red clay 2 to 6 inches thick. We gather from those facts that when the beds composing the trappean range were being originally formed, the agencies, whatever they were, which formed what are now the melaphyrs, ceased to act not only over limited but over extended areas, in once instance at least over fifty (50)

Table : Gis

Property upon, or opposite to which the measurements were made	Dip of Strata in Degrees
Number & Consecutive Widths	
South Pewabic	56
Isle Royale & Huron	55
Dakotah	55
Montezuma	55
Sheldon & Columbian	55
Pewabic	54
Douglass	[illegible]
Mesnard	53
North Star	[illegible]
St. Marys	[illegible]
Albany & Boston	52
Calumet	[illegible]
Kearsarge	42
Eagle River	31
Kingston	30
Central	
Delaware	[illegible]

a Measurement by [illegible]

‡ Scaled fr[illegible]

miles, and for periods of time long enough to allow of the accumulation of beds of conglomerate from a few to over 75 feet—in one instance over half a mile—in thickness.

In the Portage Lake District, commencing upon the eastern limit, so far as known of the range, the trappean zones have considerable thickness, which are separated by three beds of conglomerate about 50 feet thick. Then follows a zone of about 1,300 feet thick, in which there are five conglomerates from 2 to 20 feet thick. Twenty-five hundred feet of melaphyrs and amyg daloids then follow, with no known conglomerate, and then, for 5,000 feet in thickness, nine conglomerates from 200 to 1,000 feet apart, and from almost 0 to 30 feet thick occur. West of these, conglomerates and sandstones frequently occur, in aggregate thickness probably equalling that of the interstratified melaphyrs, until the wide conglomerate (slates and sandstones) is reached.

Fault in Portage Lake: Section II. stands in the relation to Section I. that it would have, supposing the Albany and Boston conglomerate were accurately projected from the Pewabic mine, with the line of strike it then has, until it intersected Section I., and then plotted off from the point so obtained, and moved down into its present position. If the Albany and Boston conglomerate were exposed upon the south shore, there would be shown a fault in this conglomerate of about 710 feet horizontal. Applying corrections for widening to the apparent faults in other conglomerates, gives from 705 to 720 feet, excepting the extreme east and west ones, which give nearly 740 feet. The discrepancies are due to the bending of the formation in crossing the lake, a source of error not easily eliminated. The numbers serve to show, however, that the actual amount of faulting is probably not far from 720 feet, the north shore having been raised or the south shore depressed, giving the effect of the north shore having been moved westward upon the southern one. If the fact that the north shore is generally over a hundred feet higher than the south shore is not wholly due to erosion, it may be in part due to this fault. The dislocation is probably made up of many lines of fracture, as several pass through the Hancock property.

CHAPTER V.

DESCRIPTIVE CROSS-SECTIONS AT PORTAGE LAKE.

N.B.—The numbers refer to the *absolute* thickness of the beds in full feet, and the sequence is from older to younger.

Cross-section I.

THIS cross-section is taken, without re-measurement, from the sketch of the exploration-trenches made by the Isle Royale Mining Co., under the superintendence of Mr. Mabbs. These explorations were filled with dirt and water when visited by the members of the Survey.

THE EASTERN SANDSTONE unconformable with the trap series, and dipping gently to the East.

? **Amygdaloid**; soft.
14. **Amygdaloid**; hard.
97. **Conglomerate.**
329. **Covered** by swampy ground.
43. **Amygdaloid**; light-colored.
40. **Amygdaloid**; light-colored.
82. **Melaphyr**; dark and rough.
82. **Melaphyr**; dark and rough.
82. **Melaphyr**; dark and rough.
82. **Melaphyr**; gray.
82. **Melaphyr**; dark and bluish; carries some fine copper.
82. **Melaphyr**; dark and bluish; carries some fine copper.
82. **Melaphyr**; gray.
46. **Melaphyr**; red.
34. **Amygdaloid.**
81. **Melaphyr**; red.
20. **Amygdaloid.**
64. **Melaphyr**; red.
64. **Amygdaloid**; red.
94. **Melaphyr**; black.
42. **Melaphyr**; red.
41. **Amygdaloid**; red.
83. **Melaphyr**; gray.
56. **Conglomerate.**

Cross-section II.

(Determined by triangulation and levelling by the Geological Survey, except where otherwise accredited.)

Hanging-wall of first conglomerate east of the "Mabbs' vein," determined by the Geological Survey.

21. "**Amygdaloid.**" From Mr. Mabbs' map of the Isle Royale Co.'s explorations.
77. "**Melaphyr**; gray." " " " "
82. "**Melaphyr**; dark." " " " "
82. "**Melaphyr.**" " " " "
2–6. "**Mabbs' vein.**" On the foot-wall barrel and mass copper in quartz and calcite. One mass weighed 2,200 lbs. The rest of the vein is stamp-work. Position determined by the Geological Survey.
62. "**Melaphyr.**"
111. "**Melaphyr**;" soft, jointed. From Mr. Mabbs' map of the Isle Royale Co.'s explorations.
147. "**Covered.**" From Mr. Mabbs' map of the Isle Royale Co.'s explorations.
20. "**Amygdaloid**; epidotic." From Mr. Mabbs' map of the Isle Royale Co.'s explorations.
31. "**Melaphyr**; dark, rough." From Mr. Mabbs' map of the Isle Royale Co.'s explorations.
39. "**Amygdaloid.**" From Mr. Mabbs' map of the Isle Royale Co.'s explorations.
148. "**Melaphyr**; dark." Float copper on the surface. From Mr. Mabbs' map of the Isle Royale Co.'s explorations.
148. "**Melaphyr**; bluish." From Mr. Mabbs' map of the Isle Royale Co.'s explorations.
152. "**Melaphyr and Amygdaloid.**" Float copper on the surface. From Mr. Mabbs' map of the Isle Royale Co.'s explorations.
23. "**Melaphyr**; dark and rough." From Mr. Mabbs' map of the Isle Royale Co.'s explorations.
17. "**Amygdaloid.**" From Mr. Mabbs' map of the Isle Royale Co.'s explorations.
12. Conglomerate. Position determined by the Geological Survey.
53. "**Melaphyr**; bluish." From Mr. Mabbs' map of the Isle Royale Co.'s explorations.
"**New Vein.**" Fissure vein of quartz, prehnite, calcite, copper. Position determined by the Geological Survey.
47. "**Melaphyr**; red, fine-grained." From Mr. Mabbs' map of the Isle Royale Co.'s explorations.
37. "**Amygdaloid.**" From Mr. Mabbs' map of the Isle Royale Co.'s explorations.
26. "**Melaphyr**; dark." " " " "
358. "**Covered.**" " " " "
1½. "Conglomerate." According to Mr. J. H. Foster, this bed is only ½ an inch thick on the north part of the Shelden and Columbian location.
Covered.

40. + **Melaphyr.**

Covered.

30. + **Melaphyr**; dark-green, mottled with purple; fine-grained to compact; slightly shimmering.

713. **Covered.**

4. **Amygdaloid**; chocolate-colored, with amygdules of delessite, of quartz, epidote, and prehnite.

16. **Melaphyr**; greenish-brown, fine-grained; contains grains of delessite. Towards the top it contains abundant amygdules of a light-green mineral resembling chrysoprase and passes into

4. **Amygdaloidal Melaphyr**; filled with amygdules of radiating prehnite. This rock passes rather abruptly into

1. **Amygdaloid**; with a dark-brown matrix. (This bed is very persistent in its character, being visible east of the Isle Royale copper-bearing bed on the "Shelden and Columbian," "Grand Portage," "Isle Royale," and "Huron" properties.)

69. **Melaphyr**; brown; fine-grained. Above the middle of the bed the trap contains round masses somewhat resembling boulders 4-8 inches in diameter, which are very amygdaloidal and have the cavities filled with prehnite. These masses increase in number towards the top of the bed. They are much harder than the melaphyr, and stand out an inch or more above the glaciated surface of the rock. Glacial furrows are sometimes forked at these harder points and continue separately beyond them. This bed passes into

15-50. **Amygdaloid**; ("Isle Royale lode;") a mixed green and brown amygdaloid; in places very epidotic with extensive segregations of quartz, calcite, prehnite, and is then richest in copper; in others, brown, with amygdules and seams of laumontite, calcite, etc., and is then poorest in copper.

18. **Melaphyr**; green; medium-grained and crystalline; contains much magnetite and a black, infusible mineral, and near the foot-wall some specks of copper. The weathered surface is dirty white and black, and exhibits magnetite and the infusible black mineral. This plane of contact is in places marked by a thin seam of epidote, which often branches out in irregular seams, and branches into the *underlying* rock.

81. **Melaphyr**; brown-green; fine-grained, contains abundant small flakes of red mica. The weathered surface is fine-grained, gray and rusty-brown from the mica.

277. **Covered.**

Fissure-vein; 6 inches to several feet; quartz and ankerite; carries considerable chalcocite and bornite. It has been struck in the "Shelden and Columbian" mine between the "Bloodgood" and "Skip"-shafts, and on both the "Grand Portage" and "Huron" properties.

38. **Melaphyr**; passing into

12. **Amygdaloid**; with amygdules of prehnite and quartz, and containing some copper.

31. **Melaphyr**; hard; fine-grained.

128. **Covered.**

? **Melaphyr**; coarse-grained and crystalline; resembles the coarse-grained crystalline melaphyr on the Dacotah location. See Cross-section IX.

Cross-section II. a.

80+ **Conglomerate**; either not present in II. or not observed by Mr. Mabbs.

Cross-section II. b.

(Determined by triangulation and levelling by the Geological Survey, except where otherwise accredited.)

Along the exploration, trenches on the Huron (formerly the Howard) property.

? **Melaphyr**; brown and green; fine-grained; the weathered surface is mottled gray green and dark red.

97. **Covered.**

? **Amygdaloid**? a brown and gray fine-grained rock, which, in places, has brown, porous portions surrounded by a yellow sandy material.

Covered.

Fissure-vein; prehnite and quartz.

47. **Melaphyr**; brown, hard, and compact, with scattered amygdules of delessite.

? **Amygdaloid**? a rock strongly resembling a conglomerate or breccia, of altered trap pebbles; the matrix contains prehnite and quartz.

216. **Covered.**

24. **Conglomerate**; in places the cement of this conglomerate is very hard, and filled with small round amygdules of calcite and laumontite. Sometimes these minerals have disappeared, and the cement then presents the appearance of a scoria. There is a thick flucan on the hanging-wall.

? **Melaphyr**; light green, with spots of dark-green delessite; soft; fine-grained.

Covered.

? **Amygdaloid**; green, compact, very hard and silicious matrix, with amygdules and segregations of quartz, epidote, and calcite; a pit was sunk on this bed.

157. **Covered.**

3.? **Conglomerate**; small pebbles of brown non-quartziferous porphyry lying in a fine-grained cement of very calcareous sandstone containing more or less epidote.

260. **Covered.**

24. **Conglomerate**; with a band of sandstone on the upper side.

255. **Covered.**

? **Amygdaloid**; ("Foster Mass vein.") ("Ancient Pit vein.")

85. **Covered.** From the notes of Mr. A. B. Wood.

5. **Amygdaloid.** From the notes of Mr. A. B. Wood.

85. **Covered.**

10. + **Conglomerate**; small pebbles of brown non-quartziferous porphyry, in a highly calcareous cement of comminuted porphyry; rich in magnetite. Carries some sheet-copper near the hanging-wall.

355. **Covered.**

7. **Amygdaloid**; light green, and filled with amygdules of quartz, chlorite, and some epidote.

150. **Covered.**

"Isle Royale" cupriferous bed.

Cross-section II. c.

(Determined by triangulation and levelling by the Geological Survey.)

(On the south part of the "Shelden and Columbian" location.)

? **Amygdaloid**; brown; soft; with irregular streaks of a compact brown material resembling hardened clay.

? **Melaphyr**; hard; fine-grained.

1. **Fissure-vein**; quartz, prehnite, and calcite.

35. **Melaphyr and Covered.**

? **Conglomerate?** grayish green rock with occasional pebbles.

Cross-section II. d.

(Determined by triangulation and levelling by the Geological Survey.)

? **Conglomerate.**

Cross-section II. e.

(Determined by triangulation and levelling by the Geological Survey.)

? **Melaphyr**; dark brownish green; fine-grained; contains considerable magnetite and fuses not very easily to a black and very magnetic globule. Towards the top of the bed it contains quartz amygdules, and then without any sharp line of separation becomes mixed with, and passes into

3-6. **Amygdaloid**; yellowish green, matrix of epidot and quartz, with quartz amygdules; this passes rather abruptly into

1. **Amygdaloid**; dark-brown, semi-vitreous matrix, with amygdules of quartz and chlorite, and seams of calcite; the weathered surface is green and brown; contains some copper.

50. **Covered.**

? **Melaphyr**; red and green; soft; contains little or no magnetite, scattered amygdules of calcite, also of quartz, sometimes carrying copper; fuses to black magnetic enamel. This bed frequently contains large and small masses of quartz and epidote, having amygdules of quartz.

? **Amygdaloid**; light green to bluish and brown green; silicious matrix, containing amygdules of quartz, with chlorite and seams of brown jasper.

Cross-section II. f.

(Determined by triangulation and levelling by the Geological Survey.)

? **Amygdaloid**; the lower half of the bed is greenish brown, hard, with amygdules of calcite, prehnite, and green-earth. The upper half is dark brown, and filled with minute amygdules of laumontite and prehnite.

58. **Covered.**

9. **Amygdaloid**; purple brown, compact and hard, with abundant amygdules of quartz; the matrix fuses with difficulty to a magnetic glass.

Covered.

? **Melaphyr**; mottled light and dark green; fine-grained.

Covered.

? **Melaphyr**; brown; very fine-grained, with some spots and seams of delessite, and occasional small amygdules of calcite.

Cross-section II. g.

(Determined by triangulation and levelling by the Geological Survey.)

? **Amygdaloid**; green; very fine-grained, and not very hard rock, fusible on the edges to black magnetic glass; probably a silicious chloritic rock.

Covered.

Amygdaloid; of varying character; in places brownish green, compact and hard, quartz epidote rock, with small amygdules of calcite; other portions are dark brown and soft, with amygdules of calcite, or calcite and green-earth.

This bed shows stains of carbonate of copper.

Covered.

Melaphyr; dark green, semi-crystalline, and slightly shimmering; contains considerable magnetite, and weathers rusty gray.

Covered.

Melaphyr; dark, almost black; compact and slightly shimmering, and tolerably soft. Near the "Capen vein" it is spotted with foliated delessite, and carries minute particles of copper; small pieces of the rock adhere to the magnet.

"Capen vein" (apparently a fissure vein). Seams of calcite, prehnite, laumontite, and fibrous chlorite, with copper in flakes and sheets. Several tons of copper were extracted from this pit.

Cross-section II. h.

(Determined by triangulation and levelling by the Geological Survey.)

On the north part of the "Shelden and Columbian" location.

? **Melaphyr**; dark green; hard, semi-crystalline, compact, near the "Capen vein" it becomes highly chloritic, and the joints are covered with smooth asbestiform delessite.

+—3. **"Capen vein"** (see description in Cross-section II. g). An adit was driven here and some sheet-copper found.

102. **Covered.**

17. + **"Ancient Pit" Amygdaloid.** (So called from the fact that a number of ancient workings were discovered on the "back" of this bed.) Portions of this bed are almost a homogeneous melaphyr, but the greater part bears a striking resemblance to a breccia, owing to the segregation of quartz, prehnite, and calcite in countless fine cracks, enclosing irregular fragments of the rock from the size of a pea to several inches in diameter. In some places the rock becomes epidotic, retaining the same structure and segregations. Some copper was found here.

168. **Covered.**

12. **Conglomerate.** (1st conglomerate east of the Isle Royale copper-bearing bed.) Some small pieces of copper were taken out of this bed.

Cross-section II. i.

(Determined by triangulation and levelling by the Geological Survey.)

"Ancient Pit" Amygdaloid; same character as in II. h.

Cross-section II. j.

(Determined by triangulation and levelling by the Geological Survey.)

? **Melaphyr.**

25. **Covered.**

? **Amygdaloidal Melaphyr**; dark green; hard; containing amygdules of prehnite impregnated with native copper.

Cross-section II. k.

(Determined by triangulation and levelling by the Geological Survey.)

14. **Amygdaloid**; bed irregularly occupied by bright green epidotic rock, abounding in grains of clear quartz, and a dark brown amygdaloid with amygdules of epidote and calcite, and of porous delessite or green-earth.

20. **Melaphyr**; green brown, fine-grained; towards the top it becomes

8. **Amygdaloidal Melaphyr**; with amygdules of delessite, and of delessite and green and rose-colored prehnite.

7. **Amygdaloid**; green brown, compact, with amygdules of quartz and chlorite, and larger amygdules and seams of green and rose-colored prehnite.

73. **Covered.**

Fissure-vein; direction north-west. Comby strings of quartz with calcite arsenurites of copper, chalcocite, and native copper.

12. **Covered.**

9. + **Amygdaloid**; this bed has been opened here by an adit, under the erroneous supposition that it forms the continuation of the Isle Royale copper-bearing bed, which really lies to the westward.

55. **Melaphyr**; brown, slightly mottled with dark green; fine-grained; it becomes green higher up and more distinctly crystalline, and near the top assumes abundant spots and grains of delessite.

5. **Amygdaloid** (probable continuation of the Isle Royale cupriferous bed); grayish green, compact, epidotic matrix, with semi-conchoidal fracture; filled with amygdules of prehnite. The upper portion of the bed, to the extent of 1 foot in thickness, is a well-defined band of dark brown amygdaloid with spike-shaped amygdules, standing at right angles to the plane of stratification, and filled with prehnite, which is frequently replaced by porous delessite.

20. **Melaphyr**; brown, fine-grained and hard; weathers dirty gray; towards the top it becomes

3. **Amygdaloidal Melaphyr**; with irregular grains of delessite, sometimes containing epidote, and amygdules of prehnite.

4. **Amygdaloid**; greenish gray, with amygdules of delessite, quartz, and laumontite. The upper portion of the bed is compact gray-green epidote, with tabular fracture filled with amygdules of quartz.

37. **Melaphyr**; passing into

6. **Amygdaloidal Melaphyr**; containing amygdules of prehnite. This passes into

1. **Amygdaloid**; brown, compact, with amygdules, and irregular masses of delessite, red feldspar, prehnite, quartz, and epidote. (Observed succession—1, delessite; 2, red feldspar; 3, quartz, epidote.)

9. **Melaphyr**; dark brown, fine-grained, weathers dirty brown gray.

20. **Melaphyr**; green brown, fine-grained, with many irregular grains of delessite, the weathered surface is brown white, pitted with rust-colored holes from the decomposition of the delessite grains. Towards the hanging-wall it becomes

7. **Amygdaloidal Melaphyr**; containing irregular small segregations of red feldspar (orthoclase ?), prehnite, and quartz. The feldspar contains casts of rhombohedrons of calcite.

3. **Amygdaloid.**

50. **Covered.**

Grand Portage Cupriferous Amygdaloid.

Cross-sections II. l; *II.* m; *II.* n.

(Determined by triangulation and levelling by the Geological Survey.)

Shafts on the Grand Portage Amygdaloid on the Shelden and Columbian, and Grand Portage locations.

Cross-section II. o.

(Determined by triangulation and levelling by the Geological Survey.)

? **Amygdaloidal Melaphyr**; dark, fine-grained matrix, with numerous amygdules of prehnite. Farther north and south this bed appears more amygdaloidal and somewhat cupriferous, and forms the so-called "Frue lode."

? **Melaphyr.**

Cross-section III.

(Determined by triangulation and levelling, etc., by the Geological Survey, except where otherwise accredited.)

? **Melaphyr.**
9. **Amygdaloid.**
8. **Amygdaloid**; with amygdules of prehnite.
8. **Amygdaloid**; (Frue "lode") bears copper.
10. **Melaphyr.**
11. **Amygdaloid**; green and hard, with amygdules of quartz.
12. **Melaphyr.**
8. **Amygdaloid**; green and hard, with amygdules of quartz.
10. †**Melaphyr.**
279. **Covered.**
? **Melaphyr.**
14. **Amygdaloidal Melaphyr**; purplish gray, fine-grained matrix, with uneven fracture, carrying grains of delessite, and large amygdules of prehnite; it contains bunches of epidote and on the hanging-wall a seam of epidote with amygdules of prehnite.
60. **Melaphyr**; gray green, hard, fine-grained, and indistinctly crystalline.
35. **Amygdaloidal Melaphyr**; gray green, and fine-grained; contains amygdules of prehnite. In some instances large amygdules of delessite show a tendency to the radiating structure of the prehnite, and in some cases the prehnite amygdules are entirely changed in the interior to delessite, the outer portion being partially intact. This bed contains many bunches of compact epidote, with amygdules of prehnite, and in the upper portion crystals of epidote are visible in the matrix.
40. **Melaphyr**; green, compact, with semi-conchoidal fracture; contains lamellæ of a red mica which resist the weathering.
Melaphyr Seam; indistinctly crystalline, containing prehnite in amygdules and impregnations.
32. **Melaphyr**; same as that below the seam.
15. **Melaphyr**; gray-green rock, with numerous red spots of mica; contains ¼ inch amygdules or grains of compact and radiating delessite. Near the hanging-wall it carries amygdules of prehnite and bunches of compact green rock, with amygdules of prehnite, epidote, and delessite. Still nearer the top it becomes
12. **Amygdaloidal Melaphyr**, with many small amygdules of prehnite.
37. **Melaphyr**; similar to the next underlying.
21. **Melaphyr**; dirty green, compact, and containing less red mica than the underlying varieties.
10. **Amygdaloidal Melaphyr**; green and hard, with amygdules of prehnite.
11. **Amygdaloid**; very hard epidotic matrix, with amygdules of prehnite and calcite.
129. **Covered.**
? **Amygdaloid.** (Position given by Mr. R. Shelden.)
79. **Covered.**

? **Amygdaloid.** (Position given by Mr. R. Shelden.)
17. **Covered.**
? **Amygdaloid,** bearing copper. (Position given by Mr. R. Shelden.)
Covered.

Cross-section III. a.

(Taken from notes furnished by Mr. A. B. Wood.)

Cross-section III. b.

(Determined by triangulation and levelling, etc., by the Geological Survey.)

32. **Amygdaloid.**
63. **Melaphyr;** gray brown, fine-grained and hard; contains scattering amygdules of delessite and much red mica.
24. **Melaphyr;** green, compact, speckled with red mica; contains some amygdules of prehnite.
Seam.
10. **Melaphyr.**
7. **Melaphyr;** same as the next underlying. Near the hanging-wall it carries abundant amygdules of prehnite.
? **Melaphyr.**
56. **Covered.**
? **Melaphyr;** green, compact and hard, with specks of red mica.
Melaphyr Seam; coarsely crystalline, greenish gray.
50. **Melaphyr;** same as next underlying the seam; weathers dirty gray; passes into
? **Amygdaloidal Melaphyr;** containing numerous small amygdules of delessite, and larger ones of prehnite, some of which last are partially changed into foliated and amorphous delessite.
10.* **Amygdaloid;** brown and green; contains a little copper.
? **Melaphyr.**

Cross-section III. c.

(From notes furnished by Mr. A. B. Wood.)

Cross-section IV.

(Determined by triangulation and levelling, etc., by the Geological Survey.)

? **Melaphyr;** gray green, hard, fine-grained, and speckled with ruby-red mica.
43. **Covered.**
Melaphyr Seam; coarse-grained and crystalline, and containing apparently feldspar, chlorite, hornblende or pyroxene, mica, and specular iron.
10. **Melaphyr;** green brown and fine-grained, with red mica and grains of delessite.
Epidote Seam.

12. **Melaphyr**; same as next underlying.
2. **Melaphyr Seam**; coarse-grained and crystalline.
8. **Melaphyr**; brown green, compact and hard, with some red mica, and weathering dirty white. Near the top it contains amygdules of prehnite and delessite.
Seam.
4. **Amygdaloid**; with epidotic matrix.
Seam.
Covered.

Cross-section V.

(Determined by triangulation and levelling, etc., by the Geological Survey.)

36. **Amygdaloid**; very epidotic, and containing ½ inch amygdules of prehnite.
36. **Jasper**; apparently stratified, and more or less epidotic. In places, highly siliceous, with amygdules of quartz; in others a soft brown amygdaloid.
? **Melaphyr**; gray green, passing into
Amygdaloidal Melaphyr; with amygdules of prehnite and of epidote.
190. **Covered.**
? **Melaphyr.**
233. **Covered.**
5. **Conglomerate.**
170. **Covered.**
? **Melaphyr**; brown, with amygdules of delessite.
6. **Conglomerate.**
24. **Melaphyr**; with much epidote.
9. **Conglomerate.**
115.+**Melaphyr**; fine-grained; brown, spotted green and red.
168. **Covered.**
Amygdaloidal Melaphyr.
? **Melaphyr**; fine-grained.
70. **Covered.**
60. **Amygdaloid**; the lower part of the bed is a confused mass of epidote-quartz rock and brown amygdaloid; while the upper and larger part of the bed is a brown amygdaloid, in which amygdaloidal masses are enveloped in a brown, hardened, clay-like material, resembling portions of the South Pewabic and Hancock beds.
133. **Melaphyr**; fine-grained.
1. **Seams** of coarsely crystalline melaphyr, containing crystals of green triclinic feldspar, delessite, and specular iron.
56. **Melaphyr.**
3. **Conglomerate**; contains rare pebbles of quartz porphyry, similar to that of the Calumet conglomerate; also traces of copper.
8. **Covered.**
10. **Amygdaloid**; green; contains segregations and amygdules of quartz, prehnite and calcite, and some copper.
114. **Covered.**

3. **Amygdaloid**; green; contains amygdules of prehnite, quartz, and calcite.
62. **Covered.**
10. ? **Amygdaloid**; ("Montezuma lode.") Brown and green siliceous matrix, with amygdules of prehnite, and quartz and some copper.
102. **Melaphyr** and **Covered.**
5. **Amygdaloid**; brown amygdaloidal matrix, with large and small irregular segregations of quartz with prehnite.
99. **Covered.**
62. **Amygdaloid**; the *lower* portion of the bed, for a thickness of several yards, is a green and more or less epidotic and siliceous amygdaloid; the *upper*, and much thicker portion of the beds, is a soft brown amygdaloid, containing amygdules of red laumontite, and irregular seams and spots of a soft white mineral (Kaolin ?)—perhaps a decomposition product of a zeolite.
127. **Melaphyr.**
?. **Amygdaloidal Melaphyr.**
60. **Covered.**
? **Melaphyr.**
60. **Covered.**
? **Melaphyr.**

Cross-section V. a.

(North ½ of Sect. 2, Town. 54, Range 34.)

(Taken in part from the notes of Mr. A. B. Wood, and in part by triangulation and levelling by the Geological Survey.)

Cross-section V. b.

(Determined by triangulation and levelling, etc., by the Geological Survey.)
Amygdaloid.

Cross-section V. c.

From notes furnished by Mr. R. Shelden.

Cross-section V. d.

(Determined by triangulation and levelling, etc., by the Geological Survey.)
Amygdaloid; epidotic.

Cross-section VI.

(Determined by triangulation and levelling, etc., by the Geological Survey.)

Along part of the road from Portage Lake to the Huron Stamp Mill.

? **Melaphyr**; hard.
44. **Amygdaloid**; the lower portion of the bed is epidotic and hard; the upper portion is a soft brown rock, abounding in laumontite, and containing patches of epidote and quartz.

37. **Melaphyr**; hard.
33. **Covered.**
28. **Melaphyr**; hard.
17. **Amygdaloid**; soft brown, and abounding in laumontite.
? **Melaphyr**; hard.

Cross-section VII.

(Determined by triangulation and levelling, etc., by the Geological Survey.)

Along the banks of Dacotah Creek.

? **Amygdaloid**; soft; brown, and abounding in laumontite.
35. **Melaphyr**; grayish brown; fine-grained; earthy fracture; soft; contains much red mica.
21. **Amygdaloid**; a very irregular mixture of bright-green epidotic rock, containing grains of quartz, with a soft chocolate-colored rock containing amygdules of delessite and calcite, and larger bunches of calcite with radiating crystals of leonhardite. The epidotic portions of the bed are impregnated with copper.
? **Melaphyr**; grayish brown; soft; fine-grained; contains much red mica.

Cross-section VIII.

Position indicated by the Hon. S. W. Hill.

1½–2 ft. **Conglomerate**; considered by Mr. Hill to be the "Houghton" conglomerate. The trench in which it was formerly exposed was filled with rubbish when visited by the Geological Survey. According to Mr. Hill, it was well filled with copper.

Cross-section IX.

(Determined by transit survey, chiefly underground, by the Geological Survey.)

? **Amygdaloid**; portions of the bed fine-grained, greenish-gray to bluish-green, siliceous matrix; very hard (strikes fire with steel), and filled with amygdules of white quartz, more rarely of calcite; portions of the bed consist of soft, chocolate-colored material, with amygdules of calcite and of green-earth. Struck in bottom of a shaft; width not seen.
? **Melaphyr**; soft; greenish brown; fine-grained with spots of delessite, and thin filmy seams of laumontite. Struck in the end of an adit. No walls seen.
120. **Concealed** on south side of Portage Lake.
11. **Amygdaloid**; fine-grained, soft green-brown matrix, with irregular amygdules of delessite and green-earth, and of delessite and quartz. The upper portion of the bed is chiefly green-brown amygdaloid, with compact siliceous matrix and amygdules of more or less crystalline quartz, of quartz and epidote, and of calcite. Forms the east end of the adit.

19. **Melaphyr**; green; fine-grained; somewhat crystalline; tolerably hard; contains minute acicular crystals of triclinic feldspar and much magnetite.

7. **Amygdaloid**; towards foot-wall, soft brown-green matrix with earthy fracture; contains amygdules of quartz coated with delessite, and of quartz and epidote. It carries fine copper. The upper portion of the bed has a compact, gray-green, and very siliceous matrix, with amygdules of quartz and of calcite.

76. **Melaphyr**; the lower portion of the bed is green and fine-grained, the upper portion contains seams of laumontite.

5. **Amygdaloid**; light green; soft, with earthy fracture; contains amygdules of quartz, of delessite, and of quartz and epidote.

6. **Amygdaloidal Melaphyr**; green brown, soft, with earthy fracture; contains spots of delessite and amygdules of green-earth and of laumontite.

8. **Amygdaloid**; soft, almost earthy, green rock in the lower part of the bed, with brown streaks; contains flakes of delessite. The upper portion is more amygdaloidal, with amygdules of quartz and epidote, and forms, in places, a very hard and compact brown and green rock, with semi-conchoidal fracture. Some copper was found in the bed, and some rich boulders of exactly similar rock, in driving the adit through the sand near by. It has been supposed to be the continuation of the Pewabic copper-bearing bed.

? **Melaphyr**; very disintegrated, and containing laumontite. Thickness not known.

106. **This ground** is said to have been cut in the next adit westward, but as this was driven below the lake level, and full of water, it was inaccessible to the Survey.

? **Melaphyr**; mottled, grayish green and dark green; contains abundant spots of delessite, and minute indistinct crystals of green feldspar. Forms the eastern end of the adit on the south side of the lake-shore road.

7. **Amygdaloid**; very hard, siliceous, compact matrix, with conchoidal fracture; brown, speckled with green; contains amygdules of quartz, of calcite, and of laumontite.

13. **Melaphyr**; soft, with considerable laumontite.

15. **Amygdaloid**; brown, shaded with green; very hard, compact, siliceous matrix, with conchoidal fracture; contains abundant amygdules of quartz, more rarely of green-earth, of calcite, or of epidote and quartz.

29. **Amygdaloidal Melaphyr**; green, fine-grained, semi-crystalline rock, mottled with dark green delessite, and containing minute crystals of green triclinic feldspar, and amygdules of quartz and delessite.

22. **Amygdaloid**; brown, semi-crystalline matrix, with earthy fracture, containing minute crystals of triclinic feldspar, and filled with irregular spots of dirty green chlorite, and amygdules of calcite and laumontite. The upper portion of the bed is siliceous and compact, with better defined amygdules of quartz and delessite, and of quartz and epidote, and in places is highly siliceous and contains seams of quartz and amygdules of quartz and of prehnite.

60*. **Melaphyr**; coarse-grained and crystalline; consists of light green triclinic feldspar, and black green delessite, in apparently nearly equal proportions. It contains considerable magnetite. It closely resembles a similar rock, west of the Pewabic copper-bearing bed, on the Quincy and St. Mary's locations.

122. **Not exposed** on the south side of Portage Lake, but very possibly the same crystalline melaphyr, since this bed is between 200 and 300 feet thick on the St. Mary's.

12. **Amygdaloid;** green, and tolerably compact, hard matrix, with amygdules of quartz and of laumontite; the upper portion of the bed has a brownish-green and rather siliceous matrix, with abundant amygdules of laumontite.

25. **Not exposed** on the south side of Portage Lake.

3. **Amygdaloid;** light green, soft matrix, with earthy fracture, containing amygdules of delessite and of quartz.

25. **Not exposed** on the south side of Portage Lake.

? **Amygdaloid;** green matrix, with amygdules of quartz, laumontite, and calcite, in places highly siliceous; often the matrix is changed to a red jasper near the amygdules.

323. **Not exposed** on the south side of Portage Lake.

Cross-section IX. a.

(Included in IX.)

Cross-section IX. b.

(Determined by the measurements of the Geological Survey.)

In the trenches east of the South Pewabic Mine.

33. ? **Conglomerate;** porphyry pebbles, with grayish-red cement. The cement effervesces somewhat in muriatic acid, and fuses partially B. B. to white highly magnetic glass with black specks.

22. **Covered.**

? **Amygdaloid;** green brown; compact, and abundantly filled with amygdules of a soft pink zeolite. The matrix is very hard, and fuses on the edges of splinters to a highly magnetic glass. It contains films of magnetic iron.

25. **Covered.**

? **Melaphyr;** gray green, fine-grained, and soft, with scattered amygdules of quartz. Is quite soft, probably very chloritic. Contains magnetite; fuses with intumescence to a dark magnetic glass.

200. **Covered.**

? **Melaphyr;** gray green; amorphous, or crypto-crystalline matrix, porphyritic, through numerous long, flat crystals of green triclinic feldspar, and grains (pinhead to pea size) of almost black delessite. Contains irregular amygdules of a soft pink zeolite (laumontite), and considerable magnetite. Mixed with the fragments of this trap are those of a more even-grained and darker variety.

65. **Covered.**

10? **Amygdaloid;** matrix the same as in next rock to west; the irregular cavities either empty or coated with delessite, or crystals of quartz. Perhaps part of next bed west.

22 ? **Amygdaloid**; blue green; hard, round, and sometimes connected amygdules of calcite or quartz, or again of a flesh-colored, soft zeolite (leonhardite?). The cavities are often bordered on one side by a dark-red jaspery substance. This red substance may be simply the result of oxidation of protoxide of iron in the bluish-green matrix, without further decomposition of the rock. It is accompanied in the debris on the side of the trench by a fine-grained, green, slightly amygdaloidal trap, with amygdules of quartz, and filled with minute green-white points.

400. **Covered.**

100. ? * **Melaphyr**; green; crystalline; light-green triclinic feldspar, and delessite in irregular grains, in apparently nearly equal amount. Contains grains of magnetite, also apparently some tabular crystals of specular iron ore. Closely resembles the bed 200–300 feet thick, in about the same position on the St. Mary's, and a similar bed on the Dacotah. Thickness not observed, but seems to be more than 100 feet. Neither wall seen.

470. **Covered.**

? **Melaphyr.**

8–12. **South Pewabic Amygdaloid**; almost uniformly chocolate brown. It is characterized by the very general presence of a fine-grained or compact brown material with conchoidal fracture, varying in hardness from quite soft to harder than steel; this substance effervesces in muriatic acid, owing to the presence of minutely associated calcite. The smallest splinters fuse on the edge, and become magnetic. This red material is so distributed as to resemble the cement of a conglomerate in which the more amygdaloidal portions resemble the pebbles, but this is merely a resemblance. The amygdaloidal portions are darker brown, often tinged with green; they are usually a few inches in diameter, and are filled with spherical amygdules of shot sizes, chiefly of quartz, calcite, delessite, green-earth, and copper; the matrix fuses to a black magnetic glass. The copper is disseminated as a coating of the amygdules, and in thin films in the cracks.

53. **Covered.**

21. **Conglomerate** ("Hancock West" conglomerate).

? **Amygdaloid.**

Cross-section X.

(Determined by underground compass survey by the Geological Survey.)

? **Melaphyr**; fine-grained green rock, with minute crystals of green triclinic feldspar, and spots of dark-green delessite and seams of chalcedony. It forms the eastern end of the easternmost adit on the "South-side" location.

22. **Amygdaloidal Melaphyr**; soft green rock, with earthy fracture, containing much delessite; it is traversed by minute threads of calcite, and filled with amygdules of laumontite.

4. **Melaphyr**; dark green, fine-grained, and hard, with minute crystals of green triclinic feldspar, and spots of dark-green delessite.

4. **Amygdaloid**; light green, rather hard and siliceous matrix, with semi-conchoidal fracture; contains amygdules of quartz and laumontite.

9. **Melaphyr**; contains laumontite.

2. **Amygdaloid**; light green, hard and siliceous, with amygdules of laumontite and quartz.
2. **Melaphyr.**
2. **Amygdaloid**; brown, speckled with green, very siliceous, with conchoidal fracture; contains amygdules of quartz and chlorite.
5. **Melaphyr**; with amygdules of laumontite near the hanging-wall.
4. **Amygdaloid**; brown and green, very siliceous, with conchoidal fracture; contains amygdules of quartz and some laumontite.
4. **Melaphyr**; dark green, fine-grained, and hard.
? **Amygdaloid**; dirty green, with earthy fracture, contains amygdules of leonhardite and of delessite, and minute specks of a soft flesh-colored zeolite. Parts of the bed are a very siliceous brown and green amygdaloid, with conchoidal fracture containing amygdules of quartz.

Cross-section XI.

(The eastern adit was surveyed by compass; the western adit by transit, by the Geological Survey; the remaining beds are given from the notes of Mr. A. B. Wood.)

On the "South-side" location.

5. **Amygdaloid**; grayish green, siliceous, filled with amygdules of laumontite.
23. **Melaphyr**; green and chloritic, becomes amygdaloidal towards the top of the bed, assuming amygdules of delessite and passing into the overlying bed.
4. **Amygdaloid**; green and hard; the cavities are either empty or filled with laumontite.
40. **Melaphyr**; dirty green, very fine-grained, with perfect conchoidal fracture, and containing much diffused magnetite. Near the hanging-wall becomes very amygdaloidal, the cavities being either empty or filled with delessite, or with quartz coated with delessite.
25. **Amygdaloid**; the matrix consists largely of a brown crystalline mineral, and contains abundant round and oval amygdules of quartz surrounded with laumontite. This rock resembles in places that of the South Pewabic copper-bearing bed, being then traversed by a chocolate-colored, very fine-grained material, which has an even earthy fracture, is easily scratched, and fuses readily to a dark-green, somewhat magnetic glass. This red material also contains irregular amygdules of laumontite enclosing quartz, and smaller flat ones of laumontite alone, so arranged as sometimes to give to the rock a laminated appearance.
? **Melaphyr**; dark gray green, and very fine-grained, with imperfect conchoidal fracture; it contains specks and threads of delessite, minute threads of laumontite and magnetite both diffused and in distinct grains.
82. **Covered.**
32.? **Conglomerate**; ("Hancock West.") From the notes of Mr. A. B. Wood.
38. **Melaphyr.** " "
Epidote Seam. " "
135. **Melaphyr.** " "

20. **Amygdaloid**; the lower portion of the bed is a fine-grained, soft rock, speckled brown and yellowish green, containing amygdules and specks of a soft, greasy white mineral, and of delessite with calcite, and more rarely of laumontite. The rest of the bed is chiefly a dark chocolate-brown rock, filled with amygdules of calcite, or of laumontite, or of laumontite enclosing calcite. Druses occur containing crystals of analcite ⅛ to ½ inch in diameter, with super-imposed crystals of orthoclase.

47. **Melaphyr.**

5. **Amygdaloid**; dark purple brown, with large irregular vesicles which are either empty except a lining of red laumontite (?), or filled within this coating with calcite or with green-earth. This bed carries some copper.

27. **Amygdaloidal Melaphyr**; soft, and containing many large and small amygdules of laumontite and calcite.

5. **Melaphyr.**

10. **Amygdaloidal Melaphyr**; soft, with amygdules of laumontite and calcite.

44. **Melaphyr**; very dark colored, hard and compact, with conchoidal fracture, and containing minute crystals of green triclinic feldspar, and very little visible delessite, some diffused magnetite, and grains of a hard, black, infusible mineral.

9. **Amygdaloid**; dark green, soft, and consisting largely of amygdules of calcite and chlorite.

155? **Conglomerate**; forms the west end of the westernmost adit on the "South-side" location.

94. **Melaphyr.**	From the notes of Mr.	A. B. Wood.
13. **Sandstone.**	"	"
100. **Melaphyr**; coarse-grained.	"	"
19. **Sandstone and Conglomerate.**	"	"
51. **Amygdaloid.**	"	"
15. **Sandstone.**	"	"
66. **Melaphyr**; compact, gray.	"	"
? **Conglomerate**; width unknown, but very thick.	"	"

Cross-section XII.

(From the measurement of Mr. Graveldinger.)

On the Douglass location.

? **Amygdaloid**; epidotic matrix.

138. **Covered.**

? **Amygdaloid.**

13. **Melaphyr.**

? **Amygdaloid.**

46. **Melaphyr.**

? **Amygdaloid.**

12. **Melaphyr.**

? **Amygdaloid**; epidotic.

60. **Melaphyr.** •

? **Amygdaloid.**

195. **Covered.**

? **Amygdaloid**; "soft."
11. **Melaphyr.**
? **Amygdaloid.**
61. **Covered.**
? **Amygdaloid**; "epidotic."
45. **Melaphyr.**
? **Amygdaloid**; light green; soft; contains amygdules of delessite and copper.
? **Jasper**; banded brown and green.
17. **Melaphyr.**
? **"Capen vein." "Mass vein."**
10. **Melaphyr.**
63. **Covered.**
? **Amygdaloid**; "Ancient Pit" bed; resembles the same bed on the Shelden and Columbian location.
168. **Covered.**
? **Conglomerate.**
525. **Covered.**
? **Amygdaloid**, in which is the Douglass mine.

Cross-section XIII.

(Determined by the measurements of the Geological Survey.)

On the North-star location.

31. **North-star Conglomerate.**
188. **Covered.**
6. ? **Amygdaloid**, resembling that of the Montezuma mine.

Cross-section XIV.

(Determined by the Geological Survey.)

At the Albany and Boston Mine.

? **Houghton Conglomerate**; rounded pebbles of medium size, of brown porphyry (non-quartziferous). The cement is, in places, calcite and laumontite (?) altered to red clay; in places, a very chloritic sand.
10. **Covered.**
38. **Melaphyr**; fine-grained, green, soft; abounds in delessite; crystals of green triclinic feldspar are visible.
9. **Amygdaloidal Melaphyr**; green brown, fine-grained, in places resembling the crystalline seams in trap west of the Isle Royale group, and carrying amygdules of delessite and others of prehnite undergoing change to delessite.
8. **Amygdaloid**; chocolate brown, finely crystalline to amorphous matrix; amygdules of prehnite changing to delessite; also of calcite.
12. **Melaphyr**; brown green, very fine-grained, soft.
Covered.

Melaphyr; greenish gray; finely crystalline.

Covered.

Amygdaloidal Melaphyr.

Melaphyr; very crystalline; contains flat thin crystals of brown and white triclinic feldspar, and grains of delessite or chlorite.

Cross-section XIV. a.

(Determined by the measurements of the Geological Survey.)

East of the Mesnard Mine.

? **Houghton Conglomerate.**

Cross-section XV.

In the open trenches on the St. Mary's location.

The positions of the conglomerates, and of the rocks lithologically described in this section, were determined by the Geological Survey; the remainder is from the notes of Mr. A. B. Wood.

Covered.

15. **Melaphyr.**
8. **Melaphyr.**
31. **Melaphyr.**
9. **Amygdaloid.**
63. **Melaphyr.**
40. **Amygdaloid.**
35. **Albany and Boston Conglomerate.**
40. **Melaphyr.**
38. **Melaphyr.**
17. **Melaphyr.**
10. **Amygdaloid.**
20. **Melaphyr.**
31. **Melaphyr.**
7. **Amygdaloid.** ("Albany and Boston Amygdaloid")
29. **Melaphyr.**
6. **Amygdaloid.**
90. **Melaphyr.**
12. **Amygdaloid.**
10. **Amygdaloid.**
9. **Amygdaloid.**
32. **Melaphyr.**
81. **Melaphyr.**
11. **Amygdaloid.** (Supposed Pewabic cupriferous bed.)
25. **Melaphyr.**
9. **Amygdaloid.**

78. **Melaphyr.**
10. **Amygdaloid.**
42. **Melaphyr.**
6. **Amygdaloid.**
65. **Melaphyr.**
4. **Amygdaloid.**
11. **Melaphyr.**
4. **Amygdaloid.**
14. **Melaphyr.**
4. **Amygdaloid.**
226. **Melaphyr**; crystalline, granular; contains light-green triclinic feldspar; mottled with dark green spots and grains of delessite.
6. **Amygdaloid.**
112. **Melaphyr**; fine-grained brown rock, spotted with dark-green amygdules of delessite, sometimes enclosing quartz.
8. **Amygdaloid**; brown matrix; amygdules filled with greenish-white, soft, decomposition product; others lined with red-stained crystals of feldspar.
40. **Melaphyr**; greenish brown; fine-grained, hard; conchoidal fracture; contains many spots of bluish-green compact delessite; also of calcite.
52. **Amygdaloid**; very fine-grained, green-brown matrix, with large amygdules of delessite enclosing quartz; others of only foliaceous delessite; smaller bird's-eye amygdules of—1st, compact delessite; and within this, 2d, a thin ring of light green chlorite (?); 3d, the centre filled with foliaceous delessite. In places, the amygdules contain more quartz, sometimes red, or, again, filled with metallic copper. Often the matrix becomes quite siliceous, the cavities being filled with small crystals of quartz, on which rest scattered crystals of white feldspar. In places, the matrix adjoining the amygdules (generally on the under side) is changed to a red jaspery substance; in such cases, many of the amygdules contain a soft, white, decomposition product.
11. **Conglomerate** (Pewabic West); reddish, somewhat calcareous cement of sandstone; containing pebbles (generally under two inches diameter) of brown non-quartziferous porphyry, which in some instances show the beginning of amygdaloidal action. Contains numerous stains of carbonate of copper.
12. **Melaphyr**; ("with orthoclase." A. B. Wood.)

Amygdaloid, the only rock identifiable on the side of the trench, is a brown, almost black, subcrystalline amygdaloid; the cavities run into each other, and are filled, some with calcite, but mostly with a soft, white clay,—a product of decomposition of some previous mineral.

8. **Amygdaloid.**
8. **Amygdaloidal Melaphyr**; grayish green-brown; medium grain, with numerous large and small spots and amygdules of dark-green delessite.
15. **Amygdaloid**; brown; tolerably compact and hard amygdaloid, with amygdules of calcite, of a flesh-red zeolite (?), and others of chlorite.
15. **Melaphyr.**
24. **Amygdaloid.**
16. **Melaphyr.**
10. **Melaphyr.**

27. **Melaphyr.**
23. **Amygdaloid.**
21. **Melaphyr.**
9. **Melaphyr**; grayish green and black green; medium-grained; contains green triclinic feldspar, and dark green chlorite.
5. **Amygdaloid.**
17. **Melaphyr.**
23. **Melaphyr.**
4. **Amygdaloid.**
54. **Melaphyr.**
5. **Amygdaloid.**
22. **Melaphyr**; the exposed rock is much affected by weather; granular; contains gray altered feldspar, delessite, and small crystals of a resinous-looking mineral resembling nepheline.
17. **Amygdaloid**; green, with branching amygdules of light pink laumontite.
7. **Melaphyr.**
12. **Amygdaloid.**
19. **Amygdaloidal Melaphyr**; very fine-grained, brown rock; hard, with conchoidal fracture; spotted dark green with amygdules of delessite; contains small spots of calcite.
43. **Melaphyr**; dirty black-green, very hard, with perfect conchoidal fracture, contains minute crystals of triclinic feldspar; also scattered spherical amygdules of chalcedony, surrounded with delessite.
13. **Melaphyr**; compact; dark.
8. **Amygdaloid** ("Hancock-vein," A. B. Wood); chocolate-brown matrix, filled with large and small amygdules of flesh-colored zeolite (laumontite ?), rarely calcite.
23. **Melaphyr**; very fine-grained; green; hard; conchoidal fracture; small crystals of green and flesh-colored triclinic feldspar.
11. **Amygdaloid**; closely resembling the South Pewabic type in form. Amygdules form a large part of the rock, and are almost wholly of laumontite.
27. **Melaphyr**; fine-grained; green; amygdules of delessite or laumontite; more rarely quartz and delessite.
32. **Hancock West Conglomerate.** At or near the bottom there is sandstone, which splits in thin layers; contains much carbonate of lime and some magnetite. There is also a finer sediment, apparently very fine arenaceous clay. Pebbles (not generally more than 1 inch diam.) are chiefly of brown non-quartziferous porphyry, with compact matrix in which are small crystals of reddish feldspar. The cement of the conglomerate is very calcareous.
14. **Melaphyr.**
5. **Melaphyr.**
16. **Melaphyr.**
43. **Melaphyr.**
8. **Melaphyr.**
24. **Amygdaloid.**

Cross-section XV. a.

(Determined by the measurements of the Geological Survey.)

Being the exposures in the eastern part of the covered drain at the St. Mary's Mine.

? **Melaphyr**; hard; dark green, speckled with red.

35. **Covered.**

6. + **Amygdaloid** ("Ragged Amygdaloid"); chocolate-brown matrix, with large cavities of irregular shape, tending to give a brecciated appearance to the rock. The cavities are partially filled in the given sequence with, 1st, analcite; 2d, calcite; 3d, orthoclase; 4th, calcite.

25. **Melaphyr**; dark greenish-gray; fine-grained; hard; exhibits beginning alteration in the form of small greenish, soft, white spots; it becomes brown and less distinctly crystalline towards the hanging-wall, and assumes veins and amygdules of delessite.

18. **Melaphyr**; brown, speckled green. Towards the hanging-wall this bed carries more grains of delessite and some amygdules of prehnite.

? **Amygdaloidal Melaphyr.**

10. **Covered.**

? **Amygdaloidal Melaphyr**; brown; fine-grained, and hard; contains scattered small amygdules of calcite and delessite.

4. **Amygdaloid**; chocolate brown; soft; contains amygdules of calcite and delessite.

? **Melaphyr**; dark gray brown; fine-grained, hard; conchoidal fracture, with minute red specks of altered delessite? At a few feet from the foot-wall the rock becomes lighter in color, and, without losing in hardness, shows decided signs of change, becoming porous near the joints. In the centre of a polygonal block the delessite is fresh; towards the joints it is changed to a blood-red, soft mineral; nearer the joints the delessite has disappeared wholly, leaving only the empty cavities.

44. **Covered.**

? **Amygdaloidal Melaphyr**; resembling the last-mentioned, but containing amygdules of calcite.

? **Melaphyr**; greenish-brown, fine-grained; hard, conchoidal fracture; contains scattered spots of calcite, and grains of delessite which show minute red specks. Higher up it contains numerous amygdules of delessite.

? **Amygdaloid**; dirty-brown, fine-grained matrix, with small amygdules of calcite and orthoclase?

Covered.

? **Melaphyr**; speckled green and pink; fine-grained and hard. Higher up it is very decomposed, and contains many amygdules of delessite and laumontite.

18. **Covered.**

? **Amygdaloid.**

35.? **Conglomerate.** ("Albany and Boston.")

25. **Covered.**

4. **Amygdaloid** ("20-foot amygdaloid"); hard, light-green matrix, with empty round cavities and round amygdules of quartz.

? **Amygdaloidal Melaphyr**; dark green; contains amygdules of delessite and minute flesh-colored acicular crystals.

15. **Covered.**

6.? **Amygdaloid.** ("**Epidote** lode.") Brown and green compact matrix, with amygdules of calcite, delessite, green-earth and quartz; often very siliceous; carries some copper.

? **Melaphyr**; dark green; fine-grained, soft; carries much delessite.

20. **Covered.**

? **Melaphyr**; brown green-gray; granular; hard; contains small spots of prehnite.

7. **Amygdaloidal Melaphyr**; gray-green; fine-grained, rich in delessite; carries amygdules of rosy and white prehnite.

? **Melaphyr**; brown green, spotted dark green with amygdules of delessite; fine-grained.

26. **Covered.**

? **Melaphyr**; similar to the last described.

4. **Amygdaloid**; chocolate brown; contains amygdules of green-earth and calcite.

8. **Melaphyr**; brown, with spots of delessite; fine-grained.

7. **Amygdaloid** (probably the "Albany and Boston amygdaloid"); violet and green; irregular hardness; contains amygdules of green-earth, quartz, and epidote.

10. **Melaphyr**; brown green; fine-grained; soft; contains many spots of delessite.

10. **Amygdaloidal Melaphyr**; green-gray; soft; the cavities are partially filled with minute pink and white acicular crystals of a zeolite.

8. **Melaphyr**; gray green; much altered; contains amygdules of delessite.

9. **Amygdaloid**; brown and green; hard matrix, containing amygdules of prehnite and quartz and some copper.

? **Melaphyr**; brown; granular; hard, with imperfect conchoidal fracture. Higher up it is gray and softer, and much altered.

27. **Covered.**

? **Melaphyr**; light gray-green, spotted with dark-green delessite; rather coarse-grained, and marking proximately the eastern limit of the coarse-grained crystalline melaphyrs.

22. **Covered.**

? **Amygdaloidal Melaphyr**; light-brown green, with spots and amygdules of delessite and quartz, and containing porphyritic crystals of a triclinic feldspar.

Cross-section XV. b.

(Determined by the measurements of the Geological Survey.)

On the "Mesnard" location.

10.? **Amygdaloid**; resembling the "Ragged Amygdaloid." (See *Section XV.* a.)

70. **Covered.**

? **Amygdaloid.**
25. **Covered.**
30.† **Albany and Boston Conglomerate.**

Cross-section XV. c.

(Determined by the measurements of the Geological Survey.)

Near the "Powder-house," at the Albany and Boston Mine.

Cross-section XV. d.

(Chiefly from the printed report of the Pewabic Mining Company for 1865.)

In the cross-cut from the "Albany and Boston Conglomerate" westward to the 70-fathom level of the Pewabic Mine.

32. **"Albany and Boston Conglomerate."** Boulders and pebbles, ¼ inch to 2 feet or more in diameter, in a tolerably fine-grained cement of green and red sand. The cement is very subordinate in quantity as compared with the pebbles; it is free from magnetite.

The predominating pebbles are of a red crystalline rock, of medium grain, of the granite family, chiefly of triclinic feldspar and quartz, with little or no mica, and some chlorite. Next in point of frequence are pebbles of a chocolate-brown porphyry, free from quartz, with a very compact matrix, in which lie numerous minute crystals of a triclinic feldspar of the same color as the matrix. This rock is the most common among the pebbles in the other conglomerates of Portage Lake.

Another characteristic variety of pebble is a dark-brown porphyry, free from quartz, with a very fine-grained base, containing crystals ¼–½ inch long, of flesh-colored triclinic feldspar, which have generally begun to change chemically.

The conglomerate carries copper, though, so far as tested, not in paying quantity.

20. **Melaphyr.**
23. **Amygdaloid** (supposed continuation of the "Epidote" amygdaloid).
41. **Melaphyr.**
8. **Amygdaloid** (supposed continuation of the "Albany and Boston" amygdaloid). Carries copper.
80. **Melaphyr.**
16. **Amygdaloid.** ("Green Amygdaloid.") Carries copper.
69. **Melaphyr.**
29. **Amygdaloid.** ("Old Pewabic" Amygdaloid.) Carries copper.
104. **Melaphyr.**
10. **Amygdaloid.** (Pewabic—Franklin—Quincy copper-bearing bed.)
? **Melaphyr.**

Cross-section XV. e.

(From the measurements of the Geological Survey.)

In the Quincy "Hill-side Adit."

East end covered; large conglomerate boulders were found in the "drift," just before meeting with rock.

32. **Melaphyr**; rather fine-grained; dirty dark and light green, with rusty brown specks; hard; tough; fracture irregular; changes into

28. **Amygdaloidal Melaphyr**; fine-grained, dirty green color; numerous dark-green delessite spots; loosely textured.

12. **Amygdaloid**; mixed green and brown; some hard amygdules in places mostly of quartz and green-earth, in others of calcite and green-earth; epidote is present, and some laumontite: some amygdaloidal melaphyr occurs near the centre.

37. **Melaphyr**; fine-grained; dark brown, mottled dirty green; hard. The lower one foot is amygdaloidal. Above, it slowly changes to

31. **Amygdaloidal Melaphyr**; coarse; dirty green, spotted with dark-green delessite; fracture very uneven.

1. **Amygdaloid**; narrow, poorly developed, hanging-wall not well defined; green; hard; amygdules of quartz and delessite, some laumontite.

73. **Melaphyr**; rather coarse, not closely textured; color, dirty green of different shades; fracture irregular; inclining toward amygdaloidal melaphyr in places. Toward the base it is dull brown and rather compact, with conchoidal fracture, while at 15 feet from the top it changes irregularly into a very compact rock; the color here is dull purple, indefinitely inclining toward, and mottled with, dull green; very hard; conchoidal fracture. Thirty-four feet horizontally from the foot-wall are two parallel seams, three feet apart; the upper one is 4 inches wide; strike about north; dip 40° east.

18. **Amygdaloid**; green; compact; hard; amygdules mostly quartz in green-earth or epidote; also calcite.

12. **Melaphyr**; very compact; color green, purplish; rather hard; elastic; highly conchoidal fracture; occasionally flesh-red triclinic feldspar crystals porphyritically imbedded; changes to

48. **Amygdaloidal Melaphyr**; in part with dirty greenish-gray base, sprinkled with small porphyritically imbedded flesh-red feldspar crystals, and numerous green delessite spots, and in part light green, with scattering brown specks, and delessite amygdules; tough; fracture uneven.

4. **Amygdaloid**; mixed green and brown; the green is hard, and usually with quartz and epidote; the green is generally softer, with calcite and delessite. Drifts of 15 feet either way near hanging-wall.

6. **Amygdaloidal Melaphyr**; base dirty greenish-gray, sprinkled with flesh-red feldspar crystals, and numerous green delessite spots; tough; fracture uneven.

5. **Seam** of laumontite, narrow.

8. **Melaphyr**; coarse-grained, rather crystalline; dark purplish-green; hard; brittle, yet tough, changes gradually into

14. **Amygdaloidal-Melaphyr**; base purplish-brown; rather fine-grained, growing coarser toward the bottom, where small flesh-red feldspar crystals are imbedded; amygdules of delessite at the bottom, of laumontite and green-earth, or delessite, near the top.

Cross-section XV. e.

(From the measurements of the Geological Survey.)

In the western half of the Quincy "Hill-side Adit."

10. **Melaphyr**; dirty brown, fine-grained, mottled with very dark delessite.

4. **Amygdaloid**; brown; hard; conchoidal fracture; siliceous amygdules of green-earth and quartz, and green-earth and calcite, and of green-earth alone. Contains some copper. Shaft sunk.

18. **Amygdaloidal Melaphyr**; green; fine-grained; crystals of green triclinic feldspar; numerous amygdules of very dark delessite; others of delessite enclosing quartz.

4. **Amygdaloid**; light green; siliceous; amygdules of quartz and calcite; some copper.

17. **Amygdaloidal Melaphyr**; very fine-grained; in places amygdules of laumontite; in others, of green-earth and quartz.

8. **Amygdaloid**; brown matrix; small amygdules of green-earth, and large ones of green-earth and calcite.

34. **Melaphyr**; brown green; fine-grained; brown triclinic feldspar and dark-green delessite.

6. **Amygdaloid**; light-green and brown; hard conchoidal fracture; very siliceous; amygdules of quartz, calcite, epidote, copper.

72. **Melaphyr**; fine-grained; green, through predominance of delessite. Some crystals of brown triclinic feldspar. Below this the rock is mottled dark and light green, and under the glass seems to consist of compact delessite, with amygdules of gray-green chlorite or green-earth, with a little quartz. It has the appearance of a feldspar-delessite trap in which the feldspar had disappeared, leaving cavities now occupied by secondary minerals. Towards the foot-wall, the rock is medium-grained, crystalline, brown, mottled with the usual dark-green delessite, and with light-green spots. It consists of crystals of brown triclinic feldspar, which are partially changed to a soft, light-green clay. Still nearer to the foot-wall the rock appears fresh; a medium-grained, dark-greenish gray mixture of gray to light-green triclinic feldspar, and dark-green delessite. Contains minute hexagonal crystals of specular iron. Resembles the coarse-grained trap west of the Pewabic lode, and on the St. Mary's.

13. **Amygdaloid**; bearing copper; shaft sunk.

46. **Melaphyr**; fine-grained, crystalline; consisting of brown triclinic feldspar and diffused delessite; contains amygdules of delessite and of green-earth.

4. **Amygdaloid**; compact and hard; green-brown matrix, with abundant small brown crystals of triclinic feldspar, and numerous amygdules of calcite, green-earth, and quartz.

31. **Melaphyr**; fine-grained; crystalline; conchoidal fracture; consists of brown triclinic feldspar and diffused delessite, with abundant amygdules (mostly spherical) of delessite.

Narrow Amygdaloidal Seam.

25. **Melaphyr**; to the eye a fine-grained crystalline rock with uneven fracture, and consisting of reddish triclinic feldspar, with delessite both diffused and in flakes and grains; general color brown, mottled and veined with dark green. Contains some magnetite. Passes towards the hanging-wall into the overlying amygdaloid.

7. **Amygdaloid**; brown, with light-green amygdules of green-earth and quartz and calcite, containing copper. Matrix contains crystals of triclinic feldspar.

24. **Amygdaloidal Melaphyr**; abounding in amygdules of delessite. In the body of this bed, there occur portions of regular amygdaloid, with amygdules of quartz and calcite with green-earth. Small cavities in the amygdaloidal trap are lined with delessite, and this with clear, slightly pink crystals of feldspar; at the hanging-wall this passes into

6. **Amygdaloid**; compact and hard; brown, mottled green, with amygdules of green-earth and quartz.

85. **Melaphyr**; fine-grained towards the foot-wall; middle of the bed medium-grained crystalline trap; crystals of green and brown triclinic feldspar; delessite near the hanging-wall, almost foliaceous from the abundance of delessite.

8. **Amygdaloid**; green brown, with amygdules of green-earth, and cavities lined with minute crystals of feldspar. Carries considerable copper.

CHAPTER VI.

UNDERGROUND CROSS-SECTION AT THE CENTRAL MINE.

Beginning south of the "Greenstone," and proceeding from younger to older.

C*

$\overline{1}$ 50 ft. **Melaphyr:** dirty green; fine-grained and soft, with almost earthy fracture; $\frac{1}{4}$ to $\frac{1}{2}$ delessite [essential], the rest chiefly triclinic feldspar much altered; contains impregnations of calcite and specks of a copper-colored micaceous mineral with a red streak. Little or no magnetite.

In Adit.......... $\overline{2}$ 18 ft. **Amygdaloid:** chocolate brown; matrix predominant; micro-crystalline, hard; semi-conchoidal fracture; contains amygdules of *prehnite;* often colored red in centre by disseminated copper; also amygdules of calcite with prehnite. This bed passes into the underlying melaphyr (C $\overline{3}$).

$\overline{3}$ 45 ft. **Melaphyr:** dark red near the top; fine-grained; uneven fracture; has scattered amygdules of delessite and of a whitish steatitic mineral; lower down this rock becomes brownish green, very fine grained and hard, with conchoidal fracture, and contains scattered irregular particles of delessite, and in places amygdules of prehnite and calcite; large *bunches* of amygdaloid occur in this rock, with brown matrix containing round amygdules of calcite and green-earth; at the bottom of the bed the rock is mottled, brown and green, the green portions being mostly delessite, while the brown are similar to the lighter spots in the greenstone, consisting of triclinic feldspar, very much fractured; the rock here contains considerable magnetite and impregnated calcite. In small veins occurs the following series: 1. Laumontite. 2. Analcite.

---

* The numbers in the first column refer to Pl. XXIII. of Atlas, where the corresponding figures indicate the points from which the specimens were taken.

	C		
90 fathom level...	4	16 ft.	**Amygdaloid**: brown; fine-grained, compact, in places hard; contains bunches of prehnite with copper, also amygdules with calcite, veins of laumontite, and very irregular cavities coated with crystals of feldspar. This passes below into C 5.
90 fathom level...	5	55 ft.	**Melaphyr**, similar to C $\bar{1}$, very fine-grained, speckled brown and green; under the glass three ingredients appear about equally divided, viz., dark green delessite, a glistening white mineral, apparently altered feldspar, and a porous, red substance, possibly an altered product of delessite; the joints are glazed with delessite.
90 fathom level...	6	15 ft.	**Amygdaloid**: brown matrix, containing amygdules of calcite and delessite.
90 fathom level...	7	40 ft.	**Melaphyr**, similar to C 5; towards the middle and bottom of the bed it becomes hard, with conchoidal fracture, and brown color predominating through comparative absence of chlorite.
Adit level.......	$\overline{10}$	15 ft.	**Amygdaloid**: brown, compact matrix, filled with numerous amygdules, and seams of a light bluish green, soft, alteration product; the same bed has similar character at the 90 fathom level.
Adit level.......	11		
90 fathom level...	12 and 13	60 ft.	**Melaphyr**, identical with C 5. This very thick bed is remarkably uniform in character, and presents the same appearance in every respect at the 90 fathom level and the adit.
90 fathom level...	14	15 ft.	**Brown Amygdaloid**, with scattered amygdules of prehnite containing copper, and others consisting of a more or less porous light-green, chlorite-like mineral, perhaps an alteration product; the vein made mass copper in this bed.
90 fathom level...	15	37 ft.	**Melaphyr**, similar to C 5; small mass here in the vein.
90 fathom level...	16	17 ft.	**Amygdaloid**: very hard, compact, brown matrix, with amygdules of prehnite and others of calcite, with green-earth.
90 fathom level...	17	45 ft.	**Melaphyr**, resembling C 5, but with perfect conchoidal fracture; vein made a mass in this bed.

	C		
90 fathom level...	18	22 ft.	**Amygdaloid**: matrix brown, in places hard; in the uppermost softer portions of the bed frequent amygdules of a white and green mineral with radiated structure, probably altered prehnite; in the harder portions frequent veins of prehnite and delessite carrying copper, also, in places, epidote, and cavities lined with feldspar.
90 fathom level...	19	5 ft.	**Melaphyr**, similar to C 5, with isolated small amygdules of delessite.
90 fathom level...	20	5 ft.	**Amygdaloid**: compact, brown matrix, with amygdules of epidote and others of red feldspar and delessite.
90 fathom level...	21	5 ft.	**Melaphyr**, similar to C 5.
90 fathom level...	22	12 ft. (22–24)	**Amygdaloid**: matrix brown, containing irregular amygdules of prehnite, more or less altered, with calcite, green-earth, and copper; passes below into amygdaloidal melaphyr, containing irregular amygdules of prehnite.
90 fathom level...	23		**Amygdaloid**, in irregular bunches of prehnite with copper, the prehnite apparently undergoing change to delessite. In this bed the feeders of the vein consist of coarsely crystallized calcite, having irregularly distributed feldspar and delessite.
90 fathom level...	24		**Amygdaloid**, with amygdules of altered prehnite and feldspar and delessite.
90 fathom level...	25	10 ft.	**Melaphyr**, resembling 5, with less delessite.
90 fathom level...	26 to 34	10 ft. 10 ft.	**Porphyry Conglomerate**, with epidote cement containing disseminated copper in both pebbles and cement. Vein in this conglomerate carries large crystals o calcite.
90 fathom level... / 90 fathom level...	35 / 36	16 ft.	**Amygdaloid**: South Pewabic and Ash-bed type; the amygdules are almost wholly calcite.
90 fathom level...	37 / 38 / 39	45 ft.	**Melaphyr**: very fine-grained; in the upper portion purplish brown matrix, thoroughly impregnated with minute irregularly shaped particles of a soft white mineral containing small isolated impregnations of calcite. The middle portion of the bed is a dark gray micro-crystalline rock. Under the glass it is seen to be somewhat impregnated with delessite, and to contain, in scattered particles, a copper-colored micaceous mineral, which gives a dark red streak. The lower portion of the bed is minutely mottled, gray and green, is fine-grained, and contains a considerable amount of the red mineral above mentioned.

C.		
90 fathom level... 40 At No. 4 Shaft...	15 ft.	**Amygdaloid:** brown matrix filled with very small amygdules of green-earth, of green-earth and calcite, and of a very red feldspar; it carries finely disseminated copper. Feeders of the vein observed in this rock consist of prehnite with younger sheet-copper and calcite.
90 fathom level... 42	68 ft.	**Melaphyr:** chocolate brown, very fine-grained, with even fracture, containing small crystals of triclinic feldspar in the matrix, and disseminated particles of the copper-colored micaceous mineral above referred to; and also isolated small amygdules of delessite at the 90 fathom level. This bed is intersected by a cross course, immediately beneath which there was found a large mass of copper.
90 fathom level... 43	20 ft.	**Amygdaloid:** green and brown matrix, with amygdules of green-earth, calcite, copper, and analcite.
90 fathom level... 44	34 ft.	**Melaphyr:** fine-grained, gray-green matrix, speckled red with the copper-colored micaceous mineral above referred to, and containing considerable magnetite; the joints of this rock are coated with laumontite.
90 fathom level... 45		**Amygdaloid.**
90 fathom level... 47		**Melaphyr:** fine-grained greenish-gray matrix, speckled red with the copper-colored micaceous mineral. Feeders of the vein as observed in this rock consist of red feldspar and calcite, with a green, talc-like mineral.
90 fathom level... 48		**Melaphyr,** resembles C 5, but harder and with perfect conchoidal fracture.
90 fathom level... 49	16 ft.	**Amygdaloid:** brown matrix, filled with amygdules of prehnite more or less altered, and of calcite with green-earth. Feeders of the vein as observed in this bed consist of prehnite filled with native copper.
90 fathom level... 51	43 ft.	**Amygdaloidal Melaphyr:** very hard, brown, fine-grained matrix, showing numerous minute crystals of triclinic feldspar, and containing isolated amygdules from ¼ to ¾ inch diameter of prehnite, apparently undergoing a change to delessite; below this passes into C 52.
90 fathom level... 52		**Melaphyr,** similar to C 5; on the 90 fathom level the vein in the members of this bed was exceedingly rich in mass copper, while at the 120 fathom level the vein was reduced in the same bed to a narrow brecciated mass, consisting of small pieces of wall rock surrounded and sustained by thin seams of laumontite.

Location	C	Thickness	Description
90 fathom level... At No. 2 shaft...	53	28 ft.	**Amygdaloid**: soft, brown matrix, filled with round amygdules of calcite and altered prehnite.
		68 ft.	**Melaphyr.**
		25 ft.	**Amygdaloid.**
		45 ft.	**Melaphyr.**
120 fathom level.. 200 ft. N. No. 2 shaft.	54	50 ft. (54–58)	**Amygdaloid**: matrix purple brown, in places very soft, almost uniformly filled with amygdules of prehnite.
120 fathom level...	56		**Melaphyr**: very hard, uniformly dark greenish gray, very fine-grained, with conchoidal fracture; exhibits under the glass scattered particles of a copper-colored micaceous mineral.
120 fathom level...	57		**Amygdaloid**: brown matrix with isolated amygdules of prehnite with delessite; approaches an amygdaloidal melaphyr; probably not a persistent bed.
120 fathom level...	58		**Amygdaloidal Melaphyr**: hard brown matrix, isolated amygdules of prehnite, delessite, calcite, and native copper.
120 fathom level... and At No. 2 Shaft...	59 and 60	68 ft.	**Melaphyr**, same type as C 5, though harder, with perfect conchoidal fracture.
130 fathom level... 75 ft. N. No. 2 shaft.	61	9 ft.	**Amygdaloid**: soft brown matrix with irregular-shaped amygdules, with prehnite, analcite, and native copper.
	62	30 ft.	**Melaphyr**: dark green, very fine-grained matrix, apparently rich in delessite.
		9 ft.	**Amygdaloid.**
	63	30 ft.*	**Melaphyr**, same type as C 5.

* This is geologically the lowest point reached.

CHAPTER VII.

GENERAL STRUCTURE AND LITHOLOGY OF THE EAGLE RIVER SECTION.

BY A. R. MARVINE.

THE width of the "Trap Range" of Keweenaw Point, measuring from Lake Superior at the mouth of Eagle River across to the unconformable sandstones forming the south-eastern border of the Point, is about six miles. Topographically it here consists of two ranges of hills, which conform in direction with the general trend of the strata and of the Point, and while rising gently on their northern, fall more steeply upon their southern slopes. Locally, these are known as the North, or Greenstone, and the South Ranges.

The former, at an average distance of two miles from the lake shore, attains a height of from 400 to 700 feet, when it falls abruptly, even precipitously, to the longitudinal valley separating the two ranges. The South Range occupies the remainder of the trappean series, and while rising to a height of 600 to 800 feet above lake level, has a more gently moulded contour than the North Range.

The drainage waters of the intermediate valley accumulate in the branches of Eagle River, and flowing east and west at the southern base of the North Range, unite at the Phœnix mine, where, turning, they pass through a break in the Range and flow northward across the formation to Lake Superior.

Except in scattered exposures, the South Range lies buried under glacial and modified drift, while the frequent exposures of the North Range, together with the channel of Eagle River, offer excellent opportunities for a geological section.

The accompanying section, in two plates, exhibits in detail the section of the rocks of the North Range taken at this point, while the geographical relations of the exposures correlated in the section are shown in the accompanying map.

The sections are numbered in Roman numerals, and their limits —with the corresponding numerals—indicated on the map. The map and sections may be still further connected by means of the lettered survey stations and the numbers of beds, commencing at the northern end, placed on each.

General Geological Structure of the Section.—In the immediate vicinity of a line drawn from the Phœnix mine to the mouth of Eagle River, the rocks are very regularly bedded, the strata having a trend of slightly more than sixty-two degrees east of north, and dipping toward the Lake, or north north-west, at an average angle of about thirty-one degrees, being somewhat steeper at the north, and less steep toward the south. Upon either side of this line, the strata, participating in the general bending of the Point, swing slowly around, until, in a mile or more, their strike differs by two or three degrees from that of the main line of section. Moreover, in leaving the section toward the east, a flattening of the strata takes place, the dip decreasing about a degree in a mile. It probably increases slightly toward the west. Besides the geographical widening of the formation toward the north, owing to this flattening of the strata, there is also a widening from the actual thickening of some beds, or from the wedging in of new ones. Even within the limits of the map the formation thickens considerably in going eastward.

This system of rocks may be divided into three groups: sandstones or conglomerates; melaphyrs, including amygdaloids; and diorites.

The latter stand alone, occurring in a single belt of 2,400 feet in width. Being much harder and more massive than the adjoining rocks, they form the highest and southern portion of the North Range. Locally, the term "greenstone" is applied to the finer varieties of the diorites, and "feldspathic trap" to some of the coarser varieties, thus distinguishing them from the melaphyrs or "traps."

Both above and below the belt of diorites, and perfectly conformable with it, are the melaphyrs. To the south, excepting occasionally intercalated narrow beds of conglomerate, they are said to extend to the limit of the range. Northward they extend to a distance of nearly 5,800 feet, or to within a quarter of a mile of Lake Superior. For the first 3,500 feet of this distance they are comparatively free from any true sandstone strata, only four sandstone seams occurring. From this point northward, however, a series of sandstones occur, interbedded with the melaphyrs and amygdaloids. These, in their aggregate thickness, amount to over one-third of that of the including melaphyrs, but, as a rule, increasing in thickness toward the north, they there dominate over them.

Resting on this interstratified series of melaphyrs, amygdaloids, and sandstones is a wide belt of coarse conglomerate extending to and under the Lake. Some miles eastward of the main line of section, this conglomerate has a width of about a mile,* a trappean belt of about two-thirds the width resting upon it, which is in turn overlaid by another wide conglomerate. The United States Lake Survey chart represents a reef about 3,400 feet off the mouth of Eagle River, which at one point rises to within six feet of the water's surface, and as it is stated to be of "trap rock," it probably belongs to the outer trappean belt. This would make the conglomerate at Eagle River from 4,000 to 4,500 feet wide.

The diorites, as before stated, form a massive group of rocks about 2,400 feet wide, or 1,200 feet thick. They are heavily bedded in strata of from 20 to 150 or more feet in thickness, and without the interpolation, so far as known, of any amygdaloidal bands. Excepting a very massive bed at the base, and a thinner one upon the top, all the interior two-thirds of the series have a coarse-grained texture, especially so as compared with the adjoining melaphyrs. This texture varies from nearly that of the coarsest melaphyrs to beds in which the component crystals average one-eighth of an inch or more square, while the specific gravities are generally greater than those of the melaphyrs, and range from 2.89 to 3.03.

These beds are composed essentially of nearly opaque, lustrous, black to greenish-black, sometimes inclined to translucent resinous-brown, *hornblende*, and *triclinic feldspar*. Two kinds may be observed: the first, in which the feldspar is white, inclining to grayish-green, and sometimes red; fuses rather readily before the blow-pipe in coarse splinters; has a specific gravity, in one bed, of 2.73; weathers less easily than the hornblende; and is probably *Oligoclase*, or Albite. Its presence gives rise to handsome gray or greenish-

---

* See Foster and Whitney's Report on the Geology of the Lake Superior Land District, Part I., Map. Washington, 1851.

gray, and sometimes mottled rocks, which may be designated as the "lighter type." In the second variety of diorite the feldspar is very brittle, hard, colorless, or inclined to green, and sometimes to resinous brown, and is transparent; fuses before the blow-pipe only in very fine splinters; weathers more readily than the hornblende; and is probably *Anorthite*. Its presence, owing chiefly to the fact that the hornblende shows through it, gives rise to dark-colored, and generally dirty, resinous-hued strata, which may be designated as the "dark type." In both types a dark green chlorite, probably delessite, occurs to a greater or less extent, occasionally nearly wholly replacing the hornblende, from which it seems to be derived, thus transforming the bed into a coarse, loosely textured melaphyr, not especially like the typical melaphyrs above and below the diorites. Segregations are quite frequent, especially in the dark-type beds. In these the feldspar generally occurs in large crystals, but irregularly, and white and opaque, like that of the lighter type, and usually imbedded in hornblende, and intersected by it. Indurated green spots of quartz and epidote also occur, and occasionally prehnite and a bronze-colored mica. In the dark type also, larger crystals of the glassy feldspar often occur. These generally contain scattered particles of the usual ingredients of the rock, often to such an extent that the crystals themselves cannot be recognized except by the reflection of the light. This tendency to a larger crystallization in the mass is not so apparent with the hornblende, except when brought out upon the weathered surface of the rock, where the feldspar crystals have been removed, while the projecting hornblende crystals, often over an area of one or two inches square, all catch and reflect the light simultaneously.

The strata of the dark type predominate in the group, especially toward the bottom, form the more massive beds, and have, as a whole, the finer texture, never reaching the coarsest. The lighter type, on the contrary, occurs in thinner beds, which generally have a coarser texture, reaching the coarsest. Other things being equal, the latter are the more enduring of the two, but their comparative thinness and coarseness of texture generally reverses this character. The beds of the lighter type generally weather rather evenly, but with a rough surface, the color being a handsome gray, sometimes mottled, and speckled with projecting black hornblende crystals. The beds of the dark type sometimes also weather evenly and rough, with projecting crystals, but generally the weathering is rugose, often very exaggerated, the depressions being rusty brown in color, smooth and free from crystals, while the projecting knobs, often an inch high, are whiter, but sprinkled with projecting black hornblende crystals, those on each knob generally all reflecting the light at once to the eye.

In one bed (No. 107) the two types are banded together in strata of from 5 or 6 inches to 5 or 6 feet in thickness, with no perfectly sharp lines of demarcation separating them. The lighter type layers were here exceedingly coarse, and their granular components could be separated quite well from one another.

The hornblende grains often showed very lustrous cleavage planes, some of which could be measured. In one of these, measured on the reflecting goniometer, the cleavage angle of Pyroxene (Augite), viz., 87° 5′, was obtained. Several gave the undoubted angle of hornblende, 124° 30′; while, in general, those which were not sufficiently perfect to measure accurately were not far from the latter angle. An analysis of this material was made which, though not wholly successful, showed conclusively that the mineral was hornblende,—the amount of lime in it not exceeding 10 per cent. Its specific gravity was 3.39,—that of the feldspar 2.73, and of the rock 3.02, showing that about 57 per cent. of the latter was feldspar and 43 per cent. hornblende.

Commencing several hundred feet from the base, there is a coarse-grained, hard, dark, resinous-hued bed. From here downward a gradual change occurs, the texture growing finer and finer, the feldspar element seeming to increase, giving a gradually increasing crystalline character to the bed, until the components finally become undistinguishable, and there is formed at the base 200 or 300 feet of an exceedingly brittle, elastic, hard, compactly crystalline aphanite, having a conchoidal fracture, and handsome clear green color, sometimes mottled vaguely with purple, and with occasional large, flat, transparent crystals of glassy feldspar imbedded in the mass. When mottled, the weathering is of a uniform rugose character, the silver-gray projections being composed of grains of glassy feldspar, sunk in which are minute crystals of hornblende, all of which on each knob, as on the larger, coarser knobs, before described, reflect the light simultaneously at the same incident angle. This shows that the rock is still of the diorite type, and that, though changed in its physical character, it has not changed materially in its chemical. The specific gravity of the homogeneous aphanite is 2.95. For 15 feet above the foot-wall of the "greenstone" its eminently crystalline character is lost, the rock looking much like some fine-grained and compactly textured melaphyrs; the specific gravity, however, still being 2.92. The lower 3 or 4 feet are broken and fissile, and are followed by the red-clay seam called the slide, separating it from the melaphyrs and amygdaloids of the Phœnix mine, or those lying below the diorites.

Commencing at the base of the uppermost bed of the diorites (No. 91), the rock is precisely similar to the purple-mottled aphanitic greenstone, and has the same specific gravity, 2.95.

It gradually changes, however, becoming at first somewhat coarser-grained, and resinous in lustre, with specific gravity 3.01, and finally a fine-grained, brownish, homogeneous rock, in all respects like some melaphyrs, and with a specific gravity of 2.89. This is separated by a distinct plane from a fine-grained, dark-bluish, semi-columnar rock, which forms the base of the melaphyrs lying north of the diorites.*

Melaphyrs and Amygdaloids.—The limits and position of the melaphyr group upon the section have already been indicated. In general characters these melaphyrs are precisely similar to those in the Portage Lake District; and the results given in the description of that district as to their general lithological characters, chemical or mineralogical composition, accompanying minerals and their paragenesis, etc., are equally applicable, in a general way, to the melaphyrs of this district. In fact, the region lying north of the diorite group is in the same horizon as the western end of the Portage Lake section, and there is not only a strong resemblance, but, so far as traced, a general accordance between them; many beds in the two being lithologically, probably stratigraphically, identical.

---

* A much more complete knowledge of these diorites may be had by referring to the detail descriptions of the typical and peculiar beds of the series in Chap. VIII., as appended. Dark type, No. 92, p. 133; light type, No. 96, p. 134; coarse, inclined to melaphyr, with large red feldspar crystals at base, No. 94, p. 134; coarse, light and dark types banded, No. 107, p. 135; dark type, changing to aphanitic, No. 108, p. 136; aphanitic, changing to melaphyr-like form, No. 91, p. 133.

The melaphyrs lying above or north of the diorites do not separate themselves into any apparent natural groups, but for convenience of reference they may be divided into three series:—(*a*), Those extending from the diorites north to the bed next below the Ash-bed, or No. 66; (*b*), from No. 66 to the Upper Falls, or through No. 45; (*c*), the remainder of the melaphyrs northward to the wide conglomerate, and containing ten intercalated sandstones. The respective widths and thickness of these three series are—(*a*), 1840 (± 925)* feet; (*b*), 1,200 (618) feet; (*c*), 2,700 (1,417) feet, with sandstones aggregating above 860 feet.

(*a*.) The most characteristic and prominent beds of series (*a*)—mostly confined to its northern 580 feet—are composed of rather coarse and not closely textured melaphyrs; tough, but not brittle, and breaking with a rough fracture. The colors are dirty light green, strongly mottled with quite well-defined spots of dark green, causing them to approach amygdaloidal melaphyrs in structure, while they have a medium specific gravity of about 2.87. Their capping amygdaloids are not very strongly developed. South of these beds finer-grained, darker, and more evenly colored beds seem to prevail, but the covering of soil renders this the least known of any part of the section. Near the base a wide bed (No. 87, 350 feet wide), similar to those at the top, occurs, below which, and at the base, is a wide (175 feet), hard, homogeneous, fine-grained, clear bluish-black melaphyr, which has a quite well-defined and striking semi-columnar structure.

The middle division (*b*) of the northern melaphyrs has at both its base and summit a porphyritic bed, composed of small semitransparent crystals of greenish-white triclinic feldspar, generally without definite form, profusely sprinkled in a very fine-grained base of purplish—or purple lightly mottled green—color, and containing considerable magnetic iron. It is a hard, elastic, brittle rock, with rather smooth and semi-conchoidal fracture.

These are the only two marked porphyritic beds among the melaphyrs of the section, and are also somewhat more brittle and elastic than any others, approaching in these two qualities the aphanitic diorites. They also have a high specific gravity, 2.93 and 2.91,

* Numbers in parentheses indicate generally *thicknesses*, or distances measured perpendicular to the bedding.

respectively. Each has a scoriaceous amygdaloid upon its summit, separated from the main bed by rather a sharp line of demarcation. Situated between these two extremes are a variety of beds, mostly abnormal in composition. Commencing at the base there is the compact dark-bluish melaphyr, with the associated scoriaceous amygdaloids called the "Ash-bed;" then a semi-columnar trap with wide, coarse amygdaloid above; then a width of curious, coarse, easily decomposing beds of very irregular structure, followed by many very thin, but perfectly bedded amygdaloids with narrow intermediate amygdaloidal melaphyrs.

The northern series (*c*), extending north to the wide conglomerate, and containing sandstones aggregating probably 860 feet in width, has, for its predominant type of melaphyr, a rather fine-grained and compactly textured rock, breaking with an irregular to semi-conchoidal fracture, and somewhat brittle and elastic. The color is generally a dark but dull green vaguely mottled purple, the latter often predominating, having in it a mottling of green, and occasionally all is green, a darker shade mottling a lighter background. Though the purple element enters largely into the colors of these rocks they are inclined to dark shades, being darker than the normal green rocks of series (*a*), while the mottling, though striking, is not so well defined or spotted, nor of such dark green color. They are also finer grained, more brittle, not so tough, nor with so rough a fracture, and show a closer texture. The quantity of magnetite in this series seems quite small, and much less than in (*a*), while the specific gravity may average lower—2.71, 2.76 to 2.89.

As exceptions to this general nature we have one light green (No. 1) and two grayish beds (Nos. 9, 10), while No. 39 is of quite exceptional character.

In amygdaloidal melaphyrs delessite is nearly always the mineral that first occupies the amygdules, the matrix or base generally remaining much the same as the accompanying melaphyr. Laumontite generally follows closely after, but north of the diorites it occurs far less than in the Portage Lake District, series (*a*) being nearly free from it, (*c*) having but little more, though considerable occurs in (*b*). Calcite and prehnite occupy the position next most important to delessite in the amygdaloidal melaphyrs. In their most developed stages in series (*a*) calcite occurs in the delessite amygdules, nearly to the exclusion of prehnite, but in (*c*), especially all through its middle parts, prehnite occurs in the amygdaloidal melaphyrs in most characteristic pink-and-salmon colored radiated amygdules, to the exclusion not only of calcite, but almost entirely of even delessite.

In the amygdaloids throughout, calcite largely predominates, but is often closely followed by prehnite (the latter generally enclosing the former), which is generally light green, semi-transparent, and botryoidal, and almost always accompanied by small amounts of copper.

Quartz follows next in importance, and those amygdaloids containing it in larger quantities are generally more or less indurated. The matrix of amygdaloids is seldom similar to the underlying melaphyr, being almost always very fine grained to compact, sometimes inclined to earthy, sometimes indurated and hard, and generally reddish-brown in color, with sometimes red crystals of feldspar porphyritically imbedded in it, while in the harder amygdaloids, the base is often green in color.

South of the diorites a few hundred feet of rock are exposed in the Phœnix mine. In the upper part of this series the melaphyrs, which all approach amygdaloidal melaphyrs, are of a rather coarse-grained, inclined to loose, texture, being rather soft and tough, and having a very uneven fracture. The feldspar is greenish-white, and occurs in elongated crystals, which are sometimes twins, and in the coarser varieties, appear spread like a network upon a background of dark-green delessite and pinkish feldspar, the delessite being separated into frequent and rather well-defined spots, while magnetic and specular iron are present in quite large quantities. In the finer-grained beds brownish-red largely prevails, with dirty white in the base, the colors, though dull, as in nearly all rocks, being well marked and in strong contrasts. At one or two points, the whole rock becomes a nearly uniform, decided, brownish-red color, and as no magnetic iron then seemed present, the color may be due to the decomposition of the iron ingredients of the original rock. In looseness of texture, and consequent rough and uneven fracture, these rocks somewhat resemble the typical melaphyrs of series (*a*), but in colors they are decidedly dissimilar. The amygdaloids are but little more than exaggerated forms of the amygdaloidal melaphyrs, the matrix not being compact but fine-grained, and of a dull brownish-red color, with but little calcite or prehnite with the delessite amygdules. Toward the south the melaphyrs become finer grained and more compactly built, harder, more brittle and elastic, and with more even and conchoidal fracture, while the color darkens, a dark green predominating, indefinitely mottled with dark dull purple. There is but little or no magnetite. The mottling spots may be .3-inch diameter, though their limits are variable and indefinite, and they may merge into one another. In part of one bed (No. 120) they reflect to the eye, when the fractured face is held at the proper angle, a glistening or sheen, which causes them to strongly resemble a similar modification which sometimes occurs near the lower part of the aphanitic diorites. The latter, however, is harder, and in texture inclines to compact crystalline rather than to fine granular, while the color is a clearer green.* The specific gravity of the former is 2.84-2.87, of latter 2.92-2.95. The amygdaloids of this part of the mine are much better developed than those at the northern end, calcite predominating in some, prehnite in others, and delessite and laumontite in the amygdaloidal melaphyrs.

One amygdaloid (No. 111), lying about between these two series, is pretty well developed, and where examined contained much copper. The prehnite of the amygdaloids, especially in this part of the mine, seems to be readily attacked by the decomposing agencies at work, changing to a soft, white, amorphous chalk or kaolin-like material, the larger amygdules sometimes having a partly unchanged prehnitic nucleus and exterior.

Certain changes which have been suggested in the preceding pages as occurring in the melaphyrs may be of sufficient interest to warrant a more detailed description.

Being of the same general nature in all, it may be convenient, in examining them, to select those features presented by the characteristic rocks of series (*a*), the coarser character of which enables their components to be more readily detected and their changes examined, referring, when occasion requires, to other beds which may differ from those under consideration.

---

* At the Allouez mine these distinctions are nearly lost, and the "greenstone" is so changed that, were it not known to be the equivalent of the "greenstones" farther northeast, it would certainly be considered as a melaphyr.

Upon breaking the rock at various points, a rather rough fracture is obtained, indicative of a rather coarse and not very closely compacted texture, while the color is decidedly a green, generally a dirty light shade, mottled or even spotted with a dark shade.

The normal form, however, seems to tend toward a uniform shade lying between the two, though the mottling is even here hardly lost. This normal form does not necessarily lie at, nor next to, the bottom of a bed, for variations from it are somewhat irregular, though generally arranged parallel with the plane of bedding; but it seems never to occur near the top and adjacent to the overlying amygdaloid.

Taking the most uniformly colored rock, the glass discloses it to be composed of small crystalline grains of light green feldspar, thickly scattered in, or rather mixed with, a dark green, soft, amorphous, generally earthy, sometimes shining mineral, which is undoubtedly a chlorite, and which, according to Macfarlane's analyses, would seem to be delessite.

Here and there, but wholly subordinated to these two principal ingredients, are shining flakes, sometimes set in a dull reddish spot, which have a brown streak and iron reaction, and which probably are specular iron, while grains of magnetite occasionally appear. Still more sparingly, small black crystals, as if of hornblende, occur in the delessite, and sometimes flakes of rubellan (?).

The feldspar sometimes assumes a pinkish hue, which, with the green, produces a subdued but decided purplish shade. This change is seldom uniform, but confined to irregular spots, giving a rude purple mottling upon the green base, an extension of which produces a purplish base with a green mottling.

This tendency, however, is almost *nil* in the coarser (*a*) rocks, though always present in the typical (*c*) rocks; in fine-grained rocks a suggestion of it is almost always present, which sometimes, by going through almost imperceptible changes, may have a dull, uniform, purplish-brown rock.

Returning to the rocks of the (*a*) series, which to the eye approach a uniform texture and color, there is found, a few feet farther on, generally above, a tendency in the delessite to segregate into spots to the exclusion of the feldspar. These spots at first are of decidedly indefinite and irregular boundary, and tend to produce the effect of mottling; but farther on they are more definite and less irregular in shape, though seldom perfectly round, till, finally, the well but not sharply defined spot of earthy-looking, dark green delessite stands on a background of nearly uniform shade or uniform mottling of shades. This background in these rocks is of a dirty light or grayish-green, and is composed mostly of feldspar, though much delessite still remains. In the change, however, the feldspar is apt to assume a pinkish shade.

It is through some such changes as these that the rock always passes in approaching the form of the amygdaloidal melaphyr, into which is an easy transition; the delessite spots becoming more definite amygdules, generally accompanied by the entrance into them of some laumontite, calcite, or prehnite.

The matrix generally remains much the same as the melaphyr below, though the feldspar is apt to become pink or red, and often porphyritically imbedded in a fine-grained mixture of feldspar and delessite, very occasionally in delessite alone. If laumontite and other minerals are present together in the amygdule, the former is always the oldest, following the delessite (see paragenetic series). The entrance of these minerals seems to be by a replacement of the delessite.

From a bed in the southern part of the Phœnix mine, containing generally delessite amygdules, many amygdules could be easily broken out intact. Some of these amygdules averaged an inch in diameter, somewhat irregular in shape, and outwardly of shining green delessite.

Upon breaking them open, however, they showed a variety of structures. One seemed to be wholly composed of what appeared like a crystalline granular mixture of white calcite and pink laumontite in a light green base. Much of this base is of dark green, as if of the original delessite filling, some of which still coats the amygdule, while the lighter green seems to contain prehnite. Laumontite crystals sometimes enclose dark delessite, while calcite sometimes encloses, in a thin coating, laumontite. The cleavage faces of the calcite particles throughout the amygdule all reflect the light at the same incident angle, showing a uniform tendency to crystallization throughout it.

In another amygdule much of the periphery was like the above, but, instead of filling the amygdule, it rapidly changed into pink prehnite, which in time became green in the interior, some spots of delessite still remaining and particles of copper being present. In a third case, the outside was much like the first, except that the light green matrix was nearly all prehnite;—this changed into a prehnitic interior like the second. Again, a portion of the periphery was composed of very small grains of delessite and laumontite, changing along an irregular boundary into a radiated laumontite core. In the accompanying fine-grained rock may be seen incipient delessite amygdules of various sizes and stages of development, some having the usual laumontite core.

The gradual development of a delessite amygdule in a melaphyr seems undoubted. From the above facts it would seem that the genesis of the laumontite was also a true metamorphic change, and took place, not by the filling of a cavity, but by direct replacement of delessite; while the incoming of prehnite and calcite would seem also to belong to the same class of changes, probably by the replacement of laumontite.

In the characteristic amygdaloidal melaphyrs of the (*c*) rocks, it has been already noticed that while a purple green mottling is nearly always present upon the rather fine-grained and closely textured rock, it seldom seems to develop, as in certain beds of the (*a*) series, into delessite amygdules. The amygdules, instead, contain a salmon-colored radiated prehnite, in appearance strongly suggestive of radiated laumontite, but which shows, in many instances, a gradual transition through vitreous pink to transparent green, non-radiated prehnite. Though there is no direct proof of the genesis of this prehnite, the facts noticed above, together with the occurrence of laumontite associated with it in a few beds (Nos. 18 and 30) would indicate that it was all derived by replacement of laumontite, which, in turn, replaced delessite. Perhaps the very resemblance of the radiated prehnite to radiated laumontite may indicate a step in this change. It seems strange, if such transition has taken place, especially when its large extent is considered, that it should not have left an easily discovered record of the intermediate steps. The beds having this character, however, contain numerous sandstones, which would readily allow of the penetration of metamorphosing agents.

Some one of the transitions from melaphyr to amygdaloidal melaphyr, as have been described, almost always occupies a more or less wide zone between a melaphyr and its overlying amygdaloid.

It also often occurs within a melaphyr, though generally in a zone parallel with the plane of bedding. But besides these cases, it is also found *cutting directly across* the bedding, in which case it accompanies a vein, joint, or fissure, from which it may extend several feet either side, the intensity of the action being greatest near the vein. The change here is just the same in kind, and often in degree, which takes place beneath an amygdaloid, amygdules of delessite and laumontite being common, sometimes with prehnite and calcite, though generally not well developed. These facts, especially the latter, point to the conclusion that the transitions from melaphyr to amygdaloidal-melaphyr are of a truly *metamorphic* character, that they were impressed upon an approximately homogeneous rock by chemical action long after it was first formed, and that the amygdules which mark this change were thus slowly developed, and are not the mere fillings of pre-existing cavities.

In passing to the consideration of the amygdaloids, it becomes more difficult to clearly distinguish between that part of their structure which existed in the original rock and that which is due to subsequent metamorphism. From what we have seen, it seems almost possible that metamorphism, by carrying the process a step beyond the formation of an amygdaloidal melaphyr, might produce an amygdaloid. The differences between the two have already been sufficiently pointed out. After the number, irregularity, contents, or general development of the amygdules in an amygdaloid, their next most noticeable characteristic, and one much less easily accounted for by metamorphism, is the almost invariably fine-grained, compact, or even earthy nature of the base, no matter how coarse the underlying melaphyr may be. But the strongest proof would seem to be in the structure of the so-called scoriaceous amygdaloids. In these, patches or balls of amygdaloidal material are associated, even surrounded by, an imperfectly stratified material, which is undistinguishable from the true fine-grained sandstones.* This association is such that it seems as if it could in nowise be accounted for by metamorphism acting on sedimentary beds, but only by supposing a peculiar mixture of the materials at the time of deposition,

* The nature of this material is set forth in speaking of the sandstones, p. 113, as well as the association mentioned, which is also more especially noticed in detail descriptions of beds No. 45, p. 125, and No. 65, p. 129.

which mixture is not such as sediments assume. In these amygdaloids, also, there is sometimes a well-marked plane of demarcation between the amygdaloidal melaphyr—or the amygdaloid, perhaps—and the scoriaceous amygdaloid which rests upon it. The fact that in sandstones which are intercalated between two trap beds the upper parts, for several inches from the hanging wall, are often changed as if by heat, while at the bottom contact there is no such change, cannot be offered as an objection to the metamorphic theory, for it would be in just such regions that metamorphism would naturally occur. But the fact that sandstone material seems to have entered amygdules near the upper part of beds covered by sandstones; that it may fill a well-defined crack extending down into an underlying melaphyr (see p. 113 and p. 119), or that melaphyr may nearly surround pebbles apparently caught up from an underlying conglomerate (p. 119); these facts, as does the peculiar structure of the scoriaceous amygdaloids above noticed, seem to point to a very different origin for the melaphyrs than a sedimentary one.

It would seem, then, that the fine-grained nature of the amygdaloids, together with amygdaloidal cavities, probably originated with the bed, though both must have been strongly affected by that metamorphism which has produced the more prominent features of the amygdaloidal melaphyrs. The amygdules must, in many cases, have been much changed, in size and shape, and subsequently filled, by this agency, while the small feldspar crystals often found porphyritically imbedded in the amygdaloidal matrix may also be due to this cause. Whenever well exposed, the melaphyrs generally exhibit, from a few inches to a foot or more from their foot-walls, a modification similar to the amygdaloidal melaphyr of the same bed, with sometimes a true amygdaloid at the very bottom. These three often shade gradually the one into the other, and, as at the top, it is impossible to accurately draw the line between what is original structure and what due to metamorphism.

As a whole, then, the *structural* features of these beds remarkably resemble those of true lavas. They have been affected, however, and to a very great extent, by metamorphism, and this metamorphism has taken place in such a manner, has so heightened and carried on the original structure, as it were, that the ordinary proof of their igneous origin, such as contact changes in adjacent sandstones, presence of amygdules, etc., fail, and it seems natural to consider this metamorphism as a *vera causa* for the whole structure. Certain extraordinary features, however, as noticed above, seem wholly incompatible with this idea, and when considered as true igneous rocks in which great and peculiar metamorphism has taken place, all the phenomena presented seem to be satisfactorily and naturally accounted for.

The general structural features of these beds long ago gave rise to the belief that they were of igneous origin, a belief now almost universal in the mining regions; while the metamorphic changes which have taken place in them since their formation are not generally recognized, if indeed suspected. These changes, however, have been both very many and very great; so great, in fact, that, as seen above, when once examined they seem almost sufficient to have developed all the peculiarities of the beds from sedimentary deposits.

The practical importance of the recognition of this metamorphism, and of a proper understanding of its methods and effects, will be apparent when it is recollected that to it is due all the economical value that the beds possess. The beds as originally formed probably contained the elements of its minerals, together with its copper and silver, more or less disseminated through their mass, as much so remains till the present day, or else they were so contained—at least in part—in overlying rocks, and in this form they could have been of no economic value; nor could any process taking place at that time have concentrated the minerals in the manner in which they now occur.

One hears of the melted copper being squeezed from the adjacent heated rock into the veins, and run into amygdules as bullets into a mould, the facts being ignored that—supposing the hydraulics of the case possible—these rocks, at the temperature of melted copper, would no longer be what they now are; and also that, at such a temperature and under a moderate pressure, a fissure, supposing it once formed, would be immediately welded solid again; while in amygdules copper is seldom, if ever, the only occupant, but occurs in them in company with minerals which are readily changed by heat, but which never show such change. The fact that chemically pure silver and copper constantly occur intimately united, but never in the least degree alloyed with one another, proves that they could never have been so placed by the action of heat, for it is well known that in melting they would have alloyed instantly. Suppose that, in an amygdaloidal cavity or vein, transparent crystals of quartz stand upon crystals or masses of green prehnite, fitting into their irregularities and touching nothing else, and that no replacement has taken place, it is clear that in such a case the quartz must have been placed upon and adapted to the form of the prehnite after the latter had been placed in its position and had assumed definite form. Prehnite, however, will almost melt in the flame of a common candle, while quartz is one of the most infusible of substances, and could not possibly have been placed upon the prehnite by the agency of heat alone without the fusion of the prehnite and the destruction of its form, to say nothing of the various other chemical effects that would be induced. Moreover, the quartz, after melting, is found to have a different specific gravity than when crystallized.

This is but one of many cases. In fact the whole manner of occurrence of, and general relations between, the many refractory and easily fusible minerals which occur in a similar way in these rocks shows that great heat has had absolutely nothing to do with arranging them as they now are. The result must be looked for in some other cause. Does chemical action supply such a cause?

Many experiments have shown that carbonic acid and oxygen held in solution in water can readily attack alkaline silicates, and that these, or similarly obtained substances, even in apparently insignificant quantities, are powerful agents in changing rock masses, provided sufficient time be allowed for them to act. Water falling upon the earth soon obtains these reagents from the atmosphere, decomposing vegetation, soil and rocks, as proved by the fact that all spring water which has travelled any distance through rock-masses always contains them. In this form, these agents are generally recognized as producing many metamorphic changes, and are known to have produced effects quite antagonistic to those produced by heat.

All the phenomena tend to prove that it is by means of some such chemical actions as these, continued through long periods of time, that the metamorphism of these beds has been effected. It is such metamorphism which has developed the amygdaloidal melaphyrs, formed segregations, modified and filled the veins and amygdules, placing in them their minerals in the present relative positions; and, in the general process, the copper, like the other ingredients, was selected from its disseminated and therefore useless condition, and concentrated in veins, amygdaloids, and conglomerates till it reached a percentage of richness that gives to the deposits an economical importance.

This action has taken place certainly not at a high temperature, and possibly at a temperature no greater than that of the beds at present, while it may have been largely aided by that electric action which chemistry almost always induces, and which is known to be active at the present day. In fact the presence of the latter is proof that chemical action is even yet going on.

This metamorphism, then, being the means by which these ore deposits have been formed, it is evident that a knowledge of its laws and methods should lead to important practical results. It is not necessary for me to say that the studies in Chapter III. go a long way in discovering these laws, and it is hoped that they may be still further developed until wholly known.

SANDSTONES AND CONGLOMERATES.—Interbedded in the northern 2,300 feet of the section are ten sandstones, which aggregate about 860 feet in width, and which, as a rule, may be considered as increasing in thickness and coarseness toward the north.

Commencing with the southernmost one and going north, we have—

Bed No.	Width in feet.	Character.
35	55	Rather coarse.
28	18	Rather coarse; inclined to conglomerate.
26	50	Very fine, shaly.
21	33	Very fine, banded, and shaly.
19	45	Coarse; some conglomerate.
17	120	Medium fine, shaly; some conglomerate.
8	170?	Rather coarse; some conglomerate.
6	280?	Coarse as far as known; much covered.
4	10?	? ?
2	282?	Coarse; considerable conglomerate; much covered.
		Above the Melaphyr occur—
−1	129	Coarse; with considerable conglomerate, coarse and fine.
−2	Over 4,000	Coarse conglomerate.

These are apparently all composed of the same materials.

The predominant pebbles of the conglomerates are of a compact chocolate-colored, felsitic rock, generally with small, lighter-colored crystals of a triclinic feldspar, porphyritically imbedded in it. Next is a coarser-grained, crystalline, hard, reddish, felsitic rock. Pebbles of quartz-porphyry occur sparingly, and occasional pebbles of a rock somewhat resembling some fine-grained melaphyrs.*

The sandstones seem made up of the same ingredients; they are generally of shades approaching brick red. All contain much calcite as an impregnation. This often occurs, filling small lenticular cavities, which may occur grouped together for a few inches parallel to the formation, and may have replaced a constituent of the rock; it also fills veins where they intersect the beds.

Where observed, the hanging-walls of the sandstones were generally smooth and gently undulating, but occasionally quite uneven, while the upper two to twelve inches were somewhat changed, being harder or softer, or lighter or darker-colored than the mass of the bed. In one instance, at the top of Bed No. 2, this change was so great as to render the sandstone difficult to distinguish from the overlying greenish melaphyr, except for the unchanged pebbles enclosed in it. The foot-walls are sometimes smooth and undulating; the surface of the underlying bed, when not an amygdaloid, as was sometimes the case, seeming to have been worn smooth, as if by attrition; or else the sandstone seemed to fill inequalities in the underlying amygdaloid.

The sandstones were not observed to be changed near the foot-wall. In one remarkable instance, a crack or fissure was observed extending down into a melaphyr, which was filled by the overlying sandstone. The conclusion is inevitable, that the melaphyr had formed, hardened, and cracked before the sandstone was deposited.†

Lying south of the sandstones there are several others scattered along the section, but which occur as mere seams of only a few inches in thickness, and generally on the surface of, and intimately associated with, the upper part of the so-called scoriaceous amygdaloids. Though these seams are generally too fine-grained to distinguish the form of their component particles, and no pebbles were observed, yet they undoubtedly are true sandstones.

---

* See description of Beds Nos. -2 and -1, pages 117 and 118.

† For this and the irregular metamorphosed hanging-wall, see figures on p. 119.

They are of the same color, and in places often, though irregularly, banded or stratified—in one instance breaking shaly, with wavy laminæ. Both contain much calcite as an impregnation, often in small lenticular cavities. Before the blow-pipe they behave like the sandstones, the harder varieties fusing with difficulty, the finer ones more easily, in fine splinters, and without intumescence, to a green-black mass. In some cases, where foreign material seems present, some intumescence occurs, but less than with the melaphyrs. Their specific gravities are the same as those of the sandstones, and lower than those of the melaphyr, ranging from 2.56 to 2.70, reaching in one case 2.80 (compact shaly).

The conglomerate pebbles are probably heavier, one of the chocolate-colored, non-quartziferous, felsitic porphyry having a specific gravity of 3.02.

The red-clay seam, a few inches thick, underlying the "greenstone" at the Phœnix mine, and known as the "slide," is the equivalent of the Allouez conglomerate, which a few miles either way along the formation expands to a thickness of 15 or 20 feet.

The "Kingston conglomerate,"—which was exposed by Mr. A. H. Scott about 500 feet south of the centre of Section 33, Town. 57, Range 31,—lies about 5,870 feet horizontally from the "greenstone," and is about 45 feet wide. This is entirely different from the preceding beds, and belongs to the quartz-porphyry type of conglomerate, as does the Calumet conglomerate. The predominant pebble is of a hard, compact, brownish-red, felsitic rock, containing quite numerous flesh-red to pearly-white crystals of orthoclase feldspar, occasionally over .25 inch in length; and grains of colorless, transparent quartz, ⅛ inch in diameter and under, which appear black from non-transmission of light. It strongly resembles the Kearsarge conglomerate exposed near Calumet, and if the mean dip of the formation between the greenstone and the Kingston conglomerate were about 25°, it would be the same bed as the Kearsarge. The dip, however, is probably steeper, so that it probably lies 600–700 feet east of the Kearsarge.

VEINS AND FAULTS.—The system of rocks described in the preceding pages is intersected by numerous veins, in many of which movement or faulting has taken place.

Two principal systems seem to exist: first, in which the trend is about N. 15° W. to N. 25° W., with a nearly vertical dip, but perhaps tending eastward; second, with trend perhaps averaging N. 16° E., dip also nearly vertical, tending eastward. Besides these, several veins have been noticed trending nearly with the formation, but having a much steeper dip.

Upon the first system more or less faulting has generally occurred, though some of the better defined veins show none. In those cases where opportunity was had to measure the amount of throw, the distance varied from a few inches to about 75 feet, the eastern side always having moved northward. In the large faults of this system, however, which are generally occupied by the larger transverse valleys, or breaks in the range, the detritus in which hides the actual fault and its amount of throw, the throw is obviously in the opposite direction,—the east side moving south,—and it is generally sufficiently great to more than counterbalance those faults which have thrown the strata in the opposite direction. The formation, which is slowly changing its trend—in going east—more and more to the southward, at some of these larger faults makes a sudden bend of a few degrees in the same direction, so that these large faults act both by offsetting and bending, in increasing the general curvature of the formation.

Faulting was not noticed in the second system, and a fault of only a few inches in one case in the third.

But movement has taken place upon the formation along certain planes of bedding separating the strata. The so-called "slide"—lying a few feet above the "Ash-bed"—is a case in point, the rocks above having slid down upon those below probably about 150 feet.

Motion also seems to have taken place upon the "slide" at the base of the Greenstone—the horizon of the Allouez conglomerate, here occupied by a clay seam—but probably to no great extent, it being said that veins are not displaced by it. Movement may have similarly occurred at other points as yet undiscovered.

The veins of the district were of course formed long after the consolidation of the beds, and probably when they were being lifted into their present position. They have subsequently been filled with the various minerals which now occupy them, wholly by infiltration and chemical, probably aided by attendant electric, action, and in a systematic and natural sequence, as shown by the paragenetic table. The more successful mines seem to have been worked in the first system of veins, on one of which the present Phœnix mine is now working. The "old Phœnix," however, on a similar vein, but farther north, was not successful. The work upon the Robbins and Armstrong lodes, both of the second system, was not remunerative in either case.

Besides vein-mining, considerable work was formerly done on the "Ash-bed," but without success. There is no apparent reason, however, why there should not exist beds in this region, either of amygdaloid or conglomerate, as near Portage Lake and elsewhere, which it would be profitable to mine for copper. A better examination of the "amygdaloidal floors" already exposed in the vein-mines might lead to profitable returns.

A system of vein and bed mining could be carried on simultaneously with considerable economy, the one helping the other: the beds generally carrying a low but nearly uniform per cent. of copper, the veins more generally containing larger but more erratic and uncertain "masses."

CHAPTER VIII.

DESCRIPTIVE CROSS-SECTION OF THE EAGLE RIVER DISTRICT.

BY A. R. MARVINE.

THE following chapter gives a detail description of the various beds which compose the Eagle River cross-section. Attention having been called in the preceding chapter to the more prominent features of the section, this chapter is intended only to be of service in prosecuting any mining enterprise, in which it would be useful to know the distances and the nature of the rocks to be passed, or of use in correlating or comparing in detail with any sections which future surveys may make in other parts of the copper region.

The initial point of measurement and numbers is at the foot-wall of the wide northern conglomerate and sandstone, lining the lake shore, or, what is the same thing, the hanging-wall of the northernmost melaphyr of the section, as exposed at the lower fall and dam just above the town of Eagle River. From this point the beds are numbered in regular order going south, while occasional larger numbers upon the left give the total *horizontal* distances of certain points, also measured from this initial point. Common Arabic numerals, also on the left, are used to express the *horizontal* widths of beds in feet, while numbers standing alone in parentheses always express actual *thicknesses* of beds, or parts of beds, measured *perpendicular* to the plane of bedding.

–2. **Conglomerate**; over 4,000 feet wide; variably bedded thick and thin; large pebbles predominating in some strata, coarse sandstone in others; latter sometimes cross-stratified, sometimes inclining to a shaly structure. Generally coarse, with predominant pebbles from 3 to 8 inches diameter, often 1 foot and more, cement being smaller pebbles and coarse sand. No angular fragments, many not perfectly rounded, those rounded affect oval form. Majority composed of (*a*) non-quartziferous porphyry; matrix, predominant color, liver to chocolate brown, compact to subcrystalline; crystals, subordinate, sometimes nearly absent, always scattering; small, generally $\frac{1}{80}$ to $\frac{1}{10}$ inch long; generally somewhat lighter colored than the base. Flesh-red crystals .1 to .5 inch long sometimes occur in chocolate-colored matrix. The crystals in these seem to be of triclinic feldspar. Subordinate to these occur (*b*) pebbles of a coarser, but compactly crystalline flesh-red felsitic rock, the light reflecting from the many small cleavage planes of the fractured crystals which form the rough broken surface. Numerous black specks of unknown nature occur (quartz?); exceedingly hard, and hence often in large pebbles. So far as seen these two varieties compose most of the bed. Very occasionally there occur (*c*) pebbles having a dull brick-red, hard, compact inclining to jaspery matrix, in which are pretty thickly scattered crystalline grains of black quartz from .02 to .20 inch diameter, together with flesh-red to pearly crystals of feldspar, probably orthoclase, from .1 to .6 long. The quartz appears black from non-transmission of light, being in reality colorless. It is invariably fractured upon breaking the pebble, and lies about flush with the fractured surface, which is uneven, and generally exhibits a rude hexagonal outline. Pebbles of a rock (*d*) very closely resembling some fine-grained or compact melaphyrs also sparingly occur. The matrix and sandstone layers are apparently composed of the same ingredients as the pebbles, small quartz grains sparingly but pretty generally occurring. Calcite, generally white, opaque, with rhomb cleavage, frequently occurs impregnating the mass, and filling the cavities and the veins where they pass into the bed.*

+ 4000

–1. **Sandstone**; generally coarse but varying from harder, thicker, coarser beds, generally more or less cross-stratified, to finer, thinner, shalier ones. Bands of pebbles occur, and much calcite is present; composition same as above; general color, dull brick red.

129
(68)

— o — *Junction*, well defined, slightly undulating, quite smooth; no contact change in either bed.

1. **Amygdaloid**, inclining to amygdaloidal melaphyr; amygdules not abundant, scattered, containing calcite, sometimes with laumontite and a light green green-earth. Matrix same as underlying accompanying **Melaphyr**, which has a medium coarse-grained texture and semi-conchoidal inclined to uneven fracture. Prevailing color a light grayish green, rudely mottled with dull flesh red. The dark green chloritic element is very subordinate; glistening black specks of specular iron occur, and much disseminated white calcite. Specific gravity, 2.94. Tendency toward amygdaloidal zones in the bed; lower two feet amygdaloidal.

85
(48)

– 85 (48) – *Junction*, very irregular. For two feet the underlying sandstone is changed and indurated, being, in places, hardly distinguishable from the overlying melaphyr, except for enclosed pebbles which are not changed.

* It is in a vein in this horizon that the impure manganese occurs, at Manganese Lake, near the extremity of Keweenaw Point.

Some pebbles rest upon the hanging-wall which are quite enclosed in the overlying amygdaloidal melaphyr.

a a—Amygdaloidal melaphyr.
s s—Sandstone with pebbles.
j j—Junction of a and s.
m s—Metamorphosed top of s, with unchanged pebbles.
p—Pebble surrounded by a.

5 (3) — 2. **Sandstone**; coarse, some conglomerate. Same as No. -1.

70 (37) — *Covered* (probably sandstone).

55 (30) — **Conglomerate** and sandstone. Same as No. -1.

148 (78) — *Covered* (probably sandstone).

4 (2) — **Sandstone**; rather coarse. Same composition as the preceding.

367 (198) — *Junction*, slightly undulating. No change or metamorphism in the adjacent beds. Extending from the junction down into the underlying

melaphyr—about eight feet being exposed—is a fissure or crack with sharply defined edges and two abrupt bends, giving widths of two and four inches. This crack is filled with sandstone similar to that above, but somewhat finer and slightly decomposed or softened. There is an appearance of irregular, but rudely curved stratification, about parallel, as a whole, with the formation.

50 (26) 3. **Melaphyr,** inclining to amygdaloidal melaphyr; medium coarse texture. Color dull reddish purple, sprinkled with numerous small green spots; former mostly feldspar, latter delessite. Some amygdaloidal portions have delessite spots of larger size, enclosing radiated laumontite.

10? (5) 4. **Sandstone.** Same as -1.

40 (21) 5. **Melaphyr,** inclining to amygdaloidal melaphyr; resembles No. 3, slightly finer,—more purple; occasional calcite amygdules, and small feldspar crystals porphyritically imbedded in the base.

175 (89) *Covered,* probably sandstone.

40 (21) 6. **Sandstone,** similar to preceding; rather coarse.

160 (83) *Covered.*

25 (13) 7. **Melaphyr,** inclining to amygdaloidal melaphyr, rather irregularly. Upper 15 feet are of dark green decomposing melaphyr. The lower 10 (5) feet are of irregular **Amygdaloid**; decomposing; the harder parts composed of a rather compact dark reddish-green base, containing small scattered flesh-red feldspar crystals, and irregular, generally elongated, amygdules of delessite, containing calcite and sometimes accompanied by small amounts of prehnite and laumontite. Enclosed in this amygdaloid are some irregular patches of fine-grained light gray melaphyr.

867 (456) *Junction,* well defined, straight; rocks adjacent rather soft, and weathering; slight stains of carbonate of copper.

90 (47) 8. **Sandstone;** upper few inches somewhat changed by the overlying rock. Same composition as the preceding sandstones. Ten feet quite consistent, followed by ten feet shaly; rest not well exposed; some conglomerate occurs.

88 (45) *Covered,* probably sandstone.

14 (7) 9. **Melaphyr;** medium to fine grained; color light grayish, from minute light-green and dark-red component spots; fracture uneven, inclined to smooth; rather hard and elastic; many small calcite cleavages glistening over the fractured face; no magnetite. Lower ten inches are amygdaloidal; calcite with some laumontite amygdules, increasing toward the foot-wall.

1059 (555) *Junction,* well defined.

20 (10) 10. **Amygdaloid;** base reddish-brown; compact, not indurated; amygdules very irregular, from .05 to .50 inch longest diameter, composing about 40 per cent. of the rock and filled with calcite accompanied sometimes by small amounts of green-earth and some prehnite for about ten feet, and then gradually changing through rather fine-grained.

Amygdaloidal Melaphyr to **Melaphyr** at fifteen feet; general color greenish-purple, composed of light-green feldspar crystals and small red specks; fracture uneven; occasional calcite amygdules.

26 (14) — 11. **Amygdaloid,** and **Amygdaloidal Melaphyr,** similar to No. 10, but not well developed.

12. **Amygdaloid,** 15 (8) feet, similar to that of No. 10; 50 per cent. amygdules.

50 (26) — **Amygdaloidal Melaphyr,** 30 (16) feet, and **Melaphyr,** 5 feet; color greenish-gray; finer grained and harder than No. 10; fracture somewhat uneven. Lower foot amygdaloidal, calcite amygdules.

13. **Amygdaloid;** narrow, 5 feet, but well developed; base fine grained; purple; amygdules mostly calcite, at top with considerable green and white prehnite and scattered specks of copper and some datolite.

40 (21) — **Amygdaloidal Melaphyr;** 20 feet; base somewhat coarser grained; dull dark green, mottled purple; fracture uneven; amygdules scattering; laumontite, sometimes surrounding a green-earth core; some of delessite and calcite.

Melaphyr; 15 feet; inclining to amygdaloidal, with delessite spots; lower foot quite amygdaloidal.

1195 (626)

14. **Amygdaloid;** 10 (5) feet; very fine-grained to earthy, reddish-brown base; amygdules irregular, not ramifying, reaching .3 inch diameter, and forming about 30 per cent. of the rock; some of calcite with delessite lining; rest partially, some wholly, filled with prehnite, some of which are further filled with calcite; ramifying seams of prehnite occur, with some copper.

30 (16) — **Amygdaloidal Melaphyr;** 20 (11) feet; medium fine grained; mottled dirty green and purple; scattering amygdules of delessite and laumontite.

15. **Amygdaloid;** 4 (2) feet; similar to No. 14, but subordinate; some

87 (45) — **Amygdaloidal Melaphyr** and **Melaphyr;** 75 feet; medium fine-grained; irregularly mottled green and purple; irregular to hackly fracture; much like melaphyr of No. 14.

1312 (687)

16. **Amygdaloid;** 25 (13) feet; mixed soft and hard, former predominating; upper part, as usual, much the best developed; irregularly associated with and changing to amygdaloidal melaphyr; the base of the amygdaloid is very fine-grained to compact; greenish to red brown; amygdules irregular and ramifying, forming 20 to 40 per cent. of the rock, mostly of calcite and prehnite, small amounts of copper generally accompanying the latter.

43 (22) — **Melaphyr;** 12 feet; medium coarse-grained; intermingled dull-green and purple ingredients, giving a mottled greenish-purple color, relieved by numerous small greenish-white feldspar crystals, which sometimes are .1 inch diameter.

120 (63) — 17. **Sandstone;** medium fine-grained; specific gravity 2.56; mostly rather heavily bedded, some thinly bedded and shaly, and some conglomerate layers; color dull brick red; not well exposed.

18. **Amygdaloid** (?); narrow; mostly covered; not well observed.

Amygdaloidal Melaphyr; 17 (9) feet; characteristic; base rather fine-grained; indefinitely mottled dull green and purple, former sometimes prevailing, giving a dark, dirty, dull-green color, and sometimes the latter, giving a handsome purple hue to the rock; fracture generally uneven, sometimes approaching conchoidal, not rough; rather tough and elastic. In this base are sprinkled amygdules reaching .5 inch diameter and averaging about .3 inch. These occur from very occasional to where they are irregularly scattered, but averaging .5 to 1 inch apart, being generally more numerous toward the upper part of the bed. These amygdules are all filled, and almost invariably with prehnite, sometimes having calcite in the interior. Very occasionally the prehnite has its characteristic green hue; but this only occurs near the centre. More often it has a flesh pink color and vitreous lustre and fracture, either radiated or not, in the latter case strongly resembling rose quartz. It is in this form, inclining to green, that it may enclose calcite, and in which it resembles the characteristic prehnite amygdules of the "Isle Royale" series of rocks near Portage Lake. But the most characteristic and predominant form that the prehnite assumes is one in which the color has less red in it than pink, being a whitish salmon color, without a vitreous lustre, while a well-developed radiated structure gives a fracture as if from interrupted cleavage. It much resembles laumontite, but is harder. The radiations diverge from a point on one side, like a palm-leaf fan, or from two, or even three, points, the systems meeting along curved lines. In a flat-sided amygdule, the radiant point will be at the corner, the rays being at first parallel to the flat side. Occasionally, immediately about the radiant point, salmon-colored prehnite occurs, changing into the rose colored variety toward the extremity; and more often a dull, earthy green band runs through the amygdule, concentric with the radiant point. The outer surface has often a thin coating of delessite; but as often the prehnite is in direct contact with the base. This contact is intimate, never perfectly smooth, but always uneven, sometimes very irregular, and never suggestive of a filled-up cavity, but rather—as in most amygdaloidal melaphyrs—of a segregation. The base but seldom exhibits a changed appearance near the amygdule, though, as a whole, it is apt to differ somewhat from the underlying melaphyr.

24
(13)

Melaphyr; 4 feet; same as the base of the above; slightly coarser.

(20) *Covered;* possibly occupied by an amygdaloid, etc., like the preceding, probably by melaphyr.

11
10
(5)

Melaphyr; like that of No. 18.

19. **Sandstone**; rather coarse; some conglomerate layers; mostly covered; neither wall observed; same as the preceding sandstones, and, like them, having carbonate of lime generally present.

±45
(24)

20. **Amygdaloid**; 10 (5) feet; soft, decomposing; mostly covered; calcite predominating. **Amygdaloidal Melaphyr**, followed by some **Melaphyr**; same as No. 18.

70
(37)

±33 (18)

21. **Sandstone**; mostly covered; very fine; some thinly banded light and dark red; in parts slightly indurated, tending toward jasper; no junction visible.

119 (62)

22. **Amygdaloid**; 10 (6)? soft; decomposing; mostly covered.

Amygdaloidal Melaphyr; 30 (16); specific gravity, one piece, 2.72; and **Melaphyr**; 79 (40); specific gravity, 2.77; same as No. 18.

14 (7)

Covered; probably occupied by an amygdaloid (No. 23) and amygdaloidal melaphyr similar to the preceding.

14 (8)

23. **Amygdaloidal Melaphyr**; 5 feet; and **Melaphyr**, same as No. 18.

17 (9)

24. **Amygdaloidal Melaphyr**; 5 feet; and **Melaphyr**, 12 feet; same as preceding, inclining to purple.

30 (16)

25. **Amygdaloid**; 20 (10) feet; some hard; matrix compact, purplish brown, with occasionally whitish-red crystals of triclinic feldspar, sometimes 2 inches long, imbedded in it. Near hanging-wall are irregular bunches of calcite, some prehnite; amygdules often reaching .25 inch diameter, and rather irregular; forming 50 per cent. of the rock near the top; mostly of calcite in thin pellicle of delessite, and sometimes enclosing green prehnite with little copper; amygdules with red lining (feldspar?) enclosing quartz, and often copper, occur in the harder parts.

Amygdaloidal Melaphyr, inclining to melaphyr, 12 (6) feet; similar to that of No. 18; amygdules small, not well marked.

±50 (26)

26. **Sandstone**; in part very fine-grained to compact texture, inclined to shaly structure; laminæ $\frac{1}{8}$–$\frac{1}{4}$–$\frac{3}{8}$ inch thick; remainder thicker bedded and coarser; specific gravity, 2.68; composition same as the preceding, having occasional black quartz grains, and the usual calcite impregnation.

1921 (1008)

Junction; smooth; slightly undulating; no apparent contact metamorphism.

79 (42)

27. **Amygdaloidal Melaphyr**; 10 (5) feet; poorly exposed; amygdules of delessite and some prehnite in a green base, which beneath becomes

Melaphyr; 15 (8) feet; rather fine-grained; irregular fracture; of dark and light green components; similar to that of No. 18, but without the purple element of the color. At 25 (13) feet from the hanging-wall it becomes amygdaloidal, with many amygdaloidal-like segregations of delessite, which are somewhat indurated but inclined to earthy; centre light green, circumference dark green, the base being strongly inclined to purple. Amygdules of quartz-like and radiated salmon prehnite also occur occasionally. Without having well-defined walls, two or three more amygdaloidal zones occur parallel with the bedding, with the greener melaphyr between. Lower 10 feet are of the darker, mottled, dirty green **Melaphyr**, somewhat coarser than the rest.

18 (9)

28. **Sandstone**; rather coarse, inclining toward conglomerate; similar to No. -1; non-quartziferous, chocolate-colored, felsitic rock, with occasional small feldspar crystals, predominating.

. . . . inclining toward amygdaloidal melaphyr; 6 (3) feet; . dark, purplish brown; fine-grained; amygdules not predominat- . mostly of some delessite with calcite and some green prehnite with . . . copper.

Melaphyr; 5 feet; medium fine-grained; like the purpler varieties of the preceding melaphyrs; foot-wall not well defined.

30. **Amygdaloid**; 4 (2) feet; hard, fine-grained, purplish-brown base; amygdules forming about 25 per cent. of the rock, irregular, but neither large nor ramifying; generally contain calcite enclosing prehnite, often the latter alone or enclosing quartz. The prehnite is usually colorless, inclining to light green, and fills also small veins in the rock. It is nearly always—the quartz sometimes—accompanied by little copper.

Amygdaloidal Melaphyr; 16 (8) feet; same type as No. 18, base somewhat coarser, though component crystals are still nearly undistinguishable; green predominates over the purple. Amygdules contain radiating prehnite, but salmon color not marked, often colorless, white; subfibrous, but lustre vitreous. Unlike No. 18, however, these are generally surrounded by laumontite, and many wholly filled with it.

Melaphyr; 33 (17); same as base above; slightly coarser and greener than No. 18.

31. **Amygdaloid**; upper foot having from a very fine-grained to compact, inclined to hard, base of dark greenish-brown color, and containing irregular and ramifying amygdules and bunches of calcite. This graduates into 18 (9) feet of softer and less amygdaloidal material, with coarser and purpler matrix, in which calcite is subordinate, and spots of dark and light green delessite or green-earth, or both, predominate. Laumontite and radiated pink prehnite also occur. This graduates into

Melaphyr; 10 (5) feet; rather fine-grained; handsome purple, irregularly mottled dark green, with uneven and often rough fracture. Same type as No. 18.

32. **Amygdaloid**; 10 (5); **Amygdaloidal Melaphyr**, 6 (3), and **Melaphyr**, 10 (5) feet, like No. 31.

33. **Amygdaloid**; well developed, calcite and prehnite, often in bunches for 1–2 feet, then less irregular for 10 feet, followed by

Amygdaloidal Melaphyr; like No. 31, but with more salmon-colored prehnite.

34. **Amygdaloid**, and **Amygdaloidal Melaphyr**, same as preceding.

35. **Sandstone**; rather coarse, similar to the others.

36. **Amygdaloid**; mostly covered, 6 to 8 feet soft, decomposing; red prehnite and delessite, followed by 2 feet of harder amygdaloid. This is inclined toward a scoriaceous character, in which occurs very irregularly banded or

41
(21)

thinly stratified material, which is wholly undistinguishable from that in some of the finer sandstones, though somewhat indurated. It often partially encloses irregular bunches or balls of calcitic amygdaloid, while it is impregnated by calcite as is the sandstones. Below this are two feet of soft, blackish green, crumbling delessitic rock, with laumontite scattered through it; crumbling shaly. A normal **Amygdaloid,** inclining to amygdaloidal melaphyr, follows, 28 feet; base dirty light grayish green; delessite, calcite.

Melaphyr; 12 (6) feet; forming base of bed; dirty gray green predominating; occasional delessite spots.

30
(16)

37. **Amygdaloid**; 4 (2) feet; dark, chocolate-colored particles in light-green base, forming grayish-green colored matrix; amygdules of calcite in pellicle of delessite; occasionally of pink and salmon prehnite.

Melaphyr; same as matrix above.

20
(10)

Covered; probably occupied by a coarse, soft amygdaloid—No. 38?—like No. 36.

27
(14)

38. **Melaphyr**; 27 (14) feet; exceptional; quite coarse-grained; fracture somewhat uneven; composed of small, but generally indistinct, crystals of flesh-red triclinic feldspar, and dark-green, to lighter bluish-green amorphous delessite; colors well marked. A soft, dark, chocolate-colored mineral is present in small quantities; no magnetite; particles of metallic copper occasionally appear through the mass, generally in the feldspar crystals; specific gravity 2.79.

150
(80)

39. **Melaphyr**; rather fine-grained; uneven fracture; mottled indistinctly and irregularly dark green and dull purple, former mostly predominating and giving rise to a dirty, dark-colored rock.

103
(54)

Covered; much very amygdaloidal débris.

95
(50)

42 and 43. Space mostly covered; two melaphyrs showing, each probably with an overlying, very amygdaloidal bed. **Melaphyrs** incline to dark color; some being nearly black or greenish black, mottled with very dark, indistinct purple.

44. **Sandstone**, with

45. **Amygdaloid**; 24 (12) feet; forming a "scoriaceous amygdaloid;" composed of irregular bunches or "bomb-like" masses of calcitic amygdaloid, 1.5 feet to .5 inch diameter, filled in with a very fine-grained to compact brick-red material, which often shows fine, but irregular, bands or lines, apparently of stratification. This material increases in amount toward the top of the bed, where it often quite encloses and surrounds the smaller, irregularly round, amygdaloidal balls. The strata-like bands are more evident when the material is in larger amounts, and they often seem to separate or open out to enclose the imbedded balls. In appearance it is undistinguishable from the finer-grained sandstones, though, perhaps,

containing more calcite, which is more often collected into small, generally lenticular, cavities, which are sometimes more numerous in wide, narrow bands, parallel with the bedding, than is the case with the sandstones.

154
(56)

Amygdaloid, or Amygdaloidal Melaphyr; 10 (6) feet; very fine-grained; purplish brown; containing small white and red crystals of felspar; amygdules of green-earth and calcite, when together, latter contains the former; very little copper. Graduates into

Melaphyr; 120 (38) feet; exceptional; very brittle and elastic, and harder than the average melaphyr, with sub-conchoidal and rather smooth fracture; specific gravity, 2.91; consists of a very fine-grained to compact purplish base, in which are profusely scattered small, semi-transparent, light-green crystals of a triclinic felspar (labradorite?), occasionally showing striation, but seldom definite form. Considerable magnetite is present, which, with small flakes of specular iron, occurs scattered here and there, often connected with a small spot of discoloration of the rock to dark brown, as if from hydration. The chloritic ingredient is occasionally segregated out in amygdules, in which flakes of copper sometimes occur. Weathered surface smooth, color gray, with felspar white. Lower 4–6 inches of melaphyr broken up, rock more compact than above, felspar crystals red, rock brown. With this, but mostly below it, are 4–5 inches of finely amygdaloidal rock, base compact; purple; red felspar crystals; amygdules mostly calcite, some green-earth.

2961
1503

Junction. The line over the broken-up material is strongly marked, while the real foot-wall of No. 44, though a true plane of separation, is not so apparent.

46.	**Amygdaloid**;	(2) feet;	**Amygdaloidal Melaphyr**	and **Melaphyr**	(27).
47.	"	(1)	"	"	(8) feet.
48.	"	(2)	"	"	(4)
49.	"	(2)	"	"	(8)
50.	"	(1)	"	"	(15)
51.	"	(2)	"	"	(7)
52.	"	(2)	"	"	(4)
53.	"	(4)	"	"	(9)
54.	"	(4)	"	"	(10)
55.	"	(5)	"	"	(10)

3100
(1630)

These ten beds, though thin, are well defined. With some variations they are much alike, and may be described as follows, taking No. 47 for the type: Commencing near the middle of one of the amygdaloidal melaphyrs—No. 1 of the adjoining section—we find a rather soft, but tough, medium fine-grained, but not closely textured rock, breaking with an uneven and rough fracture. Upon the purplish-brown base are numerous dark-green, poorly defined, delessite spots. The purple-brown base is composed of the light green or pinkish felspathic element, and small chocolate colored specks, apparently a decomposition product of an iron mineral, probably specular iron, glistening points of which still remain. Amygdules not numerous, in some beds almost wanting, generally of delessite, occasionally stellated, sometimes of calcite, laumontite, and green-earth. This grows more and more amygdaloidal in ascending, passing into an irregular, somewhat harder belt, having a finer-grained to compact matrix, with an even reddish-brown color, and amygdules of calcite and green-earth and some prehnite. This changes rather abruptly to a band—No. 2—a few inches wide, of the same base, but very

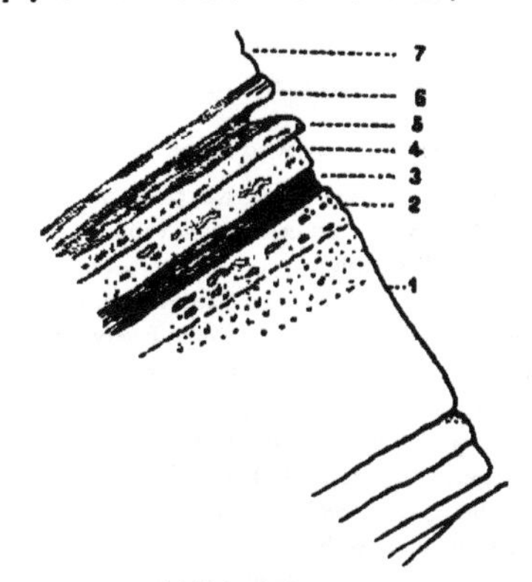

1. Amygdaloidal melaphyr.
2. Very amygdaloidal: calcite, prehnite.
3. Soft, decomposing, delessitic zone.
4. Same as 2, and top of bed.
5. Hard; not as well developed as 2 and 4.
6. Soft, fissile; delessite and laumontite.
7. Amygdaloidal melaphyr, same as 1.

amygdaloidal, green botryoidal prehnite predominating, with calcite; amygdules very irregular, reaching 5 inches long; the larger ones generally lying parallel with the plane of bedding, and mostly confined to the centre of the narrow belt. No. 3 follows, soft, easily decomposing, dirty brown, containing much delessite; about 6 inches wide; overlain by No. 4, also about 6 inches wide and like No. 2. Its top is the real "hanging-wall" of the bed, having an even, free, plane of stratification separating it from the bed above. The lower two or three inches of this upper bed—No. 5—are rather hard and somewhat like Nos. 2 and 4 below, but not so well developed. It graduates into No. 6, which is very soft, inclining to fissile, and easily decomposing, having in its lower part abundance of radiating laumontite and delessite, the latter prevailing above but decreasing in amount till about a foot or more above the "foot-wall," when a rather sudden change takes place into the melaphyr or amygdaloidal melaphyr above. The prehnite is almost invariably accompanied by small amounts of native copper. This double structure of the amygdaloids occurs in most of the beds, but is absent in some, and some of the amygdaloids are much better developed than others; while the individual distances may be somewhat incorrect.

67 (35) *Covered;* probably containing two or three beds similar to the above.

18 (9) 58. **Melaphyr.**

58 (30) 59. **Amygdaloidal Melaphyr,** varying irregularly to amygdaloid or melaphyr, and generally much decomposed.

11 (5) 60. **Melaphyr.**

185 (95) 61. **Amygdaloidal Melaphyr,** same as No. 59.

11 (6) 62. **Melaphyr,** inclined to semi-columnar structure.

These five beds form an exceptional series. No. 58 resembles the melaphyrs of the last-described series, a rather coarse, soft, easily decomposing, uneven fracturing rock of dull reddish-brown color, spotted dark green. No. 60 is finer, harder, uneven but smoother fracture; color greenish, mottled reddish brown, and of more even tint. No. 62 is very fine-grained, elastic, but tough, inclined to semi-columnar structure, and of dark blackish-green color, with occasional crystals of glassy feldspar. These three melaphyrs graduate most irregularly into the two intermediate zones, Nos. 59 and 61, the easily decomposing nature of which, however, renders them difficult of observation. Throughout there is much melaphyr similar to the enclosing beds, but seemingly mixed with it are large amounts of most irregular material, sometimes occurring in vein-like seams, or in wide irregular bands, paralled to the bedding. Some of these are composed almost wholly of coarse delessitic material, patches of which are filled with ramifying, connecting amygdules of pink laumontite and calcite, which may even become the predominating ingredients of the rock, rather than amygdules. Others, somewhat harder, have but little laumontite and more calcite and prehnite, with a little copper; but these minerals are more confined to cavities and spots, and appear more like accidental than essential ingredients of the rock, as is the case sometimes with the delessite and laumontite. Stellated groups of small quartz crystals occur in vugs. Still various amounts of compact melaphyr are frequent, balls of which may be pried from out the crumbling mass, leaving irregular cavities behind. One or two zones show a material consisting of dark glistening delessite and grains of feldspar, which crumbles away into a sand and encloses balls of harder melaphyr.

3450 (1810) *Junction, "Slide";* strongly defined, somewhat uneven. The "Old Phœnix workings on the Ash-bed" show the "Armstrong Vein," thrown horizontally along this junction about 260 feet (Merryweather's map). Considering the dip and strike of the vein, and supposing the movement to have taken place directly down the bed, it would appear that the beds above had slid down upon the beds below about 150 feet.

.5 63. **Sandstone;** seam, 1-6 inches thick; very fine-grained to compact; reddish-brown to chocolate color; occasional fine lines of stratification; undistinguishable from the fine sandstones, and, like them, melts before the blowpipe flame on their edges, without intumescence, to green-black glass; specific gravity, 2.61. Fills irregularities on the surface of amygdaloid below.

64. **Amygdaloid**; 12 (6) feet; made up of several thin beds; the upper ones very amygdaloidal; 60 per cent. amygdules, calcite, little prehnite, not very irregular, base compact, reddish gray. Near the amygdules, generally on their lower side, often only there, the base is changed, to an indurated, amorphous, chocolate-colored material, resembling the sandstone above, but harder, and intumescing before the blowpipe, like most melaphyrs.

108 (55) **Amygdaloidal Melaphyr**; 40 (20); in several beds; base varying from very compact to rather fine-grained; former even light-green color; latter same, finely mottled darker green; amygdules small, rather numerous, contain laumontite and calcite.

Melaphyr; 50 (26) feet; very fine-grained; elastic, rather brittle; rather smooth, inclined to semi-conchoidal fracture; color, dark, greenish black inclined to brown; decided semi-columnar structure.

fault

—35 —(18) 65. **Amygdaloid**; about 180 (92) feet, variable; scoriaceous; locally called the "**Ash-bed**"; upon it, and on the "Old Phœnix Vein" where it intersects the ash-bed, much mining has been done, but without ultimate practical success.* Débris on the surface and water in the workings mostly hide the bed. It seems to be composed of three or four indefinite zones or "floors," one of which is strongly marked, separated by, and intermingled with, compact melaphyr similar to that forming the base of the bed below. These zones often appear to be made up of more or less round and various-sized balls of amygdaloid, often distinctly surrounded by the melaphyr, a pretty well-defined line of demarcation sometimes separating the two, but often very irregular and not well defined. Again they are enclosed in a brownish-red, fine-grained material, similar to the medium and finest grained to compact sandstone. It is in places distinctly but irregularly stratified, but no pebbles have been observed. Specific gravity of the coarser, 2.70; of the finer, 2.66. When the melaphyr surrounds the balls, its dull bluish color sometimes changes to a reddish hue, resembling that of the supposed sandstone material, but, unlike it, containing occasional incipient delessite amygdules, and scattering red feldspar crystals. It fuses, like the parent melaphyr, before the blowpipe rather readily on the edges of large pieces, and with intumescence, while the finer sandstone-like material is not readily fused, with hardly any intumescence, the coarser being almost infusible. In one specimen the fine sandstone material was associated with small layers of ash-colored, porous, tuff-like material. The amygdaloid itself has generally a dark, chocolate-colored base, with amygdules often quite filled with delessite, but generally of calcite and the other minerals, crystallized datolite being common in veins intersecting the bed. Copper is scattered through the bed.

± 280 (143)

---

* Merryweather's map of the "Old Phœnix workings on the Ash-bed" gives this bed as about 70 feet wide, but variable. The survey measured 90 feet upon it without seeing either wall. At Copper Falls it is over 150 feet wide.

Melaphyr; 100 (51) feet; lying below the preceding, as well as associated with it. Texture very fine-grained to very compact; fracture smooth, sub-conchoidal; color, at a little distance, a soft, even, neutral, bluish tint, which inspection shows, as with most melaphyrs, to be composed of a delicate mottling of green and purplish red. Throughout, more or less definite green spots appear, sometimes increasing in number and definition, giving an amygdaloidal character to the rock. Small shining particles of copper are quite numerous, and considerable magnetic iron is present. Specific gravity, 2.98; large probably on account of the copper and iron present.

3803 (1990)

66. **Amygdaloid**; 14 (7) feet; scoriaceous; matrix brown, fine, with flesh-red feldspar crystals; amygdules small, very numerous, mostly calcite; change abrupt to

Amygdaloidal Melaphyr; 14 (7) feet; very fine purplish base, composed of green and dark-red particles, with porphyritically imbedded numerous small flesh-red crystals of triclinic feldspar, in which occasionally are flakes of copper. Amygdules are of black-green delessite, containing green prehnite and quartz, generally some copper, near the top; delessite alone near the bottom, earthy on the exterior of the amygdules, often stellated or radiated, and shiny within; specific gravity, one piece, 2.84. Changes to

88 (45)

Melaphyr; (31) feet; hard, brittle, elastic; fracture rather smooth, sub-conchoidal; color even, purplish, or lightly mottled green, with numerous small crystals of whitish-green, inclining to pink triclinic feldspar, porphyritically imbedded; considerable magnetic iron; specific gravity, 2.93. Similar to No. 45. Lower foot is amygdaloidal.

67. **Amygdaloid**, inclined to amygdaloidal melaphyr; 6 (3) feet; composed of three rather distinct layers; base very fine-grained; purplish; porphyritic, with small, flesh-red feldspar crystals; amygdules mostly delessite, some calcite and prehnite, little copper. Changes to

Melaphyr; inclined to amygdaloidal melaphyr. In descending, the feldspar changes from flesh-red to dirty white, and then to light green, composing, in a few feet, with some generally disseminated delessite, the greater part of a rather coarse-grained, dirty gray-green matrix, spotted with numerous dark-green, delessite spots, the porphyritic character no longer existing, and with but little magnetite and no observable specular iron. It is tough, non-elastic; fracture uneven and rough, giving the impression of a loose texture. Specific gravity, 2.88. The delessite spots give an amygdaloidal-melaphyr character to the rock, which exists throughout the bed.

109 (56)

68. **Amygdaloid**; fine-grained, hard; quartz, with other minerals, present.

Amygdaloidal Melaphyr; a mixture of small red feldspar crystals and much delessite, latter mostly collected into frequent spots reaching .2 inch diameter. In descending, as in No. 67, the feldspar loses its depth of shade, becomes light dirty pink and green or white, and semi-transparent, and forms, with apparently about an equal amount of delessite, a more even granular mixture of a dirty pinkish color, profusely spotted with greenish-black segregations of delessite.

Melaphyr. In descending still further, about fifteen feet from the hanging-wall, the feldspar mostly becomes of a light-green color, giving to the base a dirty gray-green hue, sometimes slightly mottled with purple, and a more compact fine-grained texture, while the dark-green delessite spots become smaller and smaller. In the fullest development of the melaphyr these spots lose their definition, forming an irregular, undefined, but evenly distributed mottling of dark green upon a lighter green background, and even this may be nearly lost. But little magnetic iron is present. Specific gravity of finest, 2.87. Fourteen (7) feet from the foot-wall a distinct head, parallel to the bedding, occurs. The lower foot is inclined to amygdaloidal, pink feldspar, larger delessite spots.

69. **Amygdaloid**; 12 (6) feet; hard; green and brown mixed; former compact, jaspery, few amygdules; latter very fine-grained; amygdules mostly of prehnite, with little copper; but often of quartz or calcite, generally free from chlorites. Softer, less developed toward bottom. At about 12 (6) feet from the hanging-wall, and along a pretty well defined plane of demarcation, is an abrupt change to

Amygdaloidal Melaphyr; (5) feet; same as in No. 68, but with laumontite accompanying the delessite near the top. Changes to

Melaphyr; same as No. 68; approaching the even-shaded variety toward the base. Thirty feet from hanging-wall is an *amygdaloidal melaphyr* zone; 25 feet from foot-wall is distinct "head" or joint, parallel with the formation, forming a fall in the river. Lower foot is amygdaloidal melaphyr.

70. **Amygdaloid**; 2 (1) feet; base fine-grained, purplish-brown; amygdules chlorite or green prehnite, with calcite and some quartz. Small orthoclase crystals resting on prehnite.

Amygdaloidal Melaphyr; 18 (9) feet; rather fine-grained; purplish and green speckled; delessite amygdules; texture rather loose; fracture rough.

71. **Amygdaloid**; (4) inches; same as No. 70, with **Melaphyr** below, same as No. 68.

Covered; amygdaloid?

72. **Melaphyr**; (25) feet; same as No. 68; rather coarse purplish base, mottled green.

Covered; amygdaloid in part?

86 (44)	73. **Amygdaloidal Melaphyr**; 20 (10) feet; base like No. 68; radiated pink prehnite in amygdules; followed by **Melaphyr**, like No. 68.
64 (32)	*Covered.*
	76. **Amygdaloidal Melaphyr**; 15 (8) feet; mostly covered; greenish; laumontite and calcite.
35 (18)	**Melaphyr**; very fine to compact, even bluish-green tint, like that of No. 65 ("Ash-bed"); débris indicates amygdaloid, probably scoriaceous.
30 (15)	*Covered.*
	78. **Amygdaloidal Melaphyr**; poorly exposed; like No. 76, followed by a
70 (35)	**Melaphyr**; very fine grained; blackish green; elastic.
125 (64)	*Covered;* the first sandstone south of the "Ash-bed" at Copper Falls may pass here.
50 (25)	82. **Amygdaloidal Melaphyr,** and **Melaphyr**; latter predominating near the base; very fine-grained; fracture quite smooth, inclined to conchoidal; similar to the melaphyr of No. 65 ("Ash-bed"), except the purple element of the color prevails, giving a dull purplish-brown color. Occasional crystals of salmon-colored feldspar occur, and few dark delessite spots. These, more particularly the delessite, increase, but irregularly, toward the summit, where calcite and quartz occur with it. This amygdaloidal tendency is not regular; often confined to irregular areas. Narrow zones of amygdaloidal action, two to six inches wide, accompany seams, cross the bed N.W. to S.E. For 12 to 18 inches from the foot-wall the bed is very amygdaloidal.
.5	83. **Sandstone**; 3-4 inches; irregularly banded; intimately associated with
	84. **Amygdaloid**; 10 (5) feet; somewhat scoriaceous; resembling No. 45, but less developed; less developed toward the bottom; occurring in patches in underlying
100 (52)	**Melaphyr**; 90 (47) feet; very fine to compact; bluish green; similar to that of the "Ash-bed." Near the foot-wall it is dull purplish, with red feldspar crystals scattered porphyritically, and spots of dark-green delessite, as in No. 82.
	85. **Sandstone**; few inches, or wanting; shaly or laminated, wavy; associated with
32 (16)	86. **Amygdaloid**; 10 (5); scoriaceous; matrix brown, compact; amygdules mostly calcite, some prehnite and quartz, little copper.
	Melaphyr; 21 (11); rather fine-grained; dirty dark-green color; uneven fracture; weathers into balls, and with a tendency to semi-columnar structure.
	Junction; marked, occupied by thin seam, but very irregular.

87. **Amygdaloid**; 13 (7) feet; pretty well developed near the top; base fine-grained, reddish brown; sometimes indurated; porphyritic with minute flesh-red feldspar crystals; amygdules somewhat irregular; quartz, enclosed in delessite; with very little copper. Graduates into an

302
(154)

Amygdaloidal Melaphyr; 45 (23) feet; composed largely of small flesh-red crystals of feldspar (triclinic) in a rather coarse base of smaller gray-green feldspar crystals and delessite, the latter being also collected into numerous prominent green spots. By the omission of the flesh-red feldspar, or by its change of color to light green, and by the more even distribution of the delessite, together with a more compact aggregation of the mass, this gradually changes to

Melaphyr; 244 (124); spotted. Like No. 68, but of somewhat more exaggerated form.

192
(97)

Covered; in part probably occupied by an amygdaloid.

90. **Melaphyr**; 172 (87) feet; exceptional; semi-columnar in structure; a very fine-grained, inclining to crystalline, rock; brittle, elastic, rather hard; fracture uneven to sub-conchoidal, and rather smooth; clear uniform bluish-black color, with faintly purple element. Considerable magnetic iron. Specific gravity, 2.89. Weathers smooth; drab color.

172
(87)

5646
(2927)

91. **Diorite**, *aphanitic;* very fine-grained, changing to sub-crystalline; rather brittle, but very tough; fracture uneven, quite rough; color greenish black, but showing through the transparent feldspar; component gives a dirty brownish tint, inclined to resinous. Near top of bed hardly distinguishable from some melaphyrs; except the tendency to a resinous hue, and lack of any columnar character, it is much like melaphyr 69. In descending through the bed, the resinous character at first increases, occasional crystals of glassy feldspar occurring; fracture irregular to hackly, and specific gravity increasing. Near the bottom it again grows finer, becoming quite compactly crystalline, the resinous and brown appearance disappearing, and a clear green color with purple mottling taking its place; purplish spots rather large, and reflecting light. The green is not the dead, nor shiny delessite green in melaphyrs. Near the bottom the rock is like the mottled crystalline "greenstone" near the base of the diorite series. Specific gravity near the top 2.89, near middle 3.01, near bottom 2.95.

187
(94)

92. **Diorite**; *dark type;* a fine-grained, glistening, crystalline aggregate of black or greenish-black hornblende, and brownish or greenish transparent feldspar, the shining of the former through the fractured edges of the latter, giving a dark dirty-brownish color, and resinous lustre to the rock. Considerable dark green chlorite (delessite?) is present, and some magnetite; tough; fracture irregular and rough; specific gravity 2.89. The bed contains irregular patches or segregations of coarser material, in which the feldspar is white, opaque, its crystals forming an irregular network on a black background of hornblende; latter often of high lustre, and

25
(13)

accompanied by, seemingly changed into, a soft dark green chloritic mineral (delessite ?). In these prehnite and a radiated bronze-colored mica are present.

93 ? **Diorite**; *dark type;* same as preceding, but coarser. Feldspar sometimes in larger crystals, or nearly square plates; cleaving, transparent, glassy, or resinous; melts before blowpipe only in fine splinters. Weathers in large flattish knobs, which are sprinkled with small protruding shiny black crystals of hornblende, all of which upon the same prominence generally catch and reflect the light at the same incident angle, as from a large but irregularly interrupted cleavage face; color of depressions rusty brown; of prominences, whiter.

25
(13)

94. **Diorite**; *light-colored type;* inclining to coarse melaphyr; a coarse, crystalline, irregular-textured bed, containing numerous segregations of coarser material composed of prehnite with feldspar, or of green indurated spots of quartz and epidote (?). The bed is composed mostly of white, inclined to red, generally elongated, crystals of feldspar, lustrous hornblende, and a soft dark-green chlorite, probably delessite, which often predominates over the hornblende and is apparently derived from it; loosely textured. In places it strongly resembles the melaphyr of bed No. 87 or No. 68. For several feet near the base (absent near station D_1) the feldspar is red, sometimes pearly, and occurs in long intersecting crystals, often over an inch long, and twinned. Sometimes the surface, and near the composition plane of twins, will be red, and the interior pearly. This feldspar here forms about 40 per cent. of the rock, more than half the remainder being chlorite; specific gravity about 2.94.

96
(48)

95. **Diorite** (observed at station D_1, absent at D_2); *dark type;* much like Nos. 92 and 93. Dark colored, resinous; hard, tough; fracture uneven, medium fine-grained, but compactly textured; some chlorite present. Weathers rather evenly and not rugose; rusty brown with protruding hornblende crystals.

40
(20)

96. **Diorite** (directly beneath No. 94 at station D_2); *light-colored type*, typical. A rather coarse-grained, but compactly textured, crystalline aggregate of white and greenish-white, opaque, triclinic feldspar, and black, often lustrous, hornblende, with occasionally a little chlorite. Color greenish gray, but the whiter and more opaque feldspar is more or less confined to certain spots, thus giving a large and handsome mottling of grayish white on a greener ground. The hornblende crystals in many of these lighter colored spots, reflect light at the same incident angle. Hard, tough, somewhat brittle; specific gravity 2.90. Weathers rather evenly, color rusty gray, mottled white, with projecting crystals of black hornblende, which are more prominent on the lighter spots.

72
(36)

60
(30)

97. **Diorite**; much like No. 92. Some chlorite and bronze mica present.

15
(8)

98. **Diorite**; similar to No. 96, somewhat darker colored, with mottling very indistinct.

Covered.

99. **Diorite**; *light type*, much like No. 96; rather fine-grained, color dark greenish-gray, not mottled; feldspar mostly greenish, hornblende not generally lustrous, considerable chlorite; fracture neither very rough nor uneven. Weathers even rusty brown, generally without projecting hornblende crystals. Has frequent hard, compact, indurated, green segregations. Thirty feet covered.

100. **Diorite**; *dark resinous type*, like No. 92 or 93, or No. 106. Quite resinous. Weathers in large but flat prominences, like No. 93.

101. **Diorite**; *lighter type*, much like No. 99, somewhat coarser.

102. **Diorite**; *dark type*, same as No. 100.

103. **Diorite**; *lighter type*, similar to No. 96, but not mottled, and with a somewhat closer structure; chlorite very subordinate.

104. **Diorite**; *dark type*, much like No. 100, but less brown and resinous. Much amorphous, and rather hard chlorite is associated with the hornblende. Specific gravity, 2.91. Weathers like No. 100.

105. **Diorite**; *lighter type*, mostly covered; unusually coarse-grained; feldspar white, opaque; hornblende, in part hard, with highly lustrous jet-like cleavage faces, but mostly dull and softer, with some changed into soft amorphous, chlorite; bands of dark diorite also occur. Weathers very roughly, with protruding snow-white decomposing crystals of feldspar; also smoothly, with little hollows, each with hornblende segregation at bottom. Bed much like No. 107.

106. **Diorite**; *dark type*, with but little chlorite; rather coarse; color very dark, but not very resinous; frequent larger crystals of transparent feldspar; excellent example of the dark type. On the more exposed points this rock weathers in a very exaggerated rugose manner; the depressed portions are rusty brown and comparatively smooth, the numerous rough projecting knobs, often an inch high, and rising quite abruptly and irregularly, are lighter colored, and dotted black with projecting crystals of hornblende, nearly all upon each projection reflecting the light simultaneously. Coarse crystalline segregations, containing larger crystals of white opaque feldspar, are quite numerous.

40 — *Junction*, a well-marked plane of separation.

107. **Diorite**; *white* and *dark types*. This bed is composed of both types of diorites in alternate strata of from 5 or 6 inches to 5 or 6 feet in thickness, separated by rather abrupt, but neither sharp nor free, planes of demarcation. The wider and predominant beds of the white type are of an unusually coarse granular character; feldspar white or slightly greenish; hornblende mostly highly lustrous, and some chlorite; specific gravity of the rock 3.03, of the hornblende 3.39, of the feldspar 2.73. In places, nearly all the hornblende is changed to a dark-green chlorite—delessite—

when, if the feldspar crystals are small, the rock resembles a coarse melaphyr; and if, as at one point, delessite occupies, at the same time, spots formerly occupied by larger hornblende crystals, it may appear like an amygdaloidal melaphyr, as the upper part of melaphyr No. 87, for instance. Again, some of the white bands tend toward a gneissoid structure, the flattish feldspar crystals being arranged approximately parallel with the bedding. In this case the feldspar and hornblende are most compactly gathered together, the former being greenish and forming 54 per cent. of the rock, the latter highly lustrous and forming 46 per cent. of the rock, chlorite being absent; fracture thin, rather smooth; specific gravity 3.02. In this gneissoid-like rock the hornblende is gathered together in frequent black spots, 4 inches or more in diameter. In weathering, each of these segregations forms a little pit, the rest of the surface being even; color, white, speckled black, the feldspar crystals protruding slightly.

83
(41)

The dark type bands are the narrower, being generally .5 to 2 or 3 feet thick; being softer than the whiter bands, they weather in, with smooth, rusty-brown surface. Rather fine-grained, tough; fracture rough; color, dark blackish, resinous. The lower stratum is of the white type, but finer grained than those above, and like No. 96, but not strongly mottled.

108. **Diorite**; *dark type;* coarse, changing to aphanitic ("greenstone"). At, and for a hundred feet from the summit this bed is, for the dark type, quite coarse-grained; color very dark greenish black, inclined to resinous; glistening highly from the large amount of transparent, resinous-hued, glassy feldspar, as well as from the hornblende ingredient, though most of the latter is not lustrous; chlorite is generally present. It is hard, and very brittle, though tough, a hard, quick blow shivering the rock with an uneven and rough fracture. Larger crystals of light resinous-hued glassy feldspar quite frequently occur, but they often contain the other ingredients of the rock scattered in them, and often make themselves known only by reflecting light from areas which seem at first to be composed of an ordinary mixture of the ingredients. The hornblende also exhibits this character somewhat, but much less than the feldspar, possibly as not having highly lustrous faces. It is well brought out on the weathered surface, however. This surface weathers in large, flattish, rounded knobs; the depressions being smooth and rusty brown, the knobs lighter colored, with numerous small, projecting, black hornblende crystals. These, often over a space of nearly two inches square, all catch and reflect the light simultaneously. In descending through the remainder of the bed, the rock, with some local reversions, gradually grows finer-grained, though more particularly does it gradually become more crystalline in its texture, and very compactly built, while the feldspar seems to increase in amount, in places apparently impregnating the rock and almost producing a crysto-crystalline structure. The change in fineness is reached in the lower 200 feet. The fracture has become much more even and smoother, the rock is remarkably elastic and brittle, while the black resinous color

825
(412)
W. of Robbins Lode.

has gradually changed to a clear, handsome dark green, sometimes vaguely mottled with an indefinite purple, from which mottling a more glistening sheen may reflect than from the adjacent rock. Specific gravity of the homogeneous aphanite 2.95. The weathering also changes; in the middle parts being smoother and whiter but less regular than at the top; the hard, white, irregular feldspar segregations standing slightly out, the small hornblende crystals being slightly sunk and mostly noticeable by reflection. In the more crystalline variety, near the bottom, the weathering is uniform again, having the surface covered with numerous small characteristic knobs, often silver gray in color, composed of grains of glassy feldspar, from which, as at the top of the bed, light reflects from hornblende crystals, but here the latter are very minute, and each is sunk slightly below the surface. The presence of these shows the dioritic character of the aphanite, and that it has not become a melaphyr. For 20 feet above the foot-wall most of the eminently crystalline character is lost, the rock becomes fine-grained to compact, of a duller green color, and resembling some fine compactly textured melaphyrs; specific gravity 2.92. The lower 3 or 4 feet of the "greenstone" is broken up and fissile.

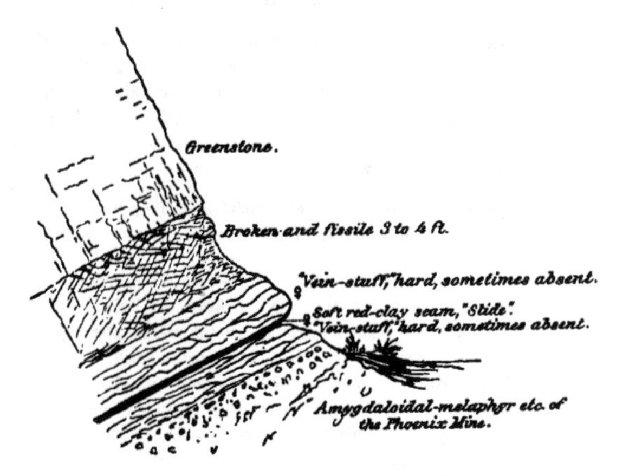

— 8090 —
± (4120)
At the Phoenix mine.

The Slide; a soft red clay seam, one to four inches thick, the equivalent of the Allouez conglomerate. Above and below it may occur several inches of hard vein-like material, red and green, with quartz, calcite, prehnite, and epidote, with some copper pretty generally scattered through it, and sometimes being present in nearly workable quantities. The Waterbury mine was worked upon the slide, but without success.

The following section is from the *Phœnix Mine;* thicknesses of beds approximate.

In the **Shallow Adit.**

109. **Amygdaloid,** narrow and indefinite, and **Amygdaloidal Melaphyr**; character obscured, apparently by the adjacent "slide" and vein. It contains much calcite, often in small veins; and scattering amygdules of delessite, prehnite, and calcite in a base similar to the underlying **Melaphyr**; rather coarse-grained, not closely textured; very irregular fracture. Composed of small but elongated crystals of reddish-white or pale-green feldspar, and very dark-green, shining delessite, which is often aggregated into mottling spots, and sometimes colored red at numerous points, probably from changing magnetic or specular iron, considerable of both being present in parts of the bed. It is coarsest at a point near the middle of the bed; whitish and dark green, spotted dark brown, with numerous incipient dark-green, delessite amygdules; colors marked. Specific gravity 2.87. Above, it is somewhat finer and greener, and at one point below nearly all of the delessite seems changed to a quite bright red, probably from change of iron minerals, no magnetite being there present. Near the foot-wall it again assumes its normal composition. Sixteen feet south of the "slide" is a narrow vein strike, with formation dip 45° S., lined with calcite in flat fundamental rhombic crystals on which rest small, pyramidal crystals of apophyllite terminated with the basal plane.

85
(42)

110. **Amygdaloid,** inclined to *Amygdaloidal Melaphyr;* matrix much as above; color brownish red; composed of small but elongated (often twinned) dirty white to transparent greenish feldspar crystals, in a reddish-brown, mottled green base; contains scattered amygdules of calcite and decomposing prehnite enclosed in delessite; graduates into **Melaphyr**; like 109, but finer, and with less magnetic iron.

48
(24)

111. **Amygdaloid,** inclined to *Amygdaloidal Melaphyr;* narrow, like 110, but finer; changing to **Melaphyr** like 110, but finer and more compact.

Section in **Sixty Fathom Level,** *Phœnix Mine,* commencing at north end and probably just above the last-described bed, or No. 111. The prehnite occurs in the amygdules like the calcite, often with calcite, enclosed in delessite, and is generally changed to a soft, amorphous, kaolin-like material. In larger amygdules, decomposition has often attacked the outside and inside, leaving unchanged prehnite between. The process of "winding," as the miners call it, goes on very rapidly in this part of the mine. The white decomposed prehnite does not effervesce with acids, and hardly differs, before the blowpipe, from prehnite. Matrix, as above, medium fine-grained; colors marked; purplish-brown, mottled green; texture not close.

165
(82)

Melaphyr; colors same, mottling fine. Near the centre of the bed is a coarse zone; feldspar in a network of whitish, elongated, often twinned crystals in a red and green base; green delessite also in frequent spots; considerable specular, some magnetic, iron; grows finer, harder, and more compactly textured in descending, becoming darker colored, green, indefinitely mottled indistinct purple. At 23 and 42 feet respectively from the foot-wall is a nearly vertical seam of laumontite.

62 (31) 112. **Amygdaloid**; quite well developed; base compact; reddish brown; amygdules irregular; near the top often united, and for a few inches below tending to form, here and there, zones of increased amygdaloidal action parallel with the bedding. While irregular amygdules often have their longer axes perpendicular to the plane of bedding, more often they are parallel with it. They contain prehnite, sometimes enclosing calcite; former often decomposed; in places, much copper occurs. Changes rather abruptly to

Amygdaloidal Melaphyr; 20 (10) feet; base fine; reddish-brown, some green; amygdules delessite, rather numerous; shades insensibly into **Melaphyr**; growing darker and more uniformly colored, being finely mottled red and green at the base, where it is fine, quite dense, and somewhat brittle and elastic.

48 (24) 113. **Amygdaloid**; 4 (2) feet; like No. 112, but not so well developed, and with but little copper. Changes rather abruptly into

Amygdaloidal-melaphyr, and **Melaphyr**, like 112, but rather darker and greener. At 26 feet north of foot-wall is a vein of laumontite 2 inches wide, approximate strike S.E., dip 75° S.W.

30 (15) 114. **Amygdaloid**; 8 (4) feet; inclined to amygdaloidal melaphyr; matrix fine, reddish or purplish; amygdules mostly delessite, but near base mostly radiated laumontite, enclosed in delessite.

Melaphyr; fine grained, compactly textured; color dark, dull purple, finely mottled darker green; becoming more compact toward base. Thin laumontite seam, with narrow accompanying amygdaloidal action, 10 feet from hanging wall.

18 (9) 115. **Amygdaloid**, nearly *Amygdaloidal Melaphyr*, narrow with **Melaphyr** below, same as 114.

24 (12) 116. **Amygdaloid**; 4 (2) feet; better developed than last; base fine-grained to compact; brown red; amygdules, mostly calcite in delessite, some prehnite. **Melaphyr** below, like 114.

57 (28) 117. **Amygdaloid**; 8 (4) feet; inclining to amygdaloidal melaphyr, and not well defined; base fine, red brown; amygdules mostly delessite; some radiated laumontite. Below is a **Melaphyr**, like that of No. 114, having at 34 and 13 feet, respectively, from the foot-wall, a tendency toward an amygdaloidal character.

40 (20) 118. **Amygdaloid**; 12 (6) feet; base compact, brown red; amygdules of calcite, with delessite exterior; irregular, inclined to large size and scattering. **Amygdaloidal Melaphyr**; 8 (4) feet; rather hard, fine, compactly textured; delessite spots in a dark purplish background. **Melaphyr**; much like No. 114; but more compactly textured, and the purplish ground being sometimes inclined to a hard metallic appearance.

119. **Amygdaloid**; 8 (4) feet; not as well developed as preceding; base greener; amygdules of calcite and delessite above, of laumontite and delessite below, changing into

115 (57) **Melaphyr**; like 114; rather hard, brittle, elastic; fine-grained, close-textured; color of brown and dark and lighter green particles, with green delessite spots. More uniform colored at base, compact, with conchoidal fracture; no magnetite. Ten feet from foot-wall is a seam; strike little east of north; dips 75° or 80° to west; laumontite and chlorite, accompanied by change of melaphyr to amygdaloidal melaphyr; amygdules of delessite and some calcite.

120. **Amygdaloid**; 4 (2) feet; base compact, earthy looking, but not soft; color reddish or chocolate; amygdules calcite, prehnite, some quartz; changes to hard **Amygdaloidal Melaphyr**; almost melaphyr, like No. 119. At 28 feet from hanging-wall is an amygdaloidal melaphyr zone, better developed than that below the amygdaloid, which changes to a **Melaphyr** like those adjacent; fine, hard, brittle, rather dark colored. 36 (18)

121. **Amygdaloid**; 8 (4) feet; base very fine; color dull red brown; amygdules of prehnite, and with little copper; changing through **Amygdaloidal Melaphyr**; 6 (3) feet; to **Melaphyr**, a wide bed; near the top inclines to brown-red color; near the middle it is a fine-grained and closely textured rock, with conchoidal fracture. The color is a clear shade of dark green, mottled purple, the dark purplish areas being larger than usual, though with indefinite boundaries, about .3 inch diameter. A spot often reflects light from all its parts at the same incident angle, causing the rock in this particular, as well as in other characters, to strongly resemble the similar modification which often occurs in the aphanitic diorite, or greenstone. It contains but little magnetite, and its specific gravity is less. Near the foot-wall the color is a nearly uniform shade of purplish brown. "*Crocker Shaft*" is about 120 feet north of foot-wall. 206 (103)

122. **Amygdaloid**; 5 (2.5) feet; not well defined; prehnitic; followed by **Amygdaloidal Melaphyr**; brown, fine; large, scattering; delessite amygdules, sometimes containing laumontite, or prehnite, or calcite, or all; sometimes in a granular mixture; extends to south end of sixty-fathom level, or 56 feet.

In **twenty fathom** level, "Bay State," a wide melaphyr, supposed to be 121, has a foot-wall at about 223 feet south of "Taylor Shaft." Below is 122. **Armygdaloid**; not well defined, with wide, chocolate-colored **Amygdaloidal Melaphyr** below, followed by a wide **Melaphyr**, like 121; color generally redder. 320 (160)

123. **Amygdaloid**; like preceding, followed by **Melaphyr**; hard, fine-grained, purplish green, dark, containing but little magnetite; extends 18 feet, reaching southern end of 20 fathom level.

INDEX.

PART III.

PALÆOZOIC ROCKS,

BY

DR. C. ROMINGER.

TO THE HONORABLE BOARD OF GEOLOGICAL SURVEY OF THE STATE OF MICHIGAN:

Gentlemen—I have the honor to lay before you the results of my investigations in the geology of the eastern portion of the Upper Peninsula of Michigan, which work was placed in my hands by the Honorable Board during the past two years.

All the facts reported upon, including the chemical analysis of quite a number of rock specimens, are derived from my own personal labors and observations. I have avoided, as much as practicable, expanding the bulk of the report; and have paid no more than the absolutely necessary attention to other branches of natural history, which, although very interesting, are not strictly comprehended under the word geology, and to which I could not make any personal contributions of value. A great portion of this material would have been a mere copy of the published records of others; and I think the public does not care so much for a compilation, as for an actual increase of our knowledge by carefully made observations, elucidating unknown or imperfectly understood features in the construction of the earth's crust.

For this object I have zealously worked. How far I have succeeded in the effort the reader may judge; and I shall feel well satisfied if he finds the picture I give worth attentive study, without having it surrounded by a borrowed glistening frame, composed of a collection of items from almost every branch of human knowledge.

Very respectfully yours,

C. ROMINGER.

INTRODUCTION.

By the Legislative Assembly of 1871, the continuation of a geological survey of the State of Michigan was determined upon, in such a manner as to divide the work into three districts, each of which was to be investigated independently by different parties.

The third district, intrusted to me, comprises the Lower Peninsula, and the eastern half of the Upper Peninsula, or that portion which is not included in the iron and copper regions. Its surface rock is exclusively composed of members of the palæozoic series; while in the other two, older crystalline and metamorphic rocks prevail.

On the Lower Peninsula only a partial reconnoissance tour has been made through *Little Traverse Bay* region.

The principal part of the time has been spent in the investigation of the Upper Peninsula, of which the Board of the Survey desired to have a complete report at the end of this year.

GEOGRAPHICAL LIMITS AND SURFACE CONFIGURATION OF THE DISTRICT.

THE Upper Peninsula of Michigan comprises an area of about 16,000 square miles, excluding the several islands belonging to the State, which add to this sum 300 square miles more.

To the north it is bounded by the waters of Lake Superior, to the south by the Lakes Huron and Michigan; the east end is bounded by the river St. Mary, which cuts from the main body three larger islands—Sugar island, St. Joseph's island, and Drummond island; the first and the last of which belong to the State; St. Joseph is Canadian land.

The western, or more accurately speaking, the southwestern boundary line between the Upper Peninsula of Michigan and Wisconsin territory, is given by the bed of the Menominee river, flowing into Green bay of Lake Michigan, and by the Montreal river emptying into Lake Superior, which bounds the remaining western portion of the Peninsula.

The so defined land lies between the 45th and 49th degrees of northern latitude, and 83° 45′ and 90° 93′ of longitude west of Greenwich.

An air line drawn from the mouth of the Menominee river to the mouth of the Montreal river is about 175 miles long; from the mouth of Montreal river to the north end of Keewenaw point, a similar line measures 150 miles; and a line drawn from Marquette to the mouth of the Menominee river amounts to about 100 miles.

These three lines inclose the Iron and Copper Districts; only the smaller portion of this area is overlaid by strata of the palæozoic series, which on the remaining eastern portion of the Peninsula, exclusively compose the surface.

This remaining portion extends about 175 miles from west to east, and is about 50 miles in width from north to south.

The surface elevation of the two western districts exceeds considerably that of the eastern district; the highest points of the eastern portion do not go beyond 400 feet, while in the two others some mountains attain a height of 800 feet, and a few a height of 1,400 feet above the lake level.

The character of these western districts is irregularly mountainous. The eastern end of the Peninsula represents an undulating high plateau, sinking in gradual descent towards the south, and more rapidly falling off towards the north. Its watershed lies much nearer to the northern than to the southern margin, and consequently the rivers emptying into Lake Superior are generally small in comparison with those entering Lake Huron and Lake Michigan. The headwaters of these rivers take their origin from numerous lakes and marshes dispersed over the plateau. None of them are navigable for anything larger than a canoe, on account of the frequency of rapids and cascades; particularly those of the north side, where almost every creek precipitates its waters over mural exposures of the Lake Superior sandstones. Nevertheless, these rivers and creeks are of high economical importance, by affording in the spring season, when the waters are high, ample power for floating the logs from the interior down to the numerous saw-mill establishments sprung up at the mouth of almost every larger watercourse, whose branches reach back into the pine lands.

The whole eastern Peninsula is an unbroken forest, having, except the lumbermen, no other inhabitants than some solitary trappers in the winter time; while, during the summer season, the fishermen take their temporary abode along the shore. Only a few of them have squatted down, and remain there all the year. Besides these there are a few migratory Indians.

GENERAL PRELIMINARY REMARKS ON GEOLOGY.

The object of the subsequent report is to give an explicit description of the above indicated portion of our State, regarding its soil, the different rock-beds composing its surface, and the various minerals associated with them. It intends to explain, as far as science will allow, the process by which all these materials were formed, and in what relation they stand to each other. It seeks to point out the different uses to which all of them are applicable in the economy of the human race, and means to aid in the understanding of the great unalterable law, which governed creation in the past, as well as it does in the present time.

In order to accomplish this in a plain and comprehensive way, I find it necessary to disregard those readers, who are already well informed, and for the benefit of those who are not so well informed, to preface these remarks by a short exposition of some elementary geological principles.

Every person is naturally inclined to inquire into the nature of things surrounding him; and one of the most obvious objects for such inquiry are the different rocks, strewn in fragments over the surface, or buried in the ground in massive blocks, or in well stratified ledges. Without being geologists, we observe granites and other rocks of crystalline structure, we notice slates, shales, sandstones, limestones, coalbeds, iron ores and other mineral substances, under various conditions.

We find, sometimes, a variety of these rocks indiscriminately mixed, in rounded water-worn fragments; at other times they exhibit a certain regularity in their position, and succeed each other in a fixed order, their layers piled up, one upon another, like the leaves of a book.

Usually, the stratified rock-beds are in horizontal position, but frequently also, we find them inclined, or standing vertical, or even reversed or twisted in all imaginable curves; ripple marks present themselves on the hardened stone-faces, as plain as those formed under our eyes by the play of waves on the shallow sand-banks of our lakes and rivers; unmistakable shells, corals, and other animal or vegetable remains are imbedded in the stone-mass, sometimes in such profusion as to constitute the principal body of the rock. All these and many other geognostical facts, present themselves to our daily observation, and there are few who are not deeply interested by them, and seek for explanations how all this originated, and how and when things assumed their present condition and form. It is the science of Geology which attempts to answer these questions.

With the increased accumulation of geognostical facts, we become more and more enabled to bring isolated observations into a system ; to see the connections in which one phenomenon stands with another, and to deduce and recognize therefrom the rules and laws which are guiding the terrestrial development.

Much has been achieved by these efforts to disclose the earth's history, but much more yet has to be learned, before we can have a clear comprehension of all the successive changes which our globe has undergone, and of which indelible traces are left to record the past. Beginning to make deductions from these our indicated daily observations, we find that the first mentioned crystalline rocks are the lowest ones we know of in the earth's crust which we have directly penetrated to the small depth of only 3,000 feet, but of which, natural outbursts and fissures have laid bare to our eyes successive layers of rock-masses, representing a depth of 80,000 feet and more.

The variety of these crystalline rocks is great; they occur most frequently in bulky masses, as if once in molten condition ; not so often in regularly stratified layers. From the existence of volcanoes, from the gradual steady increase of temperature as we go down into the interior, the hypothesis has been made that the globe was once in molten condition, and that these crystalline rocks are the cooled crust, inclosing a still fluid nucleus, at a comparatively insignificant depth.

An opposite opinion to this contends for the aqueous origin of these crystalline rocks. One of the several objections brought against the igneous theory by the so-called Neptunists, is the frequent occurrence of minerals intermingled, or in close juxtaposition, within the crystalline rocks, which in a melted condition of the mass, according to our chemical knowledge, could not have existed without combining with each other, or in some other way decomposing. The real truth is, that rocks of crystalline structure can form in both ways, and that in some instances the facts indicate an igneous origin, in others, crystallization out of an aqueous solution.

Above the crystalline rocks, if we make an ideal restoration of the series,—which, in nature, is often interrupted by elimination of one, or of a long series of successive strata,—a complex of rocks succeeds, which, by its regularly stratified condition, bears evidence of its sedimentary origin, but which has become more compact and subcrystalline in structure than unaltered aqueous deposits generally are.

These are designated as ***metamorphic rocks***, suggesting that, after their deposition, they were subjected to a metamorphosis.

Of what kind the influence was, which produced this alteration, is not accurately known. It has been supposed that the melted crystalline rocks, while breaking through these sedimentary strata, communicated to them a degree of heat sufficient to produce the metamorphism; and, indeed, the altered rock masses frequently look as if they had been subjected to a process of partial melting. We should expect, in such cases, to find the most intense metamorphism in immediate contact with the crystalline rocks; but this is not the case always; we frequently find the sedimentary beds in contiguity with the crystalline rocks, only slightly altered, while on the other hand, sometimes a high degree of metamorphism is observed in a series of strata, too far remote from the focus of heat to allow a transmission of any high degree to it, to such a distance.

The Neptunists, of course, explain the metamorphism of the rocks as dependent from the solvent action of the water, or suggest the co-operation of only a low degree of heat, and of a high atmospheric pressure. .

The metamorphic rocks are followed by a long series of rock-beds, which are all of sedimentary origin; many of them contain the remains of sea animals, and other traces of organic life.

If we had to depend alone on the visible order of stratification, and on the much more uncertain lithological characters, identifications of beds from remote localities would be difficult, and sometimes never could be made; but, by the organic remains we find included in them, we are enabled to determine with a considerable degree of precision their relative age.

In the examination of this long series of rock-beds, we find that certain animate forms have existed and flourished during certain periods, then disappeared, to give way to other new forms, which in their turn made room for subsequent new types of creation; and this coming and going of new forms is so often repeated that, according to our conception, an immense time must have elapsed while all these numerous animate forms came into existence, lived and disappeared.

By comparing the different subsequent faunas imbedded in the rock series of one place with those of a series from another place, we find that contemporaneous beds generally contain the same or similar animal forms, besides some others, which are peculiar to each locality. This gives us the ready means of investigating the stratified fossiliferous rock-beds of all parts of the world, and of assigning to them the correct place and comparative age which they have in the series of deposits. Nowhere on the globe, an uninterrupted series of all the sequence of strata and rock formations can exist. The ancient ocean which made the deposits, frequently changed its place, and in spots which emerged from the water no deposits could accumulate. Consequently, all the ocean deposits made during the interval of their emergence were wanting there, and if afterward a subsidence of the same spots took place, the newly-formed deposits would come to repose on beds of perhaps widely different, much older age, with elimination of a more or less considerable mass of strata, which in other localities separate them. Oceanic or other aqueous deposits have originally always a horizontal, or nearly horizontal position, but we find over extended districts, such deposits uplifted in every degree of inclination; we find them actually tilted over, or bent and distorted in wonderful serpentine flexures. In these disturbances we have a very palpable cause with which to explain the frequent changes in the ocean basins, to which allusion has just been made.

Even the relative time in which such disturbances occurred can be determined by observing the strata which were implicated in the process of upheaval and distortion, and by noticing those which immediately followed without being involved in the dislocating revolutionary motion.

Whether the upheavals were produced by the expansion of gases developed in the interior of the earth, and bursting out in volcanic eruptions, or by the slow contraction of the cooling earth's crust, or by the pressure of unequally distributed weight over the not entirely unyielding surface, or by other causes, are questions foreign to our present purpose, and therefore we may leave them with the few hints thus far given.

A number of other deductions can be made by wisely using the data we have found. From the nature of the deposits, and even from the fossils they contain, we draw conclusions as to the depth or shallowness of the water which deposited them; we trace the extent of ancient ocean shores by the same means; we determine by fossils whether it was fresh or salt water which made the deposit, and bring a great many other similar questions to satisfactory solution by attentive combination of what sometimes seem to be facts of small importance.

This preliminary exposition of the mode and means by which we construe a part of the earth's history, I consider sufficient to bring to the general reader of the report a fair understanding of all that is to be described in the following pages, devoted to a special description of the surface deposits of the Upper Peninsula of Michigan.

SYSTEMATICAL ENUMERATION AND DESCRIPTION OF THE STRATA COMPOSING THE EASTERN PORTION OF THE UPPER PENINSULA OF MICHIGAN.

The reports of the New York geologists have generally served, in American geology, as the basis for comparison, and in all the later geological descriptions of other States, their nomenclature and divisions of the fossiliferous strata in principal groups, have been unanimously adopted, and are also here used as the rule.

The *Palæozoic Series*, including all the deposits from the earliest time in which indubitable organic remains were discovered up to the close of the carboniferous period, is, in the New York reports, divided into the following principal groups:

I. LOWER SILURIAN.

1. Potsdam sandstone.
2. Calciferous sandstone.
3. Chazy limestone.
4. Trenton limestone, including Birdseye and Black river limestones.
5. Hudson river group, including Utica shales; by western geologists also called Cincinnati group.

II. UPPER SILURIAN.

6. Medina sandstone, a more local development in New York State.
7. Clinton group.

8. Niagara limestone.
9. Onondaga salt group.
10. Helderberg group, the upper division of which is connected with the next superimposed Devonian system.

III. Devonian System.

11. Hamilton group.
12. Chemung and Portage group.
13. Catskill group.

IV. Carboniferous System.

14. Subcarboniferous limestone.
15. Coal measures.

The Michigan district under consideration is composed of the older portion only of the above tabular series, commencing with the Potsdam sandstone, and forming an uninterrupted series up to the Devonian period, of which the upper Helderberg limestones are the highest member developed. Of later deposits nothing is known to have existed in this region before the large drift accumulations took place which now envelop almost the whole surface, and restrict the outcrops of the older formations to very limited areas.

The greatest regularity prevails in the disposition of the Palæozoic strata. They seem not to have suffered from any subsequent disturbance, and have retained their original horizontal position, with a slight dip southward, in which general direction the ancient ocean appears to have gradually receded.

We find the older strata exposed in a belt, extending from west to east, and forming the north shore of the Peninsula; moving southward from there, we successively step over newer strata, which follow each other in imbricating order, until, at the south shore, on St. Ignatz Point, or at the island of Mackinaw, we stand on the limestone beds of the upper Helderberg group.

Not all the single subdivisions made in the systematical description of the New York series can be recognized, but the principal groups are identified without difficulty. To the description of each of these a separate chapter will be devoted; but, first of all, the superficial deposits called *drift* and *alluvium*, which compose by far the greatest portion of the surface of the district, deserve our attention. From their nature and composition, the character of the vegetation in a great measure depends. From the vegetation we draw conclusions as to the agricultural characteristics of a place, and in these, the principal value of a country rests.

SUPERFICIAL DEPOSITS.

Under the name of Drift, in geology, are included immense masses of loose, rounded or water-worn rock fragments, and of finer comminuted materials, which are spread over a wide and long extended belt of country in certain latitudes of the North American and European continents.

It is an established fact that these deposits were once transported in the general direction from north to south. We find the drift material in part composed of rocks which are not found in any of the neighboring districts, but which are known to exist at variable distances north of these deposits, exhibiting the same details of structure, color, and composition, so that no doubt can remain as to the origin of the fragments.

In addition to this, the whole wide area over which the transportation of this material took place, bears unmistakable marks of the mechanical effects which the moving débris exerted on the underlying strata.

Whole formations of more perishable rock-beds, which stood in the way of the moving detritus, were broken up, mixed with the intruding masses, and carried away from their old resting-places by irresistible force.

The projecting inequalities of harder and more unyielding rock-masses were ground smooth, or deep furrows were carved into their surfaces, which exhibit yet, with exquisite distinctness, innumerable fine parallel scratches, by which this grinding and carving were effected. All these scratches point in one constant direction from which the force acted, and which varies very little from a north and south direction. In certain single instances the resistance offered by some obstacle was so strong, that the moving rock stream was deflected, and, in such a case, the local drift scratches may run in any other direction.

These phenomena are the same that have been recognized in connection with the moving rock-masses of the glaciers in Switzerland, and in other Alpine regions.

The accumulated materials of the drift, frequently like the moraines of glaciers, have no stratification, and are a heap of confused rubbish, inclosing boulders and rock fragments of such huge weight and dimensions, that the propelling power of waves and water currents, unquestionably under ordinary circumstances, was not adequate for the performance of the work.

The hypothesis of glaciers and icebergs having been the transporting mediums of the drift deposits, in consideration of these facts, is now almost universally adopted by geologists; but notwithstanding its plausibility, a great many circumstances connected with the drift formation remain still unexplained by it, and we have no definite conception of the causes which could produce glaciers of such stupendous extent. A considerable portion also of the drift deposits are not heaps of rubbish, but well stratified beds of clay, sand and gravel, exhibiting with plainest distinctness the action of waves and water currents in their deposition.

It has been said this stratified drift was the product of subsequent floods, which sifted and rearranged the materials in their present regular beds. No doubt a large portion of the drift has been rearranged and worked over by later floods; the terraces, so frequently observed around the shore line and along some river valleys of the Upper Peninsula, are confirming this suggestion. In various elevations above the lake level we see these terraces in stair-like succession ascending to the height of eighty feet, and it is probable that within comparatively recent time the waters of the lakes stood high enough to cover the whole Eastern Peninsula, and may have modified the surface; but it is equally evident that much of the stratified drift observed there, and in many other places, was originally deposited prior to the time in which the terraces were formed, and the last finishing touch was given to the configuration of the country, as we see it at present.

Much of the non-stratified coarser boulder drift, which resembles moraines, is found in the narrower channels of Lake Superior and of Lake Huron. Many of the small islands in these channels are composed exclusively of such coarser material, and shoals, paved with larger drift blocks, extend far out into the lake, endangering navigation.

The non-stratified condition of the drift in these cases is directly the result of later floods acting on the drift, which before may have been well stratified. By the action of the water, the drift barriers, separating the lakes into different levels, were broken through and all the lighter particles washed away, the current pushing with increased velocity through the newly-made outlet, and dropping on the way the heavier rock fragments and larger pebbles which we find now directly resting on the rock beds of older formations. Also, in other localities of the Upper Peninsula, on the top of higher hill elevations, composed of older rock beds, we find frequently only heavier drift boulders and scarcely any finer comminuted material spread out over the surface, which seem to be the remnant of former larger deposits. The non-stratified drift is consequently not all to be brought in connection with moraines of glaciers, and in a great many of the better exposures of the peninsula the drift is found in well-stratified layers, in which constantly a lower deposit of an impalpably fine rose-colored clay, of almost shaly structure, can be distinguished, and an upper arenaceous and gravelly portion, containing larger boulders.

These lower clay beds are exposed in the bed of Carp river in St. Martin's bay, on Manistique river, on Menominee river, Ford river, Escanaba river, and on Lake Superior shore near St. Mary, on Taquamenon river, at Sable Point, and in many other places, where they rest immediately on the older rock formations. Usually seams of gravel are intermingled with the clay deposits, or single pebbles are dispersed through the mass. On South Manitou island the lower clay beds are more sandy than elsewhere. In the districts underlaid by limestone, the clay contains a more or less considerable admixture of carbonate of lime; only in a few instances, however, the proportion of the admixture is so large as to make it unfit for pottery ware. The experiments I made with several specimens of the clay gave satisfactory results; it dries without cracking, and by burning furnishes a tough, light-yellowish colored ware, which is not fusible in moderate fire.

In the sandstone districts of Lake Superior, the clay is quite free from calcareous parts, and deserves the attention of manufacturers. From Sugar island and from St. Mary, limited quantities have been shipped to iron furnaces, to be used as ordinary clay, but the experiment is worth trying, if a proper mixture of the clay with coarse quartz sand, would not have the qualities of a good fire clay.

The thickness of the clay deposits is very variable in different localities. In the land slides on both sides of the lower part of Taquamenon river, and in the hills south of St. Mary, it is not less than sixty feet, in other places on the Lake Michigan side it is often not over ten feet.

The upper arenaceous portion of the stratified drift, is composed of alternating sand and gravel beds, with occasionally intermingled larger boulders and subordinate seams of clay.

The frequent occurrence of discordant stratification, makes the strata very irregular, and no general rule in the succession of the sand and gravel layers can be recognized; their deposition was governed by local influences of very changeable nature.

In the sandhills of Sable Point and on Point Iroquois, the aggregate thickness of this upper part of the drift formation is 300 feet. This seems, however, to be the maximum, and in most other places it is found to be less.

Over extensive areas, the drift material forming the surface is an almost pure fine quartz sand of very sterile character, congenial only to the growth of pine trees. Other districts are underlaid by the more gravelly and argillaceous portions of the drift, which form a productive soil, covered by a rich vegetation of hard-wood timber. The geologist can form here reliable conclusions on the nature of the superficial deposits, from the character of the forest trees, before he examines the soil, and the insufficient denudation of the strata in the interior of the Peninsula, makes the indications given by the vegetation frequently a welcome help to him.

The coarser gravelly constituents of the drift, and the larger boulders, represent an immense variety of rock formations and mineral species. The older granitic, syenitic and gneissoid rocks are never missing, diorites are also abundant of all shades, from the coarsest crystalline structure to the aphanitic condition, quartzites, jaspers, conglomerates, from the Huronian group, and amygdaloid and porphyritic rocks of the copper bearing series; blocks of iron ore or of metallic copper are occasionally interspersed, and quite common are also limestone and flint pebbles, containing well-preserved fossils of the Niagara group, and many Devonian species, which found their way from distant arctic regions to the shore of Lake Superior.

A very important part of the drift pebbles is taken up from the detritus of the neighboring and immediately underlying formations, which by their multitude and less worn angular form betray the proximity of the mother rock.

So we find the drift, reposing on the Lake Superior sandstones, full of sandstone blocks, and smaller fragments; further south, Trenton limestone, with its characteristic fossils, is very common. In the Niagara belt, the Niagara limestone almost excludes every other rock species, and the same thing is repeated within the limits of the Helderberg group.

Organic remains, contemporaneous with the drift period, are remarkably rare. In all the numerous exposures of undisturbed drift on Lake Superior, I did not see a single animal or vegetable fossil; but in the loose alluvial deposits, by which our present time is insensibly linked with the drift era, vegetable and animal remains are copiously imbedded.

In many of the larger river valleys, as, for instance, in the valley of Manistique river, extensive deposits of sandy argillaceous mud, mixed with vegetable débris, and interstratified with black peat-like beds of almost entirely vegetable composition, are met with. We find imbedded in them large trunks of the cedar, the birch, and other forest trees of the present day. Giant trees of the same species now grow and flourish in the soil of these deposits. The woody fibre of their buried ancestors of the cedar tribe is as well preserved as if only a few years had passed since their entombment. The bark of the birch trees looks yet as white as snow, while the more perishable wooden nucleus has almost entirely disappeared.

A similar recent deposit of vegetable débris and tree trunks of ten feet thickness, is shown on the shore of Lake Superior, in the drift bluffs of Sable Point, about two miles west of Grand Marais harbor. To a superficial observer, this deposit seems to make part of the drift; the vegetable matter reposes immediately on the lower clay beds of the drift formation. But a more careful examination reveals clearly the discordant position of the carbonaceous deposits to the adjoining strata of the drift. The vegetable substances evidently accumulated on the bottom of a deep smaller lake basin, which was excavated in the drift, and subsequently covered up by the sand falling in from the sides, until the whole of it was filled up, and made even with the surface. By the encroachment of the Lake Superior on the shore, a cross-section of this former basin was exposed in the lake bluffs, and by the perfect similarity of the covering sand layers with the adjoining sand deposits, this carbonaceous wedge, cropping out from the base of the apparently unbroken continuation of a drift bluff, seventy feet high, is very apt to deceive.

Another alluvial clay deposit, containing many vegetable débris, and large numbers of sweet water shells, planorbis, physa, cyclas, and also some helices, is exposed in the bed of Carp river in St. Martin's bay. A similar much more extensive deposit, on the Lower Peninsula, near Little Traverse bay, forms the whole bed of Pine lake, a sheet of water sixteen miles in length.

The practical value of these alluvial vegetable deposits is very small; they are mixed with too much sand and argillaceous particles to become good fuel.

In the marshy meadow lands of the Peninsula an abundance of peat is stored away for future use, and is in constant process of formation. The large demand for fuel in the production of iron has already induced a company in the iron district of Marquette to make use of the peat, and, as it appears, with satisfactory results.

BOG IRON ORES OF THE UPPER PENINSULA.

Among the more recent superficial deposits, the bog iron ores are of high practical interest. A number of localities in which bog iron ore was found were long since pointed out in the maps of the linear surveyors of the State.

In the late preliminary report of the former State geologist inexhaustible quantities of this ore were said to exist on the western tributaries of Manistique and of Taquamenon rivers, and for that reason I have given particular attention to the investigation of these districts.

Nearly all of the springs contain small quantities of iron in solution, which they deposit, if by escape of carbonic acid, or by other chemical action, the solvent power of the water for the iron should fail. The waters of all the rivers of the Peninsula are freed of the iron which they hold in solution, by passing through the extended swamps, from which the rivers originate, and the small continuous secretion of iron particles, in favorable spots, gradually increases to masses of such extent as to constitute valuable ores for the manufacture of iron. Such large deposits are, however, of rare occurrence. Usually, we find the surface coated with a crust of ochraceous mud, or the superficial sand is infiltrated and sometimes cemented into a hard sandstone band by the ferruginous matter, or concretiary lumps and nodules of a purer hard ore are sprinkled thinly over the ground, and frequently adhere to the root fibres of fallen trees. Spots of this kind, noticed by the woodmen and the surveyors, have been carefully indicated in the maps, and seem to show that the whole centre of the Peninsula is carpeted with bog iron ore ; but the quantity of the ore here alone determines its value, and not its quality. Only in very few places the ore deposits are important enough to give favorable hope for mining operations.

Among all the bog iron ore deposits, which came under my observation, I found one of the best localities on the head-waters of the Taquamenon river, which I am going to describe.

Between the head-waters of Two Hearted river and the west branches of Taquamenon river, a swampy, almost treeless, high plateau spreads out, on which, for many square miles, not a trace of bog iron is to be seen, until finally, at the southeast margin of it, the stagnant ditches of the plain collect into a tortuous channel with slow current, which, before it enters the timbered lands, is surrounded by a grassy marsh, about sixty acres in extent. Over this spot the bog iron ore is dispersed in small irregular patches, which are clear of all vegetation, excepting a few low bushes. The total of the surface occupied by the ore can be estimated to be three acres ; the greatest observed thickness was fifteen inches, but in other places it was not more than two inches, and its average thickness I would not estimate over six inches. Beyond these limits, no more ore is found in this locality.

The ore is evidently of a rich quality, but I have not found the time yet to make an accurate chemical analysis of it. Numerous other localities for bog iron, which I saw, were of no greater extent than the described deposit, and most of them much inferior to it. Also, the frequent descriptions of bog iron ore deposits given to me by the woodsmen did not indicate the existence of any more favorable spots for the ore in this district.

Taking into consideration the remoteness of these places from all passable roads, and the easy exhaustibility of such limited areas, I am of the opinion that at present little practical benefit can be derived from these mineral stores, which perhaps in the future may become valuable, after the country gets more inhabitants, and is traversed by railroads, or even only by common roads.

MAGNETIC IRON ORE SAND.

In certain localities on Lake Superior and on Lake Michigan, the beach is found thickly covered with black, magnetic iron ore sand, which has been washed out of the drift sand by the action of the waves.

The separation of the iron ore from the quartzose sand is in streaks almost perfect; other portions would require purification by artificial washing to make it fit for the melting furnace. The quantity of the ore sand is sometimes so considerable that shiploads of it could be gathered with very little expense.

HELDERBERG GROUP.

Below the drift, the next youngest of the deposits, composing the Upper Peninsula, are the limestones of the upper Helderberg group.

Their surface extent is limited to the triangular landspur, inclosed between Point aux Chenes, Point St. Ignatz and Grosspoint, in St. Martin's bay. The islands of Mackinaw, Boisblanc, and Round island are likewise composed of the same strata, which continue southward across the straits, and form a wide belt around the north end of the Lower Peninsula. The present report, however, is specially devoted to the description of the Upper Peninsula, and the Helderberg strata of the Lower Peninsula will be left out of consideration.

The thickness of this group, within the indicated district, amounts to about 250 feet. In New York State, and particularly in the eastern part of it, it has a much greater development, in which two principal divisions, an *upper* and a *lower* one, are well marked. The lower division is known to extend through Pennsylvania, Maryland and Virginia, to Tennessee. It has been recognized in southern Missouri and southern Illinois, and in the northeast part of the continent; it is largely developed in the promontory of Gaspe. Directly west from the Helderberg mountains, from which the name of the formation is derived, this lower division soon disappears, and has also in the adjoining Canadian districts, and in our State not been clearly recognized.

It is the upper division of the New York series which is represented by the strata of Michigan. The identity of the compared strata is fully demonstrable by the identity of their faunas; but a detailed parallelism between the successive layers in the different places cannot be established.

A very peculiar brecciated lime rock is considered to be the lowest stratum of our Helderberg formation in the Upper Peninsula. It is composed of a great variety of calcareous, dolomitic, cherty and calcareo-argillaceous rock fragments, mixed and thrown about through the re-cemented rock mass. A great portion of the brecciated material is distinctly recognizable as the fractured beds of the immediately underlying formation, and frequently larger rock masses, composed of a series of successive ledges, which have retained their original position to each other, are scattered through the breccia.

The lower part of the breccia rarely contains fossils; near its upper termination, fossils sometimes becomes very abundant. None of these are peculiar to the breccia; the same species continue to exist during the time in which the upper undisturbed series of the group was deposited.

The picturesque scenery of the island of Mackinaw is produced to a great degree by the bold escarpments of this brecciated rock along the shore line, and a similar character is imparted to the landscape on the landspur of the mainland of the Peninsula, as far as this formation extends. It retains, however, not uniformly, this brecciated character; in a number of places it can be observed that the bold rock masses, in their horizontal continuation, gradually are replaced by regularly stratified, less disturbed beds of dolomites, or dolomitic limestones, alternating with shaly or marly layers.

Along the lake shore, between St. Ignatz Point and Point aux

Chenes, repeatedly the precipitous walls of the breccia rocks disappear, and the hillsides assume softer, more rounded forms, on the same level, in which a continuation of the vertical rock walls would be expected. In other instances, the powerful denuding agencies of the drift period have swept away the softer, well-stratified portions, but could not destroy the more compact brecciated limestones, which now stand like obelisks, in isolated masses along the margin of the hills, or project in strong relief from their declivities, forming arches, caverns, and other specimens of natural architecture which, from popular imagination, in many instances have received fantastical local names.

The disturbing influences which occasioned the brecciated structure of the Helderberg strata were not confined to this small northern area. Similar effects, referable to the same period, can be observed on several of the islands on Lake Erie, where the lower rock beds are of a decidedly brecciated character. Also the Helderberg strata, exposed in the village of Rock Island, Illinois, are partially in brecciated condition.

The upper division of the Helderberg series in the Upper Peninsula, is made up by uneven bedded limestones, and by cherty and argillaceous layers, attaining an aggregate thickness of about 100 feet, on the island of Mackinaw, where it forms the summit elevation on which old Fort Holmes was situated. The larger portion of the island has been denuded from these higher beds during the drift period, of which thousands of granite boulders and other rocks were left behind, strewn over the rolling plateau, formed by the brecciated limestones, whose beds offered more resistance to the destructive forces. During this process of denudation, portions of the upper series were undermined, and occasionally tumbled downhill into open clefts between the breccia rocks, escaping in this way further destruction. An instance of this kind can be observed in ascending the wagon-road to the new fort. A large crevice in the breccia is seen there, replenished by a wedge of vertically standing beds of the upper fossiliferous limestones. Similar conditions are found on the triangular spur of the mainland, where several conspicuous hilltops are formed by the upper fossiliferous beds of which, at the foot of the hills, larger dislocated masses fill out the ravines.

Fossils are very abundant in this upper division and in the subjacent terminal strata of the brecciated limestone. I have stated already the conformity of the fauna in this whole upper complex. Many of the fossils seem to have been fractured and waterworn before they were inclosed by the calcareous sediment; some are compressed, and by the process of petrifaction, the finer structure of many of them has been partially obliterated. Most of the fossils are calcareous, or only partially silicified.

On comparison with the fossils of the corniferous limestones of Canada West and of the Sandusky limestones, we find the same forms in either of the deposits, testifying their equivalency.

In the following list all the species I found are enumerated:

Several species of *Stromatopora*, of which I have prepared descriptions and magnified photographic views, are not published yet.

Zaphrentis gigantea, M. Edw.
" prolifica, Billings.
Clysiophyllum Oneidaense, Billings.
Cyathophyllum rugosum, M. Ed.
Heliophyllum, not described species.
" exiguum, Billings.
Chonophyllum magnificum, Billings.
Phillipsastræa Verneuilli, variety with larger cells.
Eridophyllum Simcoense, Billings.
Cystiphyllum Americanum ?
Syringopora Hisingeri, Billings.
" Maclurei, Billings.
" nobilis, Billings.
Alveolites squamosus, Billings.
Cladopora labiosa, Billings.
" expansa, nov. spec.
Favosites hemisphericus, in many varieties, with larger and smaller cells and with variable development of perfect or compound diaphragms.
Favosites Canadensis (Fistulip. Canad., Billings).
" tuberosus, nov. spec.
" turbinatus, Billings.
" epidermatus.
Michelinia convexa, M. Edw.
And several Bryozoa.

Of shells are found:

Atrypa reticularis.
Centronella glansfagea.
Pentamerus aratus.
Stricklandinia elongata.
Strophodonta hemispherica.
" demissa.
Spirifer divaricatus.
Conocardium trigonale.

Of Trilobites: fragments of Dalmannia, of Phacops and of Proetus crassimarginatus.

The value of this limestone formation for building purposes is very limited; the upper strata are thin and not even-bedded, full of silicious veins and nodules of hornstone; the lower brecciated rocks are too heterogeneous and subject to rapid decay.

In construction of the fort the brecciated rocks have been used, but the dilapidated, rough appearance of its walls are little apt to recommend them for such purposes.

Some beds are adapted for the manufacture of quicklime, but at present only the small demand for local consumption has been supplied from them.

The marly and argillaceous constituents form a large proportion of the group, and their mixture with the superincumbent arenaceous drift material, make a very congenial soil for agriculture. By the extent of cultivated lands in this region the limits of the formation are pretty well defined.

To the west, the barren drift sand covers the surface; to the east, the broken, rocky surfaces of the Niagara group have offered to the farming settler no inducements and have remained in their primitive state of wilderness; while along the hillsides facing the lake, from Mackinaw to Point aux Chênes, many a friendly house and verdant clearing attract the attention of the voyager sailing by.

Below the brecciated limestone a series of light colored, thin, and even-bedded dolomites follow in descending order, interstratified with softer marly layers; but these succeeding lower strata are so intimately connected with the brecciated limestones, that often only an arbitrary division line can be drawn between them.

The base of the breccia, in a great proportion composed of the fractured lower strata, merely differs from them by the disturbed condition of its materials, and not by lithological characters. If in places, as has been stated above, the breccia limestone loses its brecciated character, and forms regular beds, it is almost impossible to distinguish an exact geological level.

On the east side of Mackinaw island, well-stratified, light-colored dolomites, in beds of 6 or 8 inches thickness, form an outcrop right above the water line, and are directly overlaid by the brecciated limestone.

In approaching the southeast corner of the island, in the cliffs called *Robison's Folly*, these dolomites are seen on a higher level; further on, west of the Mission House, the whole elevation of the hillside is composed of these well-stratified dolomites, in somewhat thinner ledges and interchanging with softer marly layers. The breccia rock has receded here from the margin, and seems to be partially replaced by the marl and dolomite beds, but a little distance east of the new fort, its massive rock walls come to the front again, forming a conspicuous vertical cliff, on which the southeast corner of the fort is built.

This spot offers a splendid opportunity to study the relations of the dolomites to the superincumbent breccia rock; the gradual transition of the shapeless, massive portions of the rock, into regularly stratified beds, making the horizontal continuation of the former, is here plainly laid open to the observer.

The shore line between St. Ignatz Point and Point aux Chênes repeatedly presents similar exposures. The lower well-stratified dolomites form a terrace of ten or twelve feet elevation, from which, a small distance away from the lake border, the vertical walls of the breccia limestone ascend, or on which a softer rounded hillside is built up, by a succession of regularly bedded dolomite and marly strata, replacing the breccia.

Very characteristic for this dolomite formation, in its whole extension, are tabular leaflets of calc-spar crystals, pervading certain ledges in every direction; seen edgewise, the crystals appear in acicular form. In many instances the spar crystals subsequently have been re-dissolved, and the empty spaces present themselves as narrow slits in the rock.

In one locality I found in the same strata, a piece of dolomite, representing a pseudo-morphose of a rock-salt crystal, one and a half inches in diameter. It has the shape of a quadrangular pyramid, with parallel striæ concentrically arranged over the facets, as in coarsely crystallized culinary salt.

In fresh fractures, the dolomite has a fine crystalline grain, but its weathered surface has a rough, absorbent, and mealy aspect, and as decay goes on it softens to an easy friable aggregate of microscopic, rhomboidal crystals.

Within the same horizon of the series, in which the acicular spar leaflets are most common (and this is in the upper beds of the dolomites, which frequently, also, enter into the composition of the breccia), a few species of fossils are found, which are highly interesting.

One is a large *Leperditia*, which I identify with Liperditia alta of Hall; another is *Spirifer modestus* (Hall). Both are casts which sometimes, in great numbers, cover the surface of the slabs. Except those, I only could find the indistinct cast of a *gasteropod*, and another cast of a *cyathophylloid star-coral;* all my hunting for *eurypterus*, which would be expected here, was in vain.

Leperditia alta and Spirifer modestus are known as lower Helderberg species. On such a meagre representation of a rich fauna, I would hesitate to base conclusions regarding the age of the formation; but taking also into consideration the stratigraphical sequence, surmounted above by well-characterized upper Helderberg strata, underlaid below by beds of perfect lithological resemblance to the Onondaga salt group, we may safely take the intermediate beds as contemporaneous with the lower Helderberg group.

The thickness of this dolomitic rock series, as far as exposed, I estimate at sixty feet; an accurate determination of it is impracticable. Its upper limits have been described as connected with the brecciated limestones by insensible gradations, and its connection with the gypsiferous beds below, is, for the most part, hidden in the depths of the lake or under drift accumulations.

The dolomites are generally too thin bedded to be of much practical use for building purposes and for quicklime; the large percentage of magnesia degrades the product.

A chemical analysis of a specimen from the exposures on the east side of Mackinaw, close to the water level, gives

Carbonate of lime...... 55 per ct.
" " magnesia.. 41 "

The remainder is iron, alumina and bituminous matter.

A specimen of the upper fossiliferous Helderberg limestones from Blanchart's farm gave on analysis:

Carbonate of lime.............	68 per ct.
" " magnesia.....	22 "
Siliceo-argillaceous residue, insoluble in muriatic acid...........	8 "
Iron oxyd and alumina hydrate..	2 "

ONONDAGA SALT GROUP.

In the bed of the lake, near Blanchart's farm, not many feet below the outcrops of the just described dolomites, green and red variegated marls, alternating with more massive ledges of a calcareous rock, are seen. Near Little Point aux Chênes, and on the lowlands joining the shore to the west of it, similar variegated marls with beds of shale and dolomitic ledges form the surface rock. Interstratified with these are nodular concretionary masses, and narrow bands of gypsum. The gypsum there is principally composed of a white granular plasma, through which larger brown colored crystals of gypsum are disseminated; some of it occurs also in seams composed of transversal, silky, shining fibres.

The dolomites, interstratified with the gypsum, are much more mixed with silicious and argillaceous matter than the higher dolomite beds. A specimen from the gypsum quarries of Point aux Chênes had the following composition:

Carbonate of lime..............	46 per ct.
" magnesia.........	14 "
Alumina and iron oxyd hydrate..	10 "
Siliceo-argillaceous residue, insoluble in mur. acid.............	30 "

Boiling solution of carbonate of soda dissolved out of this residue three-tenths of its weight.

Many years ago quarries were opened near Point aux Chênes for this gypsum, but are abandoned now. From the appearance of the quarries I infer, that the beds of the gypsum were too irregular, and too much of worthless material had to be handled, in order to gain it; perhaps another reason for the suspension of the work is the remoteness from the market, while the same, and I suppose a better article, is found abundantly right in the centre of the district, for which it is in demand.

The same variegated marls, as seen near Point aux Chênes, are found again underlying the St. Martin's islands; their surface is principally composed of drift material. I was informed that large white gypsum blocks are sometimes found on the beach of the islands, and a locality on the larger St. Martin's island was described to me, in which the gypsum could be seen in place, but I did not succeed in finding the spot, although I spent a whole day in hunting for it. In the bottom of the lake, twelve miles east of Mackinaw, in the circumference of Goose island, extensive patches of snow-white gypsum can be seen at the depth of eight or ten feet; the island itself is a low gravel bank, and does not show any rock exposures.

In the bed of Carp river in St. Martin's bay, half a mile above its mouth, the water runs in rapids over rock ledges, which in all probability represent the lowest beds of the gypsiferous formation. The river bank is composed of six feet of yellow sand; next below are five feet well-stratified, impalpably fine rose-colored clay; then follows a stratum of gravel two feet thick, which forms the base of the drift formation. Immediately under this gravel are ash-colored dolomites, mottled and shaded with red. They segregate in uneven rugose layers of six or eight inches in thickness, and are interstratified with thin seams of shaly and marly substance; numerous cellular cavities, filled with calc-spar and occasionally with sulphate of baryta, are seen in the rock, which has a fine-grained, somewhat earthy fracture. The strata, of which six feet are exposed, have a gentle dip to the southeast, which would bring them under the variegated marls of St. Martin's island. In the opposite direction, the first rocks met with are the Niagara limestones. The immediate contact of the two formations is, however, nowhere to be seen.

The lithological characters of this complex of strata, and their position below the Helderberg group, and above the Niagara limestone, have been to all geologists satisfactory evidence in identifying them with the Onondaga salt group of New York. Salt springs, which make this formation in New York so very important and valuable, have not been found in this district, and are not likely to be found where nearly all we have of the formation lies in the bottom of the lake, and for this same reason its thickness cannot definitely be ascertained in this region.

The Dolomites in the bed of Carp river are composed of:

Carbonate of lime..............	53	per ct.
" magnesia.........	44	"
Hydrated iron oxyd............	1	"
Insoluble matter, consisting of crystals of sulphate of baryta and bituminous substance....	2	"

The variegated marls of St. Martin's island give on analysis:

Carbonate of lime	41	per ct.
" magnesia..........	22	"
Iron and alumina hydr.....	7	"
Insoluble residue, mostly silicious sand........................	30	"

NIAGARA GROUP.

With exception of the small area, in which the Helderberg and the Onondaga groups have been described as forming the surface rock, the whole south part of the Peninsula is composed of a series of dolomitic limestones, which prove their contemporaneous age with the Niagara group, not only by numerous characteristic fossils, but also by the immediate continuity which they keep with the typical rocks of the Niagara falls.

A belt of this formation strikes from Niagara river northwestward to Cape Hurd on Lake Huron, and from there continues through the Manitou islands to Drummond's island, only interrupted by a few narrow channels of the lake.

Drummond's island is the eastern end point of the north part of our State. A small belt of its northern end only is composed of older sedimentary rocks; the remainder is all built up by the ledges of the Niagara limestone.

All the numerous islands in the south part of the bay between Drummond's and St. Joseph's islands, the south end of St. Joseph, Lime island, Round island in the St. Mary's river are underlaid by Niagara rocks, but the most part of their surface is boulder drift. On the mainland of the Peninsula, the northern limits of the formation are to a great extent conjectural.

Without considering the almost universal deep drift covering, which hides the rock beds, the uninhabited condition of the country, and the extensive swamps and impenetrable timberlands spreading over the ground make it almost impossible even to penetrate into the interior. Far less success can be expected for a geological exploration, which, however, was in many instances attempted. But to travel with such impediments for many successive days, and to return without having been able to accomplish the desired object, is a severe disappointment.

The Niagara rocks seem to extend as far north as the south shore of Manusco bay; no outcrops of rock ledges can be seen there, but the rock fragments intermingled with the drift material are mainly Niagara limestone, containing Pentamerus oblongus, and other fossils. Of the Hudson river group, which one would naturally expect to find there, not a single pebble is to be seen.

The north boundaries of the western extension of the group approximatively coincide with the centre of the Township Tier 44. In approaching Millecoquin's river, which is over fifty miles west from Manusco bay, the Niagara limestones reach further north, and are observed in outcrops as far as the north shore of the largest one of the Manistique lakes in the centre of the Town Tier 45. From there, the limits of the formation descend again in a southwest direction, pass at a distance of several miles north of Indian lake, a tributary on the west side of Manistique river, and finally run out near the mouth of Sturgeon river, in Big Bay de Noquette.

A line drawn from the mouth of this river, and prolongated until it strikes the west side of Green Bay peninsula, will indicate the direction in which the Niagara formation continues southwards. The Peninsula, and all the islands east of this line, are composed of Niagara rocks; west of it we find only older strata.

After having delineated the extent of the formation, I proceed to give the results of its special investigation and study in different localities of the district.

Commencing at the east end of the State, at Marblehead off Drummond's island, we have undoubtedly the best section which is offered anywhere to start with.

This promontory, forming the northeast end of the island, has an elevation of about 100 feet above the lake. Its sides are steep, with occasional perpendicular rock escarpments; the strata slowly dip southwards, and consequently, on the north side of the hill deeper strata come to the surface; their dip is, however, by no means so uniform that by comparing dip and distances reliable conclusions on the thickness of the formation could be derived. Undulations in the strata are frequently observed.

The summit of Marblehead hill is formed by light-colored, uneven-bedded, crystalline dolomitic limestones; their moderately thin ledges are pervaded by numerous silicious veins and nodules of hornstone, and inclose a great number and variety of silicified fossils, particularly corals. I collected from these beds:—

Halysites esharoides, Syringopora verticillata.
Syringopora, similar to S. compacta of Billings.
Heliolites interstinctus, Heliolites elegans?
Lyellia Americana, Strombodes pentagonus.
Several forms of Cyathophylloid corals, Favosites favosus.
Favosites Niagarensis, Favosites venustus, several species of Alveolites, and of Cladopora, and likewise of Stromatopora.

Of shells I found: Atrypa recticularis, Pentamerus oblongus, and a large Euomphalus, the two latter of which are generally only imperfect casts.

The thickness of these beds at the spot is about twenty-five feet; the lower ones are much less fossiliferous than the upper. Next below follows a massive, light cream-colored, obscurely stratified dolomite of coarsely crystalline grain, and projecting in perpendicular rock walls from twelve to fifteen feet high. It is filled with casts of Pentamerus oblongus, their form almost obliterated to indistinctness; or merely the cavities once containing them, and now lined with calc-spar or quartz crystals, are preserved. In the same imperfect condition, Favosites and Stromatopora are abundantly dispersed through the rock mass.

At the foot of these walls other more thinly and regularly stratified dolomite ledges of finer grain, and containing no fossils, are seen in the thickness of four or five feet. The subsequent lower strata are hidden by talus.

Some distance sideways, it may be at a ten feet lower level, a downward continuation of the section is most plainly given in a quarry, which once had been worked there, but is now abandoned.

The top of the quarry, which is about thirty-six feet above the lake, is formed by dark gray-colored, highly crystalline limestones, with silicious veins and containing many fossils. Stromatopora is most abundant; besides this, Halysites esharoides, several kinds of Favosites, Alveolites, Syringopora, Cyathophylloid Corals and Crinoid stems are found. Somewhat lower the crystalline limestones become mottled with lighter silicious spots, and at the depth of six feet from the top the crystalline structure changes into a dull, more earthy-fracturing rock of lighter yellowish color, the fossils decrease in numbers, and gradually are absent.

From there, to the depth of fifteen feet, light-colored, yellowish white, earthy-fracturing limestones, of laminated structure, continue in layers of variable thickness, which easily decay and fall into angular fragments after some time of exposure. The next lower eight or nine feet are again of crystalline structure and darker gray color, separating in beds up to two feet in thickness, while the lowest portion of the quarry exhibits again more earthy-fracturing beds of distinctly laminated structure. The sole of the quarry is about eight feet above the level of the lake.

The thicker beds in the lower portion of the quarry, including the darker colored crystalline ledges, furnish a building material ot fair appearance, but the weather seems to affect them injuriously, and the name of marble is ill-adapted to them; they have neither the lively colors of marble, nor are they susceptible of a fine polish. A small distance north of the quarry, after a short interruption of the exposures, a continuation of the section through the next lower strata can be observed by following the shore line, where successively the deeper layers are brought to light by the gentle southward dip of the whole formation. At the same time, the level of the lake intersecting the outcrops allows one to observe minutely the different undulations to which the strata are subject. The next succeeding rock-beds, below the sole of the quarry, are dark gray, bituminous dolomites, of nodular, very unhomogeneous structure. Seams of black carbonaceous matter wind themselves through the rock mass; other beds are stained with large white blotches of chert. They contain a good many fossils, principally Stromatopora and Favosites, also Halysites, Saphrentis, Orthoceras, and other remains are found, but not very well preserved.

After intersection of about ten feet of the last-mentioned strata, six or seven feet of flaggy layers follow, in various shadings of color, and of uneven nodular surface.

Next below are three feet of ash-colored, fine-grained limestones, in beds of from four to eight inches, some of which are full of fissure-like cavities, penetrating them in all directions, which, according to their shape, are the empty spaces once occupied by re-dissolved tabular spar crystals. Similar limestone specimens have been described from the lower strata of Mackinaw. Underneath these, eight feet of regularly bedded limestones, of yellowish gray color, and of dull, earthy, uneven fracture, continue the series. They inclose druse cavities filled with calc-spar or quartz, and contain the casts of an Avicula, not unlike Avicula rhomboidea of Hall, a Nucula-like bivalve, a Rhynchonella, Leperditia, and a tuberose form of Favosites. Some of the thicker beds have been quarried on a small scale, but the rock does not seem to resist the weather very well. The next succeeding rock is a dark gray-colored bituminous limestone, of unhomogeneous nodular structure; the nodules are coated with a black cuticle of bituminous shaly matter; its thickness is five feet, composed of uneven layers. Fossils are not rare, but poorly preserved; those found are Favosites, Leperditia, Rhynchonella, of apparently the same form as those in the strata above; besides these, casts, similar or identical with Murchisonia subulata; some Nucula-like bivalves and a fragment of Orthoceras have been found.

Finally, light-colored absorbent limestones, separating in thin slabs, with uneven conchoidal surface, make the end of this section; their exposed thickness is ten feet. Some of the layers are harder, giving a ringing sound under the hammer; others softer and easy weathering, in angular fragments. The lowest layers are darker, and full of ramified, fucoid-like maculæ, the only distinct fossil I noticed in them is a Leperditia.

By this time we have followed the shore to within one-quarter of a mile from Pirate harbor, in Sitgreaves bay, and the rocks are lost under the drift, or disappear in the bottom of the lake.

On the opposite side of the bay, only about one and a half miles distant, the shore is literally covered by hard, compact, loose limestone slabs, containing silicious concretions, part of which are evidently obscured specimens of Stromatopora; also the cast of a large orthis was found.

I could not find an outcrop of these layers; but the abundance and unworn condition of the loose slabs leave no doubt about their close proximity *in situ*. The lake bottom at this spot is formed of a different kind of rock bed—a blue arenaceous limestone, with intercalated shales, which, by the immense number of chaetetes stems they contain, indicate the upper part of the Hudson river group; while the loose limestone slabs, which undoubtedly form the next layer above. in lithological characters, more resemble the dolomites of the Niagara series, on the other side of the bay.

Further west, on the islands at the north end of the channel between Drummond's and St. Joseph's island, similar hard dolomitic rocks are exposed, in which the casts of a large orthis, similar to Orthis Occidentalis, are the only recognizable fossils. A critical consideration of these beds, and the lower members of the above given section through the Niagara rocks, I postpone for a subsequent chapter.

On the west side of Drummond's island other interesting sections of the Niagara group are observed on Quarry Point, and in another locality about one mile north of it, in Town 42, Range 6 east, Sections 18 and 19, where, also, quarries have been opened.

In the latter place, the quarry itself comprises twenty-five feet of strata; its sole is about sixteen feet above the level of the lake, and its upper edge forty feet below the highest hill elevation, which is formed by massive dolomite ledges, containing numerous casts of Pentamerus. The next lower beds are not well exposed, but from about ten feet above the margin of the quarry, down to its sole, all the ledges are remarkably even, and of more or less distinctly laminated structure. The upper beds, in and above the quarry, are generally not thick enough for a building stone; they have been used to supply the limekilns erected in front of the quarry. The lower ten feet above the sole bed are in layers from twelve to thirty inches thickness, which break in fine, square-edged blocks; the coarser grained ledges seem to be preferable; the middle bed exhibits the before-mentioned laminated structure in a high degree, and not only may be split in this direction in slabs of any desired thickness, but unfortunately also separates in such laminæ after a short exposure to the winter frosts. This is probably the reason why quarry and limekiln were both abandoned several years ago. Some of the coarser grained blocks left in the quarry have withstood during this time the effects of five or six winters, and are sound yet; but these beds alone would scarcely remunerate for the necessary expense and labor to get at them. The blocks of laminated structure left there may, by the stroke of a small hammer, be shattered into innumerable thin slabs, like roofing slate.

From the bottom of the quarry, down to the edge of the water, the dolomite, retaining the same lithological characters, continues in thinner ledges, and disappears then in the lake channel.

Throughout this lower complex of strata, scarcely any fossils are noticed; only a few calyces of a Zaphrentis could be seen projecting from one of the layers in the quarry.

The rock beds of the Niagara group allow a subdivision in three well-marked sections. It is exclusively a limestone formation. The lower section is always very regular and even-bedded, composed of comparatively thin layers of a fine crystalline grain, or with a dull, more earthy fracture. In composition, most of the strata are dolomites; only a few layers are found to be a pure limestone. Fossils are rare in it. The middle division is made up by more massive highly crystalline dolomite ledges, which usually contain a large number of the casts of Pentamerus oblongus, and some ill-preserved corals. The third upper division is a series of thin, uneven layers, with intermixture of much silicious matter with the dolomite mass, and of seams and nodular concretions of hornstone. In this upper division, also, the greatest abundance of fossils is found.

The south part of Drummond's island is exclusively composed of the two upper divisions. Their outcrops are not in high bluffs, as on the north and west side; the southern dip of the strata brings the rock ledges down to the water level, which crown the hill-tops at the northern end.

Only a few of the rock cliffs have an elevation of 20 feet; the higher background of the south shore is formed by the drift hills, densely overgrown with forests.

At the southeast side of the island, the massive rock masses of the middle division form the surface rock of the flats adjoining the lake. The rock could be easy quarried there, and is of sufficient durability for ordinary rough constructions; but for such uses there is little demand in this region, and for finer masonry the structure is too seamy and unhomogeneous.

The upper thin-bedded, fossiliferous strata have no economic value, but the profusion of fossils imbedded in them excite the highest interest of the palæontologist.

The environs of the old English fort, in Whitney bay, have long been celebrated for their abundance of beautifully preserved silicified fossils, particularly corals; but this much-hunted place has been largely spoiled of its treasures by frequent visitors, and no new ones are brought to the surface, since the industrious settler, and with him the plough, have retired from this once flourishing military station.

On the opposite side of the St. Mary's river, along the hill-sides of Point Detour, the numerous drift terraces, surmounting each other in succession, are almost exclusively composed of the broken-up confused layers of the upper Niagara limestone, which in spots are seen yet retaining their undisturbed position on top of the Pentamerus limestone, which composes the body of the hills, and is in several places denuded from its drift covering.

In this accumulation of rubbish, fossils are crowded in astonishing numbers, far surpassing the Drummond island localities. Here is scarcely any room left for the rock mass; it is all one heap of silicified corals, and a part of them preserved as perfectly as if the animal only had recently ceased to live in the corallum.

The fossils found in this locality are enumerated in the subjoined list:

Favosites favosus.
Niagarensis.
venustus.

These three forms are the closely allied representatives of a peculiar type in the Favosites tribe, by which peculiarity Hall was induced to separate them from Favosites, under the name of Astrocerium.

Spinules, projecting in longitudinal rows from the interior of the tube walls, and a similarly spinulose or granulose surface of the transverse diaphragms, with more or less developed syphonal depressions in their circumference, are characteristic of all the three species; but these characters are so variable, that we find all degrees of transition; from almost smooth, only slightly granulose, tube cavities to such as are decorated with rows of long projecting spines. The marginal depressions of the diaphragms may give the tubes the appearance of a star-cell, while in other portions of the same specimen scarcely any such depression is perceptible. The convexity of the diaphragms is a distinctive mark of Favosites favosus from F. Niagarensis or Gothlandicus; but we find all gradations from convex to perfectly level diaphragms, represented in the specimens. If we consider the size of the tubes, which appears to be an excellent character for distinction of species, we are lost altogether. We find specimens having all the particular qualities of Favosites favosus, with tubes six millimetres wide, and of every possible intermediate size, down to specimens with only 1½ millimetre tube diameter. This latter small-tubed form connects the species immediately with the still more minute form of Favosites venustus, which is only a miniature edition of Favosites favosus.

The almost unlimited variability in the size of the calyces, in specimens which otherwise are perfectly alike, is observed in all the Zoantharia tabulata, and in the majority of Zoantharia rugosa.

Other Favositoids found in this locality are:

Favosites reticulatus, M. Edw.

Cladopora multipora, Hall.

laminata, n. sp.

Several species of Alveolites not nearer determined.

Two species of Thecia and one of Striatopora.

Halysites catenulatus, in all possible gradations of tube size and chain mashes.

Syringopora verticillata, Goldf.

similar to compacta, Billings.

and several other not accurately determined forms of Syringopora.

Heliolites interstinctus, Waheenb.
megastoma, McCoy.
Lyellia Americana, M. Edw.
papillosa, nov. spec.
Strombodes pentagonus, Goldf, in a great variety of forms with larger and smaller calyces.
Strombodes mamillaris, Owen.
Zaphrentis Stockesii, M. Ed.; and several other not determined forms of Zaphrentis.
Omphyma verrucosa.
Amplexus Shumardi (Cyathoph., M. Edw.); and a great many other forms of Cyathophylloid corals, to be subordinated in the genera Streptelasma, Heliophyllum, Eridophyllum, Cyathophyllum and Cystiphyllum.
A species of *Columnaria*, and a *Tetradium*, are of particular interest, as not known before to occur in the Niagara group.
Chaetetes and other Bryozoa are not very common.
Stromatoporas, in many different forms, are among the most abundant fossils in the formation. In a not published manuscript I have described and figured one form as *Stromatopora minuta*, another as *Str. vesiculosa*.
Except the Corals and Sponges, only a few other fossil forms are found. They are: Atrypa reticularis, Pentamerus oblongus, a peculiar Stricklandinia, Orthis flabellulum, a large Euomphaloid shell, and casts of other gasteropodes.
Several forms of the problematic fossil Huronia, and of Discosorus conoideus are found quite frequently; and Crinoid stems, common everywhere where fossils are, are not missing.

From Point Detour westward to the mouth of Pine river, almost continued outcrops of the middle, rarely of the upper fossiliferous division of the Niagara group, offer themselves in the low landspurs or in the islands and shoals surrounding the shore. Some distance back from the shore, where the land makes its first terrace-like ascent, a string of low rock bluffs can be all the way followed, but the higher elevations are all covered by the drift.

The rock character remains uniformly the same; it is a hard crystalline dolomite, in massive ledges, full of irregular, impurer silicious seams. The dissolving action of the lake water on the exposed rock-faces excavates the purer limestone portions, and the more resisting silicious seams project in high relief.

Fishermen frequently bring up such eroded pieces with their nets, and admire them as curiosities, comparing them with all sorts of natural objects, which similarity frequently only can be seen by persons gifted with a livelier fancy than naturalists generally have.

By some explorers, the above-mentioned rock bluffs, exhibiting about fifteen feet of the massive beds of the middle Niagara division, have been noticed in a locality in Town 42, Range 1 east, Sect. 21 ectr., and the discovery of a *great marble quarry* was announced. After careful inspection of this place, I express my opinion to the numerous inquirers about the locality in the following short sentence: It is a good common building stone, not better than in any other locality, where the central ledges of the Niagara group are exposed, and a great deal more out of the way than many others.

The islands of the Cheneau group are all underlaid by a nucleus of this rock formation, but in most of them it is totally enveloped with boulder drift and with débris of the fractured rock ledges.

In the hills surrounding Pine river and the upper branches of Carp river off St. Martin's bay, the massive pentamerus rock frequently comes to the surface, but it offers no peculiarity which would require a separate description.

Going further west, along the shore of Lake Michigan, for a space of fifteen miles, nothing but bluffs of drift sand are encountered, after having passed the Helderberg outcrops from St. Ignatz Point to Point aux Chênes.

In the shoals of Manitou payment, the Niagara rocks make their first appearance again, and continue from there, with only short interruptions, to form the surface rock of all the low lands and projecting spurs along the lake shore, as far as Summer island, at the entrance of Big Bay de Noquette.

The hills coming up to the shore or ascending further inland are built up by drift deposits, and only in deep ravines and the channels of creeks, could an erosion down to the rock beds occasionally be observed.

Epouffete Point and the surrounding low lands are only four or five feet above the water level; the exposed ledges are very rugged, cellulose dolomites, containing quite numerous ill-preserved fossils or casts of them. Most common are Pentamerus oblongus, Halysites esharoides, and Syringopora verticillata.

In the surroundings of Millecoquin's river, beds somewhat higher in the series, and containing more and better preserved silicified fossils, underlie the first drift terrace, having an elevation of about twenty feet above the lake. Most of the drift material is only the shattered fragments of the ledges in place, mixed with a few erratic boulders and gravel. The river is the outlet of a lake, situated on top of the drift plateau, whose elevation above Lake Michigan I estimate to be about 100 feet; its distance from it is six miles. In the creeks entering Millecoquin lake, and in the bed of its outlet not far from the lake, the rock ledges of the Niagara group, containing Pentamerus, Catenipora, and Favosites, are exposed under drift banks of 30 and 40 feet height.

Travelling from Millecoquin lake, in a northwest direction, to the Manistique lakes, we pass over undulating drift hills overgrown with splendid hard-wood forests. In the beds of creeks, or along the slanting hill-sides, the Niagara limestone frequently comes to the surface, well characterized by the usual fossils, Pentamerus, Halysites, Favosites, and other corals. The same rock is exposed on the west shore of South Manistique lake. The environs of North Manistique lake exhibit also outcrops of limestones in the northwest corner of the lake, not far from the exit of Manistique river.

The rock there is a light-colored gray dolomite, in uneven beds, generally not over five or six inches thick, enclosing some fossils; I found Zaphrentis, Favosites, a Leptæna similar to subplana, a Conularia, and Crinoid joints. The Conularia differs from the Trenton and Hudson river group species; the other few and imperfectly preserved fossils are not sufficient to entitle to any conclusions on the age of the strata, but Favosites are not known to occur in any lower position than the Clinton group, and the rock character is so similar to the well-identifiable Niagara limestones of the neighborhood, that I feel inclined to consider the age of these limestones as upper Silurian. The limits of the Hudson river group cannot be far off from there, pebbles containing Ambonychia radiata, Murchisonia bicincta, and other Gasteropodes similar to Murchisonia gracilis, are not uncommon along the beach, but an outcrop

I could not find; heavy drift deposits form the principal part of the lake borders.

In the interior of the Peninsula very few opportunities are offered to see extensive sections through the strata, and my companions were not woodsmen enough to risk a further advance, so I returned to the lake shore.

West of Millecoquin's river, the Niagara rocks are not well disclosed before reaching Seul Choix Point, where the massive dolomites project from the lake in a string of reefs, and form the surface rock of the low landspur, in places overlaid by the thin-bedded more fossiliferous upper strata. The fossils, which in the eastern localities usually are found silicified, are here and in the more western outcrops partly calcareous, rarely exhibiting structure as well preserved as in the silicified specimens. Halysites esharoides, Syringopora verticillata, Strombodes pentagonus, Zaphrentis Stockesii, several species of Favosites, of Alveolites, Cladopora, and of Stromatopora are found in great abundance; also several species of Huronia, supposed to be the siphuncules of Orthoceratites, but probably of altogether different nature.

Two miles west of Seul Choix Point, which is a very important fishing station, the rocks have disappeared again from the shore under a belt of low barren sand hills, thrown up by the present lake, and perhaps with co-operation of the gales acting on the loose sand and transporting it in the air. Subsequently, in some low marshy headlands and in the shoals of the lake, the rock beds are found again denuded. After entering the mouth of Manistique river, I had for the first time after leaving Drummond's island, a good deal more extensive section of the Niagara group laid open before me.

At the saw-mill, a dam is erected on the rock ledges over which the river used to run in a small cascade.

From the water-level below the dam, to the level above, twelve feet of strata are exposed. Lowest is a hard crystalline blue and yellowish spotted dolomite, above it are two feet more of similar strata, which contain numbers of Stromatopora and Pentamerus oblongus, also a specimen of Discosorus conoideus was found. Above follow four feet of a hard brittle limestone, weathering white, but interiorly of a bluish color. It splits in slabs of uneven surface, and has a conchoidal fracture. On the surface of the beds Pentamerus oblongus, long stems of crinoids, and fucoid remains of two different kinds are seen. One of them resembles closely Fucoides cau-dagalli, also specimens of Dictyonema gracilis and of Favosites are inclosed in the rock.

About a quarter of a mile east of the river, in a ridge of sixty feet elevation, and extending alongside of the river for many miles up stream, a continuation of the section is exposed. Above the rock ledges, supporting the dam, similar beds continue for several feet, then follows a coarsely crystalline, unhomogeneous, easy weathering dolomite, containing large numbers of half silicified corals and casts of Pentamerus. It forms the surface rock of a slowly rising belt around the hill, and its thickness cannot be accurately ascertained. Higher up the dolomite becomes almost entirely replaced by flinty layers, which contain a great abundance of silicified fossils, but also not in very good preservation. The following collections were made: Strombodes pentagonus, Halysites esharoides, Syringopora verticillata, Lyellia Americana, several species of Favosites, and of Stromatopora, Atrypa reticularis, a species of Rhynchonella, and fragments of Orthoceras. Of these flinty layers four or five feet are deposited.

Above them are four feet of compact dolomitic limestones of uneven bedding, laid open in a quarry, which furnished the stone to fill the cribs in the construction of the mill-dam. For other building purposes the stone would not answer. The upper part of the hill is formed by drift, and the top layers of the quarry are found polished by drift action. The bed of the river is, for many miles upwards, formed by flat rock ledges of hard dolomitic nature, over which the water rushes in rapid current. The rocky river bed continues up to Indian lake, whose east side is lined with cliffs from ten to fifteen feet in height, which belong to the Niagara limestone series, containing Pentamerus, Stromatopora, and in the upper beds many fucoid-like stems. On the east shore of Indian lake a tract of high land, overgrown with fine beech and maple trees, is inhabited by Indians, and the ruins of some larger log houses, erected there by the white men, indicate that in former times this was an important trading-post. Towards the north end of Indian lake, extensive marshlands are spread out, in which nothing, but here and there dispersed patches of rusty mud, or a small bog-iron deposit, attracts the interest of a geologist. Also, in ascending the west branch of Manistique river, up to the centre of Town 44, not a single rock exposure could be found; the river has carved its bed in the drift, which frequently forms embankments forty and fifty feet high. On the east side of the main river, a ridge of Niagara limestone strikes in a northeast direction through Towns 42, and 43, R. 14, and terminates on the west side of South Manistique lake.

The rock beds composing this ridge are partly even-bedded thick layers of a dull earthy fracturing limestone, partly in thinner uneven beds, more crystalline in structure. Fossils are rare, only fragments of Cyathophylloid corals and specimens of a Rhynchonella have been observed. Well-opened sections are not seen in this forest land where, before the sight of rock can be had, a thick blanket of moss has to be peeled off from their surface. By the owners of the saw-mill I was informed that the rock ledgesextend from the river bed out in the lake for quite a distance. Step by step, with the increase of the depth, lower strata come out in the bottom, until, at forty-two feet, the rock beds disappear at once, and clay replaces them.

As we leave Manistique river, and sail west, the thin, hard, bluish colored limestone slabs, forming the upper surface ledges at the mill-dam, are found exposed in the shoals of the lake, on a land tongue in Town 41, R. 16, Sect. 22. A few compressed shells of Pentamerus were the only fossils found in them. In several places, west of this landspur, the proximity of the strata is indicated by accumulations of little water-worn, angular limestone slabs along the shore; but regular outcrops are only found after having passed Point of Barques. On Point of Barques, and east of it, two successive drift terraces are almost entirely composed of very fossiliferous larger rock fragments, the unbroken ledges of which cannot be far from the surface.

The lake throws out thin-bedded, blue-colored limestone slabs, filled with calcified or half-silicified fossils; the fossils in the limestone fragments of the superincumbent drift bluffs are all silicified, and, like the calcified specimens below, represent the same species, which have been enumerated as found in the upper Niagara division at Point Detour, on St. Mary's river. Some of them are very beautifully preserved.

In Town 39, R. 17, Sect. 17, the rock ledges project about one foot above the water, and underlie the lowlands along the lake, which are overgrown with mixed timber of small size. In Town 38, R. 18,

Sect. 4, the rocks project about ten feet, in uneven slabs; the upper portions of which contain only a few Pentamerus shells; the lower beds, intermingled with more argillaceous matter, are rich in fossils, identical with those thrown out by the lake on the east side of Point of Barques. I suppose this to be the place called by Desor *Orthoceras Point*.

Huronia vertebralis is found there in numerous specimens, but, like the other fossils, not in very good preservation.

Onwards to Point Detour, which forms the eastern boundary of Big Bay de Noquette, the rock exposures continue almost without interruption, but the ledges do not rise much above the water level, and the whole country is very low land. At Point Detour the cliffs are about eight feet high, and are composed of the same crystalline, massive limestone beds which form Seul Choix and Epouffete Points. Casts of Syringopora verticillata and of Pentamerus oblongus, besides indistinct Stromatopora specimens, are almost the only fossils found there.

South of Point Detour, Summer island presents bold perpendicular rock walls around its circumference, which project much higher than the cliffs of Detour Point.

Little Summer island, situated north of the large island, is nearly all drift; on its northern end, and on a small island close by, the rocks are at the surface. As we sail into Big Bay de Noquette, the upper fossiliferous limestones are everywhere close under the surface in the low, rounded bluffs extending up to Elliot's harbor, only covered by a thin coating of drift.

On the north side of Elliot's harbor perpendicular rock walls begin to emerge from the water; at first only in cliffs of ten feet height. Further north, in quick succession, lower and lower beds come up under the gently southward dipping strata; and, after the short distance of only three miles, at *Burnt Bluff*, the perpendicular escarpments are already sixty feet high; above which, in a less abrupt manner, the series of rocks continues, or forms in places a second perpendicular escarpment of the same height; so that at Burnt Bluff an uninterrupted section, of at least 125 feet of strata, can be observed. The rock escarpments continue further north, without rising any higher. In an indenture of the coast line, perfectly surrounded by the high cliffs, an iron furnace has been erected, melting Lake Superior ores shipped from Escanaba. At Garden

Point the rock walls are already considerably lower, and near the head of Bay de Noquette they are lost out of sight under the red clay banks and gravelly sand masses of the drift.

The lower fifty feet at Burnt Bluff are very evenly and thinly bedded dolomitic ledges of a smooth, fine-grained conchoidal fracture. Except the casts of a Murchisonia-like gasteropod, and some nodular masses of Stromatopora, I did not observe any fossils in them. Next above are succeeding more thickly stratified crystalline dolomites, containing numerous spherical lumps of Stromatopora. Within this horizon lenticular masses of snow-white granular calc-spar of the grain of Italian marble, but softer, are intercalated between the beds. At the same height, a band about two feet thick of a cellulose brecciated rock, with the rock fragments all incrusted by calc and dolomite spar of yellow color, is found. This band is shown in the whole length of the exposure; at Burnt Bluff it is about sixty feet above the water; in the bluffs a mile south of this place it is found down at the water's edge. The next higher strata, above the last-described series, are thick-bedded crystalline dolomites, alternating with thinly and unevenly bedded nodular limestone slabs. Several of the layers contain silicified corals; and, in the upper beds, Huronia vertebralis and Discosorus conoideus, with an abundance of casts of Pentamerus.

The summit of the perpendicular cliffs is formed by even-bedded limestone slabs, fifteen or twenty feet in thickness; and above them, in the receding portions of the hill-top, still twenty or thirty feet more of silicious limestone ledges, with nodular seams of hornstone, are found. All these layers contain a great number of silicified, but poorly preserved fossils, of the usual kinds found in this upper division of the Niagara group.

The lower even-bedded limestones at the iron furnace are used as flux in the melting process, and as ordinary building stone, as far as there is any demand for it in the locality. Neither there, nor in another neighboring place, did valuable building stones, worth trans-shipment, come under my observation. On the west side of the bay, near the mouth of Sturgeon river, the lowest beds seen in the section at Burnt Bluff seem to continue. Ledges of them are visible in the shoals surrounding the headlands on the north, and on the south side of the mouth of Sturgeon river. The rock specimens from both sides of the bay do not differ lithologically.

CHEMICAL ANALYSES OF DIFFERENT ROCK SPECIMENS FROM THE NIAGARA GROUP.

In all the analyses I selected homogeneous pieces, avoiding such portions in which visible impurities were intermingled.

Names and Localities.	Carbonate of Lime.	Carbonate of Magnesia.	Allumina et iron oxyhydrat.	Residue Insoluble.	Remarks.
	Per cent.	Per cent.	Per cent.	Per cent.	
1. White crystalline dolomite from the middle division of the group from the Lake Huron shore, one mile east of Pine river	56	40	1	1.5	Insoluble residue, consisting of clear quartz crystals.
2. Light-gray colored dolomite from the middle division at Point Detour on St. Mary river	56	43	Traces.	1	
3. Dolomite forming the top stratum of the quarry at Marblehead, Drummond's island	62	33	2	3	Cloudy bituminous residue, with some quartz granules.
4. Upper part of Marblehead quarry of laminated structure with absorbent earthy fracture	54	39	4, most alumina.	2	Residue bituminous and silicious.
5. Lower beds of Marblehead quarry, Drummond's island	58	32	2	7	Residue siliceo-argillaceous.
6. Limestone from the section of Marblehead with *acicular cavities* (*vide* description above)	95	1	2	2	
7. Limestone from the Marblehead section, 30 feet below the acicular limestone, described as nodular bituminous limestone, containing fossils.	94	2	1	2	Residue bituminous, and some quartz granules.
8. Lowest beds in the Marblehead section	52	35	2	9	Siliceo-argillaceous residue.
9. Loose slabs on the west side of Sitgreaves bay, immediately above the Hudson river group strata, Drummond's island.	52	38	3	6	Silicious residue.
10. Quarry point on the west side of Drummond's island. Quarry stone of laminated structure	60	32	1	3	Silicious residue.
11. Sole bed of the same quarry of more crystalline structure than specimen No. 10	59	38	1	1	
12. Lowest strata at Burnt Bluff of Big Bay de Noquette	56.6	39	1	2.5	

The white, lenticular, marble-like masses, imbedded between the lower strata of the Burnt Bluff section, in Bay de Noquette, are almost chemically pure carbonate of lime.

CLINTON GROUP.

Intermediate between the Niagara group and the Hudson river group, in the New York series, a complex strata is found intercalated, which is characterized by peculiar lithological features and by a number of peculiar fossils. These strata are known under the names Clinton group and Medina sandstone.

Also, in the last geological reports of Michigan, the Clinton group is mentioned as a distinct member in the series of strata in our State, and the lower thirty-two feet of Dickinson's quarry, on Marblehead, off Drummond's island, have been pointed out as being the upper portion of this group, and the beds exposed on the lake shore between this quarry and Pirate harbor, as the lower portion. In proof of this assertion, only three not nearer described fossils—an Avicula, a Murchisonia, and a Leperditia are mentioned, all of which, as far as my experience goes, are only found in the lowest portion of the section, while in the quarry, which altogether does not comprise much over thirty-two feet of strata, only Niagara fossils are found, or at least such kinds which are common to both formations. The three mentioned fossils are only imperfect casts, not allowing an exact determination, and scarcely justify important deductions to be based on their occurrence. But, allowing to these meagre representatives the importance given to them, the indicated limits of the Clinton group had to be lowered considerably. I do not doubt that a portion of the lower strata is contemporaneous with the Clinton group, because the lithological changes from the upper portion of the Hudson river group to the typical Niagara group are so gradual that no important alterations in the ocean bed can have taken place during all this interval, which would have caused the elimination of this group; but, for the same reason, I do not admit a division line to be drawn at random where an uninterrupted continuity is indicated by nature. The Clinton and Medina groups of New York are evidently littoral deposits, in which the amount of sedimentary material is much greater, and in quality more variable, than in the deep ocean deposits, going on in quiet uniformity in the western countries. There was no cause for a sudden change in the fauna, while the local conditions near the shore altered frequently, and, consequently, also, a local change in the fauna was induced by these alterations.

A critical examination of the faunas of the Niagara group and of the Clinton group, demonstrates that a great many of the fossils of the Clinton group were living during the Niagara period, and those which were not are frequently found only in certain restricted localities, standing in some relation with the character of the rock, or are stragglers of the vanishing Hudson river group fauna; as, for instance, Orthis lynx. A further proof of the uninterrupted succession of deposits in this western country, from the end of the Hudson group period to the Niagara time, I see in the continued existence of Halysites esharoides, and its different varieties, through all the concerned strata. A similar continuity of the successive deposits during this era, we learn from the descriptions of the Canadian survey of the island of Anticosti, where the richness of the fauna in all the different horizons is peculiarly favorable for making comparisons regarding the duration of each species of fossils. We find there the faunas of the Hudson river group and of the Niagara group intimately linked, the one gradually disappearing, the other slowly developing itself.

The summary result of my studies of the Clinton group in our State, in the States of Ohio, Indiana, and the more western States, is, that during that time, in which the littoral deposits of New York State, and their organic contents, frequently changed their character and were accumulating to mighty masses, contrasting with the deposits above and below, in the West, the deposits went on in slow uniformity, without any sudden changes in the fauna or in the deposited materials. In consideration of this fact, I see no reason why we should uphold here an artificial line of demarcation where nature has made none.

HUDSON RIVER OR CINCINNATI GROUP.

It has been mentioned, that in the shoals surrounding the mouth of Sturgeon river, on the west side of Bay de Noquette, ledges of limestone are seen, which are lithologically the same as the lower beds of the opposite shore, on Burnt Bluff. Further inland, about eight miles in direct line from the shore, we find the Sturgeon river running in rapids over ledges of a dark bluish-gray arenaceous limestone, containing Streptelasma corniculum, stems of Chaetetes,

Crinoid stems, fragments of Isoteles gigas, Streptorhynchus filitextus, Murchisonia bicincta. The banks of the river are formed by drift deposits thirty feet high. In going up stream, we see the rapids continue, and successively other rock-ledges come to the surface, and form part of the river banks. These are of a darker, bluish-green color, more arenaceous, and interstratified with arenaceous marls. Of fossils, they contain a great many Chaetetes stems, and a coral named by Hall, Sarcinula obsoleta (more likely it is an Eridophyllum). The cavities of the stems have generally lost their organic structure, and are filled with calcspar, and occasionally with violet-colored fluorspar. Besides these, are found specimens of Halysites, in no way much differing from the Niagara species, fragments of Bryozoa, Isoteles gigas, Streptorhynchus filitextus, Crinoid stems. The exposed rock series amounts to about fifteen feet in thickness at this place. In a locality four miles higher up the river the same strata are met with again; the outcrops are so limited and disconnected that little of information can be collected about the structure and sequence of the strata, for which reason I returned from there to the mouth of the river, in anticipation of finding better opportunities along the shore line. I am informed that, twenty-seven miles from the mouth of Sturgeon river, twelve miles above the place from which I returned, other more considerable rapids with rock exposures, are found in the river, which probably are of older Trenton age.

The whole Peninsula, separating Big Bay from Little Bay de Noquette, is underlaid by the Hudson river group. Its east shore is low, and has few rock exposures, but numerous limestone blocks strewn along the beach, indicate, by the fossils they contain, the nature of the subjacent rocks. Leptæna alternata, Zygospira recurvirostra, Orthis testudinaria, Isoteles gigas, are the common fossils.

On a landspur four miles east of the lighthouse at the entrance of Little Bay de Noquette, in Town 38, Range 21, Sect. 8, six feet of uneven and thin limestone beds, interstratified with arenaceous shales, form an outcrop along the shore. They contain numerous stems of Chaetetes ramosus, Orthis testudinaria, Orthis lynx, Leptæna alternata, and other fossils. The interval from there to the lighthouse is filled out by a sandy, low drift beach. On the lighthouse point the rock beds emerge again, and from there, continued outcrops for ten miles up Little Bay de Noquette, line the west side of the Peninsular spur, and form conspicuous vertical bluffs of from 12 to 50 feet in height. At the lighthouse, the elevation of the strata above the water is only a few feet. The uppermost beds, not more than three feet in thickness, are light-colored silicious limestones, full of flint nodules, and inclosing a great number of fractured shells, of the same kinds as those found in the beds below, which are blue-colored argillaceous limestones, alternating with blue shales, and perfectly crowded with fossils. The whole thickness of the exposed strata, amounting to about 60 feet, is composed of these alternating beds of shales and limestones. At times the shales prevail over the limestone, other times the case is reversed.

Notwithstanding the immense abundance of fossils, well-preserved specimens are rare, and also the number of species is comparatively small.

The following list gives the names of the specimens found:

Streptelasma corniculum, rare.
Chaetetes ramosus, and other species, very abundant.
Protaræa vetusta, rare.
Tetradium cellulosum, rare.
Stictopora, spec. not determined.
Orthis lynx.
" plicatella.
" testudinaria.
" occidentalis.
Rynchonella increbescens.
" small variety.
Zygospira recurvirostra.
Leptæna alternata, in a smaller and in a larger variety.
" sinuata.
Streptorhynchus filitextus.
Ambonychia carinata.
Modiolopsis modiolaris.
Avicula demissa.
Cyrtodonta, sp. not determined.
Indistinct casts of Gasteropods, and of Orthoceratites.
Isoteles gigas.
Crinoid stems, and a specimen of a sessile Crinoid not generically determined.

The lowest beds seen in these outcrops are a less argillaceous kind of limestone slabs than those above, and are almost entirely composed of Chaetetes ramulets. In the bank, the upper argillaceous limestones appear solid enough to be used for a filling stone in the cribs of aquatic dams, and have been quarried for this purpose; but the experiment proved to be a failure; the rock rapidly decomposed into a soft pulp, and was swept away. Opposite Escanaba, the bluffs have attained their maximum elevation; further north they begin to get lower again, and two miles south of Squaw point all the older rock beds have disappeared under the drift.

About six miles further north, in Bill's creek, a branch of the Whitefish river, 12 feet of blue shales, and a few intercalated limestone beds form, for a short distance, the embankment of the creek.

The shales contain a large number of Lingula shells of suborbicular form, with delicate concentric lines of growth and a faint radial striation from the umbo to the periphery. The dimensions of some perfect specimens are two centimetres from the umbo to the front, and seventeen millimetres in transverse direction. Of other fossils there were only noticed compressed specimens of an Orthis, small casts of Gasteropods, and, quite abundantly, the small casts of a Nucula-like shell (Cleidophorus), covering the surfaces of the slaty laminæ, together with branchlets of a Stictopora and of a Trematopora. The intercalated seams of limestone are an agglomeration of Chaetetes stems. These outcrops are perfectly isolated. All the surrounding country is covered by drift, their central position between the Trenton limestones of Whitefish river and the Hudson group of Bay de Noquette indicates them to be of lower position than the Hudson group strata exposed in the shore line of the bay. The similarity of these beds with the shales representing the Hudson river group in the lead regions of Illinois and Iowa is very obvious, and the differences existing between them and the more eastern strata would find their explanation in the somewhat older date of the western prevalently shaly deposits. South of Bay de Noquette, the continuation of the belt formed by the Hudson river group is found on Wisconsin territory. It makes a few outcrops on the west side of Green Bay Peninsula. On the mainland, near the end of the Bay, on Duck creek, quarries have been opened in this formation, which furnish large blocks of a stone, which is worked into window sills, door steps, etc., but I notice too many argillaceous seams in the blocks, which must impair their durability. Some of these layers are very rich in large shells of Leptæna, Orthis occidentalis, Orthis lynx, etc. Of the eastward continuation of this belt of the Hudson river group, through the Michigan Peninsula, we have very little definite knowledge. I found in a few places of the interior rock fragments mixed with the drift, which contained Ambonychia radiata and other characteristic Hudson river group fossils, but a well-disclosed outcrop of the formation is not known to me east of Bay de Noquette, before reaching St. Mary's river. The absolutely regular geological construction of the east part of the Peninsula makes it certain that a belt of the formation traverses its centre, in the direction from west to east; but the drift has exclusive possession of the surface in this region, and the watercourses creeping over the swampy plateau of the centre have not dug their channels deep enough to reach the rock beds. No doubt a minute investigation of this strip of land would lead to the discovery of some favored spots, where glimpses of the rock beds could be caught, but this would require more than the labor of a whole working season, with no prospect of any practical or scientific acquisition, equivalent to the expense and the unusual difficulties connected with a travel, confined to the central portion of an uninhabited forest, where the remoteness from all supplies is an impediment more serious than can be imagined by those who have not had experience of backwoods travelling.

The first disclosures of the Hudson river group, near the east end of the upper Michigan Peninsula, are found on the Canadian island St. Joseph; but also there, most of it, is covered up by deep drift deposits. Finally, along the north shore of Drummond's island, a series of good exposures allows the study of the formation. The rock differs little from the equivalent beds at Little Bay de Noquette; it is an argillaceo-arenaceous limestone of blue or greenish color, with intercalated beds of shale. If we take into account the beds seen twenty feet under water, in the lake bottom, an aggregate thickness of about fifty or sixty feet of strata comes to view.

The strata of these eastern localities likewise abound in fossils. Some are the same in the east and in the west, but one kind may be very common in one place while it is rare in the other, and a number of species are found only in the eastern localities, and there are inverted. For instance, at Drummond's island, Favistella stellata (columnaria), and a coral related to the genus Lyellia, and identical with Calapœcia of Billings, are found very abundantly, while not a single specimen of either of them is to be found at Bay de Noquette. Streptelasma corniculum and Rhynchonella increbescens are very common at Drummond's island, and only few small specimens of both can be seen in the west. At Drummond's island the following collections have been made:

Streptelasma corniculum (Zaphrentis Canadensis, Billings).
Columnaria stellata (Favistella of Hall).
Calapœcia Huronica.
Chaetetes, several species of it.
Stictopora, not determined.
Stromatocerium rugosum, in large laminated masses, but with very obscure finer structure.
Protaræa vetusta.
Crinoid stems.
Rhynchonella increbescens, large variety.
Zygospira recurvirostra.
Leptæna alternata.
Orthis lynx.
Orthis plicatella.
Orthis occidentalis.
Streptorhynchus filitextus.
Modiolopsis modiolaris, and a number of other indistinct casts of bivalves and gasteropods.
Various species of Orthoceras, one quite large kind of the type of Endoceras proteiforme, with large conical siphons of the shape of gigantic Belemnites. Others have small central siphons.
Isoteles gigas, and fragments of another kind of Trilobite.

The extent of the Hudson river group on Drummond's island is restricted to a strip of land a few miles in width, and its approximate demarcation from the Niagara belt will coincide with a line drawn from the fond of Vermont harbor, in Portaganissing bay, across the island, in southeast direction, to the centre of Sitgreave's bay. The interior of the island, through all this district, is a low swampy forest, in which no outcrops of any extent can be expected. The principal outcrops are confined to the different promontories on the northeast side of the island, between Sitgreave's bay and Reynold's bay. The rock beds are in stairlike offsets, extending far out into the lake, and can be seen yet at a depth of 30 feet; their dip is slowly southeast.

The economical value of the rock series is small. For building purposes, and for lime burning, the ledges are too argillaceous and sandy. A specimen of the calcareous shaly beds from Bay de Noquette gave, on analysis:

Carbonate of lime..............	36	per ct.
" magnesia..........	18	"
Alumina and iron oxydhydr.....	4	"
Insoluble argillaceous residue, mixed with some quartz sand, and with iron pyrites........	42	"

TRENTON GROUP.

This group has in its western extension a very analogous development, in lithological as well as in palæontological characters, with the Trenton strata of New York. It underlies a considerable part of the Upper Peninsula. A belt of it, coming up from the head of Green Bay, enters our State at the mouth of Menominee river, and forms the whole shore line, up to the head of Little Bay de Noquette. At Menominee, the belt has a width of 12 miles; at Escanaba, its extent from east to west is from 20 to 25 miles. Near Grand rapids of Menominee river, the river is left to the west, and the formation ascends with its western limit to the southwest corner of Town 42, R. 25. It is intersected by the Escanaba river, in the centre of the west side of Town 43, R. 24. From there it bends eastwards, crossing the Escanaba and Negaunee railroad a short distance north of Centreville, and comes up to a mile south of the north end of Town 45, R. 22. In an easterly direction its limits are found again near the outlet of Mud lake into Whitefish river. Further east, for a long distance, the formation has not been observed. At Carp river, six miles south of Waiska bay, on Lake Superior, for the first time, the

Trenton strata are met with again, and a few ledges of it are seen on the top of the calciferous formation at West Anebish rapids; from there it strikes under the drift across Anebish island to the island of St. Joseph. The west side of St. Joseph's island does not show the formation, except in a large accumulation of drift boulders, composing the hills near St. Joseph's village.

On the east side of the island the Trenton strata are quite extensively exposed along the shore. An outlier of the Trenton limestone forms also the top of Encampment d'Ours island, and of several other islands in the north channel of Lake Huron, all of which are Canadian territory.

The southern limits of the formation, and its junction with the Hudson river group, are very imperfectly known. Even at the shore line, where fine outcrops of the two joining formations are to be seen on the opposite shores, the places of contact are either in the bed of the lake or hidden under the drift. In many localities, through the centre of the Peninsula, the Trenton strata can be identified from west to east, but the exact line of its southern boundaries is as little known as are the northern limits of the Hudson river group. Also in the eastern exposures on St. Joseph's island, the contact of the two adjoining formations has nowhere been observed.

In the bed of the Menominee river, below the mill-dam near its mouth, 10 or 12 feet of hard, crystalline, gray-colored dolomitic limestone form the surface. The uppermost beds are polished by drift action; otherwise the ledges are of a rough, uneven surface, containing druse cavities lined with spar crystals. The rock is durable for ordinary purposes, but not homogeneous, and not in blocks sufficiently large to be useful for finer masonry. It contains some fossils, of which the following specimens have been collected: Lingula quadrata, Leptæna camerata, Streptorhynchus filitextus, Murchisonia major, Buccania expansa, Trochonema umbilicata, Maclurea (large casts three inches in diameter), Conularia Trentonensis, Dictyonema (a species with very delicately-reticulated fronds expanding from a transversely-wrinkled hollow cylindrical stem, with a shining carbonaceous surface; also indistinct specimens of Chaetetes frondosus and Crinoid stems are included in the rock.

Above the mill-dam no more rock-ledges are visible before arriving at the next rapids, which are six miles above the mouth of the river. The outcrops there are all confined to the river bed; only few ledges project from the banks. The rock is a thin and uneven-bedded arenaceous limestone of blue color, weathering yellow, interlaminated with shaly seams. Numerous fucoid branches, similar to Buthotrephis succulens (Hall), stand out in relief from the surface of the ledges; the other fossils found are Orthis testudinaria, Streptorhynchus filitextus, Rhynchonella plena, ramulets of Stictopora, Phænopora multipora, Esharopora recta, and other Bryozoa, Schizocrinus nodosus (stems), Streptelasma corniculum, fragments of a Calymene. These appear to be inferior in position to the Trenton strata at the mouth of the river, and represent the middle horizon of the Trenton series. From this locality to the Grand rapids, a distance of 15 miles, no more rock exposures are seen on the river. At Grand rapids, the ledges of the calciferous formation are the surface rock, and will be considered in the following chapter. The Trenton formation keeps itself east of the Menominee river, but Cedar river, with all its ramifications, and Ford river for almost the whole length of its course, are within the Trenton area. Numerous good outcrops are noticed along these river beds, but the Escanaba river flows over exactly the same rock strata, and offers by far the best and most continuous sections through the whole series. I restrict myself, therefore, to an accurate description of the Escanaba section, to avoid tiresome repetition of one and the same thing.

At the mill-dam, one mile above the mouth of Escanaba river, the highest beds of the section make an outcrop in the east bank and in the channel of the river. The rock is in ledges from four to ten inches in thickness, partly an impure limestone with silicious seams, partly of crystalline dolomitic character. The following fossils are found in this place: Leptæna sericea, Leptæna alternata, Zygospira recurvirostra, Orthis testudinaria, Orthis pectinella, Murchisonia gracilis, Murchisonia major, Trochonema umbilicata, Conularia Trentonensis, Echinoencrinites anatiformis, Poteriocrinus, similar to alternatus, Heterocrinus, similar to H. simplex of Hall, Chaetetes petropolitanus, and ramose forms of Chaetetes.

The more crystalline rock beds have been quarried in this place to supply local demands, but the rock is not clear enough of argillaceous seams to make a good building material. About eight feet of the strata form the exposure. A half a mile above the dam the same rock beds form the shallow river bed and its low banks. The strata dip slowly down in the direction of the stream, and by ascending the river the next lower strata are soon found at the surface, and, shortly after, they rise in the river banks into vertical rock walls twelve feet high. This second division is composed of thin-bedded nodular limestones, intermingled with irregular seams of argillaceous and cherty matter. They are called in Hall's report "*wedge-shaped limestones*," and form for several miles onward to the lower Falls, the surface rock of the river valley. Fossils are not abundant, but everywhere specimens of Leptæna sericea, Leptæna alternata, Zygospira recurvirostra, and fragments of Isoteles and of Orthoceras are to be found. The lower Falls, six miles above the mouth of the river, run over the ledges of this wedge-shaped limestone, making a descent of twelve feet in three offsets. The lower beds at the foot of the falls are thicker and more regular than those above. Underneath these, a series of shales, alternating with arenaceous limestone slabs, are denuded in the river channel, representing a third division of the section, characterized by its prevailing shaly constituents, and, in certain layers, by a great abundance of fossils, in numbers as well as in variety of kinds. The beds exposed below the Falls are not so fossiliferous, and contain only Fucoid branches and Chaetetes petropolitanus. Above the Falls the wedge-shaped limestones forming the surface soon to disappear under drift accumulations, which now make the embankments of the river for the distance of a few miles. The first rock ledges met with again, in ascending further up the river, are the wedge-shaped limestones, and above them more thick-bedded dolomite ledges stand out from the upper portions of the embankments. The strata are subject to frequent undulations, causing repetitions in the sections ; but as we ascend the river the erosion steadily becomes lower. At the big bend of the river, in the south part of Town 41, Range 23, some thicker rock beds, full of silicious concretions are found at the base of the wedge-shaped limestones, and then follow, in the descending order, the before-mentioned alternating beds of shales and arenaceous limestone slabs, and remain for several miles onwards the surface rock of the valley. The thickness of these strata amounts to about thirty feet. In Section 17 of the above-mentioned township, the finest outcrops of the fossiliferous beds are found. The fossils are nearly all nicely preserved. An enumeration of them is given in the subjoined list:

Plants.—Paleophycus rugosus, Licrophycus Ottawaensis, and several other Fucoids.

Sponges.—Stromatocerium rugosum.

Corals.—Streptelasma corniculum.

Bryozoa.—Chaetetes petropolitanus, Chaetetes ramosus, and several other forms, of which one is interesting enough to be described here. It is of small nummiform shape, with conspicuous solid dots, formed by closed tubes and closed finer interstitial cells. These solid dots in some specimens project like warts, and are surrounded by a depressed polygonal area, which gives the surface a striking similarity with a compound star-coral. Perfectly identical specimens also occur in the Trenton limestone of Canada, near Ottawa river, and are preserved in the collections of the Geological Survey in Montreal.

Phænopora multipora. Phænopora n. sp.

Coscinium flabellatum, Stictopora ramosa, and two other not determined species; Ceramopora, n. sp.; Arthroclema pulchella, Billings.

Crinoids.—Schizeerinus nodosus, Comarocystites punctatus.

Lamellibranches and Gasteropods.—Nucula levata and casts of other bivalves; Murchisonia major, Buccannia expansa, Trochonema umbilicata, Cyrtolites compressus.

Brachiopods.—Leptæna sericea, Leptæna alternata, Streptorynchus filitextus, Orthis pectinella, Orthis testudinaria, Orthis tricenaria, Orthis lynx, Zygospira recurvirostra, Rhynchonella plena, Crania Lælia, Pholidops Trentonensis.

Articulata.—Isoteles gigas, Illaenus Americanus, Cheirurus pleurexanthemus, Calymene senaria, Phacops callicephalus, Encrinurus sp. indic.

Cephalopoda.—Several imperfect specimens of Orthoceras.

The next lower beds presenting themselves in the section below the fossiliferous shales and limestones, are seven feet of light-colored subcrystalline limestones, with an absorbent earthy fracture, which are followed by a few beds of dark-blue crystalline limestone, filled with crinoid stems and the shells of a Cyrtodonta; and these are underlaid by a series of thin-bedded limestones, of uneven, nodular surface, and with silicious veins and concretions, which in appearance scarcely differ from the higher beds designated as *Wedge-*

shaped limestones. They contain some fossils which do not extend into the higher beds; but many others of them are found in the lower as well as in the higher strata. In this lower wedge-shaped limestone the following fossils were collected: Columnaria alveolaris, Tetradium fibrosum, Streptelasma corniculum, Stromatocerium rugosum, Chaetetes in several ramose species, a tubular Bryozoon, described by Billings doubtfully under the name of Stromatopora compacta; Stictopora ramosa, Phænopora multipora and another species, Orthis testudinaria, Orthis tricenaria, Streptorynchus filitextus, Atrypa bisulcata, Rhynchonella plena, Cyrtolites compressus, Pleurotomaria and casts of other Gasteropods, fragments of Orthoceratites, Asaphus, Illænus, and Leperditia Canadensis. The thickness of these lower wedge-shaped limestones is 15 feet, very nearly the same as the upper series has it; their principal exposures are at and below Oak falls, in a narrow cañon with vertical walls 20 feet high, through which the river rushes with great velocity for the length of a mile. The next lower beds seen in the base of these cliffs are thick-bedded, hard crystalline dolomites, eight feet in thickness, and at the foot of the falls, which make a vertical descent of eight feet, still lower rock-beds come to the surface, of the thickness of six feet. These are limestones of a greenish ash color, with smooth conchoidal fracture, and full of silicious veins and nodular concretions, which, on the surfaces exposed to the dissolving action of the water, stand out in high relief. Some of the ledges are almost entirely composed of Cyrtodonta shells and their casts.

Three miles above Oak falls, another cascade of five feet is formed by the river; the rock ledges continue to be of the same geological horizon. Columnaria alveolaris, Tetradium fibrosum, and Stromatocerium rugosum, are here the most obvious fossils. Analogous lower Trenton strata are seen in numerous places upwards on the river. In Town 42, Range 24, Section 16, a light-colored brittle limestone, with smooth conchoidal fracture, and shattering in uneven slabs, contains specimens of Endoceres proteiforme, Ormoceras tenuifilum, Orthoceras multicameratum, Tetradium fibrosum, Chaetetes, Stictopora, Leptæna alternata, Rhynchonella plena, Cyrtodonta, Isoteles gigas, and other fossils. These beds are strikingly similar in rock character and in fossils to the upper brittle limestones of St. Joseph's island, which hereafter will be described. In the north end of Town 42, the river begins to cut down into deeper dolomitic strata, which I consider to represent the calciferous formation. The greenish-gray limestones, containing many Cyrtodonta shells, and mentioned as the lowest strata seen at the foot of Oak falls, are found in this more northern locality, resting immediately on the above dolomitic ledges of the calciferous group.

The aggregate thickness of all the Trenton strata exposed in the section given by the Escanaba river I estimate at 100 feet. Frequent repetitions and interruptions in the section make an accurate measurement impracticable. East of the Escanaba river, on the height of the drift plateau, near Maple ridge station, and at Centerville, on the Escanaba and Negaunee railroad line, the Trenton strata come very close to the surface, and are repeatedly denuded in the railroad cuts. Near Maple ridge a quarry has been opened to supply the foundation stones for the buildings of a newly erected iron furnace at Escanaba. In the quarry about ten feet of the strata are denuded; they are of a crystalline dolomitic structure, but contain too many argillaceous seams to be recommended as building stone. The top ledges in the quarry are all ground level by drift action. Of fossils the following species were found in the rock: Murchisonia major, Leptæna sericea, Leptæna alternata, casts of Orthoceras, Conularia Trentonensis, Echinoencrinites anatiformis, Dictyonema growing out from a transversely wrinkled hollow cylindrical stem, and forming a fan-like reticulated expansion. These beds, like the analogous beds on the mouth of Menominee river and of Escanaba river, represent the highest portion of the Trenton group. The same strata form the bed of Day's river for several miles upwards from its mouth. The lowest ledges seen in the river are silicious limestones, which on their surfaces, exposed to the dissolving action of the water, are peculiarly excavated, so as to bring out the more resisting silicious portions in high relief. The upper more thin-bedded and more argillaceous ledges are covered with stems and disjointed heads of Echinoencrinites anatiformis, besides Leptæna sericea, Orthis testudinaria, Streptorhynchus filitextus, Rhynchonella plena? and stems of Chaetetes. The rock beds opened in the railroad cuts near Centerville are lower in the series, and represent the shaly fossiliferous strata, below the wedge-shaped limestones of the Escanaba river section.

Whitefish river, with all its ramifications, lies within the Trenton area. The lower part of the river, below the mill-dam, has no outcrops of rock ledges. Above the mill-dam, the water, by its artificial retention, has overflowed the lowlands surrounding it, and expands into a long, narrow pond, two miles in length. Above this pond the water is found running in rapids over coarsely crystalline dolomite ledges, containing silicious veins and druse cavities filled with rock oil. Of fossils are found Crinoid stems grown into the rock mass, Orthis testudinaria, Leptæna camerata, Zygospira recurvirostra. The river appears to flow, for a long distance, constantly over rock ledges of the same kind, without cutting deeper into the strata. In various places, as high up as Town 43, Range 20, whenever I struck the river, I found it invariably flowing over the same kind of rock beds. The river valley is an impenetrable cedar swamp, and for this reason I had to follow the road over the adjoining drift hills, and only from time to time I found a chance to descend to the river. On the borders of a small lake, in Town 43, Range 20, Section 2, the bed of the river is formed by hard dolomitic ledges. In the drift deposits forming the banks of the lake, a large proportion of the material is made up by angular blocks of a dolomitic limestone containing a number of fossils, which indicate the upper division of the Trenton group, as represented in the bed of Day's river. Echinoencrinites is very common, besides Leptæna sericea, Orthis testudinaria, Orthis pectinella, Calymene senaria, Streptelasma corniculum, and various Bryozoa. On the west branch of Whitefish river, in Town 44, Range 21, the lower Trenton strata, containing fossils characteristic of the Birdseye limestone, are, in various places at the surface, and are likewise found higher up in the east branch, near its exit from Mud lake, whose basin is underlaid by the calciferous sandrock.

East of Whitefish river the swampy condition of the centre of the Peninsula prevented me from penetrating in this direction. The opportunities to study the rocks are too rare in such districts, and after having repeatedly experienced how small the results of such forced excursions are in comparison with the expense of time and money, I desisted from a detailed survey of this part of the interior, which would have required a whole season's work, and a larger body of assistants than I had to dispose of. Near the east end of the Peninsula I found the Trenton formation again in the bed of Carp river, south of Waiska bay, and in the hills striking from southeast to northwest, through the diagonals of Towns 45 and 46, Range 1 west. The light-colored limestones found there contain Crytodonta Huronica, Vanuxemia inconstans, Orthis tricenaria, Leptæna alternata, Crinoid stems, Streptelasma corniculum, and branchlets of Stictopora, and evidently represent the lower horizon of the Trenton group. At no great distance from these limestones, conglomeritic dolomite strata, with streaks of coarse-grained calcareous sandstone, come to the surface, which belong to the calciferous formation. In the west Anebish rapids the calciferous sandrock is overlaid by some ledges of an impure argillaceous limestone, containing Chaetetes petropolitanus, Stictopora, Streptorhynchus filitextus, Cyrtodonta, Murchisonia gracilis, and Orthoceras proteiforme, all fossils of the Trenton group. Anebish island is entirely covered by drift, and in order to get a fair opportunity to study the Trenton formation as it is in this eastern end of the State, we have to intrude on Canadian territory, where instructive outcrops of the formation can be seen at the northeast shore of St. Joseph's island, on Encampement d'Ours island, and on several other smaller islands in the north channel of Lake Huron.

The base of Encampement d'Ours island is formed by the quartzites and slates of the Huronian group, projecting with vertically erected ledges in bold cliffs from the shore, and in isolated knolls from the lake channel. On a part of the island these vertical ledges are overlaid by horizontal strata of a light-colored, soft, sometimes conglomeritic sandstone, attaining the thickness of 100 feet, or even more. In the northeastern portion of the island this horizontal sandstone is conformably overlaid by a series of shales and limestones 60 feet thick, which betray their Trenton age by an abundance of characteristic fossils.

The lowest beds of this limestone formation are prevalently arenaceo-calcareous shales of a dusky green or bluish color, and containing numerous fossils.

The middle strata are thin-bedded nodular limestones, with shaly intercalations, also of darkish color, like the strata below, and equally abounding in fossils.

The upper strata are light-colored brittle limestones, with conchoidal fracture, splitting in uneven, wedge-shaped slabs by exposure or under the stroke of the hammer. They are likewise well stocked with fossils.

At a first glance the three indicated subdivisions of this series appear to be well defined by the occurrence of fossils peculiar to each section; but after hunting over the ground, one will learn that certain kinds of fossils in beds prevail over other kinds, but that only a small number of them are peculiar to a certain horizon, and that nearly all are to be found in any one of the layers from top to base.

The following collections were made:

1. *In the lower argillaceous beds:*

Rhynchonella plena, Leptæna alternata, small variety; Cyrtodonta Huronensis, Cyrtodonta subtruncata, Vanuxemia inconstans, Matheria tener, Pleurotomaria Eugenia, Orthoceras multicameratum, Orthoceras granulosum, n. sp.; Cyrtoceras, Stictopora ramosa, Chaetetes ramosus, small specimens of Columnaria, and a Bryozoon of nodose glandular form of a most delicate tubular structure, provisionally named by Billings *Stromatopora compacta*, to which genus the fossil has no relation.

2. *From the middle nodular limestones:*

Stictopora ramosa, Phænopora multipora and another not determined species. Chaetetes in ramose forms, Tetradium fibratum, Crinoid stems, Lingula quadrata, Leptæna alternata, Rhynchonella plena, Zygospira recurvirostra, Trochonema umbilicata, Pleurotomaria Eugenia, Cyrtodonta subtruncata and other species of Cyrtodonta, Orthoceras granulosum, Cheirurus pleurexanthemus, Phacops callicephalus, Bathyurus spiniger, Asaphus platycephalus, Leperditia Canadensis, Beyrichia Logani.

3. *From the light-colored upper limestones:*

Several forms of Fucoids, Paleophycus and Buthotrephis, Tetradium fibratum, Columnaria alveolaris, Chaetetes stems, Stictopora, Schizocrinus nodosus (stems), Orthoceras proteiforme, Orthoceras vertebrale, Orthoceras Huronense (Billings), Orthoceras of elliptical form with eccentric Sipho, Cyrtoceras Huronense, Cyrtoceras Isodorus, Leptæna alternata, Zygospira recurvirostra, Pleurotomaria subconica, Pleurotomaria rotuloides, Subulites elongatus, Subulites vittatus, Murchisonia bicincta, Belerophon bidorsatus, Ambonychia amygdalina, Modiolopsis gesneri, Vanuxemia inconstans, Cyrtodonta Huronensis, Cyrt. subtruncata, and other species, Asaphus platycephalus, Illænus crassicauda, Phacops callicephalus, Cheirurus pleurexanthemus, and Leperditia Canadensis.

The strata composing the Landspur forming the north-east end of St. Joseph's Island, are identical with those of Encampement d'Ours. They offer a still better opportunity for observing every single successive stratum of the series, fifty feet of which are laid open.

Lowest, partly below the water level, dark blue or greenish colored arenaceous limestones, interstratified with sandy shales, are seen to the thickness of six feet. They contain Fucoid branches and stems of Chaetetes, but the most obvious fossils enclosed are numerous well-preserved specimens of Orthoceras and of Cyrtodonta. I collected:

Cyrtodonta Huronica, Cyrtod. subtruncata, Vanuxemia inconstans, Pleurotomaria Eugenia, Orthoceras proteiforme, Orthoceras multicameratum, Orthoceras granulosum, Orthoceras tenuifilum, Orthoceras vertebrale, several species of Cyrtoceras, and one species of Orthoceras, similar to multicameratum, with preserved shell, in which the difference of colors is yet distinctly perceptible. Reddish brown longitudinal stripes alternate with narrower uncolored interstices on the surface of the shell. A similar specimen, with much finer and narrower, but equally distinct, colored stripes, I found in the drift deposits of Ann Arbor.

The next layers above, six or eight feet in thickness, are nodular sandy limestone slabs, alternating with shaly and marly easy decomposing layers, of dark greenish or bluish color. Most conspicuous among their fossils are the immense numbers of Rhynchonella plena, almost entirely composing the rock mass. With them are found Leptæna alternata, small variety, Streptorhynchus filitextus, Orthis subæquata, Stictopora ramosa, Phænopora multipora, Coscinium flabellatum, Chaetetes ramosus, Tetradium fibratum, (Stromatopora ?) compacta, covering whole ledges with its glandular nodules, and several forms of Fucoid branches.

The third succeeding group of strata in the ascending series are 10 feet of light-bluish gray argillaceous limestones, with smooth conchoidal fracture. They contain fucoid branches, Stictopora ramosa, Chaetetes ramosus, Streptelasma corniculum, Leptæna alternata, Zygospira recurvirostra, Cyrtodonta Huronensis, Vanuxemia inconstans, Ambonychia amygdalina, Modiolopsis gesneri, Ctenodonta nasuta, Illænus crassicauda, Asaphus platycephalus, Cheirurus pleurexanthemus, Leperditia Canadensis.

Above these limestones are a few thick ledges of a brown-colored, sandy-looking, tough dolomite, with large fucoid branches of fasciculated form (licrophycus) projecting from their surfaces in high relief; and after them follow light colored, brittle limestones, with smooth conchoidal fracture, which separate on exposure into thin, uneven, wedgelike, interlaminated slabs or smaller fragments. The top of the embankments, near the shore, is formed by these limestones, which have a thickness of about 12 or 15 feet. Their fossils are: Orthoceras (Endoceras) proteiforme, Orthoceras tenuifilum, Orthoceras Huronense, Orthoceras vertebrale, Cyrtoceras Huronense, Cyrtoceras isodorus, Lituites, and several other not determined Cephalopoda; Vanuxemia inconstans, Cyrtodonta Huronensis, Cyrtodonta Canadensis, Ambonychia amygdalina, Modiolopsis mayeri, Pleurotomaria rotuloides, Pleurotomaria subconica, Murchisonia bicincta, Subulites elongatus, Subulites vittatus, Belerophon bidorsatus, Leperditia Canadensis, Illænus crassicauda. Some distance back from the shore, on a little higher level, several feet of crystalline limestone strata are found deposited on the top of the just mentioned strata, which contain numerous specimens of Columnaria alveolaris, and of Tetradium fibrosum, besides some other fossils met with in the strata below.

Southeast of the location just described, along the shore of another promontory of the island, an outcrop of five feet of rock-ledges projects from the water, which in all probability are the next higher strata to the above given section. Their contact is not to be seen; the two outcrops are separated by a bay three miles wide, and on the shore their junction is hidden under drift deposits. Dark gray limestones, with a rough uneven surface, and containing numerous flinty concretions, form the outcrop. Of fossils, Receptaculites occidentalis is peculiar to these beds, and is found quite common. Other fossils of this place are: Leptæna alternata, Streptorhynchus filitextus, Orthis tricenaria, Orthis testudinaria, Rhynchonella plena, Pleurotomaria subconica, Lituites undatus, Leperditia Canadensis, Phacops callicephalus, Cheirurus pleurexanthemus, Illænus Conradi, Illænus crassicauda, Bathyurus, Streptelasma corniculum, Chaetetes, Crinoid stems, and a great many fucoid branches spread over the surfaces of the rock ledges.

Four miles north of Drummond's island is Sulphur island, formed by a protrusion of the Huronian Quartzites, overlaid by fossiliferous limestones, without the intervention of sandstones, which separate the Quartzites from the limestones on Encampement d'Ours and on St. Joseph's island.

In the Canadian reports these strata have been identified with the base of the Hudson river group. Before making any suggestions on the age of the strata, I will describe the locality, and mention its fossils.

The Quartzite rocks in their present elevated position seem to have formed an island or submerged reef in the Trenton ocean ; we find now the rounded water-worn edges of the old Quartzite cliffs covered up by limestone beds, and the fissures between the erected ledges filled with angular fragments of the Quartzites and with rounded water-worn pebbles of granitic rocks, which were thrown into the fissures by the breakers, and subsequently cemented into a limestone breccia by the calcareous mud of the Trenton sea. The ledges of the limestone abut inconformably against the more abrupt rock-walls of the Quartzite, while they adapt their bedding to the moderately inclined surfaces, covering them like a cap and slanting off towards the lake, parallel with the inclination of the underlying surfaces. This dip does not seem to be produced by a subsequent elevation of the island, but is due to the surface attraction, which arranged the sedimentary molecules in laminæ parallel with the surface on which they were deposited. We see at least no signs of a subsequent dislocation of the contiguous rock-masses. During the Drift period a part of these limestone strata were broken up and swept away, or were mixed with the drift-masses, which now cover a considerable portion of the island. The lowest beds, seen to the thickness of eight feet, are dark blue arenaceous limestones, weathering rusty brown, or, in the proximity of the Quartz cliffs, have the nature of a coarse breccia. Fossils are very abundant, in particular Chaetetes ramosus and Chaet. petropolitanus, Stictopora ramosa, Phænopora multipora, Stems of Glyptocrinus and Schizocrinus, Orthis tricenaria, Orthis pectinella, Rhynchonella increbescens. Less frequent are Orthis testudinaria, Orthis subæquata, Orthis subquadrata, Orthis lynx, Leptæna alternata, Streptorhynchus filitextus, Zygospira recurvirostra, Discina, Cyrtodonta subtruncata, Ambonychia suborbicularis, Columnaria alveolaris, Streptelasma corniculum, Stictopora recta.

Succeeding above are six feet of light-colored limestone ledges, mixed with much silic'ous matter and deposited in irregular easily splitting beds like the wedge-shaped limestones of the Escanaba section. They contain large specimens of Orthoceras proteiforme, Chaetetes petropolitanus in larger hemispherical masses, Chaetetes ramosus, Leptæna sericea, Leptæna alternata, Orthis testudinaria, Rhynchonella increbescens and Streptelasma corniculum. Large blocks of this same rock are mixed with the drift, covering the hills near the village of St. Joseph, and are in places so densely crowded that they could be taken for regular outcrops of the formation. Besides the fossils mentioned in the Sulphur island strata, some other species are found in the blocks of St. Joseph. These are Receptaculites occidentalis, Orthis plicatella, Orthis pectinella, Illænus crassicauda, and fragments of a Calymene.

The next following higher strata of Sulphur island are compact, hard, somewhat arenaceous limestones of a dusky yellowish color, and in moderately thin beds. All their fossils are silicified. Leptæna sericea, Orthis testudinaria, Orthis lynx, Rhynchonella increbescens, Camerella hemiplicata, Streptelasma corniculum, and Crinoid stems, are the usual forms. The thickness of these upper beds is six feet.

Glancing over the complex of fossils found in the Sulphur island strata, we notice them all to be Trentonian forms. Many of them are likewise found in the Hudson river group, but the general similarity of the fauna is much greater with the Trenton group than with the Hudson river group. The place which I would assign to these strata in the series would be above the limestones of St. Joseph's island, containing Receptaculites occidentalis, and if this be correct, we would have not far to go to reach the lower limits of the Hudson river group. The difference of opinion, then, whether we connect these beds with the Trenton strata or with the Hudson river group, is of small importance. The continuance of so many Trenton forms through the Hudson group period, proves their intimate connection.

An isolated spot, covered with limestones of Trenton age, has, since 1847, been discovered fourteen miles west of the Bay of L'Anse. The exact locality is in Town 51, R. w. 35; the deposit extends through several sections, 13, 14, 23, 24. Dr. Jackson considered these limestones as of upper Silurian age, because he believed he recognized among the fossils Pentamerus oblongus, a decidedly upper Silurian fossil. But Prof. Hall, in examining the specimens collected by Foster and Whitney, declared all the fossils found there to be lower Silurian forms, and my own examination of the place and its fossils brings me to the same conclusion.

From a branch of Otter creek I ascended the hill range, capped with the limestones, from its northwest side. The foot of the hill is all drift on the surface, and even in the deeper ravines the sandstones, which must form the nucleus of them, make no actual outcrops on this side, but loose fragments of the red and variegated sandstones are plentifully mixed with the drift. As we come near the top of the elevation, the limestone is seen to form a long line of vertical escarpments around its upper circumference. The rock walls rise to the height of thirty feet, and at their foot, in the inclined plane, covered by talus, about twenty feet of limestone strata are found underlying, which gives to the series in this place a thickness of fifty feet. The strata appear to be very nearly horizontal.

After climbing over the top of the vertical cliffs I proceeded, still slowly ascending, in a southeast direction, until I arrived at the opposite declivity of the hill range, from whence an extended view over the country is to be had. The limestone ledges are found there dipping to the northeast, in an angle of thirty degrees, and sometimes greater. I could not find any positive facts explaining the causes which produced this inclined position of the limestones, but from all circumstances it appears to me more probable that there was an underwashing and sinking of the strata during the drift period, rather than an actual upheaval of earlier date.

The thickness of all the limestone strata together will not be much below 75 feet. Their general character is dolomitic, partly silicious.

Fossils are found through the whole series, but nearly all are obscure casts. Only in the upper strata well preserved silicified specimens are found. In the following list all the fossils found by me in the locality are enumerated.

In the lower ledges, casts of Bivalves and of Gasteropods are numerous, but not well enough preserved for determination; the same is the case with fragments of Orthoceras and Cyrtoceras. I have identified: Orthis occidentalis, Orthis testudinaria, Orthis similar to Pectinella, Orthis lynx, Rhynchonella increbescens, Leptæna alternata, Leptæna sericea, Lingula quadrata, Pleurotomaria lenticularis, Subulites similar to elongatus, Murchisonia major, Buccania, Ambonychia orbicularis, Cyrtodonta subtruncata, Nucula levata ? larger than Hall's specimens, Streptelasma corniculum. The valve of a Brachiopod, similar to the dorsal valve of Orthis occidentalis, but with the hinge line extended ear-like, and exhibiting an internal septum like the ventral valve of a Pentamerus. A specimen of this kind may possibly have induced Jackson to mistake it for Pentamerus oblongus.

CHAZY LIMESTONE AND CALCIFEROUS FORMATION.

Below the well characterized Trenton strata, and reposing on the Lake Superior sandstones, we find, over the whole extent of the Peninsula, a series of calcareous or arenaceo-calcareous beds, which hold the place of the Chazy limestone and the calciferous formation of the Eastern States. We cannot distinguish two different formations, with different faunas, in the West, where all the fossils ever found are three or four species of shells, and those generally in imperfect condition. But we can see a plainly expressed typical similarity between the fossils of the Eastern and Western localities. Also the lithological characters of the compared rocks are in perfect general correspondence, so that we can safely consider our Western strata as the equivalents of the two named groups of the New York system. The greatest observed thickness of the formation within the district is near 100 feet, but usually it is not found in so large a development.

Its general rock character is that of a coarse-grained sandstone, with abundant calcareous cement, in alternation with pure dolomitic or sometimes Oolitic beds. But sometimes the arenaceous character is less obvious, and the dolomites wholly prevail.

In the lower portions of the group, some beds are found in brecciated condition;—an observation which has also been made by Prof. James Hall in several other localities of the west. A special description of the different localities in which the formation was observed will make further general remarks superfluous.

Starting from the southwest portion of the district, we first meet this group at the Grand rapids of Menominee river. The rapids are nearly three miles long, and in the bed and the banks of the river a succession of about 25 feet of strata can be observed in the following ascending order:

1. The lowest seen rock is a white coarse-grained sandstone of different degrees of hardness, with interstratified seams of arenaceous shales. The surfaces of the ledges are plainly ripple-marked. Numerous angular limestone fragments and pieces of shale are inclosed in the upper sandstone layers. The sand granules composing the rock are small perfect quartz crystals, with glistening facets and sharp unworn angles. Mixed with the quartz crystals are numerous dispersed oolith globules. Portions of the sand-rock have scarcely any calcareous cement, and are easily friable; others are hard, and are rich in calcareous cement. Exposed thickness, five feet.

2. Hard dolomitic limestones and Oolite beds, mixed with a greater or smaller proportion of quartzose sand granules. Thickness, four feet.

3. Fine-grained argillaceo-arenaceous limestones, banded with red stripes, or variegated with irregular red blotches, in thin and even-bedded layers of a dull, earthy fracture, glistening from the admixture of delicate mica scales to the rock-mass, which evolves a strong bituminous smell under the stroke of a hammer. Its thickness is three feet. I found some irregularly dentate, dark brown, leaf-like bodies in the rock, which are decidedly of organic origin.

4. Compact dolomites, partly arenaceous, partly of oolitic structure, two feet.

5. Nodular limestones of peculiar concentrically laminated structure, much resembling irregularly-contorted, nodular masses of Stromatopora; but the nature of the laminated masses is concretionary, and not organic, in its origin. Frequently, portions of the laminated rock are composed of differently-colored, alternating layers of chalcedony, equal in beauty with the nicest Band agates. Thickness, three feet.

6. Fine-grained crystalline even-bedded limestones, in thin layers, with argillaceous partings, four feet. These disappear under the drift.

The next rock exposure in the river is twelve or fifteen miles further down stream, and exhibits the middle portions of the Trenton group. Also, up the stream, the rocks are soon lost from sight. At White rapids, sandstones similar to the lower beds at Grand rapids form the river bed, and in the river banks numerous blocks of the red blotched bituminous limerock, No. 3 of the Grand rapids section, are found mixed with the drift, indicating the extension of the limestone formation on the river for a good distance north of Grand rapids.

In the drift forming the embankments of Menominee river, above the Grand rapids, calcareous rock fragments have been found containing Trilobite remains, which belong to the Primordial fauna. It is most likely that these rock fragments belonged to the calciferous rock series, and not to the Potsdam group, as is generally supposed. I shall speak again of this subject farther on.

When describing the strata of Escanaba river valley, I closed the section with the greenish-colored silicious limestones containing many Cyrtodonta shells, which I assumed to be the lowest Trenton strata. In Town 43, Range west 25, Section 24, to which point my description extended, we see next inferior to the Cyrtodonta beds, dark grey-colored crystalline Dolomites. A part of them is in uneven slabs, containing numerous carbonaceous scaly fragments, which seem to be of vegetable origin. Other beds are of a more massive kind, and inclose thin seams of coarse-grained calcareous sandstone. Still lower, more light-colored crystalline dolomite ledges, mottled with yellowish-white more porous dots, succeed in the series. They form the surface-rock at the forks of the river, where the west branch enters the eastern trunk. Underneath these, follow even-bedded ledges of a light-colored earthy fracturing limestone, striped or dotted with irregular dark-red blotches, which are in all respects like the stratum No. 3 of the Menominee section. The lowest visible beds are brecciated limestones; the limestone fragments composing the breccia are of the same material with the cementing rock mass. The fragments are rounded on the corners and of a smoothened surface, as if mixed in semi-plastic, half-indurated condition, with the softer calcareous mud which forms the matrix of the rock. A green pigment of ferruginous composition frequently coats the limestone pebbles, and pervades in thin seams the remainder of the rock-mass.

The limestone pieces found in the drift of Menominee river and inclosing Trilobites (Dikelocephalus), are of exactly the same appearance with these brecciated strata. This brecciated limestone disappears under the drift, which forms the bed of the river for several miles further north, when we meet with the Granite. In the drift of the river bed, we find a great many angular arenaceous and partly conglomeritic limestone blocks, which contain well-preserved specimens of Lingula antiqua, and also casts of Ophileta levata. There is little doubt that these arenaceous limestones form the next succeeding stratum under the brecciated limestone, but I could not find the undisturbed rock-beds in contact. Eight miles northeast from this place, close to the 45th milestone, the Escanaba and Negaunee railway cuts through a low ridge, formed by soft friable sandstone. This ridge, which forms an isolated outlier on the drift plateau, is capped with calcareous sandrock ledges, representing the calciferous formation. The rock contains small cavities, which once seem to have been filled with organic contents, but I could not find a distinct determinable form. Further eastwards, on the headwaters of Chocolate river, the older strata are hidden under heavy drift deposits. On the heights surrounding the upper course of Laughing Whitefish river, the drift covering is only thin, and the calciferous sandrock, which crowns all the higher elevation in this district, is frequently seen at the surface. The calcareous sandrock, which is perfectly identical in appearance with the ledges covering the ridge cut through by the Escanaba railroad, contains here the casts of a small Pleurotomaria. The falls of the Au Train river, 12 miles east of Laughing Whitefish river falls, are entirely formed by beds of the calciferous formation, which here attain a thickness not far from 100 feet. After rushing over a series of rapids, and making a descent of 30 feet within a short distance, the river tumbles over an inclined plane, with stair-like projections, 40 feet deeper; other perpendicular falls from five to eight feet in height follow, and all the rocks exposed within this space are of highly calcareous nature; in some of them the arenaceous constituents are not more than 15 per cent. of the rock-mass.

Lower down, the softer whitish sandstone layers of the Lake Superior sandstone form the river bed for a while, but I did not notice the junction of the calcareous sand rock with the underlying sandstones in the narrow and sometimes inaccessible ravine through which the river hastens to lower levels. Fossils were not observed. Following the river above the Falls, backwards to its source from Mud lake, the rock beds of the calciferous formation are constantly found near the surface of the valley, which the river has carved through the drift covering of the high plateau of the Peninsula.

Also Mud lake, and the series of connected lakes and marshes from which Au Train river makes its exit on the north side, and Whitefish river on the south side, have their basins eroded through the drift down to the older rock beds. The level of the lakes lies from sixty to seventy feet below the top of the plateau.

In the preliminary report of the former State Geologist, the discovery of an ancient outlet of Lake Superior into Lake Michigan, was announced as existing in this locality. It is true, the waters of Au Train river and of the Whitefish river, take their source from a series of connected lakes and marshes, in which the watershed of both rivers is so little defined, that the direction of the wind may sometimes predispose the water to flow one way or the other. This watershed lies, as has been mentioned before, in a valley eroded into the drift, to the depth of perhaps seventy feet, but the sole of this valley is at least 300 feet above Lake Superior, and the river has to descend in numerous cascades and rapids through a tortuous narrow ravine, before it reaches the lower levels of the indenture of the land, formed by Au Train bay.

If we suggest a former connection of Lake Superior with Lake Michigan through this narrow superficial scratch, we may as well at once go a little further with our suggestions, and let the waters rise a few feet higher than otherwise would have been necessary, and we have the connection perfect, the whole Peninsula is then under water.

The calciferous formation is traceable from the falls of Au Train river eastwards, to the highlands above Munising furnace in Grand Island bay, and from there to the heights south of Pictured Rocks, along the head-waters of Miner's river and Chapelle river. Near Munising furnace, the top ledges of the ravine in which the furnace is erected are formed by an arenaceous dolomite, which contains traces of fossils.

Only a small distance back from the margin, on the same little brook which precipitates itself into the ravine, the next sequent higher strata are opened in a quarry. These are thin bedded, blue colored calcareous sandstones, with shaly partings, and with indistinct fucoid-like relief forms covering their surface. They are charged with considerable admixture of iron pyrites, and for this reason are of little durability.

Four miles east of Munising furnace, on their coaling station No. 3, by well-digging and other artificial excavations, similar calcareo-arenaceous strata became exposed, which contain the casts of Pleurotomaria Canadensis in tolerably good preservation.

A part of the layers in this locality is a fine grained dolomite, very little contaminated with impurities; other layers are of oolitic structure; in the centre of every oolite granule is a small quartz grain.

The principal mass of the exposed ledges is of more sandy nature, and resembles the calcareous sandstones in the quarry above the furnace.

Similar fossiliferous limestones are reported to occur on the northwest portion of the summit of Grand island, but I did not visit the spot.

East of the Pictured Rocks, near the head-waters of Twohearted river and Taquamenon river, I had no opportunity to observe the calciferous formation, or indeed any outcrops of rocks at all. I met with the calciferous formation for the first time again south of Waiska bay, in a locality mentioned before, while speaking of the Trenton group. The brecciated character of the rock in this place was mentioned.

We find more extensive denudations of the calciferous formation in the West Anebish rapids. The rock there is a hard crystalline dolomite, in ledges of variable thickness, from six to twelve inches, and divided by vertical fissure cracks, crossing each other almost in regular right angles, into large quadrangular slabs. It appears to be a durable building stone, and could easily be quarried on the west shore of Anebish island, where it is not much covered up by drift, and sufficiently above the water level to be accessible for working. The fossils of this locality are crinoid stems, Rhynchonella plena, casts of Pleurotomaria Canadensis, and of a Subulites, besides stems of Chaetetes. On the east side of Anebish island the same kind of rock is accumulated in loose slabs along portions of the shore line, but no ledges could be seen.

Further east, on the island of Encampement d'Ours, the rock is of more arenaceous nature, and differs from the underlying Lake Superior sandstone only by a greater proportion of calcareous cement. The formation is well exposed on the eastern declivity of the hill-top, crowned with the fossiliferous limestones of the Trenton group. Right below the shales, forming the base of the Trenton strata, half way up the terminal limestone terrace, a small quarry is opened in the calcareous sandstones of yellowish white color, and of middling coarse grain, which, in certain seams, contain numerous Lingula shells and stems of Chaetetes. The Lingula is twice as long as wide ; its greatest width is below the centre, and a faint radial striation from the umbo to the periphery is perceptible. Lingula mentelli of Billings is the nearest form to it.

A number of rock specimens from the above described lower Silurian strata have been chemically analyzed. The results are given in the subjoined list. The small quantities of iron and alumina contained in almost every limestone have not been separated ; and also the analysis of the insoluble residue, after treatment of the rock with muriatic acid, was neglected. The nature of the residue could in many cases be determined by the eye, because the rocks were not pulverized before dissolving them, and the intermingled insoluble minerals could be much easier recognized.

[illegible]	[illegible]	[illegible]	[illegible]	[illegible]
[illegible]	Per cent.	Per cent.	Per cent.	Per cent.
[illegible]	[illegible]	21	4	42 Argillaceous, with [illegible] quartz sand and iron pyrites.
[illegible]	52	38.5	3	5.5 Silico-argillaceous.
[illegible]	88	4	1	6.4 Silicious.
[illegible]	55.8	21	2.4	20.8 Quartz sand with some clay parts.
[illegible]	90	3	1	4.6 Silicious.
[illegible]	92	2	1	5 Silico-argillaceous.
[illegible]	51	38	2.5	7 Quartzose.
[illegible]	89	2	1	8 Quartzose.
[illegible] group on St. Joseph's Island	82	3	1.5	13 Argillaceous.
Lowest sandy beds on St. Joseph, containing Orthocerata and Cyrtocerata [illegible]	47.3	2.5	1.6	48.6 principally coarse quartz sand.

Name and Locality of the Specimen.	Carbonate of Lime. Per cent.	Carbonate of Magnesia. Per cent.	Iron and Alumina, Hydrat. Per cent.	Residue, insoluble in Muriat. Acid. Per cent.
Dolomite forming the top stratum of Sulphur island.	47	38	3	12 Quartz sand.
Calciferous formation, Grand Rapids of Menominee river, lowest strata above the sandrock of Oolitic structure.	54	42	1	2 Silicious.
Variegated limestone, stratum No. 3 of the Grand Rapids section.	42	33.6	1	23 Silicious sand, with dark, bituminous, cloudy sediment.
Upper strata at Grand Rapids of Menominee river.	45	35	2	18 Quartzose.
Arenaceous limestone, with Lingula antiqua, from Escanaba river.	50	33	2	14 Quartz sand.
Lowest brecciated limestones, near the forks of Escanaba river.	49	32	2.5	15 Silicious.
Dolomite from the forks of Escanaba river.	47.1	37	0.8	15 Quartz sand.
Falls of Au Train river, calciferous sand rock.	47	36	2	15 Quartz sand.
Mud lake, three miles south of Au Train river falls, calciferous sand rock.	42	34	0.7	23 Silicious, finer comminuted dust.
West Anebish Rapids, calciferous dolomite.	52	40	2	6 Quartz sand.
Calciferous strata on top of the ravine near Munising furnace, Grand Island Bay.	49	40	5	6 Quartz sand.
Calciferous strata from coaling station No. 3, of Munising furnace, pure dolomite stratum.	53	39	4	3.7 Quartz sand.

LAKE SUPERIOR SANDSTONE.

Of the palæozoic rocks, comprising the Upper Peninsula of Michigan, it remains for me yet to describe the sandstone formation, of which the larger portion of the south shore of Lake Superior is formed.

The age of this sandstone has been a disputed question. Jackson and Marcou considered it as being of *Triassic age*. Others made it contemporaneous with the Potsdam sandstone, and still another shade of opinion took it as contemporaneous with the Chazy or the Calciferous formations of the Eastern States. Recently, the almost forgotten first idea of its Triassic age, has found an advocate again in Mr. Bell, of the Canadian Survey. There is nothing to support this latter opinion, except a vague similarity of rock characters between the Lake Superior sandstones and the sandstones of the Connecticut valley and other Eastern localities, or with the variegated marls and sandstones of the European Keuper. The lower Silurian age of the Lake Superior sandstone is unequivocally proved by its stratigraphical position. In its whole extent it is visibly overlaid by calcareous ledges, containing fossils peculiar to the Calciferous formation, or, in other cases, by the Trenton limestones.

The recognition of a separate rock-series, identifiable with the Calciferous formation, at once nullifies the other mentioned opinions of Geologists, and leaves no choice but to see in the Lake Superior sandstone the equivalent of the Potsdam sandstone.

Except the stratigraphical position, and the lithological similarities existing between the Potsdam and Lake Superior sandstones, in proof of the identity of the compared sandstone formations, some fossils are enumerated, said to be found in the Lake Superior sandstones; but these are only of a relatively affirmative force. There is no record of any instance in which recognizable fossils were found in situ in the Lake Superior sandstones.

The Lingula, mentioned as occurring in the sandstones of the Taquamenon river, was found in a loose fragment of a highly Calcareous sandstone, mixed with the drift pebbles near the shore of Taquamenon bay. Lingulas of similar form are found in the Calciferous sandstones of the Escanaba river and in the Calcareous sandstones of Encampement d'Ours island. The Calcareous nature of the specimen from Taquamenon river likewise points to these higher strata. Those forming the bed of Taquamenon river are sandstones with silicious cement. Mr. Murray found near Marquette the cast of a Pleurotomaria, which was identified by Mr. Billings with a Calciferous species, and which probably originates from the Calcareous beds overlying the sandstones some distance east of Marquette; the exact locality where the specimen was found is not stated. Much stress has been laid on the occurrence of *Dicelocephalus*, in loose arenaceous rock-fragments, found by Mr. Desor in the bed of the Menominee river, not far above Grand rapids. Prof. J. Hall makes the statement that the rock-piece containing the Trilobite is eminently Calcareous. At the indicated locality, or not far from it, I found in the drift water-worn limestone pebbles, which enclosed apparently the same Trilobite species as those figured in Foster and Whitney's report, besides many fractured shells of Lingula. The nature of the rock is that of a fine-grained brecciated limestone, pervaded with seams of a green pigment, and perfectly resembling the brecciated limestones forming the lowest bed of the Calciferous formation in the Escanaba river section. I have not the least hesitation in connecting my rock specimen with the Calciferous series, and not with the Potsdam sandstone. Very likely in all the three recorded cases the fossils had nothing to do with the lower sandstones, which, up to the present time, have frustrated all my efforts to discover fossils in them.

The thickness of the Sandstone formation is difficult to ascertain. Its lower portions are so intimately connected with the sandstones and conglomerate beds of the copper-bearing Trappean series, that I could draw only an arbitrary division line between the two groups, which would swell the thickness of the sandstone group to many thousand feet, while east of the Copper range, the whole sandstone series reposing on the Huronian and Granitic rocks does not exceed the thickness of 300 feet. Leaving, for the present, the sandstones west of the Copper range out of consideration, the formation east of the range can very appropriately be divided in two sections, an upper and a lower one. The upper section is composed of light-colored, almost white, sandstones of generally soft friable nature. The lower section is intensely red colored by iron pigment, and contains various hard compact ledges, which are valuable building stones. West of Marquette, only the lower section of

6

the group is developed; east of it the heights are formed by the upper division; the lower has exclusive possession of the shore, as far as Grand Island bay. East of Grand Island bay, the upper division sinks down to the level of the water, and only in limited spots the lower red-colored strata come to the surface.

In giving a detailed description of different localities, in which the sandstone is well exposed, I shall have occasion to communicate the results of my investigations with more rapidity.

ENCAMPEMENT D'OURS.

As we approach from the west, on the northwest side of this island, a steeply ascending hill, densely overgrown with hard wood timber, becomes conspicuous. The lower 15 feet above the water are formed by erected beds of a red Quartzite of Huronian age, above which horizontal layers of coarse-grained white sandstone project in mural escarpments. About 10 feet above the base of the sandstone, some harder, but partly conglomeritic, beds are noticed, in which a quarry has been opened, from which the material for the erection of a dwelling house, in St. Mary, was taken; the quality of the rock is not very good. The remainder of the hill, which is about 100 feet high, is built up of regularly stratified, soft, and partly conglomeritic sandstone ledges. The upper strata may belong to the calciferous sandrock, but at the time of my visit to the spot I was not aware of the fact, which only subsequent examination of the east side of the island clearly verified to me.

Eastwards, high cliffs of Quartzite and Slate rock, with vertically erected ledges, form the shore portions of the island, and seem to be the surface rock over a large portion of the more elevated central parts. Not far from the northeast end of the island, where a now deserted farm was located, a road leads up the hill to a plateau, elevated 60 or 70 feet above the lake. The shore is formed by Huronian Quartzites and slates, the platform is superficially covered by drift sand, but is underlaid by a soft white sandstone of Potsdam age. On this platform a second steep hill elevation, also 60 feet in height, reposes, which at its base is formed by the above mentioned Calcareous sandstones, containing Lingula mantelli and Chaetetes stems. A similar outcrop of the sandstones, overlaid by the Trenton strata, is described in the Canadian reports; on the north side of St. Joseph's island, two miles south of Encampement d'Ours. Along the northwest end of St. Joseph, and also eastwards, near the described outcrops of the Trenton formation, heavy boulder drift hides the sandstones from view, and on the shore only the bold cliffs of the Huronian Quartzites are denuded.

SUGAR ISLAND.

The Huronian Quartzites forming the Canada shore on the north side of the narrow lake channel, and a number of small islands, compose also the south part of Sugar island, but are only in a few places visible at the surface, which is formed by sandy drift material, and has a comparatively low elevation above the lake. Towards its northern end the ground gradually rises and attains in its highest portions an approximate height of from 100 to 120 feet. This highest part of the island is all composed of drift, inclosing an abundance of large metamorphic boulders and of more angular sandstone blocks of the Potsdam group which underlies the north part of the island. On its east side, six miles south of Church's Landing, a drift terrace surrounding the base of the island is almost entirely composed of large sandstone blocks, intermingled with a smaller proportion of large metamorphic boulders. The unworn condition of the sandstone blocks, and their crowded condition, indicate the proximity of the undisturbed ledges, which are not seen here, but on the north side of the Landing, a small distance west from the saw-mill, under a similar accumulation of sandstone and metamorphic blocks, the actual strata are found well denuded. The rock is a coarse-grained violet-colored, or, in part, variegated, hard sandstone, with silicious cement. Some of the ledges are disposed to split into thin laminæ, and are frequently of irregular discordant stratification; other beds break in thick massive blocks and promise to be an excellent durable material for coarser masonry work. By these sandstones the upper terminus of the lower division of the Potsdam group is represented.

SANDSTONES OF ST. MARY.

In the falls of St. Mary, and in the channel of the ship canal, the lower division of Lake Superior sandstone is exposed. In excavating the new canal, (at this time in construction,) the following strata are penetrated : First, several feet of coarse boulder drift, under which follow five feet of thin and even-bedded hard sandstone slabs of a light drab color, or sometimes variegated with red. Below them, alternating strata of argillaceous fine grained sandstones and beds of sandy shales are cut through to the depth of 20 feet.

These lower beds are of dark red color, or in places variegated with white, in stripes parallel with the stratification, or in irregularly dispersed blotches and dots. Many of the ledges are plainly ripple-marked. When fresh quarried, some of the sandstone layers appear to be good building stones, but are found to decay very rapidly on exposure to frost and weather. Some harder ledges of the upper thin-bedded series are more durable, and have been used in part to build up the inclined protection walls of the canal, and frequently find application for building up stone fences. The before mentioned more massive blocks of sandstone near Church's Landing, on Sugar island, are the next higher stratum to the rock beds of St. Mary, and loose boulders of it are found plentifully mixed with the drift of the place. A similar belt of argillaceous red sandstones is found on the opposite Canada side, surrounding the foot of the higher mountain range of crystalline and metamorphic rocks. Above the Falls the red sandstones soon disappear under the drift, which, for a good distance westwards, takes undisputed possession of the shore line, and shows itself in Point Iroquois in a thickness of over 300 feet. *Salt Point*, near the Indian reservation in Taquamenon bay, offers the first outcrops of sandstone after this interruption. The land there is a low, marshy spur, surrounded by low cliffs not over four feet high, and by shoals, exhibiting the flat sandrock ledges. We find here the soft, white sandstone of the upper division, with very irregular discordant stratification, and full of ripple marks, indicating the shallowness of the former ocean-bed in this place.

At the west end of this spur, on the farm of Mr. Tipple, the sandrock comes out from the side of a drift terrace rising some distance back from the shore. The higher hills further inland are all drift covered, but I am informed that occasionally the sandrock makes limited outcrops from their sides. The drift in this district is a mixture of clay and gravel, and its fertility is indicated by the vigorous growth of hardwood forests on the heights south of the shore.

SANDSTONES OF THE TAQUAMENON RIVER.

The lower part of Taquamenon river creeps in meandering course through its drift-lined bed to the lake. Fourteen miles above its mouth, to which distance it is navigable for a Mackinaw boat, it is found running with swift current over sandrock ledges, and a short distance farther, we see it coming down with noisy haste over a rocky stairway, with very irregular steps, from one to seven feet in height. Another arm, which branches off from the main river above the Falls, comes in from the east side, and leads us to the foot of a second series of cascades, which, after several smaller leaps, make at once a vertical descent of twenty feet. The entire descent the river makes in the Falls, I estimate to be fifty feet. The exposed rock-ledges are white sandstones, composed of glistening, little water-worn, small quartz crystals, cemented by silicious matter, and of sufficient hardness to be used as building stones, but nearly all the ledges are too thin for the purpose, and made very irregular by the frequency of discordant stratification. Of fossils I could not find a trace, ripple marks are very common.

Four miles above this locality the river forms other falls of 40 feet perpendicular height, and above the falls, for the distance ot half a mile, it makes in rapids a descent of 20 feet. The rock-beds are light-colored sandstones of a soft friable nature, and of various grain, interstratified with conglomerates and arenaceous shales. Thin seams of intensely red-colored micaceous shales, or similar streaks of bright green tints, are occasionally wedged in between. The strata of the upper falls, doubtless, follow those of the lower falls in regular ascending succession, and the thickness of the whole series will amount to about 120 feet. A mile above these second upper falls, all rock-ledges have disappeared under the drift, which on the east side of the river ascends in hills from 70 to 100 feet in height. The west side is lower, and extends in swampy undulating high plains.

The river has here scarcely any current, and is very deep. I had struck the river twelve miles further up, and came down on it on an extemporized raft of cedar trunks lashed with bark. The slightest head-wind would move our craft, provided with a blanket for a sail, in a retrograde direction, if we did not double our efforts to pole it downwards, and by the entire absence of a current at the entrance of a sidearm we came into an embarrassing dilemma how to find our way, and really ascended the sidearm in the belief that we were descending the main river.

West of Taquamenon river, for the long distance to the Bay of Grand Marais, neither in the shore line nor in any of the river beds outcrops of the sandstones could be discovered. This whole district is overlaid by sandy drift, and its forests are almost exclusively pine. In a creek entering the southwest side of Grand Marais, the lower red sandstones give occasion to small cascades; and six miles further west, at the foot of the drift hills of Sable point, the shore is lined with sandstone escarpments from 15 to 20 feet in height. Intensely red-colored, or red and white, variegated arenaceous shales alternate with red sandstones and conglomerate beds of very irregular, discordant stratification, and with ripple-marked surfaces. No rock of any practical value is to be found there.

Passing around the cape at the mouth of Hurricane river, the sandstones have again disappeared under the drift, which forms the shore for eight or ten miles westwards, when we come to the renowned vertical rock-walls, the Pictured Rocks, which for many miles face the lake in an ever-changing variety of grotesque forms, into which the soft, easily-disintegrating rock was moulded by the battering waves and the winter frosts of centuries.

I admired the silvery water streams leaping over the rocks into the lake, and listened with awe and delight to the splashing sound of the breakers, reverberated from large sublacustrine caverns; but I looked in vain for the brightly-painted rock-walls from which the name Pictured rocks is derived. What I saw in this kind was of rather doubtful beauty.

The vertical walls of the Pictured rocks rarely exceed the height of 70 or 80 feet. From the top of these, in receding steps or in gradually sloping ascent, the hills rise to the height of 150 and 200 feet, but not all of the rock is Sandstone; the hill-tops bear a considerable coating of drift. The aggregate thickness of the Sandstone strata I estimate to be 120 feet. The Pictured rock series is exclusively composed of light-colored, almost white, soft sandstone beds, with intermingled conglomeritic seams. The conglomerate pebbles are principally a milk-white or a translucid Quartz; bright red Jaspers of banded or sometimes oolitic structure; black or black and white banded Hornstone and other varieties of Quartzose rocks; together with some Feldspar and Granite. The cement of the Sandstones is a kaolin-like absorbent substance. The vertical cliffs are formed by massive soft Sandstones of a pinkish white color, with interstratified seams of Conglomerate, or with single pebbles dispersed through their mass. The higher strata are more thin-bedded slabs with argillaceous partings, some of them quite indurated, but the prevailing part being very soft, exhibiting much discordant stratification, which is also observed in the lower strata. For building purposes none of the layers are suitable.

GRAND ISLAND BAY.

In the recess of Grand Island bay, at Munising furnace, the Sandrock escarpements have lost their imposing character, the hill-sides become gently rounded, and only in ravines now and then vertical rock walls are met with. The furnace is built up in such a ravine at an elevation of about 50 feet above the lake, leaning against a vertical cliff of a soft white Sandrock, 75 feet high, which is overlaid by ledges of the arenaceous dolomite, mentioned in the description of the calciferous strata.

On the landspur forming the opposite side of the bay, owned by Mr. Paul, at the water's level, fine-grained dark red or variegated sandstones of argillaceous, easily-decaying character, and alternating with softer shaly layers, project to the height of ten feet. They are overlaid by 15 or 20 feet of massive, thickly-stratified, coarser-grained sandstones of pale bluish red color, or mottled with lighter and darker-colored specks.

A quarry has been opened in them, and the stone used in the construction of the two neighboring furnace buildings. A part of this rock is of fair quality; but the frequent false bedding and the easy fissility in these directions, together with numerous bits of shaly matter mixed with the rock-mass, greatly impair its usefulness. This same rock, with the underlying strata, composes the base of the rock escarpments of Grand island, which compare with the Pictured rocks in grandeur. The top of the island is formed by the entire series of the upper light-colored soft sandstones, with some super-imposed layers of the calciferous formation. The two lower islands west of Grand island, Au Train island, and the land-spur forming the east shore of Au Train bay are formed by the lower red-colored and variegated division of the Lake Superior sandstone. The lower course of Au Train river and the Au Train lake are surrounded by a low, marshy land, covered by loose, fine drift-sand.

Beyond the lake the river bed begins to rise rapidly, and occasionally exhibits denudations of the sandstones, which form the nucleus of all the highlands there. Higher up the river more extensive rock exposures present themselves, which, as already described, belong to the calciferous formation.

LAUGHING WHITEFISH RIVER.

One of the most instructive localities for the study of the Lake Superior sandstone formation is around the mouth and along the course of the Laughing Whitefish river. It offers a connected section through the whole thickness of the formation, as developed in this portion of the country. Commencing at the east end of Whitefish point, the following section can be observed in ascending order, as follows:

1. Alternations of thin-bedded, hard, often micaceous, sandstone slabs, with arenaceous shales. The sandstone slabs are in beds of from half an inch to one inch in thickness, but often splitting in paper-thin laminæ. Their surface is perfectly even, others are of undulating, rippple-marked surface. Some beds are white, or blotched with red and white, or red and white laminæ alternate. The inter-stratified shales are prevalently red or spotted with round white dots. Their visible thickness amounts to 25 feet, but is probably much greater, as the lower strata continue downward in the bottom of the lake.

2. A fine-grained, more or less argillaceous red sandstone, in

layers from one to three feet in thickness, with seams of red shale twelve feet thick. This sandstone is of an even, agreeably red color, only rarely spotted with white; it has a silicious cement, is sufficiently hard, and can be worked well with the chisel or the saw; but it is not in all localities alike, and becomes in places so argillaceous as to be worthless. A company opened a quarry on this point, at very great expense, but suspended operations after severe loss. The rock did not hold out in uniformly good quality, and too much worthless material had to be handled. I found a considerable lot of large blocks piled up on the dock, which were exposed there to the severity of the winter frosts, and were partly enveloped in ice even at the end of May; they were not seriously damaged by this exposure.

3. A hard, coarser grained, red or speckled sandstone, in heavy ledges, up to four and five feet in thickness, and amounting in the aggregate to fifteen or twenty feet. This rock forms high cliffs around the shore line, and is the surface rock of the landspur and of all the lower elevated hills on both sides of the river, as high up as Whitefish lake, which is partly surrounded by cliffs of the same rock. In places the regularity of the ledges is much disturbed by discordant stratification, and also streaks of conglomerates run through it. I consider it a very valuable rock. It is close to the surface, which, for miles in extent, is actually paved with huge blocks of it, which after an exposure of centuries have remained perfectly sound. Their great hardness and coarse grain does not qualify them for the same purposes as the softer sandstones of the Marquette quarries; but for strong, heavy masonry, much exposed to hard usage, as for bridges and canal structures, I think it would serve an excellent purpose, and is worth the closer attention of the professional builder.

Above the Whitefish lake, in whose cliffs, fifty feet high, this coarse-grained sandstone, and the before mentioned lower beds, No. 1 and No. 2, are exposed, the hills rise 120 feet above the river bed, and, in their declivity, the subsequent higher strata are exposed.

4. Light-colored, middling soft sandstones in thick ledges, with seams of Quartz pebbles, followed by a few feet of a dark-red, coarse conglomerate. Thickness not accurately ascertained.

5. A series of thin-bedded, soft, whitish sandstones, each layer

separated from the other by a narrow seam of bluish shale. Thickness, from 75 to 100 feet. It forms the steeply inclined declivities of the hillsides, and is crowned with massive soft white sandrocks projecting in vertical walls 50 feet high, which form the *sixth uppermost* member of the section. Right above, we find the strata of the calciferous sandrock, with casts of Pleurotomaria. The Whitefish river precipitates itself in a beautiful cascade over this upper vertical portion, strikes then the thin-bedded sandstone series below, and shoots, fan-like, expanding itself laterally over their highly inclined surface to the bottom of the ravine-like valley, where we find an outcrop of the conglomerate, No. 4 of the section.

To the west of Whitefish river, in the hilly country at the head waters of Chocolate river, this upper division of the sandstone group is hidden under the drift. The lower red-colored division is exposed in various places between Whitefish point and Marquette, on the shore, and in the valley of Chocolate river.

SANDSTONES OF MARQUETTE.

A locality on the shore, two miles south of Marquette, where the sandstones in their contact with the Huronian Quartzites can be seen, has been previously described in Foster and Whitney's report on the Lake Superior district. We find here vertically erected white Quartzite beds of the Huronian group projecting into the lake, which have preserved their granular sandstone structure, and are distinctly ripple-marked. They are surrounded by brown sandstone and conglomerate ledges, horizontally abutting against them. The sandstones, which are of very irregular discordant stratification, closely adapt themselves to all inequalities of the cliffs, which exhibit under the sandstone covering a rounded water-worn surface, indicating their long exposure before they were enveloped by the sandstones.

The Potsdam deposits seem to have formed a continuous belt all around the Huronian mountain district, which must have been an island in the ancient ocean. But a part of these deposits has been washed away again, and only in protected situations have patches of the rocks resisted destruction in places where the denuding forces had freely acted. One of these patches, surrounded by Diorite and Slate hills, we find within the city limits of Marquette, in a small side valley, at the lake front of which the Marquette gas works and an iron furnace have been erected.

The stratification of the ledges in this favored recess is more regular than usual, and in the quarries opened there an admirably fine building material is obtained. About 30 feet of the rock series are exposed to view, the ledges slowly dipping away from the hills towards the lake. In Mr. Wolf's quarry, the lower beds are a uniformly dark-brown colored sandstone, of middling coarse grain, easily worked, in banks of six feet thickness, which are capable of being split into thinner blocks. Very few vertical fissures interrupt the continuity of the ledges, and there is no difficulty of getting blocks of any desirable dimensions. The upper strata of the quarry are somewhat more broken up, have a coarser grain, and are sometimes conglomeritic ; their color is less uniform, but they are still.a very useful material, which supplies the wants of home consumption, while the larger fine blocks go abroad to distant larger cities in the west.

Only a few steps from this quarry, on the opposite side of the little valley, another quarry is opened, in essentially the same rock-beds, but of not near as valuable a character. Their stratification is much less regular. We see fine substantial ledges quickly thinning out and replaced by thin-bedded, worthless slabs, and conglomeritic layers are more abundant on that side of the valley.

The sound blocks, I think, are equal in both quarries, but the most advantageous situation is decidedly Mr. Wolf's quarry. Comparing these strata with the beds of other localities, I consider them in age contemporaneous with the strata No. 1 and 2 of the Laughing Whitefish river section. They are below the hard, massive, coarse-grained ledges, which form the surface rock of Whitefish point. The former extension of the sandstones over all the lower levels of the shore line is made evident by the great quantity of sandstone fragments mixed with the drift deposits covering these spots. In places, undisturbed sandstone strata may be hidden under this drift, but the generally great thickness of this covering would be a serious impediment for the quarrymen, even if such places were known.

PRESQUE ISLE.

North of the mouth of Dead river is another instructive point to study the sandstone formation in its relation to the older subjacent rocks. This landspur is formed by a protrusion of peculiar rock-masses, differing considerably from the rock-beds of the Huronian group in the vicinity. Lowest is a black, unstratified, semi-crystalline magnesian rock, resembling a half-decomposed basalt or a highly ferruginous serpentine. It forms considerable cliffs at the north end of the spur;—more to the south we find it overlaid by a more light-colored, once-stratified rock, which is involved in the upheaval, with its ledges bent and broken up in great confusion. A network of sparry veins pervades the rock-mass in every direction with jasper, quartz crystals, chlorite, asbestus, iron and copper pyrites, besides a number of other minerals copiously intermingled. The principal rock-mass, which is found in all forms, from compact crystalline to an absorbent, earthy condition, is chemically a Dolomite. A specimen of the more compact kind, and of flesh-red color, gave on analysis:

Carbonate of Lime................	55	per cent.
" Magnesia............	35	"
Iron oxyd Hydrate, with little Albumina......................	5	"

The remainder was intermingled crystals of foreign mineral species.

On the south portion of Presque Isle this dolomite is inconformably overlaid by a conglomerate and succeeding sandstone layers, which are identical with the sandstones of the Marquette quarries. The sandstone strata some distance off from the protrusive rocks are nearly horizontal. In immediate contact with them, they have a considerable dip, corresponding to the convexity of the underlying surface. It is possible that the strata were slightly uplifted after their deposition, but I am more inclined to explain the existing dip as an adaptation of the sediments to the surface on which they were deposited. The conglomerate beds at the base are five feet thick, and contain numerous fragments of the underlying dolomite rock and of their inclosed Jaspery minerals. In the sandstones above, which I have stated to be the same as the Marquette quarry rock, we find also wedge-like seams of conglomerates interposed. Some of the beds are in thick banks of apparently useful quality, others are more thin-bedded and alternate with shaly layers. The exposed faces of the better sandstone ledges are somewhat fissured, but we may believe that the more protected interior portions are of better quality, and may be successfully quarried. The partly argillaceous nature of all the Lake Superior sandstones makes them subject to be injured by heavy frosts, but in a more or less degree this is the case with all porous sandstones, and of our Michigan building stones I know of none which excel them in good qualities.

Although it is outside of the scope of my investigations, in going over this ground, some interesting facts regarding the Huronian rocks presented themselves to my observation, which I may be allowed briefly to mention.

The Diorites interstratified with the Huronian schistose rocks in the environs of Marquette, and particularly at the Light-house point, are of an evidently intrusive character. At the mentioned point, we see seams of a coarsely crystalline Diorite intercalated between the ledges of the schistose rock, which are connected among themselves by transverse bands, cutting across the strata of the schist, and forming in this way a network, in which smaller or larger portions of the schist are entangled, with retention of their bedding direction. The ends of the entangled pieces, and the edges of the exterior schist strata accurately correspond to each other. To explain this by intrusion of the liquid Diorite between the ledges of the schist and into the existing fractures, seems to me most natural. On Pic-nic island, the Diorites are of eminently

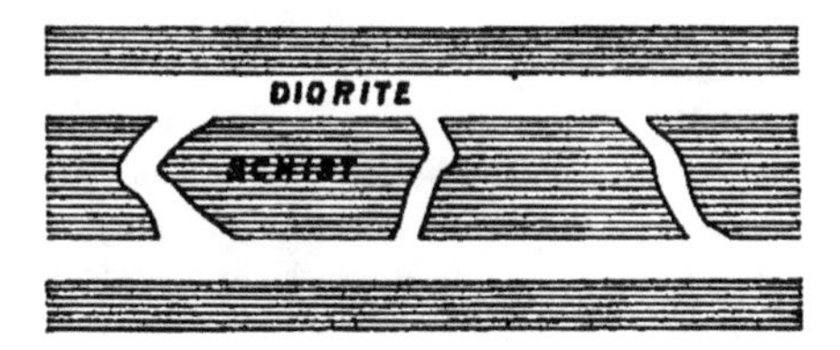

coarse grain, and inclose in the style of a breccia, numerous pieces of Huronian schists, and larger red granite masses. The granites at the mouth of Dead river are also pervaded by bands of Diorite, which is identical in appearance with the Huronian Diorities of the neighborhood.

SANDSTONES AT GRANITE POINT.

Eight miles north of Presque Isle, the granite sends a prominent spur into the lake, which is overlaid by about fifty feet of a soft friable conglomeritic sandstone of dark red color. The surface of the granite in contact with the sandstone is partly decomposed, and such portions can be crushed by the pressure of the hand into a granular sand mass. The lower portions of the sandstone contain also a large proportion of such decomposed granite particles. The sandstone is of thinly laminated structure, interstratified and intermingled with much argillaceous matter, and is of no practical value whatever. I observed there some of the light discolored round dots, so often noticed in these lower strata, which exhibited the centre perfectly bleached, and from there gradually darker and darker rings followed each other towards the outside. At the same locality I saw small vertical fissures extending for several hundred feet over the exposed flat rock-ledges. The rock mass joining the fissure was for half an inch or an inch perfectly discolored.

The removal of the coloring matter in this case plainly came from the easier access of solvents or deoxydizing gases through the fissure ; and the white dots of the sandstones and shales in every other mentioned locality find their explanation in such a process of subsequent discoloration. In the bed of the creek, which enters the lake at Granite Point, at the saw-mill situated one-half a mile back from the shore, and having about thirty feet higher elevation, red sandstones of middling coarse grain form outcrops seven or eight feet high. This sandstone seems to be the same with stratum No. 3 of the Laughing Whitefish river section. It is in moderately thick ledges of sufficient hardness for a building stone, but much shattered at the surface, and in the interior of an excavation made for the water-wheel the rock is not free from cracks and fissures. A number of outcrops of the sandstones line the shore all along the distance west to the Bay of L'Anse, in which generally the lower shaly portions are exposed in the thickness of from forty to fifty feet. The inconformable super-position of the sandstones over the erected edges of the Huronian slates has already been mentioned.

On the west side of the bay, similar rock escarpments, composed of shales and thin-bedded variegated sandstone layers, ascend in vertical mural walls from the lake, and above them coarse-grained thick-bedded sandstones and conglomerates come out in the terrace some distance back from the shore. Some of them are in sufficiently thick and solid ledges to be used as building stones.

At the lighthouse on the entrance to Portage lake, thin-bedded, variegated sandstone ledges, with ripple-marks and frequent discordant stratification, project a few feet above the water, and ascend northwards at a rapid rate, so that we find them a quarter of a mile further in this direction, 20 feet above the level of the lake. Underneath them, a red argillaceous fine-grained sandstone layer seven feet thick rises at the same rate, and soon after its emergence from the water, is found many feet above it. As it goes on towards the north, it thins out again, and the lower strata, which are perfectly similar to the beds above the red stratum, unite then in one uniform succession of thin-bedded variegated sandstones, without the intervention of the thick stratum, which, where it is best developed, has the qualities of a good valuable building stone, equal in all particulars with the quarry stone in stratum No. 2 of the Laughing Whitefish river section. I have not followed the further extent of the shore-line of Keewenaw Point, considering it the territory in charge of another member of the Survey ; but, in order to inform myself of the existing relations between the Sandstone group and the Copper-bearing rocks, I went through Portage lake across to the other side of the Peninsula.

The sandstones lining the eastern shore of Keewenaw Point extend approximating to the centre of the Peninsula, retaining their horizontal position, and also their lithological characters to such a degree that the different strata can be parallelized without difficulty with those of the more eastern localities. Near the centre, the horizontal sandstone ledges are found at once abutting against the uplifted edges of a different rock series—the *Copper-bearing rocks*—which form the most elevated central crest of the Peninsula. The strike of this upheaved rock series is in conformity with the shape of the Peninsula, from southwest to northeast. The abrupt edges of the strata look to the southeast, and their dip is in the opposite direction, under angles variable from 40 to 70 degrees. Without intending to enter into a closer examination of the structure and composition of the Copper-bearing series, I describe it in general terms as composed of mighty conglomerate beds, connected with sandstone ledges, exhibiting perfectly plain ripple-marks, which demonstrate their aqueous sedimentary origin, alternating in often-repeated sequence, with powerful seams of crystalline or semi-crystalline rocks, which are comprehended under the collective name of Trap, but are of a very variable character and composition. The thickness of this formation is very considerable, and I think is rather under-estimated at 10,000 feet. The inconformable abutment of the Lake Superior sandstones against the Trappean series, is in several places near Houghton plainly to be observed. One place is on the property of the Isle Royale Company, in Town 54, Range west 33, Section 6, where the top of a ravine is formed by mighty conglomerate beds, inclosing pebbles of a porphyritic nature, besides fragments of a shaly, well-stratified sandrock, and of amygdaloid trap; they dip under a high angle to the northwest, and form the terminal point of a line, on which the company, for the length of a mile, systematically had opened exploring ditches at close distances, to get accurate information of the succession of strata within this interval. Immediately against the faces of the westward inclined projecting conglomerate beds, the horizontal ledges of Lake Superior sandstone, of much lighter color than the sandstones connected with the conglomerates, are seen abutting in the bed of the small creek, which runs through the ravine. The same inconformable abutment can be seen in a creek entering Torch lake, near the stamp works of the Calumet and Hecla mines, and on the railroad line coming down from the mines to the stamp works.

At the last mentioned locality, the sandstone is of pink color, speckled with round white spots, perfectly resembling the speckled beds on St. Mary's river. A large patch of horizontal sandstones overlies unconformably the trap rocks on top of the hills near Houghton, on the Sheldon and Columbia property. I am not absolutely certain whether it came there as a huge drift mass, or whether it is the remnant of deposits which were there in their original position; but I am inclined to the last opinion. On the west side of the Trap range, half a mile south of Portage canal entry, large outcrops of only slightly inclined sandstone strata border the lake shore, and continue southwestwards as far as the eye can reach. Along the space of about a mile, 200 feet of strata come to the surface. The uppermost are thin-bedded argillaceo-arenaceous layers; below these follow light colored sand rocks in thick ledges, which are quarried for the purpose of filling the cribs built out into the lake for protection of the entry; under them again more shaly and thin-bedded layers follow; and the lowest exposed beds are dark, fine grained, hard sandstones of laminated structure, in beds of five and six feet thickness, which are susceptible of being split into thin, even slabs of any desired thickness. This rock is also used for the above-mentioned purpose, but could be quarried in large, fine blocks, which would serve a better purpose. The strata are very frequently ripple-marked, and exhibit discordant stratification.

The white and red banded, or spotted appearance, so common in the series of the east side, are also observed here, and the geological horizon of these layers cannot be far below the eastern deposits. For the distance of ten or twelve miles eastward from there all the surface rock is sandstone, but the forest covering of the country does not allow us to follow across the series. A few miles west of Houghton, about a mile west of the South Pewabic stamp mills, dark, blackish brown sandstones of fine grain, intermingled with micaceous scales, and quite hard, compose the hills. Their dip is about 35 degrees to the northwest, and a succession of such layers continues as we go eastwards to the South Pewabic stamp mills, where apparently lower strata having the same strike and dip are largely exposed.

They are in beds of various thickness, and alternate with sandy shales full of ripple marks. Next below follow conglomerate beds, some of which are composed of granules not larger than mustard seed up to the size of a pea; they have a very abundant Zeolithic cement (Laumonite).

Other conglomerate beds are very coarse, with pebbles, some of which are bigger than a man's fist. The pebbles are of porphyritic character, and a good proportion of Trappean rocks and pieces of sandstone and shale are intermingled.

Laumonite and calcspar crystals likewise make part of the conglomerates, which immediately rest on crystalline trap rock. All these beds, which must amount to several thousand feet, are in conformable superposition; and the suggestion, which however is not perfectly demonstrated, is, that such strata, with gradually decreasing dip, succeed in a westerly direction, and connect in uninterrupted comformable series with the sandstones forming the western shore line.

The rock character of all the sandstones of the west side of the Trap range is throughout of much darker ferruginous tint, and mixed with a greater proportion of cementing substance than the rocks of the east side. The red Zeolithic mineral exclusively forming the cement of the finer-grained conglomerates at the South Pewabic stamp mills is also, in the much higher beds near Portage canal entry, distinctly recognizable as an admixture to the sandstones. These upper beds of the west side seem to be lower than any stratum of the east side, but from their almost horizontal position it seems highly probable that they follow the strata in conformable succession; and, as the beds near the South Pewabic stamp mills, which undoubtedly make part of the copper-bearing series, seem to be their conformable continuation in the descending order, an uninterrupted serial connection between the Trappean copper-bearing deposits and the Lake Superior sandstones is obvious. The discordance of the strata on the east side of the axis of elevation, and their conformability on the sloping west side, finds its explanation in the hypothesis of a gradual submarine upheaval of the Trap range, in its subsequent rupture, and the final emergence of the western margin from the water, while the eastern portion of the fissured earth's crust remains submerged. The deposits, which on the west side continued to accumulate with undisturbed regularity on the gradually diminishing slope, had to meet with the abrupt edges facing the east side in discordant horizontal position; and if we further suggest a following subsidence of this eastern portion, we can explain why, so close to the Trap range, on the east side of it, none of the lower beds of the series are found superimposed on the Huronian slates. These were submerged at the time that the later horizontal strata were forming.

The large area of Michigan territory to the south and southwest of the Huronian mountains, which is partially overlaid by sandstones of Potsdam age, forms part of the iron and copper district investigated by other parties. It appears, however, that in the northwestern portion of this area only the older red-colored members of the series are developed, while in the southeastern part, along the ramifications of the Menominee river, principally the upper light-colored sandstones of the Pictured rock series are found. At the Breen iron mines, in the Menominee district, the vertically-erected edges of the ore beds, and the accompanying slate strata, are overlaid by horizontal sandstone ledges, which in their immediate contiguity with the hæmatites are colored red by the admixture of ore particles; but a few inches higher, the sandstone has the usual whitish color of the upper Lake Superior sandstone division, and is composed of sharp-angled quartz crystals—a peculiarity also of the upper beds. In a great many places of the surrounding district, similar horizontal sandstone strata are found superimposed on the Huronian series, which is there the principal surface rock, and presents for the lithologist an astonishing variety of rock species, particularly of the Hornblende series.

The accurate limits of the extent of the sandstones over this region are not easy to be delineated; we find them dispersed far westwards in outliers. South of White rapids, on Menominee river, the formation strikes westward across the State of Wisconsin to the Mississippi, where in some localities many fossil animal remains are found imbedded, of which the strata in Michigan seem to be destitute. The few instances in which fossils are recorded from our beds are probably referable to the calciferous strata, as has been intimated previously. It seems, however, that the fossiliferous sandstones of the Mississippi valley correspond to the upper light-colored division of the Lake Superior sandstones, and that the lower red-colored division is older than any fossil-bearing strata in the West. Nearly all the limestone deposits of the Upper Peninsula are restricted to the palæozoic rocks in the east part, and their usefulness for technical purposes has incidentally been mentioned in the description of the respective strata. But the Huronian group is not absolutely without limestones, as we find considerable beds of it interstratified between the Slates and Quartzites of this formation, which have the qualities of a variegated marble. Many inquiries have been made to me for information about these marbles, the description of which will make part of the Report on the Iron District. But as they came under my observation during a reconnoissance trip through these districts, and are, as building materials, allied with the limestones which I had to describe, I will say in a few words what I know about them.

These limestones appear to be one of the upper members in the series. South of Marquette, below the mouth of Carp river, a very interesting section is offered in the hillsides facing the lake shore. As we go from north to south first we find Quartzite beds dipping under a high angle southwards; these are followed by a series of schists and slates dipping in the same direction; and still further heavy limestone beds, interstratified with shales, succeed regularly. We find in a higher position large limestone beds presenting themselves in a horizontal position, but bent in many short flexures by a lateral compression. After having passed these we find at once the same series of strata, in inverted order, dipping northwards, which had a southern dip only a few steps north of this place. We have the centre of a synclinal axis of elevation before us in which the central limestone beds, by the pressure acting on them from both sides, were pushed together into serpentine flexures.

An ideal section given below will illustrate the described conditions.

Similar limestone deposits are seen near Morgan furnace, but I have not had occasion to make a close examination of the locality. In the Menominee district the limestones seem to be connected with the slates and quartzite beds of the upper division of the Huronian series.

These limestones are found in sufficiently thick layers to allow the quarrying of large blocks; their color is flesh-red or variously variegated with white and other shades of red; also cream-colored, almost white masses are found. They are of a compact subcrystalline grain, and very hard, susceptible of a splendid polish. Their chemical composition is dolomitic, with a very little of other impurities in selected portions of the ground-mass. But in nearly all the localities this purer ground-mass is pervaded by a dense network of coarse quartzose seams, which spoil the rock for ornamental purposes. On the Sturgeon river the limestone strata are less contaminated by these silicious seams, but are penetrated in all directions by fine fissure-planes, which, at the stroke of a hammer, open themselves, and cause the rock to break into angular fragments, much beyond the calculated effect of the stroke, so that it is difficult even to procure hand specimens of regular shape. Such rock sawed into slabs, as marbles generally are, would, as I suppose, be shattered into pieces before it came out of the hands of the marble cutter. It may be that some layers of better quality may be found in the Menominee district, but those which I saw I should not recommend for use.

An analysis of a specimen from Sturgeon river, of flesh-color and free from silicious seams, gave:

Carbonate of lime..............	61	per ct.
Carbonate of magnesia	34	"
Hydrated oxyd of iron and manganese...	1	"
Silicious matter................	0.25	"

After having described the geological structure of the eastern part of the Upper Peninsula of Michigan, it may be proper, to look back over the district and examine its advantages for the practical uses of the State.

We see at once the greater value of the western districts, which rival any other part of the world in mineral wealth. But I think, also, that the eastern part of this Peninsula has enough of less striking sources of prosperity, to make it a precious part of our country, and a desirable home for many. Already of high importance are the extended pine lands, which furnish us with excellent lumber, and will do so for a great many years to come, particularly if more care should be taken to prevent the fires, which every year destroy thousands of acres of the finest forests.

Not less valuable are the widely extended hardwood lands, stocked with excellent timber for fuel and other purposes, to which, up to the present time, comparatively little attention has been paid, but which soon will be in demand with the rapid increase of our iron industry. The ample water-power of the rivers, the limestone for the flux, sandstone for the buildings right at hand, and the lakes an open street to all parts of the country, are advantages not found everywhere.

With the clearing of the hardwood lands, a large area of fertile land will be opened for tillage. It will, perhaps, be said, that the severity of the climate is an objection to the cultivation of these fertile lands. It is true, the winters are much longer there than in the south part of our State, and corn and wheat prove to be very uncertain crops; but potatoes, oats, and grass in particular, grow as finely there as farther south, and the few farmers who have settled here and there seem to be doing very well. The mining districts are always an open, profitable market for their products; the lumber and fuel trade is a considerable additional income to them, or gives them employment if they seek it; and those near the lake have a rich harvest in the fisheries. With the increase of the inhabitants in these districts the facilities for all will be improved, and perhaps the time is not far distant when this neglected forest country shall justify the application of our State motto: "*Si quæris Peninsulam amœnam, Circumspice.*"

INDEX.